POLYMER ADHESION
Physico-Chemical Principles

Ellis Horwood Series in
POLYMER SCIENCE AND TECHNOLOGY
Series Editors: T. J. KEMP, University of Warwick
J. F. KENNEDY, University of Birmingham

This series, which covers both natural and synthetic macromolecules, reflects knowledge and experience from research, development and manufacture within both industry and academia. It deals with the general characterization and properties of materials from chemical and engineering viewpoints and will include monographs highlighting polymers of wide economic and industrial significance as well as of particular fields of application.

POLYMER ADHESION
Physico-Chemical Principles

V. L. VAKULA Dr.Sc.(Chem)
Head of the Chair of Physical and Colloid Chemistry
Institute of Applied Biotechnology, Moscow, USSR

L. M. PRITYKIN Cand.Sc.(Chem)
Professor of Physical Chemistry
Civil Engineering Institute, Dnepropetrovsk, USSR

Translator
ALEXANDER V. VAKULA
Institute of Organometallic Compounds, USSR Academy of Sciences

Translation Editor
ERIC CATTERALL B.Sc., M.Sc., Ph.D.
Honorary Research Fellow
Department of Chemistry, Coventry Polytechnic

ELLIS HORWOOD
NEW YORK LONDON TORONTO SYDNEY TOKYO SINGAPORE

First published in 1991 by
ELLIS HORWOOD LIMITED
Market Cross House, Cooper Street,
Chichester, West Sussex, PO19 1EB, England

A division of
Simon & Schuster International Group
A Paramount Communications Company

Printed and bound in Great Britain
by Redwood Press Limited, Melksham, Wiltshire

British Library Cataloguing in Publication Data

V. L. Vakula and L. M. Pritykin
Polymer adhesion: Physico-chemical Principles
CIP catalogue record for this book is available from the British Library
ISBN 0–13–662990–3

Library of Congress Cataloging-in-Publication Data available

Table of contents

Preface

The period following the publication of the first (Russian) edition of this book in 1984 was marked by further progress in the study of the physical chemistry of polymer adhesion: the variety of materials investigated has been substantially expanded, experimental instrumentation improved, new techniques introduced, and the theoretical treatment of the phenomenon has been further advanced. The number of publications concerning the topic has been steadily increasing. In this situation we had but only one course of action to keep pace with the modern state of the art and that was to follow Lewis Carroll's Alice who had to run faster and faster in order to stand still. Striving not only to stand still, but to push ahead, we tried to move even more rapidly. Also, we had the opportunity to ensure that our work still meets the challenge of the time. Moreover, some of the aspects have gained an additional, modernistic sophistication.

These were the reasons that encouraged us to continue our work on the book. The manuscript was thoroughly revised, several mistakes and irrelevant expressions were omitted, the 'key' trends were revealed and clarified and the text has been substantially expanded by inclusion of data published after 1983.

Having undertaken this task we considered it appropriate to present the monograph to the English-speaking specialists in the field of adhesion. The all too well understood desire to get our foreign colleagues acquainted with the contribution of Soviet scientists to the field of adhesion encouraged us in our resolution, as we believe that being, according to Dahlquist, an interdisciplinary science, adhesion cannot in principle avoid being international with regard to the company of the 'invisible college' of investigators working in the field.

We wish to acknowledge the invaluable help of Dr Alexander V. Vakula in translating the book so well and preparing the English manuscript, without which the English Edition of the book would have been impossible. Thanks are due also to Dr E. Catterall the translations editor for this work in making it even more readable; as also to the publishers Ellis Horwood of Chichester for their help and patience during the preparation, in which tribute I include Professor T. J. Kemp Selector of Ellis Horwood Series in Polymer Science and Technology.

V. Vakula
Moscow Medical Polymers Research Institute
February 1991

L. M. Pritykin
Research Institute for Chemical Reagents and Superspun Chemicals
Dnepropetrovsk,
U.S.S.R.

Introduction

One who has no knowledge of anything but chemistry is sure to escape the true understanding of it per se.

Georg Christoff Lichtenberg

Among fundamental physical phenomena rarely can one be encountered that, like adhesion, displays such a degree of generality attracting the close attention of researchers working in diverse branches of science and technology. The variety of examples of adhesion found in nature, added to the sophisticated character of the phenomenon proper, necessitates complex investigations. The success of such studies can be provided only through the combined efforts of experts in physics, physical chemistry, organic synthesis, mechanics and technology. On the other hand, these very same factors are the source of specific problems and difficulties which one is bound to encounter when investigating the nature and characteristics of adhesion.

In this connection we would like to make some general remarks which we believe are necessary to precede the treatise.

Formation of the adhesive joint starts when its elements, an adhesive and a substrate, are brought into contact. The rapidly (in the case of a Newtonian-liquid adhesive) or slowly (for polymeric liquids) spreading adhesive wets the substrate, leading to molecular contact between the adjoining surfaces. Depending on the degree of interfacial interaction the bond ultimately formed demonstrates certain strength properties which reflect the ability of the bond to withstand the effect of external factors.

Such a simplistic view of the physical pattern of polymer adhesion must require, it would appear, a simple theoretical description of the phenomenon. However, application to the real examples observed both in nature and technology meets considerable difficulties. 'Uniformity of nature' — an attractive slogan on science's banner appears to be a fruitless objective for definitive investigations. (As pointed out by Sir George Thomson, 'The "uniformity of nature" is a phrase that one frequently hears. It has been well said that it was never discovered in a laboratory . . .' (*The Inspiration of Science*, London, Oxford University Press, 1961, p. 15).)

In fact, the multiplicity of stages and the factors involved in the phenomenon of adhesion is such that it is usually examined piecemeal, the accent on distinct aspects of the problem being determined by the immediate objectives or even the personal interests of the investigator. The case fits the situation philosophized upon by Francis Bacon, who wrote of knowledge 'being steeped and infused in the humors of the affections' (*The Advancement of Learning*, Oxford, The Clarendon Press, 1891, 1.3). This produces a degree of pessimism regarding attempts to formulate general modes of adhesive joint formation and performance (behaviour). Hence, though the complex analysis of adhesion as a complete phenomenon has been lacking, a number of specific hypotheses and concepts have been conceived. The authors of these tried to reduce the intricate pattern of adhesive processes to particular characteristics observed in studies of a limited number of examples — sometimes selectively chosen — with the objective of explaining this or that aspect of the problem by the effect of separate factors.

Hence, considering the beginnings of the theoretical concepts of polymer adhesion,[†] it is not hard to see that the mechanical concept relates both the phenomenon and the ultimate effect to the development of the substrate–surface microprofile [2], adsorption theory relates it to the sorption of adhesive [3], the chemical concept to the formation of interfacial valency bonds [4], diffusion theory explores the compatibility effects within the zone of adhesive contact [5], the rheological theory associates the phenomenon with strengthening of boundary layers of the contacting phases [6], while the microrheological theory places the accent on the penetration of adhesive via viscous flow into the microdefects of the substrate surface [7]. The electron concept reduces the problem to that of the electrical double layer formation of the surfaces in contact [8], the electrorelaxational theory attempts to consider the behavioural specificity of polymeric materials [9], and the molecular theory is actually the expansion and development of the adsorption theory [10].

As a result, the literature on adhesion is vast. Selected bibliography concerning only measurement of techniques and their associated problems comprises 280 references [11], the number of proceedings of various meetings reaches 150 references [12], and even an incomplete assessment of the number of reviews varies from 275 [13] to 365 [14] reaching a figure of 640 according to the latest reports [15]. At first glance this situation could be regarded as clearly a very positive one. However, one must remember a shrewd observation of Balzak who considered that the piling up of facts was evidence of helplessness. In essence, we meet here what is called in France an *embarras de richesse*. The continuing expansion of the research front and the complication of applied problems provoked by the growth of technology, widens the gap between theoretical presentations and the factual material that actually makes up the basis for the former. The classical Occam's razor, *Entia non sunt multiplicanda praeter necessitatem* ('Entities are not to be multiplied beyond necessity'), seems to be purposely devised to refer to discarding the compilative approach to the theory of adhesion as the sum of particular concepts. This is why we assume that a variety of sometimes mutually incompatible concepts appears to be obvious evidence of the lack of a unified physically non-contradictory theory [16].

---

† Here we refer to the major early publications only, and not the more recent discussions of the corresponding ideas.

We believe that today the science of adhesion is experiencing a formative period quite normal for precise disciplines. It is now at the stage when it becomes necessary to transfer the focus from compiling, systematizing and interpreting experimental data on the basis of particular concepts to generalizations of a wider scope, which make it possible to examine the general features of adhesion and to analyse their manifestations. The necessity is even more urgent as, according to Albert Einstein, experimental observations can be given adquate explanation only by relying on a theoretical background. The obvious need for such generalization is to interpret the observed phenomenon from one unified standpoint. According to the methodology in physics there are two approaches capable of providing such interpretation, namely thermodynamic and molecular–kinetic, each of which treats different aspects of the problem while supplementing each other.

Hence, it would be quite natural to analyse the phenomenon of polymer adhesion within in the scope of thermodynamics, which would give a general picture of its interpretation in energetic terms. Similarly, within the framework of the molecular–kinetic concept which provides the possibility of discovering the distinct mechanism of adhesive joint formation with regard to the chain structure of macromolecules.

In spite of the fact that not all aspects of the phenomenon of adhesion have been investigated with sufficient thoroughness to allow for the best general conclusions we believe that our undertaking is not untimely. Firstly, during the last decade modern experimental techniques and instrumentation have been applied in the field of adhesion producing much more reliable data than before, the range of materials examined has been substantially extended and both, in turn, have provided stronger evidence for the experimentally observed consistencies. Secondly, broad developments in the study of the phenomenon of adhesion as a whole, as of its distinct components, can be related to the general progress in the physical chemistry of polymers, in the theory of surface phenomena, and of solid physics, including that of polymer solids. As applied to adhesion, revealing and extending corresponding analogies seem to be obviously beneficial. We have ventured to put this as one of the major distinguishing objectives of the treatise, taking theoretical and experimental data from related branches of science and thus allowing the examination of adhesion as though from the outside.

In accordance with the stated propositions, in the two major chapters of this book the thermodynamic and molecular–kinetic approaches to an analysis of the consistencies of polymer adhesion are treated. The general concepts, including a consideration of the significance of the phenomenon of adhesion and the terminological ambiguities and descriptions, are considered in the first chapter. In the final chapter we attempt to show how the basic physico-chemical concepts can be applied to the problem of increasing the adhesive ability of polymeric materials.

Let us consider the structure of the monograph in greater detail.

Two problems need to be elaborated on when considering the thermodynamics of polymer adhesion. These are the formation of the adhesive contact and the interaction between the polymers after contact has been established. The first one involves the application of the basic concepts of thermodynamics to adhesion (section 2.1.1), predominantly interfacial processes, of which wetting and spreading are clearly most important (section 2.1.2). However, formal treatment requires to be supplemented by the analysis of interfacial interactions between polymeric phases. Hence, in section 2.2.1 the problem of assessing the energy characteristics of solid surfaces, mainly those of polymeric solids, is discussed. These concepts appear to be of prime importance in the studies of adhesion as soon as the significance of processes taking place at the interfaces between the elements of an adhesive joint is understood. Application of these concepts to real materials is described in section 2.2.2 in which adhesive interactions between polymers are discussed with regard to the type of interfacial forces.

Within in the framework of the molecular–kinetic approach to adhesion identification of the difference between the properties of a polymer in the bulk and in the surface layers has fundamental significance; and this problem is examined in section 3.1.1. As the characteristics of macromolecular compounds are represented as a function of the distance from the geometric surface of the object, reliable evaluation of the parameters of the relevant surface layers provides for the success of subsequent analysis of interfacial interactions between polymers. In section 3.1.2 the main emphasis is on the effect of macromolecular flexibility and mobility on adhesion processes. On the other hand, at the macroscopic level rheological effects are of primary importance, and these are treated in section 3.2.1. The ultimate contact area between the phases (this aspect is treated in section 3.2.2) is determined by a number of factors which are discussed in section 3.2.3.

The account of the theoretical concepts of adhesion will be incomplete if one does not try to deduce from these the possible ways for solving certain applied problems. Of course, engineers constructing and exploiting various types of adhesive joints (glued and welded, hermetically sealed junctions, coatings, composites etc.) prefer to use lists of practical data. This tendency to reduce the results of theoretical investigations to tabular form is probably as old as the Bible. Beginning the discussion of adhesive binding Cagle [17] reminds us of a pessimistic Biblical verse: 'He that teacheth a fool is like one that glueth potsherd together.' The approach to physico-chemical studies as though it were an all-tastes-satisfying compilation of food recipes would be superfluously pragmatic. It neglects the perspectives in adhesives technology determined by the synthesis of new materials, development of new technologies and processes. As regards the relationship between the chemical nature of polymers and the strength of their adhesive joints, we do not anticipate that it could acquire a sufficiently rigorous form, bearing in mind the complex and varied character of the phenomenon of adhesion. We consider it more appropriate to respond to practical requirements by applying the results of thermo-dynamic and molecular–kinetic approaches to adhesion for the purposes of predicting and controlling the adhesive ability of various materials. Hence, relying on the unified basis described in section 4.1 the trends in the synthesis of polymers for adhesives and the formulation of adhesive compositions are discussed in section 4.2.1, while substrate pretreatment procedures are considered in section 4.2.2. At the same time, one should bear in mind that, because of the lack of certain thermodynamics and molecular–kinetic data, these sections give an essentially qualitative description of the subject contrary to the earlier sections in which we attempt to maintain an adequately rigorous level introducing some new parameters and improving the calculation procedures of the existing ones.

To conclude with we would like to emphasize two principle points.

Firstly, the intensity of adhesive interaction, as well as the effect caused by various factors are usually judged by examining the ultimate result of the interaction, i.e. by referring to the overall strength of an adhesive joint. This approach leads to the situation analogous to that when a prosector guides the activities of a physician and, hence, might be referred to as anatomical. Its true description was given by Goethe:

Wer will was Lebendigs erkennen und beschreiben,
Sucht erst den Geist herauszutreiben,
Dann hat er die Teile in seiner Hand,
Fehlt, leider! nur das geistige Band.[†]

The ultimate characteristic, sometimes unfairly called the adhesive strength, is determined by the loading conditons and the geometry of a system, by the existence of internal stresses and deformative properties of the bonded materials, by the character of joint fracture etc. At the same time, quite obviously the strength of adhesive joints appears to be largely determined by the efficiency of interfacial interaction if examined from the standpoint of physical chemistry and not solely as a concept pertaining to rupture mechanics. At the present state of the art corresponding contributions to the overall property can be evaluated only on a qualitative level, hence, the assessments of the intensity of adhesive processes rely on the tests of joint strength. We make use of these only to the extent necessary to interpret the physico-chemical regularities.

Secondly, one should be careful in extrapolating certain theoretical statements taking care not to exceed the limits of their correctness. The danger of overstating formalization was emphasized by Van't Hoff as long ago as in 1891 in his closing lecture at the Congress of Dutch physicians and natural scientists. Sixty years later another Nobel laureate, Professor C. N. Hinshelwood, wrote: 'It seems to me specially important in modern physical chemistry to be clear and honest about fundamentals. This is not so easy as it sounds. Some of the current working notions are expressed in words which easily become invested with a more literally descriptive character than they deserve, and many young chemists — this is my impression at least — are led to think they understand things which in fact they do not. Something simple and direct seems to be conveyed by words such as "resonance" and "activity", which is not legitimately conveyed at all. By certain descriptions which is it easy to give, one is reminded of Alice: "Somehow it seems to fill my head with ideas — only I don't exactly know what they are". Many of the mathematical equations which serve important technical purposes in the modern forms of theoretical chemistry are of a highly abstract kind, but they have acquired a dangerous seductiveness in that they clothe themselves rather readily in metaphors. Occasionally it is salutary to regard this metaphorical apparel with the eyes of the child who surveyed the emperor's new clothes" [18]. The concepts constituting the theory of adhesive phenomena were developed on a certain factual basis, hence each further step beyond this basis requires special verification that the emperor has his clothes on.[‡] Otherwise the theory would become discredited, and, in the final analysis, it would provide neither a true understanding of the phenomena nor any predictive value for the theory.

We are aware of the fact that at the present state of the physical chemistry of polymer adhesion some of the problems cannot be treated except in the most general way. Moreover, having undertaken one of the first attempts to present a unified generalization of a

---

[†]   He who would know and treat of aught alive,
      Seeks first the living spirit thence to drive:
      Then are the lifeless fragments in his hand,
      there only fails, alas! the spirit-band.
           (Goethe, *Faust*, translated by Anna Swanwick)

[‡]   The attempt to describe adhesive interaction in terms of quantum mechanics [19] provides an illustration of the fruitlessness of this approach.

vast and sometimes heterogeneous body of information, we are also aware that some of the statements may lack precision. It is the reader who must decide whether this adds to or detracts from the book. We would be grateful to receive readers' observations concerning the contents proper, the structual composition of the book or any other matter pertaining to the subject.

Having taken the classical approach to an 'Introduction' involving a justification of the need to publish the book, describing the premises which prompted the work and the underlying basic principles but including timely provisos and apologies, we are pleased to arrive at the least arguable part of the introduction, viz. to the acknowledgements. The monograph compiles the results of our own investigations and the invaluable contribution of our co-workers. Most of the problems elaborated on in the book were frequently discussed with many Soviet scientists in physics, physical chemistry, mechanics, synthesis and processing of polymers. Naturally, they do not bear any responsibility for the book; however, by their criticisms, they have made a valuable contribution which we duly acknowledge. We are also grateful to some of our foreign colleagues for providing the opportunity to discuss with them their latest results. We are unable to list all of them, but we are especially grateful to R. Good, H. Kleinert, L.-H. Lee, K. Mittal, V. Raevskii and W. Wake. And finally we would like to single out the role of S. S. Voyutskii, a founder of one of the most productive scientific schools, to different generations of which the authors belong. The memory of this brilliant man has been with us throughout the work.

1

General concepts of adhesion

1.1 THE ROLE OF ADHESION IN NATURE

In one of the first issues of the periodical *Interdisciplinary Science Reviews* founded in the 1970s, Dahlquist published an article entitled 'Adhesion: An Interdisciplinary Science' [20]. The paper presents an eloquent treatment of the broad spectrum of problems involved in the study of adhesion and related technological processes. The phenomenon of adhesion involves a surprisingly large variety of materials which may be conventionally categorized as technical and biological. The specific features of the formation and behaviour (performance) of the former are treated throughtout this entire book. In this chapter we confine ourselves to the most general information concerning this subject, while particularly elaborating on the specificity of adhesion as manifested in biology.

The phenomenon of adhesion concerns the interaction of condensed phases, and yet is the fundamental requirement for the production of heterophase systems. Of these, composite materials are claimed to be most important. Their significance is beyond that of a simple example of the role of adhesive processes, and is best reflected by the thesis: 'the history of mankind is the history of materials' [21]. The choice of the distinct components of such systems is determined to a much greater extent by their propensity for adhesive interaction than by their separate intrinsic characteristics. A large amount of data, largely experimental but also theoretical has been compiled in this field. At the same time, the development and application of composite materials, are to-day only in the early stage, and further progress in this direction is related to the development of the concepts of adhesion. These concepts, at present, are derived mainly from studies of polymeric materials (as their characteristics are easier to alter when examining their relationship to adhesive properties) and glass-fibre reinforced plastics, though composites based on metals and synthetic inorganic materials are equally important.

Rubbers are amongst the most common of the traditional composite materials. The complexity of their service properties is largely determined by the reinforcement effect provided, in the first instance, by adhesive interaction between the polymer matrix and the carbon filler. As the characteristics which describe the interaction are defined so the range of filled systems will increase, involving plastimers and thermoelastoplasts as well as elastomers. Additionally, the search for new fillers or their modifiers could also be of great promise.

The two processes traditionally relying on a knowledge of the characteristics of adhesion are bonding (cementing) and the application of protective coatings. In fact the modern view on adhesion is based almost exclusively on the studies of adhesive joints and paint coatings. The effective application of welding [22] and metallizing [23, 24, p. 55, 86] are also dependent on the adhesive properties of polymers.

Adhesion is inevitably involved when considering friction of solids [25], particularly those which are polymeric [26, 27]. Rigorous treatment of the problem is rather difficult as it is not easy to choose an adequate model [28]; however, in a number of cases semi-quantitative relationships between the frictional force and the adhesive properties of the samples were established [29]. Modern tribology relies on the concept that the overall friction coefficient is a sum of the deformational and the adhesive components and while the former is defined within the framework of continuum mechanics, the latter is determined according to the concepts of adhesion [30, 31]. This is, apparently, the only basis for the study of polishing, of wear and the frictional transfer of metals and related problems [32, 33]. Metal-to-metal gripping is supposed to be directly associated [34] with the energetics of interfacial interaction both at ambient and at elevated temperatures [35].

The role of adhesion in various material bonding technologies, e.g. welding [36], (including diffusional welding [37]), soldering [38] and the production of clad metals by rolling and blasting techniques [39, 40], cannot be overestimated. Adhesive properties of melts govern their interactions with slag and flux [41], and consequently determine the choice of distinct regimes of metallurgical practice.

Less evident, but still as important, is the role of adhesion in a variety of modern technologies such as powder metallurgy [42], flotation, impregnation, printing arts, grouting, and anti-erosion soil protection, production of reinforced constructions, metal lubrication [32], aerosol filtration, board production, lithographic printing [43], and in many other processes listed, for instance, in a brochure by Zimon [44]. All of these technologies, each intrinsically specific, have the same common background of the characteristics of adhesive interaction.

Consideration of many of the concepts of adhesion as stated in the 'Introduction', provides for different degrees of progress in various industries. There have been certain instances when authors have come to contradictory conclusions even concerning major problems. For instance, on the one hand, optimum fine polishing of the substrate surface during mechanical prebonding treatment was claimed to be essential, whereas, on the other hand, a rough well-developed surface was also proposed as being better; the contribution of the adhesive component in polymer friction is possible only under high vacuum etc. This is why today, within the framework of general concepts, analysis of the nature of adhesion using the properties of interacting products is so important.

Such analysis is beyond the aims of this monograph; however, we believe that it could be initiated by the study of adhesion in biological materials. Evolutionary development might thus be taken into consideration. In fact, according to Trinkaus [45] the adhesive properties of external cellular membranes determine the main steps of an organism's development — cell differentiation, histo- and organogenesis. Adhesion is claimed to be the factor providing for the ultimate oneness of biological systems [46].

As a consequence, an apparent growth of interest in the study of intercellular inter-actions in terms of adhesion has been observed recently [47–49]. A corresponding analysis of the biophysical characteristics of the cells' surface [50, 51] assists one in the solution of a variety of problems ranging from marine growth on the submerged parts of ships[†] [53] to the formation of cancer tissue [54].

The effect of adhesive interaction between isolated cells is demonstrated, for example, in that in the region of dense adhesive contact membrane conductivity is several orders higher than that of the 'free' membrane not involved in adhesive interaction [55]. Thermodynamic parameters of this region are essentially altered also, these changes are quite significant, e.g. in the region of intercellular contact the phase transition temperature of lipids is more than 160° above that of the unaffected membrane [56].

These facts have provided a background to the adhesive–condensational mechanism of membrane coalescence [57]. Membrane coalescence starts with the formation of dehydrated contact between the membranes, which is plausible in the presence of calcium ions [58]. However, for coalescence to occur the integrity (intactness) of the bilayer must be damaged. Here once again calcium ions are involved. Being adsorbed at the charged functional groups of lipid molecules and neutralizing the overall charge of the membrane, these cations form a 'chess-like' lattice of charges. The heads of lipid molecules are strongly attracted to each other in the non-polar environment of the region of dehydrated contact. This is what determines the drastic change in the thermodynamic parameters of the system described above. As a consequence, at temperatures of 293–313 K the molecules of the external monolayer contributing to the contact zone may form a crystalline structure, whereas other molecules of the bilayer remain in the liquid state. Hence, stresses arise in the external monolayer of the vesicle that lead to its deterioration. At the start of membrane coalescence inner volumes of vesicles are separated by only the bilayer, i.e. a three-laminar structure is formed and monolayer coalescence occurs. Subsequent development of this structure may lead to rupture of the bilayer separating the inner volumes of the cells and, consequently, to their total coalescence. This mechanism is in accordance with the experimental data concerning the effect of calcium cations on the regularities of the process described] 59, 60]. Other polyvalent cations, as well as monovalent, at sufficiently high concentrations, are able to impose the same effect as Ca^{2+} [61].

The substances responsible for these interactions are the protein and glycoprotein complexes, usually acting, like antibodies and lectins, quite specifically [62]. This fact is important when intending to regulate adhesive interaction, which can even be totally blocked by treating the cells with complexing agents [48], glycoproteins [63], and proteolytic enzymes [64]. Hence such treatment procedures can be very effective in disintegrating the tissues into separate cells.

On the other hand, some researchers suggest that the main factor which provides the cells' stability against adhesive interaction is the structural–mechanical barrier [65]. This viewpoint is supported by the observation that the specificity of the coalescence mechanism of different cells is due to steric effects in membranes [66]. In particular, this

---

† It is relevant to note that fouling of the Trans-American oil pipe-line from the inside due to micro-organism growth inflicts a 1 million dollar loss per day on the operator [52].

approach makes it possible to account for the fact that plasmatic membranes are not entirely liquid-crystalline, but comprise domains of different microviscosity separated by diffusional barriers, i.e. these membranes are heterogeneous [67, 68]. Such phase segregration predetermines the difference in adhesive ability of distinct surface areas: 'liquid' domains are inactive with regard to other cells, however, the 'solid' domains are sufficiently strongly associated [69]. The significance of structural factors in intercellular adhesive interactions becomes even more evident as soon as one becomes aware that the cell surface is highly developed (due to the outgrowths of appendages, pili, cilia, microscopic bristles and so on). These factors are most clearly seen in microorganisms. In fact, these have specific organelles, called the adhesines, that provide their attachment to the host surface. In lacto bacteria these organelles are of a 'lash-like' form, up to 1 μm long and 0.02–0.06 μm in diameter, in Candida bacteria they form a zone of filaments 0.05–0.08 μm thick in which the filamentous structures are alternated with the granules 2.5–3.0 nm in size with fimbriae 1–3 μm long, and in *Escherichia coli* adhesines have the form of fimbriae or fibrilles 1–7 μm long and 2–7 nm in diameter [52]. It is the adhesive function of adhesines that determines the behaviour of microorganisms in cenose and the ability of heterotrophic microorganisms to lead a parasitic way of life.

These facts lead one to conclude that intercellular interactions can be investigated with those techniques which are conventional in the studies of adhesion in technological products. Measurements of contact angles performed for various bacterial cultures [70, 71] demonstrated that their values were determined by the lipid composition of the outer cellular membranes. This parameter appeared to be quite sensitive to the presence of antibodies and can thus be applied to the analysis of phagocitosis [72]. Relying on the values of contact angles, surface free energies can be estimated [70] giving a quantitative assessment of the ability of cells to adhere to low energy polymers [73, 74] and to antibodies [75].

As a characteristic of adhesive ability, the surface energy of biological materials provides an important tool for uncovering the basic principles of intercellular interactions. Sharma [76], for instance, has found that within a series of polymers their biocompatibility could be correlated with the variation of surface energy [76]. At the same time, adhesive interactions in biological systems were shown to lead to conformational changes in peptide chains [77], hence, care should be taken when making use of the surface parameters of contacting materials. Modelling of adhesive interactions in biological systems usually involves glass and sufficiently simple polymeric substrates, e.g. hydrophilic–hydrophobic copolymers of 2-hydroxyethylmethacrylate [78], on the one hand, and polypeptides, bacterial cultures [71], blood cells [71, 79] and living tissues [80] as adhesives, on the other.

The two latter cases are, literally, of vital importance, as adhesive interactions of blood cells and live tissues with a substrate determine the choice of thromboresistant. materials and of those suitable for implantation as grafts, respectively.

In fact, thromboresistance (as evaluated by the blood clotting time) and surface energy of polymeric substrates were shown to be smoothly correlated [81]. Adhesion of platelets results in their destruction, subsequently the contents of intracellular granules (serotonine and adenine nucleosides initially) are discharged into the blood stream and may function as adhesives stimulating platelet aggregation. Along with this process and simultaneously with it, contact blood clotting factors are adsorbed at the surface and activated to initiate a cascade of enzymatic reactions resulting in active thombin being liberated into the blood. This proteolytic enzyme catalyses a specific hydrolysis of fibrinogen and stimulates further platelet aggregation (autohesion) and adhesion [82]. On the other hand, it is not only energetics that determine the thromboresistance of the material but also the topography of its surface [83]. The latter is clearly liable to affect the hydrodynamics of blood flow [79]. To perform as thromboresistant matter the material has to be inactive with regard to the blood components, or, if this is impossible, it should not cause activation of contact blood clotting factors and platelets or it should adsorb only those proteins that take part in thrombus formation (thrombosis). Materials that may cause lysis of thrombi resulting in soluble products also tend to display reasonable thromboresistance. Platé and Valuev suggest that if at least one of these four conditions is fulfilled, the material can be considered as hemocompatible [82].

The same premises of adhesion origin determine progress in one of the most rapidly growing fields, viz. creation of materials for endoprosthetic purposes. The experience accumulated in the field is so vast, and successes have been so tremendous, that here it will suffice to emphasise the basic character of adhesion as the phenomenon underlying the choice of metal, glass, ceramic, and, initially, polymeric materials intended for use in internal organs.

This choice cannot be confined to the recognition of interrelationships between the *in situ* performance of the material and its characteristics even if these are adequately specified as, for instance, biological inertness, strength, or durability for bone grafts. Neither is it confined to providing the ability to bond various tissues using adhesives [84, p. 120]. This choice is much more fundamental in character, as adhesion determines biocompatibility of the implant with the body in the broadest meaning of the term [85, 86]. Understanding this interrelationship stimulated the development of a wide variety of endoprosthetic materials that included materials other than polymers. Polymers as a class do indeed comprise most inert materials; however, restricting oneself to them would lead one to neglect the experience of evolution. In fact, the majority of nature's projects involves actively interacting objects. This observation has led to the creation of the next generation of biomaterials comprising bioglasses and bioceramics. The first contain elements of bone tissues, i.e. sodium, calcium, phosphorus whose atoms partially substitute silicon atoms of the matrix. The second are based on calcium phosphate as the basic constituent for the reason that calcium phosphate is a natural constituent of bone tissue. As a result these materials are tightly bonded with the tissues of the body via natural 'cementing' processes [87]. They do not provoke any inflammatory reactions, and are not rejected by the tissues. In spite of the fact that the use of artificial implants in modern surgery is quite impressive and is continually growing,[†] the investigations in the field are performed to a large extent on a semi-empirical basis lacking a sufficient recognition of the effects of adhesive origin. We believe that extensive and systematic application of the laws of adhesion to the studies of general biological problems may appear to be the lever of Archimedes that may give a major and novel impetus to biomaterials science.

[†] According to the information of the Department of Science and Technology of Japan, world-wide in 1986 artificial blood vessesl were implanted in 750 000 patients, heart valves in 300 000 patients and artificial bones and joints in 7 000 500 patients. These figures are predicted to increase by 3.2, 3.0 and 2.0 times, respectively, by 1990. In the USA alone the annual production of synthetic mammary glands reached 125 000 items.

When making a choice of chemotherapeutic drugs for targeted application one should take into account that the adhesive ability of external cellular membranes may be modified to ensure the affinity necessary to fit that distinct application. However, with rare exceptions, this approach has been neglected. In a few of the studies that came to understand the prospects of this approach, the direct relationship between the chemical nature of the modifier (as a model of a drug) and adhesiveness of the cell structures subjected to modification was elucidated. As such modifiers of cell adhesiveness, Weiss, for instance, suggested colchicine [88], actinomycine, cycloheximide, pyromycine [89] and sialic (salivary) acids [90].

We believe that in regard to biological subjects one of the most interesting applications of the concepts of adhesion, though a localised one, is offered by carcinogenesis. The corresponding approach, distinct from the more common genetic and molecular–biological ones, makes it possible to get to the terminal links of the causality–consequential chain of events resulting in the growth of cancer tissue. What differentiates the cancer cells from the normal ones is not their chemical composition but the magnitude of the negative charge and surface topography of the external membranes [91]. On the other hand, the number of intercellular contacts decreases during tumour growth [92], resulting in decreased total interfacial energy. The difference in adhesiveness of the normal and cancer cells was directly demonstrated by Coman [93]; the applied mechanical stress to tear cancer tissue was shown to be an order of magnitude lower than that to tear normal tissue.

Relying on these data and taking into account that at physiological temperatures membrane lipids are in the liquid–crystalline state the 'drop'-model of the cell has been proposed [54]. According to this model the wholeness of the cell is defined by the balance between the surface charge and the surface energy of external cellular membranes. Analysis performed on the basis of the concepts of adhesion demonstrated that the direct physical premise for uncontrolled tissue growth characteristic of malignant tumours and metastatic spreading lies in that the balance is shifted towards the prevalence of the negative charge over the surface energy. This approach offers a description of carcinogenesis in terms of adhesion that is consistent with the results of molecular–biological and oncological studies. Moreover, it suggests physically sound ways to regulate the neoplasm growth up to its termination by increasing the adhesiveness of external membranes of lesioned cells [54].

Complexity of biological subjects makes unambiguous interpretation of the modes of their adhesive interaction quite difficult. However, the progress of bio-organic chemistry and physical chemistry of polymer adhesion is though to be able to provoke a mutually stimulating influence on both. This viewpoint has been illustrated above, as applied to the solution of the purely biological problem of carcinogenesis. On the other hand, during evolution nature produced a number of designs, surveyed in bionics, that are of major importance in the development of the theory and practice of adhesion phenomena as they are demonstrated in man-made technology.

In a popular review calling for use to be made of nature's experience relating to adhesion, Matsumoto gives a list of examples [94]. Natural adhesives were shown to be beneficial in that they are universal in action, adhesive bond formation takes a very short time and there is no need for either heating or increased pressure for bonding to occur [95]. According to the nature of the major constituent, Evdokimov [95] lists four types of natural adhesives, terpenes (e.g. excretions of 'Macrotermes' termites that provide the concrete-like strength of impressively large clay structures), lipids (e.g. fly's glue providing the bond, up to 5 MN/m^2 strong, in fractions of a second), hydrocarbons (e.g. excretions of Balyanus' mussel that form water-resistant and very strong bonds reaching 80–85 MN/m^2 for bonds between metals, and 150–160 MN/m^2 for joints between a mussel's half shells, gum karaya, gum tragacanth), and peptides (like the bees' propolis excretions or the serine excretions of spiders which produce structures of no less than 30 MN/m^2 strong within seconds by means of an adhesive layer 2–3 μm thick). He suggests [96, 97] that such examples of adhesion be attributed to an independent field of bionics [98].

Analysing the above examples and others in nature, of which a quite impressive illustration for instance is a snail climbing up the smooth low-energetic surface of poly-tetrafluoroethene without leaving a visible sticky track, one should note a feature which is general for all of these systems, i.e. no covalent bonds are formed and the strong adhesive interaction (joint) is due to the attainment of the maximum surface area of interfacial contact. Adhesive tapes are only a pale reflection of this approach in man-made technology. Bioadhesives, however, may additionally structurise rapidly and reversibly, thus providing a succession of immediate joining–disjoining cycles finally leading to the formation of very strong structures. It has to be accepted that modern approaches to the formulation of adhesives are quite different from those developed in nature. In this regard, analysis of nature's experience may be especially valuable. More-over, we believe that as the chemical nature and structure of biological materials is immeasurably more diverse than in man-made technology, the 'centre of gravity' of theoretical studies of adhesion in the long view should move from technological to biological areas.

Summing up the role of adhesion in technology and biology, one can see that it cannot be overestimated. The phenomenon of adhesion implies interaction between any materials more complex than molecules. The essence of the diverse processes involved in formation of the adhesive bond, in the broadest meaning of the term, can then be con-fined to a limited number of relatively clear physical concepts. On this path there unavoidably exists the need to establish the links between different branches of natural sciences and technology, as well as for an understanding of the fundamentals of adhesion.

1.2 BASIC CONCEPTS AND DEFINITIONS

Within the scope of the ideas under consideration that of adhesion obviously occupies the central place. The previously mentioned interdisciplinary character of the science of adhesion, as well as the diversity of materials capable of adhesive interaction, lead to unavoidable discrepancies in defintions. Quite obviously, the subject of the monograph provides the need to clarify this matter in order to avoid ambiguity.

Terminological contradictions are most perceivable in applied texts [99]. However, even speciality treatises are not free from ambiguities in interpreting the terms. For instance, Berlin and Basin define adhesion as a bond and also as a phenomenon [10]. In this respect the views of Zimon are quite illustrative; as regards disperse particles his definition of adhesion is purely tautological [100]. Speaking of liquids, he defines adhesion as an interaction [101], while referring to coatings he defines the term (i.e. adhesion) as a bond, as an interaction, and as a phenomenon, all these definitions being listed on one page [102] which one should admit is rather extreme. Nevertheless in a monograph dealing with polymer adhesion to metals [103] adhesion is interpreted as an energy (p. 82) and as a strength (p. 237); such expressions as 'the strength of adhesion' (p. 15) and 'adhesion undergoes a phase transition' (p. 84) are used.

In encyclopedias [104–108] and dictionaries (including such authorities as *Webster's New Collegiate Dictionary* (Springfield, 1973) and *Oxford English Dictionary* (London, 1975)) the definition of the term is confined to tracing its etymological origin to the Latin *adhaesio*. In spite of the fact that defining by using the synonym seems to be false practice, such an approach has been exercised in a number of publications, whose authors interpreted adhesion as fixing [34], sticking [109, p. 9] etc.

As sticking implies formation of some kind of a bond, corresponding definitions have gained certain usage. Even when adhesion is referred to as a bond [107, 110] a number of implications of the term have been used, e.g. adhesion has been treated as a certain initial step ('conception of the bond' [106, 111]), as a process ('progressive growth in time' [109, p. 9]), and ultimately as the final product, i.e. the strength of an adhesive joint ('resistance to the breakdown of contact' [109, p. 17]) [112, 113]. In some cases the concept of 'instantaneous interaction', meaning the ability for bond formation, has even been involved [114]. There is a certain freedom with regard to what is to be considered as the objects that are being bound with the said bond. These have been assumed to be either the surface layers [104], or the bodies as a whole [115]. The bond itself is usually treated as the molecular one [10, 116], although it is clear that depending on the nature of contacting phases and contact conditions the energy of the bond may reasonably exceed the energy of molecular interactions. Definition of adhesion as of a force [117] ('interlocking force' [118], 'molecular force acting across the interface' [119], proportionality coefficient in the equation for the friction force [120]) is also inadequate as adhesion and the strength of an adhesive joint are undoubtedly distinctly different concepts. This fact is reflected in ASTM Standards [121].

Hence, it is wrong to restrict the meaning of adhesion to that of a bond, or sticking, or force (strength), or ability etc. Adhesion is the phenomenon that is clearly indicated under certain conditions for certain objects and leading, ultimately, to a definite result. As has already been noted, any system more complex than a molecule may be involved in adhesive interaction as it is only at this level that the macroscopic characteristics of condensed phases may be introduced. The distinctive feature of adhesive interaction is that it involves the molecules constituting the surface layers of condensed phases. Summing up, adhesion is a surface phenomenon leading to the formation of a new system, an adhesive joint displaying a complex of inherent characteristics, determined by both the bulk properties of the adhesive and the substrate and the existence of the interface between them.

The last definition, cited for instance by Kaelble [116], needs, we believe, to be complemented. Regardless of the prehistory of joint formation the adhesive joint performs as a single system. Its ability to transfer mechanical stress from one phase to the other is an important feature. This condition suggest that both phases are brought together to distances that allow for the physical and chemical forces to be established. It also characterizes the principal difference of adhesive joints from other heterogeneous systems such as, for instance, mechanical suspensions.[†] Depending on the value of the

---

[†] Strictly speaking, the ability to transfer mechanical stresses between phases exists in concentrated mechanical suspensions as well. The nature of this transfer and the distribution of the stresses in such systems, however, are basically different from those intrinsic to adhesive joints.

ratio between the number of adhesive (a) and substrate (s) phases (a : s) adhesive systems are classified as cement (glue) bonds at a : s < 1 and as composites at a : s = 1, the latter being distinguished according to the type of wetting as either filled systems (immersion wetting) or coatings (contact wetting). This classification is convenient for applied purposes; it should be remembered, however, that as regards the physico-chemical viewpoint the formation and performance of all of the listed types of adhesive systems are governed by the common laws of adhesive interaction between the contacting phases.

When examining the conditions leading to adhesive joint formation, one should distinguish between macro- and micro-processes. The first objective is the establishment of molecular contact by bringing the surfaces of the condensed phases together to the distances necessary for the establishment of interfacial bonds resulting from the development of the microprocess of interfacial interaction. The energy spectrum of such bonds, in principle, is unlimited. This division, suggested independently by Pritykin [122] and Good [123], is not a mere convention since the phenomenology of the macro- and micro-processes is different. The former are described in terms of wetting and spreading taking into account rheological factors, while the latter involve the formation of molecular, hydrogen, valence, or ionic bonds.

Summing up the concepts discussed above and generalizing the different approaches, [1] we define adhesion as the phenomenon of binding together the surfaces of condensed phases brought into contact with each other.

The concepts that are not related directly to adhesive phenomena are surveyed below in the corresponding sections. We believe it appropriate to conclude this chapter by defining what the elements of the adhesive joint are, these definitions being more related to the etymological implications than to those of physical origin.

To denote the elements of an adhesive joint two terms, the adhesive and the substrate are used. Their etymology is obvious; both are derived from the Latin, *adhaesio* and *substratum*. The term glue, which is widely used in the English-language literature, relates to the Greek κολλα (κολλαω — glueing) and γλίνε. The first of the synonyms indicates the common Indo-European root. For instance, in modern German literature the terms *klebstoff, kleben, klebrigkeit* etc. were derived from the Middle–Low German *helen* (to stick) and *klei* (viscous sludge, clay), while *leim* and *lehm* are from the Latin *lutum*. The second synonym reveals the origin as the Latin *gluten, glutimo* (to glue up), and *glus* (a glue); in modern usage it refers predominantly to natural objects (glue), in biology, in particular, it can be traced in the term agglutination.

From the technological standpoint the distinction between the terms glue and substrate is quite conventional, as one and the same product may perform as any of the joint elements. Linguistic inhomogeneity of these terms is avoided by using one-root international definitions adhesive and adherend derived from adhesion. It seems more reasonable to use glue–substrate and adhesive–adherend combinations of terms; however, the actual usage is often a matter of traditions and not the philology.

The last term requires comment. By analogy, the phenomenon of adsorption, closely related to adhesion, and which involves sorption of adsorbate on to adsorbent, Dukes [124] suggested denoting the object interacting with the adhesive by the term adherent. Mittal [125] argues that this approach contradicts the rules of English grammar and suggests that to express the meaning the term adherate be inferred from the verb to adhere [126]. We believe that this term is related rather to that of the adsorbate and correspondingly to the adhesive.

Regardless of the outcome of this discussion there has arisen the necessity to unify international terminology. We assume that the terms adhesive and adherend be used to designate the idealized (theoretical) elements of an adhesive joint, while the terms glue and substrate be retained to refer to the real (technological) objects.

2

Thermodynamic approach to polymer adhesion

Thermodynamic analysis of polymer adhesion provides the unified basis for a more general treatment of the phenomenon with distinct regard for the processes of interfacial contact formation and adhesive interaction between the contacting objects. We recall that thermodynamics deals with thermodynamic systems represented as a material continuum of certain dimensions and describes them in terms of parameters of state and parameters of processes. An important role in thermodynamic analysis belongs to the concept of characteristic functions of the parameters, regarded as equilibrium potentials when referring to the characteristic functions of the parameters of state and kinetic potentials if the parameters of processes are involved.

2.1 FORMATION OF ADHESIVE CONTACT BETWEEN POLYMERS

2.1.1 Major thermodynamic characteristics of adhesive interaction

A surface characteristic of the condensed (either liquid or solid) phase may be expressed, according to the concepts of statistical physics, as some specific parameter of the bulk multiplied by the measure of surface protraction, i.e. as a certain surface excess. Let Y be any arbitrary extensive thermodynamic parameter and y the magnitude of the corresponding intensive parameter of the matter in the bulk with reference to a single molecule. Then the surface energy excess is given as the difference

$$\Delta^s Y = Y - Ny \tag{1}$$

where N is the number of molecules constituting the phase. This definition is of the most general in character; it is free of the necessity to define the particular location of the interface, as it does not imply that any specific model of the solid has to be chosen. Nevertheless, its experimental verification (by .means of calorimetric studies, for instance) allows for theoretical interpretation within the framework of either a continual [127], or an atomistic (in a quasiharmonic approximation) [128] model.

Such an approach, traditionally ascribed to Gibbs, does not take into account the function of the distance from the surface [129] of a geometrical interface assuming that extensive thermodynamic parameters are constant up to the boundary. However, this assumption is true only as a zero-order approximation and it has to be reconsidered already for one- [130], two-, and three-phase liquid systems [131]. For any parameter X, e.g. internal energy, free energy, entropy, number of moles and masses, corresponding excesses can be determined as

$$\Delta^s X = X - (X_1 + X_2).\tag{2}$$

According to Guggenheim, however, the surface layer is a region of certain finite thickness in which interfacial forces are not at equilibrium (Guggenheim's approximation). Such regions are formed when at least two phases intereact in the following combinations; liquid–gas, solid–gas, solid–liquid, solid–solid, and liquid–liquid, each denoted in abbreviated form as la, sa, sl, ss, and ll; a, s, and l being for air (which is the most common for a gas), solid, and liquid, respectively. It should also be borne in mind that the Gibbs' approach may scarcely be applied to the studies of real heterogeneous systems as here the layers of one phase between the interfaces may be so thin that the boundary layers overlap and excessive pressure, with regard to the adjoining phase, may arise within the layers (tending to wedge the adjoining phases as introduced by Derjaguin [132]). Let us elaborate on this problem in more detail.

A variety of approaches to the study of interfacial layers relying on statistical physics involves as a background the basic equations defining the state function (3) and the probability distribution function within the large canonical ensemble (4)

$$Z = \sum_N \frac{(j)^N}{N! \, h^{3N}} \left\{ \int \exp\left(-\frac{i}{2MKT}\right) di_x \, di_y \, di_z \right\}^N$$

$$\left\{ \int \exp\left(-\frac{E - \mu N}{KT}\right) dV_1 \dots dV_N \right\}\tag{3}$$

where i, i_x, i_y and i_z are the impetus of a microparticle and its projections on to coordinate axes, M is the mass of a microparticle, E is the total potential energy of interparticle interactions within the system, $dV_1, \dots, dV_N$ are microvolumes occupied by microparticles, j are the increments of vibrational and rotational motions, μ is the chemical potential of a microparticle (J), N the number of microparticles constituting the system, T is temperature in degrees K and K and h are the Boltzmann and Planck constants, respectively;

$$\omega = Z^{-1} \exp\left[-(KT)^{-1}(H_N - \mu N)\right]\tag{4}$$

where

$$H_N = T + E + \varphi; \quad T = \sum_{N=1}^{N} i^2/2M; \quad E = E(\mathbf{r}_1 \dots \mathbf{r}_N),$$

$$\varphi = \sum_{N=1}^{N} \varphi'(\mathbf{r}_N),$$

where $r_1 \ldots r_N$ are the radius-vectors of microparticles and $\varphi'(r_N)$ is the potential of the external force field affecting a particle. Restricting attention to two-particle interactions only, the interactions between three and more particles being neglected ($N \geqslant 3$), one can express the surface energy of a flat liquid–liquid interface via a binary distribution function in the following way

$$\sigma = 0.5 \int_{-\infty}^{\infty} dZ \int_{r_{12}} f_2(z_1, r_{12}) E(r_{12}) \frac{(x_{12})^2 - (z_{12})^2}{r_{12}} dr_{12} \qquad (5)$$

where $E(r_{12})$ is the interaction energy of the two particles 1 and 2, $r_{12} = r_1 - r_2$, $f(z_1, r_{12})$ is the binary distribution function, $x_{12} = x_1 - x_2$ and $z_{12} = z_1 - z_2$.

The rigorous classical treatment described (3)–(5) was recently complemented by the so-called quasi-thermodynamic analysis [133]. Within the framework of this approach the most acceptable averaging of various parameters, initially the intensive ones, is performed. For instance, the averaging procedure utilized in the quasi-thermodynamic analysis of the density profile of matter within the interfacial region is achieved by neglecting the orientation and length of the molecules along coordinate axes. Subsequent analysis involves Bakker's equation [134] which in the general form is given as follows [135]

$$\Delta \psi - \psi / k_1^2 = 4\pi k_2 \, \rho(x, y, z) \qquad (6)$$

where ρ is the density, ψ is the specific potential of interparticle interactions (J/kg) and k_1 (m) and k_2 (m^3/kg s^2) are the coordinate–independent constants. The $\rho = \rho(x, y, z)$ function, according to its definition, is a smooth one and, in the case under consideration, is reduced to a one-variable function of the distance from the interface $\rho = \rho(x)$. These considerations being taken into account, the thickness of the interfacial region between a pure liquid and its vapour appears to be determined only by the surface energy of a condensed phase and the corresponding densities

$$d = 8\pi^2 \sigma / RT(\rho_1 - \rho_2) \ln (\rho_1/\rho_2). \qquad (7)$$

The result is in agreement with the experimental data, e.g. those of Hey and Wood [137] concerning the thickness of the interfacial layer as a function of the surface energy of the liquid in a vapour–liquid system.

The approach described is supposed to complement the classical concept of Gibbs by extending its major limitation to the practical application of the thermodynamic treatment of adhesive interaction between contacting phases, polymeric in particular.

From a thermodynamical point of view the major difference between a molecule in the bulk phase and one at the interface is attributed to the non-symmetry of the force field affecting the molecule. In fact, three forces impose their effect on a molecule at the interface, while at least four forces affect a molecule within the bulk. Quantitatively this difference is expressed in energetic terms; hence surface energy quite naturally becomes the most important thermodynamic characteristic of the surface of the condensed phase.

For liquids surface energy can be easily evaluated from the data on enthalpies of vapourization. For a liquid metal, for instance, surface energy contributes 15% to the heat of vapourization, the error of determination being within the limits of 8%. In the case of solid metals sublimation energy is used to the same effect with the corresponding contribution of 16% [138].

One should bear in mind, however, that a rigorous interrelationship between the heat of vapourization and surface energy is quite complex, even for the most simple liquids, and may scarcely be capable of experimental verification [139]. Therefore, it appears sensible to make use of the most general concepts, viz. the concept of cohesion energy

$$E_{coh} = (\Delta H_v - RT)/V \qquad (8)$$

(ΔH_v — heat of vapourization) which is taken as the excess of the potential energy of a liquid over the potential energy of the ideal vapour [140]. Hence,

$$-E_{coh} = \Delta H_v - p(V_a - V_1) \qquad (9)$$

as $V_a \gg V_1$ and $pV_a = RT$ for ideal gas. The knowledge of boiling point is, in principle, sufficient to assess ΔH_v, for instance, Hildebrand suggested a semi-empirical expression

$$\Delta H_v = 0.02 \, T_b^2 + 23.7 \, T_b - 2950. \qquad (10)$$

The approach provided by the theory of thermodynamic congruence [141] seems to be more firmly based. According to the expanded law of corresponding states, the dimensionless heat of vapourization is a function of the criticial temperature T_c

$$\Delta H_v^* = \Delta H_v/RT_c. \qquad (11)$$

Then, the dimensionless surface energy is given by the following equation

$$\sigma^* = \sigma/(KT_c p_c^2)^{1/3} = N_A^{1/3} \, V^{2/3}/RT_c \qquad (12)$$

where N_A is the Avogadro number, and p_c the critical pressure. These equations explain the existence of a linear dependence between σ^* and ΔH_v^* [142] observed for 11 aliphatic, alicyclic, aromatic and diene hydrocarbons. Passing from ΔH_v^* and σ^* to the macroscopic characteristics by making use of the right group in expression (12), and taking into account the specific molar volume V [84, p. 13], it may be noted that, there is a linear correlation between σ and ΔH_v of the compounds listed. A more precise description of the behaviour of condensed phases at close-to-critical conditions was given on the basis of scaling concepts [143]. However, like most of the approaches referred to, these are only at the beginning of their application to high molecular mass compounds.

Considering the factors that would make it possible to overcome this serious drawback let us now discuss the physical implications of the fundamental thermodynamic characteristics of a surface.

Imagine an interface between the condensed phase and the vapour resulting from either evaporation or sublimation, on which a series of parallel 'notches', as deep as several interatomic distances, is formed so that it forms a closed circuit. To maintain a mechanical equilibrium force has to be applied to the atoms in the direction along the conventional 'notches'. Clearly, the force acts in the direction normal to the contour through the 'notches' and tangential to the liquid–gas interface at the point on the contour. The magnitude of these forces acting on the contour of 'notches' multiplied by the length of the contour represents the surface energy of a liquid often called the surface tension (the concept introduced by Leonardo da Vinci). This term is used to designate the stresses in the surface layer of a non-elastic body. Surface tension is defined as a vector normal to the portion (section) of the line drawn through the 'notches' and tangential to the interface between the phases. The modulus of this vector has the dimension of dynes cm^{-1} or Nm^{-1} (in SI).

In the case of two fluid masses the concept of surface tension as of a tangential force applied to a unit length of the contour confining a certain portion of the interface is equivalent to that of surface energy. Prigogine [144], however emphasizes that the equivalence is true only for one-component fluid systems in which there are no manifestations of adsorption, i.e. accumulation of mass at the surface does not take place.

Assessment of the surface energy of solids presents a much more sophisticated task. For a molecule to be removed from the solid surface then the bonds linking it to the neighbouring molecules and to those constituting the layer underneath have to be ruptured. Then [145]

$$\sigma = F + A(\mathrm{d}F/\mathrm{d}A)_{\mathrm{T}} \tag{13}$$

where A is the surface area and F is the surface free energy. While for liquids $A(\mathrm{d}F/\mathrm{d}A) = 0$ and consequently $\sigma_l = \sigma_s$, for solids the values of F and $A(\mathrm{d}F/\mathrm{d}A)$ are of the same order.[†] Therefore, the concept of surface tension of a solid is deprived of any physical sense and has to be substituted for that of surface energy. In spite of the fact that this conclusion is now beyond any doubt [146] since eq. (13) has been verified experimentally, nevertheless the difference between the two concepts is frequently ignored even in the specialist literature [101].

The origin of such terminological ambiguity is difficult to trace as it was Gibbs, the founder of the theory of equilibrium of heterogeneous substances,[‡] who emphasized the difference between the two quantities: 'As in the case of two fluid masses we may regard σ as expressing the work spent in forming a unit of the surface of discontinuity — under certain conditions which we need not here specify — but it cannot properly be regarded as expressing the tension of the surface. The latter quantity depends upon the work spent in stretching the surface, while the quantity σ depends upon *the work spent in forming the surface* (italicized by Gibbs: authors). But when one of the masses is solid, and its strains are to be distinguished, there is no such equivalence between the stretching of the surface and the forming of new surface' [129, p. 257]. Note that the last statement clarifies the validity of eq. (13).

The common reason for these concepts being confused follows from the fact that both have the same dimension (dynes cm^{-1} = erg cm^{-2}). However, in distinction from the surface tension, which is a vector, the surface energy is a scalar whose dimension is dynes (cm/cm cm). Analysing σ_s into normal (n) and tangential (t) components of the volumetric pressure tensor p with respect to the z-axis, Bakker derived the following equation [134]:

$$\sigma_s = \int_{-\infty}^{\infty} (p_n - p_t)\, \mathrm{d}z. \tag{14}$$

p_t can be expressed in terms of the internal pressure of the bulk phase p_u, concentration C_i^σ and chemical potential μ_i^σ of the surface layer [149]:

[†] One cannot omit the case when a special choice of equimolecular interface in a one-component systems results in $A(\mathrm{d}F/\mathrm{d}A) \to 0$.

[‡] It was once said that it was easier to use Gibbs' formulae than to comprehend them [148].

$$p_t = p_u + 0.5 \sum_i C_i^\sigma \mu_i^\sigma \tag{15}$$

In the last equation the chemical potential is very difficult to account for [150]. Gerbacia and Rosano [151], having modified Bakker's approach by taking into account the interphase distance r and the concentrations of surface layers of both phases 1 and 2, derived from eq. (14) the following

$$\sigma_s = r^3 \int_{-\infty}^{\infty} \sum_{1,2} k_{12} \left(\frac{dC_1^\sigma}{dz} \right) \left(\frac{dC_2^\sigma}{dz} \right) dz \tag{16}$$

where k_{12} is the positive coefficient of interfacial interaction. This result is in accord with the analysis of eq. (14) elaborated in terms of the quasi-thermodynamics of interfacial zones [136], i.e. when the forces arising from collisions of the microparticles constituting the system are taken into account, these forces are neglected in classical approaches [134, 151].

The surface layer of a solid is characterized as being under elastic stress. This state is reflected by the concept of surface stress γ which is defined as the sum of internal forces keeping the surface layer in an equilibrium position [145]. The stresses in the surface layer result from the fact that the molecules within it are subjected to forces directed towards the bulk of the phase. The quantity γ is a scalar and has the dimension of surface tension. For a solid both quantities are usually identical, although the thermodynamic analysis of surface stress is rather complicated when there is adsorption in the system [152]. However, strictly speaking, the identity is specific for the geometric interface only, while for the entire system γ and σ are different. To emphasize this fact Rusanov [153] suggested that γ be taken as the true surface tension and σ as the conventional surface tension; while Scherbakov [146] prefers to distinguish surface tension as being of types I and II. Both characterizing the processes of surface growth; but type I is attributed to the processes of interface growth due to the disintegration of the solid into parts, the newly formed surface being identical to the initial one, whereas type II is due to surface deformation.

The above discussion takes one back to the major problem of the physics of surface phenomena that is the non-equivalence of the concepts of surface energy and surface tension. The latter acknowledges the mechanical (a two-dimensional tensor γ) and the thermodynamic (σ) definitions [154]. They may be related to each other via the following expression

$$\sigma = \gamma + A^{-1} \sum_i \overline{\mu_i M_i} \tag{17}$$

which may be derived from eq. (13) by giving the variation of free energy in terms of excesses of the chemical potential and mass $\overline{\mu_i M_i}$ for the immobile components of the ith solid body. As it has been already stated, for liquids $\sigma = \gamma$. For solids $\sigma \neq \gamma$, except for the specific case when $A^{-1} \sum_i \overline{\mu_i M_i} = 0$.

In fact, in liquids, since there is no long-range ordering, the immediate localized surrounding of atoms does not depend on the conditions at the boundary of the system, while in the case of solids the opposite situation applies [155. Therefore, in liquid systems in surface stress may be treated as the isotropic surface tension, whereas in solids it produces compensatory stresses even in the bulk of the sample being particularly manifested in small particles [156] and in multi-component systems such as alloys [157].

These views are also valid in the thermodynamics of irreversible processes [158]. Moreover, they are strictly prerequisites in terms of mechanics. Actually, equations of state make it possible to relate the tensor of surface stresses to specific thermodynamic potentials reduced to a unit surface area (with surface energy in particular), and on this basis eq. (13) is easily derived [159]. However, it is not these implications that need to be discussed at this point, nor even the fact that it is true for materials that are most common, namely crystals [146] and polymers [160]. What is really worth discussing is the physical meaning of the inequality $\sigma \neq \gamma$.

This may be seen via an analysis of the tensor character of surface tension and of the general relation between surface energy and the thermodynamic surface tension as half the trace of the surface stress tensor [161]. Calculations performed within the framework of this approach demonstrate, for instance, that for alkali halide crystals e.g. sodium and potassium chlorides, the increment of the vibrational component to the surface tension at $T = 0\,\mathrm{K}\,(\sigma = \sigma_0 - \sigma_K)$ is around 2–3%, whereas for surface energy it amounts to 40% [162]. Consequently, the difference between surface energy and surface tension is due to the different mobilities of the particles. As a mechanism for such processes Kornblit and Ignatiev [163] suggested the migration of dislocational groups [163]. The validity of this mechanism was confirmed in computations of the surface energy based on determining the work of dislocation formation as a function of the geometric parameters of dislocations, elastic shear modulus and Poisson's coefficient. Then the surface energy is also a function of the elastic modulus, Poisson's coefficient, and the value of Burgers' vector. For 17 metals the computed values of the surface energy were found to be in adequate agreement with the experimental data. In the general case the surface energy of solid bodies is given by the difference

$$\sigma_{sa} = \sigma_0 - \sigma_s' \tag{18}$$

where σ_0 is the energy required to remove an atom from the surface and σ_s' is the energy required to move an atom within the surface to the position from which it is to be removed off the surface (the said migration taking place, for instance, via the above dislocational mechanism). The calculations performed produced $\sigma_s' \approx 0.5\sigma_{sa}$ for ionic crystals, while for the crystals of noble gases $\sigma_s' \approx 0.01\sigma_{sa}$ [164], which is just what one could anticipate.

Let us now examine the major question of how the magnitudes of the bulk and surface energies of a condensed phase are related to each other [165]. Relying on the most general considerations concerning non-symmetry of the force field at the surface, one can assume $\sigma^s/\sigma^v > 1$. For instance, for non-associated phases, according to Frenkel [166] the potential energy excess is

$$s_E = 0.5\,E_{12}\,N_{(1)}^s(N^s - N^v) \tag{19}$$

where E_{12} is the potential energy of the interactions between molecules 1 and 2, $N_{(1)}^s$ the number of molecules per unit surface area of a solid and N^s and N^v the number of the neighbour-molecules from the surface and the bulk of the phase, respectively. Consequently, surface excesses are not merely an abstract concept. According to expression (2), at the surface of a condensed phase there always exists an excess number of molecules whose specific value $\Gamma = \Delta^s N / A$ describes mass 'accumulation' at the surface often described as adsorption. In other words, under any conditions, except those of vacuum, the surface is always enriched with molecules of the environment. Thermodynamic treatment of the problem [167] involves basically the common statistical physics approach regarding adsorbate as a two-dimensional analogue of three-dimensional systems; the adsorbed molecule is assumed to be localized in the vicinity of an active site at the surface of an adsorbent [168]. The mathematical procedure within the framework of this approach is considerably simplified.

Then, to the three parameters characterizing the adsorbent, namely temperature T, surface concentration C^σ (Γ in the case under consideration), and chemical potential μ (more precisely the activity, i.e. the exponential function of the chemical potential $\exp(\mu/KT)$), the fourth is added, i.e. the intensive variable which describes the difference between surface energies of the pure adsorbent and of the resultant two-component system. In analogy with the common equation of state, relating pressure, molar density and temperature, for the adsorbed layer the corresponding expression will acquire the form

$$\pi = - \int_{p=0}^{p} (\Gamma/KT)\, \mathrm{d} \ln p \qquad (20)$$

where K is the Boltzman constant. According to the definition

$$\pi = \sigma^s - \sigma_0^s. \qquad (21)$$

Expressing π via the quantities commonly measured under equilibrium conditions, Hill has obtained the following equation [169]

$$\pi = RT/A V_m \int_{p=0}^{p} (V_a/p)\, \mathrm{d}p \qquad (22)$$

where V_a and V_m are the molar volumes of the adsorbate at the surface and at normal conditions, R is the Gas constant. Despite the energetic treatment of π resulting from eq. (21) this parameter is inherently of molecular–kinetic origin. This is required since it is directly proportional to the mobility of molecules in the surface layers of an adsorbent [170] and inversely proportional to the activation energy for the migration of the adsorbate molecule from one active site at the surface to the other [171]. It is clear that these ideas, together with those determining eq. (19), are also true with regard to adhesive interaction.

To complete this survey of the basic thermodynamic parameters let us list some major functions involving these parameters. For the bulk of the phases the thermodynamic potentials may be written in the form of Gibbs function (Gibbs free energy, free enthalpy)

$$G^v = E^v - TS^v + pV = F^v + pV = \sum_i \mu_i N_i^v, \qquad (23)$$

the Helmholtz function (Helmholtz free energy)

$$F^v = E^v - TS^v = G^v - pV = \sum_i \mu_i N_i^v - pV \qquad (24)$$

or the Gerring function

$$\Omega^{v} = E^{v} - TS^{v} - \sum_{i} \mu_{i}N_{i}^{v} = F^{v} - \sum_{i} \mu_{i}N_{i}^{v}. \tag{25}$$

For systems comprising interfaces these functions are as follows

$$G = E - TS + pV - \sigma A, \tag{26}$$

$$F = E - TS, \tag{27}$$

$$\Omega = E - TS - \sum_{i} \mu_{i}N_{i} = F - G = \sigma A - pV. \tag{28}$$

Thermodynamic surface functions may be easily obtained as the difference between corresponding equations of the two groups (23)–(25) and (26)–(28) resulting in expressions of the type of eq. (2):

$$G^{s} = F^{s} - \sigma A, \tag{29}$$

$$F^{s} = E^{s} - TS^{s}, \tag{30}$$

$$\Omega^{s} = E^{s} - TS^{s} - \sum_{i} \mu_{i}N_{i}^{s} = F^{s} - G^{s}. \tag{31}$$

It is a much rarer case when the concept of surface stress is included in thermodynamic analysis. Thermodynamic potentials of solid surfaces involving γ are given by expressions similar to (23), (24), (26), and (29) and were subjected to a thorough examination by Rusanov [153] and Balmer [172]. Note that according to definition γ is given by a two-term expression relating it (γ) to the work required to change the surface area and to the deformative factor [155], the second being different from zero only for elastically deformed bodies, e.g. polymers.

Besides basic thermodynamic concepts there are also other parameters, e.g. contact angle, that are essential for the description of adhesion processes. However, due to the fact that they are directly derived from experimental measurements they are worth examining in subsequent sections.

Let us now survey the major consequences of the thermodynamic approach to the phenomenon of adhesion.

Creation of a new surface involves certain work which for an isochoric–isothermal process is equal to $\sigma\Delta A$. Consider the interaction between a solid and a liquid. Neglecting the shape-dependence of the volume of a liquid one may derive from eq. (23) the following

$$U = TS - pV + \sum_{i} \mu_{i}N_{i}. \tag{32}$$

It is then easy to obtain

$$U = T(S_{sl} - S_{s} - S_{l}) + \mu(N_{sl} - N_{s} - N_{l}) + \sigma\Delta A. \tag{33}$$

Dividing eq. (33) by ΔA one comes to the general relationship

$$U_{sl} = TS_{sl} + \sigma + \mu\Gamma \tag{34}$$

where U_{sl} is the maximum total energy excess in the surface layer as compared to the phases and S_{sl} is the corresponding entropy excess. By analogy with eq. (19) expressions may be derived for the maximum potential energy excesses in the surface layer both on the solid side (substrate)

$$\Delta^s E = 0.5 \, E_{sa} N^s_{(1)} (N^s_s - N^v_s) + U_{sl} N^s_s \tag{35}$$

and of a liquid (adhesive)

$$\Delta^l E = 0.5 \, E_{la} N^l_{(1)} (N^s_l - N^v_l) + U_{sl} N^s_l. \tag{36}$$

This thermodynamic approach makes it possible to elaborate on at least two problems which are essentially important in the physical chemistry of adhesion. These are the relationship between surface energy and adsorption, and the pattern of the energy spectrum of the adhesive joint. Let us examine these problems in more detail.

At any moment τ the total surface energy U^s of a system comprising an adhesive and a substrate is given in the form of eq. (34)

$$U^s = TS^s + \sigma_s + \mu\Gamma \tag{37}$$

where S^s is the surface entropy. As at $\tau = 0$, there is as yet no adsorption at the substrate surface and the initial value of the total surface energy of a substrate is correspondingly

$$U^s_s = TS^s_s + \sigma_s. \tag{38}$$

The energy conservation law for the case under consideration can be written as follows

$$U^s \pm c\Delta T = U^s_l + U^s_s \tag{39}$$

where c is the heat capacity (the sign depends on whether the adhesive joint is heated or cooled in the temperature range $\Delta T = T - T_0$). Then combining these equations one obtains

$$(\sigma_s + \sigma_l) + T(S^s_s + S^s_l) = c\Delta T + \sigma_s + TS^s + \mu\Gamma. \tag{40}$$

On the other hand, the entropy gain due to the fact that the molecules are engaged in adhesive interaction should be generally governed by the magnitude of the ratio between the mean frequencies of thermal vibrations [166]

$$\Delta S = \Delta N_{sl} \ln (\bar{\nu}_l / \bar{\nu}_s) \tag{41}$$

where ΔN_{sl} is the number of molecules additionally incorporated within the interfacial region due to adsorption and which may be given by the following

$$N_{sl} = -k_r \Gamma. \tag{42}$$

Hence the relationship combining Γ and σ may be derived.

Zilberman, who suggested the approach described, obtained the following dependences: for the isothermal formation of adhesive joints

$$\Gamma^{max}_{\Delta T=0} = \frac{\sigma_s + \sigma_l}{-K_r T \ln (\bar{\nu}_l / \bar{\nu}_s) + \mu} \tag{43}$$

and for non-isothermal formation

$$\Gamma^{max}_T = \frac{\sigma_s + \sigma_l}{\mu - K_r T \ln (\bar{\nu}_l / \bar{\nu}_s)} + \Delta T \frac{c + S^s_l + S^s_s}{\mu - K_r T \ln (\bar{\nu}_l / \bar{\nu}_s)}. \tag{44}$$

Eq. (43) indicates that even when there is no chemical interaction between the phases, due to the restrained mobility of the adhesive molecules in contact with the substrate, the first member of the denominator increases more rapidly than the numerator. According to eq. (44), increasing the temperature results in a non-linear growth of adsorption. An important inference from both equations is that Γ versus σ dependence deviates from a linear one.

Let us now elaborate on the second of the problems stated above, namely the one concerning the energy spectrum of the adhesive joint. U_{sl} defined according to eq. (34) as the total energy excess of the interfacial zone over the energies of each of the contacting phases provides the parameter that determines the feasibility of adhesive interaction on a fundamental thermodynamic basis.

The total surface energy of an adhesive U_l^s (or a substrate U_s^s) comprises the energy of interactions within the bulk of the phase U_l^v (U_s^v) and the energy $U^{s'}$ which is due to the non-symmetry of the force field at the surface. The first component is determined by the nature of interacting molecules being measured by the work required to pull apart the molecules within the phase. The second component is measured as the work required to separate the molecules when they are affected solely by the non-symmetric force field. The total interaction energy resulting from molecular contact between a solid and a liquid U_{sl} comprises the energy of attractive interaction in the volume of each phase and the energy of interaction governed by the symmetry gradient of the force field in the inter-phase zone. The first component is measured as the work required to remove the molecule of adhesive (substrate) from out of its (adhesive/substrate) bulk. The second component is measured as the work that should have been done to separate the molecules of both phases provided that their interaction is due solely to the non-symmetry of the force field at the surface.

The interfacial layer resulting from adhesive interaction is characterized by the force field of its own and by the total energy U_{sl} whose value may either exceed the specific cohesive energies of each of the elements of the system (E_s^{coh} and E_l^{coh}) or may have some intermediate value. If two separate parts belonging to one phase are brought into molecular contact resulting in autohesive interaction (in this case $E_1^{coh} = E_2^{coh}$) then $U_{sl} > E_1^{coh}(E_2^{coh})$ and the total energy of the interfacial zone exceeds the energy of cohesive interaction in both phases U_{sl} reaching its maximum at the geometric interface between the phases. The physical meaning of this quantity may be interpreted by analogy with the cohesive energy density [140] as the specific energy of adhesion.

The concepts described make it possible to examine the energy spectrum of an adhesive joint comprising two substrates and a layer of adhesive between them. Fig. 2.1 shows a schematic diagram of a typical spectrum. The diagram, as it is, allows for a rough comparison of the energies pertaining to the elements constituting the system. The detailed discussion of this spectrum should involve questions concerning the existence and the effect of boundary and transition layers; these are examined in section 3.1.1. At this point it is sufficient to emphasize that the form of the spectrum, i.e. the existence of maxima and minima and the mode of energy variation within the interfacial zone, is governed by the ratio of the cohesive energies of both substrates to the adhesive layered between them.

Having described the role of U_{sl} in adhesive interaction from the phenomenological point of view let us now proceed to discuss the thermodynamic background of the relationships. It is easy to convince oneself that the quantities in eq. (34) are not just

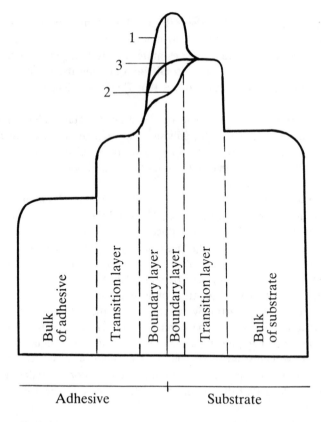

Fig. 2.1 – Energy spectrum of adhesive joint: 1, $U_l^s \ll U_s^s$; 2, $U_l^s \leqslant U_s^s$; 3, $U_l^s < U_s^s$.

mathematical abstractions, as Gibbs presumed. In fact, if the pattern of intermolecular interaction potential and the radial distribution function are known the gradients of the parameters discussed may be determined and the planes may be chosen so as to define the range that would 'contain', say, 80, 90 or 99.9% of the isochoric–isothermal potentials values of the corresponding bulk phases [173].

To conclude this section let us now elaborate on the major problem concerning the limitations of the thermodynamic approach in describing polymer adhesion.

When speaking of the merits and demerits of thermodynamics both the generality and the rigour of this discipline are referred to initially. Thermodynamics tends to reflect the ultimate state of a system, regardless of the distinct mechanism via which it (state) has been reached. In this respect thermodynamical methods are formal, however, they also make it possible to reveal analogies in describing phenomena that are quite different, and in particular those comprising adhesion [174]. Of the two main functions, general theoretical and applied, Rusanov [175] considers the latter to be the major application of thermodynamics today. It involves relevant dynamic methods being used to assess, from experimentation data, the various properties, surface ones in particular, of materials. It is quite naturally implied that because of its generality thermodynamics cannot pretend to provide an exhaustive description of examined materials.

To eliminate this drawback the equation of state of the system should be included in the thermodynamical mathematics. This equation of state could probably have been obtained within the framework of statistical mechanics [159]. According to the modern interpretation of the theory of corresponding states the functional form of the free energy of a system presented in reduced variables is determined by the pattern of the intermolecular interaction potential and by the type of statistics controlling the system [176]. Hence, such 'supplementation' of thermodynamics offers the required degree of generality along with a more substantial interpretation of the computational results [177]. Indeed, for classical systems that are characterized by the same type of this potential of the equation of state the diagrams of state should also be universal provided that the list of variables is retained. Different polymer characteristics may be used for such variables.

For instance, Verkin et al. [178] used the molar packing coefficients of polymers to obtain a number of fundamental relations between certain thermophysical parameters. In particular, this approach enabled the derivation of a universal diagram of states of a polymer in different phases that satisfactorily fitted experimental data [179]. As another parameter of the bulk though a less fundamental one, the volume itself may be used in the equation of state. For instance, Doolittle [180] suggested that the equation of state be represented as a function of the specific energy versus the difference $(v^{1/3} - v_o^{1/3})$, where v and v_0 refer to the specific volume and specific 'occupied' volume of the system, respectively. The dependence gives a fine fit of the computed values to those observed for all thermodynamic properties of various liquids, including polymeric adhesives: the average standard percentage error for 1034 points relating to the 'region of regular performance' of a liquid is 0.32%; temperature dependences of specific volume, energy and entropy were found to be linear. For the 'region of irregular performance' of a liquid the increasing contribution of cohesive forces probably has to be taken into account. For polymer melts (a quite common class of adhesives) the equation of state was suggested which related surface energy to internal pressure in the liquid [181]. Having been thus modified it yielded, in particular, a sufficiently precise assessment of σ versus temperature and molecular mass dependences [182]. Another illustration of the approach is the use of criticial temperature and critical pressure to calculate dimensionless surface energy [84, 142] (see p. 29). Analogous solutions were derived relying on the concept of free volume. On the basis of this concept, despite the fact that the latter is less rigorous than those above, the well-known equation of Williams—Landell—Ferry was generalized and the viscosity and stress relaxation in organic and organic glasses were reliably computed [183].

These issues are fully applicable to polymer adhesion. In many cases thermodynamics appears to provide the only opportunity to describe the behaviour of various systems, particularly when there is no information concerning the mechanism of the distinct adhesive interaction between the elements comprising a system. The major restrictions to this approach may be eliminated by supplementing it with the methods of statistical mechanics. On the other hand, mathematical models ostensibly in general must involve the parameters of molecular–kinetic origin (e.g. rheological) and these are treated in section 3.2.1. The situation is that which perfectly matches the Cont principle 'not or, but and'. Of the two approaches to adhesion phenomena, thermodynamic or molecular–kinetic, neither can be considered to have the prior claim. It is only when combined to supplement each other they provide for a thorough description of the formation and performance of polymer adhesive joints.

2.1.2 Thermodynamics of wetting and spreading of polymers during their adhesive interaction

Wetting is defined as a phenomenon of spontaneous (or forced as when the process is carried out deliberately) diminution of the energy of a system which occurs when its components interact. Actually, it may be exhaustively treated in terms of thermo-dynamics (at least in the case of low-molecular mass compounds for which there is no need to account for rheological effects). In the simplest and most common case such a system is characterized by the value of the angle between the surface of a drop and that of a solid, the so-called contact angle. For systems at equilibrium (equilibrium wetting), according to Dorsey [184], this angle is designated as θ, and for non-equilibrium systems φ (advancing and receding of a liquid showing advancing and receding contact angles). When examining surface phenomena and adhesion in particular, mainly contact wetting is considered. In this case interaction involves three phases, and not two as in the example of immersion wetting where a solid is fully submerged in a liquid.

Consider a system comprising three homogeneous phases whose surfaces intersect along one line (Fig. 2.2). One of the phases is a lens-shaped liquid and its dimensions are such that the gravitational force does not substantially affect the form of the interfaces (surfaces of discontinuity). When the temperatures and chemical potentials are constant and equal, and the volumes of the phases do not change on reversibly varying the position of the surfaces of discontinuity, the general condition for equilibrium of a system has, according to Gibbs [129], the form

$$\sum_i \sigma_i \, dA_i + l \sum_i \sigma_{i+1} \, dT_{i+1} = 0 \qquad (45)$$

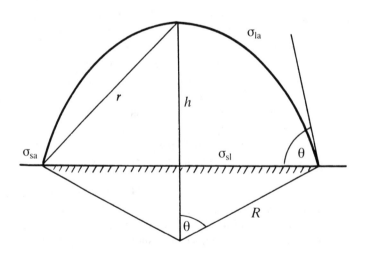

Fig. 2.2 – Wetting of a solid surface by a liquid.

where l is the perimeter of wetting, σ_{i+1} and dT_{i+1} are the increments of the respective parameters corresponding to dislocation of any element of the line l and in the direction normal to it within a plane tangential to the ith surface. If the surfaces of discontinuity form a continuous plane the condition for equilibrium at the perimeter is given by the expression first suggested by Young [185]

$$l \sum_i \sigma_{i+1} \, \mathrm{d}T_{i+1} = \sigma_{AB} - \sigma_{BC} - \sigma_{AC} \cos = 0. \tag{46}$$

The condition for equilibrium (45) may be represented as

$$\mathrm{d}F/\mathrm{d}A_{BC} = 0. \tag{47}$$

Considering the changes of surface energies at the interfaces between the three phases

$$\mathrm{d}A_{la}/\mathrm{d}A_{sl} = (\sigma_{sa} - \sigma_{sl})/\sigma_{la}. \tag{48}$$

According to Summ and Goryunov [186], the area A of the base of a wetting drop may be substituted for the radius r. Then for the system shown in Fig. 2.2

$$\mathrm{d}A_{sl} = 2\pi r \, \mathrm{d}r \tag{49}$$

and

$$\mathrm{d}A_{la} = 2\pi(r \, \mathrm{d}r + h \, \mathrm{d}h). \tag{50}$$

Taking into account that, on spreading, the drop does not change its volume, i.e. $V = \pi(3r^2h + h^3)/6$, one has

$$\frac{\mathrm{d}V}{\mathrm{d}r} = \frac{\pi}{2} \left(r^2 \frac{\mathrm{d}h}{\mathrm{d}r} + 2hr + h^2 \frac{\mathrm{d}h}{\mathrm{d}r} \right) = 0 \tag{51}$$

and

$$\mathrm{d}h = \frac{-2\pi r \, \mathrm{d}r}{r^2 + h^2}. \tag{52}$$

From the geometric considerations regarding the system presented in Fig. 2.2 (the OAB triangle) it follows that

$$(R - r)/R = \cos \theta. \tag{53}$$

Then

$$\mathrm{d}A_{la}/\mathrm{d}A_{sl} = \cos \theta \tag{54}$$

and the condition for equilibrium acquires the form of Young's equation

$$\frac{\sigma_{sa} - \sigma_{sl}}{\sigma_{la}} = \cos \theta. \tag{55}$$

This equation can also be derived from the general concepts of mathematical simulation theory. Within its framework quantitative evaluation of variations of the parameters of state of material continua which accompany different physico-chemical processes can be performed by constructing equations balancing moment and the moments of moment. For a one-dimensional continuum (a line) such equations of balance are written as the condition of a discontinuous jump at the boundary line dividing the phases [187, 188]. Neglecting the moments one has

$$n_{sl}\sigma_{sl} + n_{sa}\sigma_{sa} + n_{la}\sigma_{la} - \frac{\partial \gamma_L}{\partial \gamma'} \, l - k_L \gamma_L t = 0 \tag{56}$$

where n and t are the unitary vectors normal and tangential to the line of contact, γ_L is the linear stress tensor, γ' is its component ($\gamma' = n_j \gamma_j$), and k_L is a coefficient. When the system is axially symmetric this equation acquires the form analogous to that of the equation independently derived on the basis of the different concept of Rusanov which is treated below [189] (77):

$$\sigma_{la} \cos \theta = \sigma_{sa} - \sigma_{sl} - \frac{\gamma_L}{r} \cos \varphi \qquad (57)$$

where r is the radius of the base of the drop as in eqs (49)–(52), φ is the angle between n and t vectors, i.e. the tilt angle of the la interace at the edge where three phases meet. At $\varphi = 0$ eq. (57) is further simplified to the relationship previously reported [190]

$$\sigma_{la} \cos \theta = \sigma_{sa} - \sigma_{sl} - \gamma_L/r. \qquad (58)$$

In the case under consideration, i.e. of a one-dimensional material continuum, the γ_L tensor can be neglected. Then the basic equation (56) is reduced to give a vector equation for the Davydov–Neuman triangle, while eq. (58) leads to eq. (55) [159, p. 94].

The parameters of eq. (55) have deep physical meaning. In conformity with eq. (27) they may be represented as partial derivatives of Helmholtz free energy:

$$\sigma_{sa} = (\partial F / \partial A_{sa})_{T,\mu_i}, \qquad (59)$$

$$\sigma_{la} = (\partial F / \partial A_{la})_{T,\mu_i}, \qquad (60)$$

$$\sigma_{sl} = (\partial F / \partial A_{sl})_{T,\mu_i}. \qquad (61)$$

Nevertheless it should be remembered that eq. (55) was derived using certain assumptions which exclude some aspects of the performance of real adhesive systems ('There are some subtleties with respect to the physical chemical meaning of the contact angle equation . . .', claims Adamson [112]).

Eq. (55) refers to the state of hydrostatic equilibrium at the interphase boundaries in a three phase system. Quite obviously it does not hold true during spreading [191].

In fact, a drop of polymeric adhesive on a solid surface is rarely spherical; as a rule it asymptotically tends towards the substrate surface [192]. Hence, macroscopic (θ) and microscopic (ϑ) contact angles should be distinguished (Fig. 2.3). The former obey Young's equation, while to calculate microscopic contact angles one has to take into account the capillary effects due to the radius of curvature r and which cannot be neglected [193] when the drop is small enough. In the general case θ and ϑ are interrelated by the following relationship

$$\cos \theta = \cos \vartheta \frac{\sigma_{la}}{\sigma_{la} - r(d\sigma_{la}/dr)}. \qquad (62)$$

Distinguishing macro- and microscopic contact angles makes the thermodynamic analysis of three phase systems a much more complicated task [194]. It is however a quite general observation that the mode of the surface free energy change apparently fits the pattern of the rising curve of the hyperbolic function [195]. Fig. 2.4 depicts, for example, surface free energy as a function of macroscopic contact angle, it being assumed that $\theta + \vartheta = 180°$. Both of these characteristics are free from the effects relating to the geometry of a system and to the influence of the gravitational field. If interfacial inter-

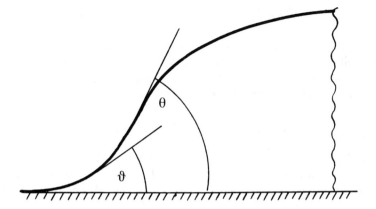

Fig. 2.3 — Micro- (ϑ) and macroscopic (θ) contact angles for a liquid wetting a solid.

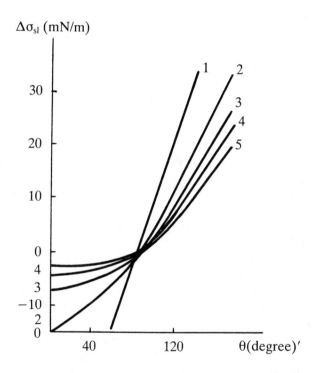

Fig. 2.4 — Interfacial energy for polytetrafluoroethene wetted with a liquid with $\sigma_{la} = 50$ mN/m versus macroscopic contact angle for various magnitudes of microscopic contact angle ϑ: 15° (1), 30° (2), 45° (3), 60° (4), and 90° (5).

action between a liquid and a solid surface results in σ_{sa} being greater than that of the 'unperturbed-by-the-substrate' state, then $\vartheta > \theta$ when $\vartheta > 45°$ and $\vartheta < \theta$ when $\vartheta < 45°$ [196].

It is important to note that σ_{la} increases when an initially spherical drop acquires a 'flattened' shape. However, the formation of a new interface which occurs on wetting causes σ_{la} to decrease. This decrease is less than the work of adhesion by the value of σ_{sa} [197].

Besides, this, when using eq. (55) to describe a system comprising a liquid–gas interface and a solid spherical particle immersed in the liquid one begins to understand the necessity for geometrical considerations to be taken into account. Huntsberger [198] gives the following relationship between θ and h (depth of immersion of a particle of radius r into the liquid)

$$\cos \theta = (1 + 2r + h)/r. \tag{63}$$

Clearly, the effect is magnified when the particle is non-spherical.

In reality one always has to take account of the adsorption of a wetting liquid (adhesive) by the surface of a solid substrate which decreases σ_{sa}. Introducing π from eq. (22), one obtains

$$\cos \theta = (\sigma_{sa}^\circ - \sigma_{sl})/\sigma_{la} = (\sigma_{sa} - \sigma_{sl} - \pi)/\sigma_{la}. \tag{64}$$

Hence, at thermodynamic equilibrium the wetting perimeter borders the already wetted surface (as Frenkel describes it — the spreading drop 'layers a path underneath itself' [166]). Interaction of a liquid with a solid substrate, as seen by high speed cinephotography of 1000–1500 frames per second, initially results in a compressed disk-like form of the drop (liquid advancing) which is subsequently changed to a spherical one[†] (liquid receding) the whole process showing a decaying pulsating pattern [199]. Consequently, each subsequent step of interphase interaction involves either discrete (unfused disks) or continuous (a film) fluid coating over the solid surface rather than the solid surface (unwetted) proper.

In the case of low-energy hydrophobic surfaces [201], e.g. of polyethene and polytetrafluoroethene, this effect is apparently negligible as for these with water $\pi \approx 0$. For high-energy surfaces (e.g. quartz whose surface is actually always covered with a monolayer of water) the extent to which the effect is manifested is difficult to assess as there is no means by which the precise value of σ_{sa} can be determined; the measured value is, as a rule, lower than the real one due to adsorption. However, quite reliable data were obtained for a glass substrate. According to interferometric measurements, the sorption of oxygen, nitrogen and argon on its surface at 79–90 K was almost linearly related to the parameter π [202]. As an example, Fig. 2.5 shows the dependence of π on the length of a sintered glass sample (that is in its turn a function of adsorbed water) [203]. As can be seen, the dependence is linear.

The effect of an area developing in front of a spreading drop, i.e. of precursive film formation, was first observed by Hardy [204]. He attributed it to condensation of the liquid from the vapour phase around the spreading drop. Hence the effect was assumed to be characteristic of volatile (with low b.p.) liquids. However, this assumption was subsequently rejected as similar films where shown to exist in all cases, even when there was no vapour fraction at all [205]. Within the framework of the scaling concept the thickness of the precursive film is determined by the viscosity of a polymer, the

---

† A rigorous description of the shape of an axially symmetrical drop on a solid support was derived from the theory of perturbations with the aid of the first and the second order solutions of the differential equation of the second order [200].

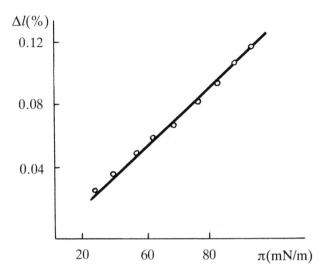

Fig. 2.5 – Surface pressure π versus variation in length of a porous glass sample accompanying water adsorption.

magnitude of the contact angle in the system, and the geometry of the boundary layers of the wetted solid substrate [206].

Since real surfaces of solid bodies are characterized by a somewhat developed topography the last factor to be considered has major significance, with regard to most of the aspects involved in the wetting process. Surface roughness is usually evaluated as the ratio between the actual surface area and its projection on to a horizontal plane. The defects of polymeric surfaces have a rather diverse shape varying from cylindrical and cone-like to trapezoidal [207] (the effect that this factor produces on the process of adhesive joint formation is considered in a three-component Vasenin's equation [208] by endowing discrete values to the power index). The shape of the defects is of prime importance when the rheological aspects of adhesive joint formation (section 3.2.1) and the manufacturing technology of adhesive bonds (section 4.2.2) are treated. However, it is seen even at the earliest stages to be involved in adhesive joint formation, namely during wetting of the substrate by the adhesive [192, 209]. For instance, a polytetra-fluoroethene sample thoroughly defatted with absolute ethanol and subsequently kept for a day under dry nitrogen gave the contact angle with water of 108° when its surface had been polished with aluminium, whereas the contact angle was 124° when the substrate surface was an unpolished sample produced by compression moulding of the granules or powder [210].

Taking into account the highly developed relief of the substrate surface it is natural to believe that the wetting rate should be dependent on the mutual position of the front of the spreading liquid and the profile of the defects (e.g. grooves) on the surface of the solid. Indeed, a study of the spreading of mercury over a zinc surface carved with grooves of triangular shape (112°) 170 μm deep [211] confirm the suggested dependence (Fig. 2.6). One can see that as the distance between the grooves (arranged in a staggered configuration) diminishes the spreading rate of mercury decreases due to the mechanical impediments to a moving liquid. In a detailed analysis two stages of the process were distinguished [212] — a short term (hundreths of a second) stage corresponding to

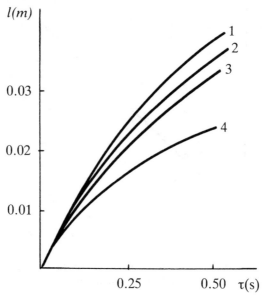

Fig. 2.6 — The distance between the moving front of a liquid and the wetting perimeter of a drop of mercury on a zinc surface as a function of time. 1, Single isolated groove, a system of parallel grooves; 2, staggered arrangement of grooves with the distance between the nodes 0.02 m; 3, ditto with the internodal distance 0.01 m; 4, ditto with the internodal distance 0.005 m.

formation of a 'sessile' drop and long term (tens of seconds) stage corresponding to streaming of a fluid from out of the 'sessile' drop and along the grooves crossing the perimeter of the base of the drop. This effect is especially easy to perceive when the grooves do not cross each other and the liquid spreads parallel or perpendicular to them (see Fig. 2.7). This case is one of the most common. With this case, as with the simpler ones of staggered arrangement of the grooves [211] the spreading of the liquid along the groove and across it [213, 214] were given a kinetic description.

The general condition for a liquid at equilibrium on a rough surface is described by Wenzel's equation [215]

$$\cos \theta_r = k_r \cos \theta \qquad (65)$$

where k_r is the surface roughness coefficient. This relationship can be derived from thermodynamic considerations [216]. Examining the surface roughness of PTFE samples subjected to cyclic thermal treatment, Dettre and Johnson [217] discovered that surface roughness varied in a manner inversely related to the advancing and receding angles with water (Fig. 2.8). However, it is only the advancing angles that fit the relationship (65); receding angles do not usually obey thermodynamic dependences [218]. In general, surface roughness of the substrate tends to reduce the values of θ at $\theta < 90°$, whereas the opposite effect is found at $\theta > 90°$ [194]. Künzer and Bonart [219] examined the wetting of substrates that were different in chemical nature but possessed similar surface relief as they were prepared as polyacrylate replicas of steel samples. Their conclusion was that the adhesive joint strength was better correlated to the surface roughness than to θ.

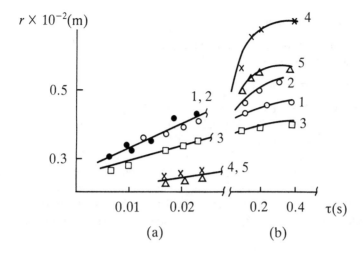

Fig. 2.7 – (a) Spreading of ethanol (1–3) and mercury (4, 5) drops on a zinc surface parallel (1, 4) and perpendicular (3, 5) to the grooves of triangular profile with the angle at the top 112° and 170 μm deep; (b) shows the spreading on a polished surface.

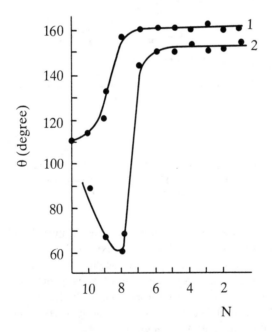

Fig. 2.8 – Contact angles of water with polytetrafluoroethene versus surface roughness of PTFE samples. 1, Advancing; 2, receding.

This inference is quite reasonable as the $\beta(\cos\theta_r - \cos\theta)$ term from Wenzel's original form of eq. (65) [215] (β is the Wenzel coefficient) is constant and, according to Kawasaki [220], identical to the k_w coefficient of the relationship relating the interfacial energy σ_{la} to the mass M of a liquid drop, the width l of the wetting path on the plane tilted to the angle ψ_w at which the drop slides down the plane

$$M g \sin \psi_w = k_w l \, \sigma_{\text{la}}. \qquad (66)$$

This relationship was proposed by Bikerman [221] who inferred it from a consideration of the well-known equation of Frenkel [222]

$$M g \sin \psi_w = 2 r_w W_{\text{Ad}} \qquad (67)$$

where r_w is the radius of the sliding drop, and W_{Ad} is the thermodynamic work of adhesion discussed later (p. 59). The validity of eq. (66) was verified by Johnson [223] and Olsen *et al.* [224]: however, at this point we consider it more important to stress the validity of the analogy between 'Wenzel's difference' and k_w. Saito and Akagawa [225] established that eq. (66) is true only when the wetted surfaces are hydrophilic (polyvinylchloride, cellulose diacetate) in which case the k_w versus $\sin \psi_w$ dependence is characterized by the constant slope coefficient. For hydrophobic surfaces (e.g. polystyrene), on the contrary, k_w remains constant while $\sin \psi_w$ is increasing (Fig. 2.9).

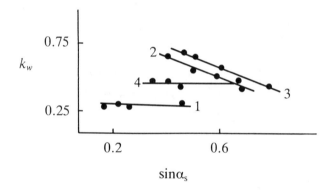

Fig. 2.9 – Coefficient k_w of eq. (66) versus the sliding angle of a fluid drop sliding down the tilted surface of polystyrene (1); Poly(vinyl chloride) (2); cellulose diacetate (3); and cellulose triacetate (4).

Roughness and the surface energy of substrates have a mutual and interrelated effect on the strength of an adhesive joint. This is to be expected since they are involved in the same functional relationships. For example, the roughness of a sand-washed steel surface, evaluated as the height of microprojections (automatically produced using a diamond indentor with $d = 2 \, \mu m$), and which corresponded with the maximum surface energy (sessile drop method) was found to be $10-11 \, \mu m$. The strongest bonding to the substrate was obtained at a height (roughness) of $9.8 \, \mu m$ [226].

Besides the mechanical heterogeneity, chemical inhomogeneity of real solid surfaces has to be taken into account also. Consider the solid surface comprising microscopic areas of two types. The surface energies at the solid—liquid and solid—gas boundaries can be written as follows:

$$\sigma_{\text{sl}} = A' \sigma'_{\text{sl}} + A'' \sigma''_{\text{sl}}, \qquad (68)$$

$$\sigma_{\text{sa}} = A' \sigma'_{\text{sa}} + A'' \sigma''_{\text{sa}} \qquad (69)$$

where A' designates the fraction of the total area occupied by the microscopic areas of a specified type. Substituting eqs (68) and (69) into eq. (55)[†] one has

$$\cos \theta = A'(\sigma'_{sa} - \sigma'_{sl}) + A''(\sigma''_{sa} - \sigma''_{sl})/\sigma_{la}. \tag{70}$$

Considering that

$$\cos \theta' = (\sigma'_{sa} - \sigma'_{sl})/\sigma_{la} \tag{71}$$

and

$$\cos \theta'' = (\sigma''_{sa} - \sigma''_{sl})/\sigma_{la} \tag{72}$$

(σ' and σ'' are the equilibrium contact angles corresponding to the homogeneous surface regions of each type), the expression for $\cos \theta$ can be transformed to

$$\cos \theta = A' \cos \theta' + A'' \cos \theta''. \tag{73}$$

Or in a generalized form

$$\cos \theta = \sum_{i+1}^{N} A^i \cos \theta^i. \tag{74}$$

According to Good [228]

$$\sum_{i+1}^{N} k_r^i = 1. \tag{75}$$

On the other hand, the movement of the molecules of liquid at relatively short distances along the surface with randomly distributed microdefects is described, in the first approximation, by the same functions for both the rough and the chemically inhomogeneous substrates [206], i.e. regardless of the nature of the microdefects.

Then, in conformity with these views, the inhomogeneity of polymer surface composition must result in different advancing and receding contact angles. In high precision measurements Penn and Miller [229] discovered that, for low-energy surface fragments (low σ_{sa}), advancing and receding angles equal each other ($\theta_+ = \theta_-$) when σ_{la} exceeds a certain threshold value σ_{la}^c ($\sigma_{la}^c > \sigma_{la}$); however, when $\sigma_{la}^c < \sigma_{la}$ then $\theta_+/\theta_- > 1$. Here, the magnitude of σ_{sa} (either high or low) governs the efficiency of relaxation of the advancing and receding angles, which, as is shown in capillarographic investigations [230], is greater in the case of low σ_{sa}. This effect is attributed to the different orientational ability with regard to the surface [231] of certain types of polar side groups on macromolecules. In fact, hysteresis in contact angle measurements correlated quite satisfactorily with the polarities of polymeric substrates [231]. For example, with the length of side chains in copolymers of hydroxyethyl methacrylate and sodium sulfoalkylmethacrylates [232].

Even more delicate effects determined by the influence of the substrate surface have to be considered when examining the role of wetting in adhesive joint formation. For

---

[†] A more rigorous thermodynamic treatment [227] shows that the generalization of eq. (55) taking into account the effect of solid surface inhomogeneity must involve the interaction between the three interfaces in the proximity of the three-phase borderline.

instance, σ was found to be a function of the radius r of the surface of a spreading liquid, the analytical expression for it being quite complex [233]

$$\frac{d \ln \sigma}{d \ln r_1} = \frac{\dfrac{2\Delta r}{r_1}\left[1 + \dfrac{\Delta r}{r_1} + \dfrac{1}{3}\left(\dfrac{\Delta r}{r_1}\right)^2\right]}{1 + \dfrac{2\Delta r}{r_1}\left[1 + \dfrac{\Delta r}{r_1} + \dfrac{1}{3}\left(\dfrac{\Delta r}{r_1}\right)^2\right]} \tag{76}$$

where Δr is the radius variation of the equimolar surface; at $r_1 = 0$ its magnitude reaches several A. Generally the $\Delta r(r)$ function is not known. Its pattern, however, may be derived by the methods of molecular dynamics [234]. For solids a solution to this problem is reasonably simplified as, in this case, the theory of elastic contact [235] that relates adhesive forces to the radius of curvature and the surface energy of the substrate [236] can be applied. By cleaving mica into separate plates of definite curvature, the plates having been premodified with calcium stearate or hexadecyltrimethylammonium-bromide, Israelashvili *et al.* [237] obtained σ values for mica that demonstrated the validity of this approach.

There is another factor to point out that has not yet been given adequate consideration. This is the substrate deformation resulting from wetting by the adhesive. This is especially so with elastomers. The major difficulties arising in treating the phenomenon are related to the fact that the magnitude of the ultimate contact angles (those at equilibrium) have to be considered, as well as the equilibrium conditions on the whole. They also involve the effect the presence or absence of gravitational fields [238, 239]. Construction of empirical dependences proves to be the simplest route to describing related effects; for instance, the authors of reference [240] proposed an empirical formula connecting the surface energy of elastomers with their deformational characteristics. A theoretical description of such effects was elaborated by Rusanov [153, 189, 238, 239, 241, 242] which takes into account the forces which act from the liquid phase on to a solid surface. One of these forces is associated with excessive hydro-static pressure, its magnitude being $\pi \Delta p A^2$ as applied to the surface area of πA^2 (Δp is the pressure difference between the inside and the outside of the drop). The second is the force arising from surface tension which is applied to the perimeter of wetting and equals $2\pi A \sigma_{la}$.

The normal and tangential components to σ_{la} are $\sigma_{la} \sin \theta$ and $\sigma_{la} \cos \theta$; however, the latter is affected by σ_{sa} and σ_{sl} to give the resultant force which is reflected in the following equation

$$\sigma_{la} \cos \theta + \sigma_{sl} - \sigma_{sa} = \sigma_{la} \Delta \cos \theta \tag{77}$$

where $\Delta \cos \theta$ is the difference between contact angles relating to the initial ('unperturbed') and deformed substrate surfaces. The complete solution is quite complicated, though it may be simplified for certain valid situations. For example, when there is no gravitational field the normal deformation of a solid substrate at the central point of the area underneath the wetting drop is given by the following equation

$$\xi_n = \frac{2(1 - \nu^2)\sigma_{la}}{E}\left[\sin \theta + \frac{1 - 2\nu}{2(1 - \nu)}\Delta \cos \theta\right] \tag{78}$$

where ν is the Poisson modulus and E is Young's modulus. Analysis of eq. (78) reveals that the shift of the central point does not depend on the dimensions of the drop; at $\sigma_{la} < 1$ N/m and $E > 10^4$ Pa $\xi_n < 10^{-3}$; for the solids with $E \approx 10^{10} - 10^{11}$ Pa ξ_n is negligibly small. For a large flat-topped drop subjected to gravitational field

$$\Delta p A = \rho\, g h A = A \left[2\sigma_{la}\, \rho g(1 - \cos\theta) \right]^{0.5} \qquad (79)$$

(ρ is the density of the liquid, h the height of the drop) the shift becomes dependent on the dimensions of the drop. Hence, the real deformational effects in substrates are detectable only when Young's and Poisson's moduli are small [243]; this is the case with polymers. According to the theory [244] these effects are especially noticeable in less cohesively strong materials, viz. the gels [245]. For hydrogels of 2-hydroxyethylmethacrylate-co-amino (or diethylamino)ethylmethacrylate $= 8:2$ with $E \approx 10^4$ Pa an error exceeding $2°$ in contact angle measurements arises when the deformational factor is neglected [246].

These various factors having been taken into consideration, the very concept of the contact angle has now to be put more precisely. As is seen from Fig. 2.10, the measured quantity does not represent the one to fit eq. (55) as a variable. Strictly speaking this fact serves as a fundamental background for hysteresis in wetting of these solids with a uniform homogeneous surface [247], e.g. metals [248] not to mention polymers [229, 248–251]. The expressions for initial and ultimate values of the contact angles corresponding to the initial ('unperturbed') and strained surfaces (θ_0 and θ_∞, respectively) can be easily derived [189] from the condition of mechanical equilibrium [241] of a drop sessile to the substrate and subjected to an external force field

$$\cos\theta_0 = \left[\pi E(\sigma_{sa} - \sigma_{sl}) + \sigma_{la}\,\Delta\rho g r\right] / \sigma_{la}(\pi E + \Delta\rho g r)^{\dagger} \qquad (80)$$

where r is the radius of the drop and $\Delta\rho$ is the difference between densities of a liquid and the surrounding medium (air). Fig. 2.11 exemplifies this difference between θ_0 and θ_∞ (expressed according to eq. (77) versus $\Delta\cos\theta$). It depicts relative deviations of the initial and ultimate contact angles obtained on wetting of elastomers from the value of θ deduced from eq. (55). A generalized linear tension κ that takes account of the deformation was introduced [189] to describe the relationship between these parameters

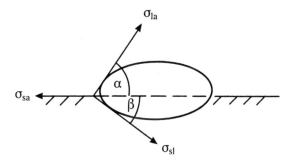

Fig. 2.10 Resultant of the three forces acting when a solid deformable surface is wetted by a liquid.

† A coefficient equalling 3 is introduced into the second term of the numerator and denominator of the formula when calculating $\cos\theta_\infty$.

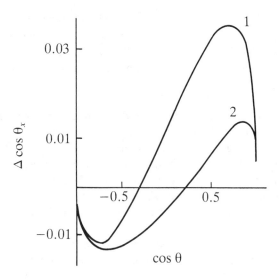

Fig. 2.11 – $\Delta \cos \theta_\infty$ (1) and $\Delta \cos \theta_0$ (2) versus cosine of the contact angle with elastomers corresponding to equ. (55); $r/d = 10^6$; $\sigma_{sa}/\sigma_{sl} = 2$; $3\sigma_{la}/2\pi Ed = 10^3$.

$$\cos \theta_\infty = \cos \theta_0 - \frac{1}{\sigma_{la}} \left(\frac{\kappa}{r_1} - \frac{\partial \kappa}{\partial r_1} \right) \tag{81}$$

(r_1 is the radius of the perimeter of wetting). For macroscopic drops [153]:

$$\frac{1}{\sigma_{la}} \left(\frac{\kappa}{r_1} - \frac{\partial \kappa}{\partial r_1} \right) \approx \frac{\sigma_{sa}}{Er_1} \tag{82}$$

and, consequently, the second term on the right side of eq. (81) may be neglected if the elastic modulus is sufficiently high. It has been reported however that in crosslinked elastomers the growth of the conventional equilibrium modulus leads to a decrease of θ [252]. The general solution that includes the effect of uniform elongation of the substrate on the magnitude of the contact angle, can be obtained [153] by differentiating eq. (55) which accounts for the concepts of surface stress and surface tension

$$\frac{d \cos \theta}{d \ln A} = (\gamma_{sa} - \gamma_{sl} - \sigma_{sa} + \sigma_{sl})/\sigma_{la}. \tag{83}$$

Wl.en the load is applied in a perpendicular direction the right-hand side of eq. (83) has to be multiplied by $(1 - \nu)$.

Recalling the effect of the hysteresis of wetting, three major causes that contribute to its realization can be recognized: roughness of the substrate surface, its chemical and structural inhomogeneity and the presence of surface-active or film-forming compounds dissolved in the liquid. The first of these has been examined most thoroughly. With somewhat casual simplicity, but very descriptively de Gennes [206] defined it as 'anchoring' of the line of three phase contact at the defects of a solid surface. Surface roughness (rugosity) offers an explanation for the advancing and receding angles shown in Fig. 2.8;

their difference is not less than $10°$ if the surface has not been specially treated. Quantitative analysis of the situation reveals that local disturbances (perturbations) of the line of wetting can be described by the Fourier integral, so that there should be some critical threshold concentration of the surface defects giving rise to the appearance of hysteresis [253]. The separate effect of each of the factors listed causing hysteresis can be shown experimentally only in rather specific model systems. For instance, in the case of hydrophilic copolymers of L-glutamic acid and L-leucine surface roughness and surface heterogeneity were distinguished, by measuring the hysteresis of contact angles in buffer solutions, as dependent on pH, substrate composition and composition of the solvent from which the substrate had been cast [254].

By examining eq. (81) one more valid conclusion can be inferred that also cannot be deduced from the Young equation. According to its physical meaning κ is related to the work of deformation on the line of three phase contact [189] and, consequently, its magnitude for an anisotropic solid body is a function of the molecule's position on this line even at constant σ_{sl} and σ_{sa}. In other words, the work of substrate deformation carried out by the applied force will be different in different directions and this is reflected in changes of κ and, correspondingly, θ. Such an extent of contact angle anisotropy was observed experimentally in studies of the wetting of uniaxially strained elastomers [255, 256], a class of materials which, when loaded, display orientation of molecular chains. For instance, elongation of butadiene-acrylonitrile copolymer SKN-18 (17–20%) acrylonitrile content) by 150% results in an almost two-fold rise in surface energy attributed to molecular orientation [257, 258] and for polyethene and polymethylmethacrylate σ_{sa} increases by 6 and 7 nN/m, respectively [258]. For polyvinylalcohol and polytetrafluoroethene anisotropy of the contact angle was shown to result from the elastic deformation of the substrate surface layers, the effect was enhanced when the substrates had been first subjected to shear strain that had provided for a better orientational ordering of macromolecular chains [259].

Good *et al.* [260, 261] subjected the dependence of advancing (θ_+) and receding (θ_-) contact angles on PTFE deformation to a more detailed study; the difference between θ_+ and θ_- was used to measure the extent of hysteresis:

$$H = \theta_+ - \theta_-. \tag{84}$$

In full accord with the views described they discovered perceivable anisotropy of contact angles, the wetting drop stretching in the direction of substrate extension. When the substrate deformation was varied the θ_+ and θ_- values were liable to vary also;[†] the first, predominantly in the direction perpendicular to the orientation axis (Fig. 2.12) and the second in the direction parallel to it (Fig. 2.13). Hence, elongation of the substrate results in an increase in $H_\perp$ and a decrease in $H_\parallel$ [260]. This effect does not depend on the morphological features of the polymer surface and is mainly associated with the orientation of molecular chains arising as a consequence of substrate deformation. These observations are in full accord with the results of [259].

[†] One cannot ignore the case when only one of the two types of the contact angles varies. For instance, a constant advancing angle was observed on wetting of poly(chloro-*p*-xylylene) coating of ceramic fibres with octane. Hence, hysteresis was a function of θ_- only [262].

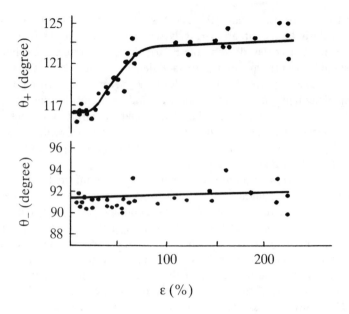

Fig. 2.12 – Advancing and receding contact angles with PTFE as dependent on the strain of the PTFE sample; the direction of advancing and receding is perpendicular to the strain direction.

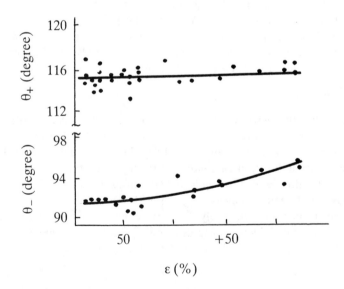

Fig. 2.13 — Advancing and receding contact angles with PTFE as dependent on the strain of the PTFE sample; the direction of advancing and receding is parallel to the strain.

It is quite natural to presume that the hysteresis of wetting observed to follow from deformation of the substrate may be attributed to the associated changes in roughness.[†] The principal question arising in this regard is how can one discriminate between the contributions to the overall H value of roughness, characterized by Wenzel's coefficient β, and of energetic non-uniformity of the substrate surface. Transforming the original form of eq. (65) in terms of Good's approach, for a horizontal surface we arrive at the following expression

$$\cos \theta = (\cos \theta_+ + \cos \theta_-)/2\beta. \tag{85}$$

By analogy, Wolfram and Faust suggested the expression for an inclined surface as

$$1 - \cos^2\theta = (\cos \theta_- - \cos \theta_+)/\beta_e \tag{86}$$

where Wenzel's coefficient is substituted for the parameter β_e characteristic of the energetic non-uniformity of the surface [265]. By determining β and β_e they were able to distinguish the effect of each of the factors. In the studies of the hysteresis of wetting of polyethene, polystyrene, polyvinylchloride, PTFE, polypropene and polycarbonate β and β_e were shown to be interrelated linearly [265].

On the other hand, one may easily assume that the magnitudes of advancing and receding angles must be governed by the roughness profile, i.e. within the framework of the 'groove model' discussed above, by the angle at which the groove's walls diverge. This conclusion was verified experimentally by Bayramli [266]. His experiments involved a sliding vertical aluminated sapphire rod with diamond-scratched microgrooves on the surface. θ_+ values were found to be determined by the maximum slope of the wall of a microdefect, and θ_- by the minimum one. These experimental results are in good agreement with the theoretical dependences [267].

Anisotropy and hysteresis of wetting of rough substrates are phenomena of principal importance with regard to their effect on adhesion. Let us elaborate on these in greater detail.

As we have already mentioned, de Gennes ascribes hysteresis to the 'anchoring' effect of microdefects on the propagation of the line of three-phase contact [206]. This concept seems to be entirely applicable to the 'grooves model' inherently exhibiting high anisotropy. However, this is the case only when the grooves are parallel. If the line L crosses the grooves at a certain angle φ_L (Fig. 2.14a) it can move, according to Cox [268], without sustaining any 'anchoring', as was experimentally confirmed by Mason [269]. Indeed, at $\varphi_L = 0$ the line tends to move from one ridge to another parallel to it (Fig. 2.14b). Obviously, such massive (extensive) and simultaneous motion is very unfavourable thermodynamically, as it implies that a high energy threshold has to be overcome. The actual mechanism of this process is assumed to be the following: at some point of contact the line L_1 jumps to the neighbouring ridge by forming a loop, the tip of the loop being the point of the line resting on the second ridge. The loop then

---

† Some authors deny the idea of such influence [263]. There are several reasons for taking this viewpoint and the experimental difficulties in measuring contact angles accurately and reliably are not least among them. In fact, the magnitudes of contact angles on real solids with developed microrelief are affected even by the size of a drop. For instance, for water drops 3–20 mm in diameter resting on a PTFE surface different methods lead to $\theta_+ = 89°$ and $\theta_- = 108.5°$. However, when the size of the drops diminishes the results produced by different procedures tend to deviate significantly [264].

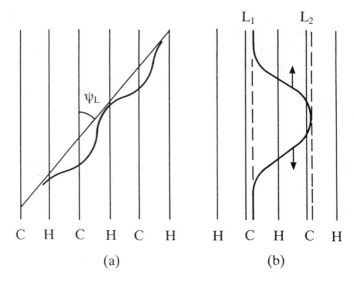

Fig. 2.14 – Line of three-phase contact moving over the rough substrate surface with micro-projections C and microditches H. a, Continuous propagation without intermediate anchorings; b, successive propagation with loop-like anchorings.

straightens out until the entire contact line moves on to the position L_2 on the neighbouring ridge. In this case the energy barrier is determined by the modes of loop formation and straightening, but not by the size of microdefects. According to de Gennes [206], this conclusion holds true for infinitely long grooves or for those forming a closed circuit.

For a thorough description of the features of wetting hysteresis it is obviously necessary to extend the study to examining different shapes of the line of three-phase contact [270]. With this intention one can make use of the approach described by de Gennes [253]. To the microdefect on the substrate surface we ascribe a certain perturbation function $h(x, y)$ localized around a certain point x_d, y_d (Fig. 2.15). There

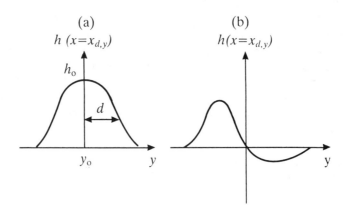

Fig. 2. 15 – Microdefects of a substrate surface. Microdefect, having a smooth structure, is localized around (x_d, y_d) point. a, Chemical inhomogeneity; b, mechanical non-uniformity.

may be more than one equilibrium position of the line L in the proximity of this micro-defect and in certain cases (Fig. 2.16) the line may get 'anchored' over the latter. When far from the defect the form of the line coincides with that of the straight line $y = y_L$. It is only in the vicinity of the defect $(x = x_d)$ that the line is perturbed to acquire $y = y_m$. At the top of the microdefect the line $x = x_d$, $y = y_d$ is at equilibrium resulting from the action of two forces, the one imposed by the ridge (causing the perturbation) and the other an elastic force tending to return y_m to the unperturbed coordinate of the y_L line. If the microdefect is of Gaussian form, the first force is also described by a Gaussian distribution function

$$f_1(y_m) = (2\pi)^{0.5} h_0 \, d \exp\left[-\frac{(y_m - y_d)^2}{2d^2}\right].$$ (87)

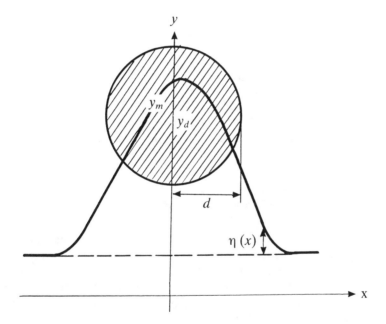

Fig. 2.16 — 'Anchoring' of the line of three-phase contact at the microdefect of the sub-strate surface.

The second force is accepted as obeying Hooke's law

$$f_2 = k_L(y_L - y_m).$$ (88)

In this expression the elasticity coefficient of the line of contact is determined by the 'long-distance cutoff' l_{max} and the size d of the microdefect as follows

$$k_L = \pi\sigma\,\theta^2/\ln(l_{max}/d).$$ (89)

The condition for equilibrium is then expressed via the following relationship

$$k_L(y_m - y_L) = f_1(y_m),$$ (90)

for which the solution may be found graphically, as is shown in Fig. 2.17. It is easy to demonstrate that at small values of h_0 there exists only one root y_m for any y_L, and hysteresis is not observed. However, when h_0 exceeds a certain criticial value there exist three roots within a distinct interval of y_L values and hysteresis becomes quite probable. In other words, when the perturbances are small, and $f_1(y_m)$ is a smooth function with a finite derivative $f'(y_m)$ there is no need for the surface to be ideal in order to determine the contact angle reliably. It is important only that the deviations from ideality do not exceed the threshold value, around 100 nm according to the assessments of [206]. When the microdefects are smaller the energy barrier between equilibrium positions of the line of three-phase contact can be overcome due to thermal fluctuations. When microdefects are not continuous but stepwise the features described are invalid, and wetting hysteresis is exhibited even at very small values of the perturbation function h [271].

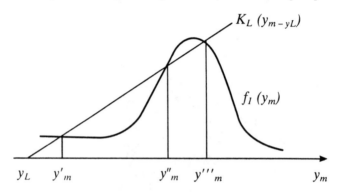

Fig. 2.17 – Equilibrium positions of the 'anchoring' point ($y = y_m$) of the line of three-phase contact at the surface microdefect.

It is necessary to emphasize that, on the whole, the thermodynamical description presented is derived on the basis of an analysis of the classical system depicted in Fig. 2.2. This system, however, simulates the behaviour of a non-deformable substrate only. Hence, as is stressed by Rusanov [238], the basic relation (55) reflecting its behaviour is a scalar one. In a more general case the drop produces a mechanical disturbance of the substrate surface (see Fig. 2.10). Then [272]

$$\sigma_{sa} = \sigma_{la} (\cos \alpha + \sin \alpha \; \mathrm{ctg} \; \beta) \tag{91}$$

and the task of determining the substrate surface energy is essentially simplified. Indeed, expressing each of the three surface parameters (energies) as corresponding vectors one comes to the following relationship for the equilibrium condition

$$\sum_i \left(\sigma_i \bar{t}_i + \frac{\partial \sigma_i}{\partial \varphi_i} \bar{n}_i \right) = 0 \tag{92}$$

where $\bar{t}_i$ and $\bar{n}_i$ are the tangential and normal unit vectors of the ith boundary, and $\partial \sigma_i / \partial \varphi_i$ is the spinning moment of σ_i arising from the dependence of the surface energy upon the orientation direction of the substrate [273]. Projecting eq. (92) on to horizontal and vertical directions, on the assumption that interfacial surface energy is anisotropic (i.e. $\partial \sigma_{la} / \partial \alpha = 0$) one has

$$\sigma_{sa} = \sigma_{la} \cos \alpha + \sigma_{sl} \cos \beta - \frac{\partial \sigma_{sl}}{\partial \beta} \sin \beta \tag{93}$$

and

$$\sigma_{la} \sin \alpha = \sigma_{sa} \sin \beta + \frac{\partial \sigma_{sl}}{\partial \beta} \cos \beta. \tag{94}$$

Assuming the last members on the right-hand side tend to zero (which is incorrect only for metals) one obtains eq. (91). This inference holds true for systems of the type depicted in Fig. 2.10.

The general ideas examined concerning the effect of deformational factors on the wetting pattern of the substrate are particularly important in regard to the thermo-dynamics of adhesive joint formation. In real cases deformation of the substrate affects the magnitude of the contact angle quite noticeably. For instance, a 30% elongation of polyesterimide [274] decreases the θ value by 5.5° when the wetting liquid is water, and by almost 7° in the case of less polar liquids with $\sigma_{la} = 61-62$ mN/m (Fig. 2.18). At the same time, a 20% deformation of the substrate results in a 20% and 15% rise of the strength of adhesive joints produced with the aid of polyester-epoxy and polyester-cyanurate adhesives, respectively (Fig. 2.19).

The considerations discussed provide grounds for doubting the validity of applying Young's equation to the thermodynamic analysis of the adhesion of solids. In fact, the 'three forces' model that is usually involved to derive eq. (55) is intrinsically contradictory as two forces acting along one line cannot be balanced by the third force acting at an angle to the first two [275]. In addition, eq. (45) holds true for liquid surfaces only, as the properties of solids are dependent on dT_{sa} and, consequently, the magnitude of $\sum \sigma_i dA_i$ is determined by σ_{sa} which reflects the extension of the wetting perimeter. Then, in agreement with the condition stated above, $dF/dA_{BC} = 0$, the wetting lens is not at equilibrium, i.e. $\sum \sigma_i dA_i \neq 0$.

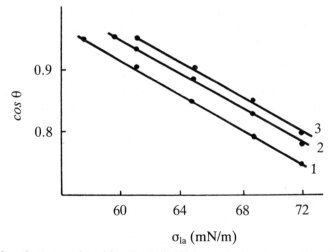

Fig. 2.18 – Contact angles with polyesterimide versus surface energy of wetting liquids. 1, non-deformed polyesterimide sample; 2, 3, samples with the deformation degree of 10% and 20%, respectively.

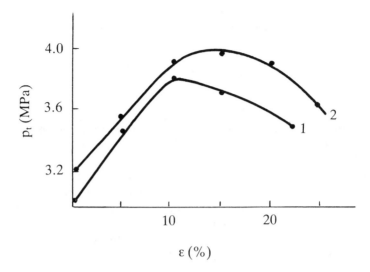

Fig. 2.19 – Tensile strength of the adhesive joints of polyesterimide versus substrate deformation ratio. 1, polyester cyanurate adhesive; 2, polyester–epoxy adhesive compositions.

On the other hand, when examining the applicability of Young's equation Frumkin demonstrated [276] that the parameters of eq. (55) are characteristic of the adsorptive layer at the surface of a solid but not of the solid surface itself. This thesis was later further developed by Bikerman [275]. Then, taking into account the extremal criteria of thermodynamic equilibrium, eq. (55) acquires the following form [190]:

$$\sigma_{sa} - \sigma_{sl} = \sigma_{la} \bigg/ \left[1 + \left(\frac{dz}{dx} \right)^2 + \left(\frac{dz}{dy} \right)^2 \right]^{1/2} \tag{95}$$

where x, y and z are as in the equation for the free surface of a drop and define the horizontal (x), oblique (y) and vertical (z) axes. Consequently, the relationship representing the physical meaning commonly ascribed to Young's equation should have been derived by the methods of capillary hydrodynamics [278] rather than thermodynamics [277]. The effects related to sorption of the components of adhesive compounds by the substrate surface have been demonstrated to affect the free energy at the interface. Fig. 2.20 shows the corresponding changes [195] resulting from the interaction of the solid–liquid interface with microparticles of cubic and spherical form.

Similar considerations lead one to doubt the most common modification of Young's equation suggested by Dupré [279], who introduced the concept of the reversible (thermodynamic) work of adhesion.[†]

$$W_{Ad} = \sigma_{sa} + \sigma_{la} - \sigma_{sl} \tag{97}$$

[†] This concept was first used by Young [185] who excluded σ_{sl}, which was difficult to measure, by simplifying eq. (55) to the form

$$\cos \theta = (W_{Ad}/\sigma_{la}) - 1. \tag{96}$$

(a)

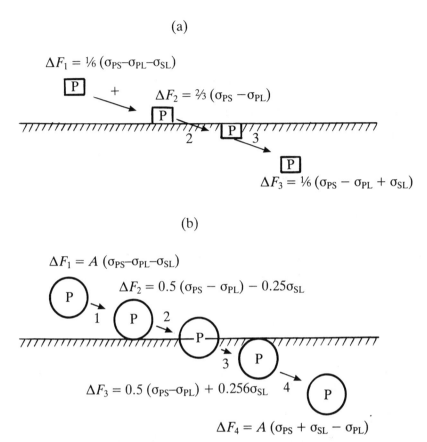

(b)

Fig. 2.20 – The pattern describing the changes of the free energy accompanying the transfer of cubic (a) and spherical (b) particles from the liquid phase L into the solid one S (A is the interphase area of the particles with the medium).

Indeed this last relationship is expected since the change of the free energy of a system accompanying its transition from one state to another via an isothermal process is given as the difference between the free energies of the system in these states. In other words, the work of adhesion may be defined as the free energy required for the reversible equilibrium separation and removal of the phases to infinitely large distances when T is const. and p is const. Combining eqs (55) and (97) we obtain the following expression for the work of adhesion (cf. footnote page 58)

$$W_{Ad} = \sigma_{la} (1 + \cos \theta). \tag{98}$$

Equation (98), however, takes no account of the processes accompanying phase transition i.e. evaporation, dissolution etc. Thus, the Dupré–Young equation is valid only for those systems in which sorption of components at the interface and the changes in surface tension and chemical potentials of the phases in contact can be neglected. Strictly speaking equations (55) and (98) are authentic only for isothermal–isobaric wetting of a low compressible and insoluble macroscopic body by the massive medium. Adamson believes [112] that Young's equation pertains to the contact angles at the microscopic level whereas those actually measured are the macroscopic values of θ whose magnitude is determined by the geometry of a system and by gravitational effects.

In fact the presence of the sorbed impurities, e.g. water vapour at the substrate surface, requires that at least two values of contact angle be considered. For instance, to give an adequate description of the interaction of an elastomer with a glass surface (θ_1) which is actually always 'covered' with hydroxy groups (θ_2) eq. (98) has to be modified to involve the respective quantities referring to both types of surfaces [280]:

$$W_{Ad} = (\sigma_{sa} + \sigma_{la} \cos \theta_1)(1 + \cos \theta_2). \tag{99}$$

This equation is essential when seeking to assess the spreading coefficient [281]. For the most general case Gent and Schultz [282] suggest that instead of the basic equation (97) the following relationship be introduced

$$W_{Ad}^l = \sigma_{sa}^l + \sigma_{la}^l - \sigma_{sl}^l \tag{100}$$

where the index l refers to adhesives (liquids). Then the newly introduced concept pertaining to the work of adhesion (eq. (100)) and the traditional one (according to eq. (98)) are interrelated in the following way

$$W_{Ad}^l = W_{Ad} - \Delta \tag{101}$$

where

$$\Delta = \sigma_{la} (\cos \theta_1 + \cos \theta_2). \tag{102}$$

Designating the contact angle corresponding to interphase segregation as θ_0 one can obtain the following relationship

$$\theta/\theta_0 = W_{Ad}^l/W_{Ad} = 1 - \Delta/W_{Ad}. \tag{103}$$

The task of deriving it is justified in that Δ appears to be more susceptible than W_{Ad} [283] to variations in nature of the substrate surface covered with adsorptive layers. This conclusion is verified by the data presented in Fig. 2.21. The data refer to the fracture tests of adhesive bonds between butadiene–styrene elastomer and polyester film performed in various liquid media [282].

It should also be borne in mind that equations (55) and (98) do not involve parameters related to molecular mass and this significantly complicates the task of describing the behaviour of polymeric adhesives. The problem is treated with greater sophistication in the next section but here we confine ourselves to noting that this drawback in the approach discussed is eliminated by introducing technological parameters. For instance, it was suggested that a 'wetting index' be used as a technological parameter to describe the ability of a polymer to wet a solid [284]. This index was defined as the ratio between the diameter of a wetted and the initial surface of the sample before it has been melted in an inert medium and in vacuum. It is important that, though empirical in origin,[†] this index was found to be related to the surface energy of a solid substrate [284].

[†] This conventional index is related to polymer viscosity [285], however it is more sensitive to the spreading pattern of a polymer [286]. But what makes the wetting index a really useful implement is that it is related to the molecular mass of adhesive [284] and its molecular mass distribution [285].

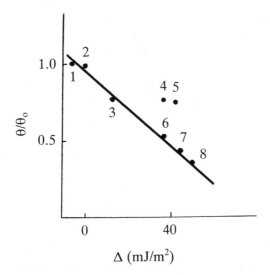

Fig. 2.21 — θ/θ_0 versus Δ pertaining to the fracture of adhesive joints between butadiene–styrene elastomer and Mylar film. 1, fracture in water; 2, in air; 3, in 10% ethanol solution; 4, in ethylene glycol; 5, in formamide; 6, in 50% methanol solution; 7, in ethanol; 8, in butanol.

The question arising in this connection is what makes the Young and Dupré–Young equations so popular for the thermodynamic analysis of adhesion systems? It cannot be attributed merely to 'seductive simplicity' or 'lucky chance' [287]. We believe that the main reason lies in an incorrect interpretation of the corresponding parameters, viz. the experimentally determined surface tension is identified by the majority of authors with surface energy of a solid. The tensor of surface tension σ_{ik} is related to specific surface free energy $\bar{F}$ by the following expression involving the tensor of substrate deformation ξ_k and the Kronecker symbol δ_{ik}:

$$\sigma_{ik} = \delta_{ik}\bar{F} + d\bar{F}/d\xi_k. \tag{104}$$

It would be now easy to demonstrate that the Young equation bears the energetic meaning[†] but not the 'force' one as Young himself claimed.[‡]

Thus any speculations concerning adhesive phenomena based on the equations of Young and Dupré–Young cannot be considered to be unequivocally sound. The approach relying on eqs (55) and (98) is valid predominantly for the most simple model systems [289] whose behaviour is not subject to side effects. This conclusion is even more justified when encountering attempts to interpret adhesive bond formation in terms of contact angles, and these, as is known, are determined, along with factors of thermodynamic origin, by rheological ones also.

---

† Strictly speaking such treatment is based on the assumption that $\bar{F}$ is constant; a fact that does not follow from thermodynamics. Then the physical meaning of eq. (55) is determined mainly by the probable types of virtual displacements characteristic of solid bodies [238].

‡ It was even attempted to correlate the work of adhesion with 'ideal' strength of adhesive joints; the suggested relationship $P_{id} = 3.079\,W_{Ad}/z$ involves the equilibrium distance z between the phases, usually around 0.5 nm, and the coefficient resulting from the operation is $3^{1/2} \times 16/9$ [288].

For epoxy–polyamide compositions it was established, for instance, that variation of the filler content affects both θ and adhesive viscosity in the same manner [290]. Besides this it is necessary to distinguish static θ_s and dynamic θ_d values of contact angles; these are measured by methods [291] some under equilibrium and others under non-equilibrium conditions [292]. Quantitative relationship between these quantities was found only for the simplest cases identified with capillary wetting [293]

$$(\cos \theta_s - \cos \theta_d)/(\cos \theta_s + 1) = tg\,[4.96\,(\eta V/\sigma)^{0.702}] \tag{105}$$

where η indicated viscosity. This is just the case for which, obviously, the rheological character of spreading is manifested quite clearly. This approximation allows the evaluation of static contact angles θ_s in polymer–liquid systems relying on experimentally measured θ_d values. According to Jiang, Oh and Slattery [292] the discrepancy between the values of θ_s produced by eq. (105) and those observed experimentally constitutes 5.9% for the system involving di(2-ethylhexyl)sebacinate and polytetrafluoroethene; 29.7% for n-octane–polytetrafluoroethene, 13.9% and 15.4% for poly hexamethyleneadipamide with water and methylene iodide, respectively.

However, in certain cases by eliminating some limitations it becomes possible to relate the strength of adhesive joints to the magnitude of the contact angle. This suggestion should be valid for systems which are free from the influence of kinetic effects, i.e. for those that are formed via the interaction of a solid with a permanently tacky adhesive. It was shown that tackiness may be considered as 'instant' adhesion [294]. This approach provides explanation for the linear dependence between conventional (relative) tack and θ [295] (Fig. 2.22).

In spite of these limitations it is still tempting to express surface energies of contacting phases via interfacial energies. The simplest of such attempts is attributed to Antonoff [296] who demonstrated that when $\sigma_{sa} > \sigma_{la}$

$$\sigma_{ls} = \sigma_{sa} - \sigma_{la} \tag{106}$$

and when $\sigma_{la} > \sigma_{sa}$

$$\sigma_{sl} = \sigma_{la} - \sigma_{sa}. \tag{107}$$

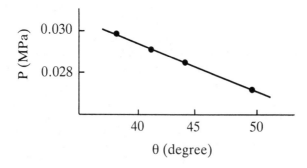

Fig. 2.22 – Tackiness of KS-1 adhesive to ceramic substrate versus contact angle.

These relationships were obtained in studies involving shear measurements of a viscous medium at the surface of a solid, hence they are valid when $\theta = 0$ and, as Antonoff has himself stressed later [297], when there is no adhesive interaction between the phases. Strictly speaking eqs (106) and (107) hold true only for ideal liquids [298] and two-phase liquid systems displaying positive spreading coefficients. This is confirmed by the measurements of interfacial tension at the interface between water and either hexane or petroleum solutions of polyisoprene (SKI-3 elastomer) [299]. However, when one of the phases is solid, Antonoff's rule is true only for substrates in the temperature range around their melting points, as was demonstrated by Mitchell and Elton [300] for acetophenol, diphenyl ether, octa- and hexadecane with water.

Such contradictions are easy to understand as soon as one reconsiders the geometry of the system predicted in Fig. 2.2. As is known (see, for example, [134]), interfacial tensions and angles adjacent to the contact ones are related to each other as sides and angles of the Neumann triangle [301] for which the following inequalities are true

$$\sigma_{la} < \sigma_{sa} + \sigma_{ls}, \tag{108}$$

$$\sigma_{ls} < \sigma_{sa} + \sigma_{la}, \tag{109}$$

$$\sigma_{sa} < \sigma_{la} + \sigma_{ls}. \tag{110}$$

When any of these relationships is converted to an equality, i.e. when the largest of the three surface tensions equals the sum of the other two, the Neumann triangle degenerates into a line and the liquid completely spreads over the substrate surface.

This is the exact condition that determines the rules eqs (106) and (107) which in this case are true within the entire region of equilibrium in three-phase states and, consequently, they can no longer be considered as some sort of approximation. However, the liquid phase trapped between the other two can also acquire the form of either a lens or a bead. In such cases relationships (108)–(110) are strict inequalities leaving no possibility for Antonoff's approach [302].

At the same time applicability of Antonoff's rules (eqs (106) and (107)) to the study of adhesive interactions cannot be considered as having been completely exhausted. To illustrate this the approach developed by Summ and Abramzon [303] should be noted. Substituting eq. (106) into eq. (97), they obtained simple expressions

$$W_{Ad} = 2\sigma_{la}, \tag{111}$$

$$W_{Ad} = 2\sigma_{sa} \tag{112}$$

for the cases $\sigma_{sa} > \sigma_{la}$ and $\sigma_{sa} < \sigma_{la}$, respectively. The dependence of the work of adhesion, calculated from eq. (98), versus the surface tension of liquids wetting various polymers was indeed observed to be rectilinear, thus proving the validity of eqs (111) and (112) and where the inflexion point on these dependences corresponds to $\sigma_{sa} = \sigma_{la}$. Hence the adhesive interaction of two substances and the cohesive interaction within that phase with the minimum cohesion were concluded as being equal to each other [303]. If both contacting phases are polymers their behaviour, according to Wu [304], does not follow Antonoff's rule.

To attain a more complete agreement between calculated and experimental data on the wetting of solids attempts were undertaken to modify eq. (55). The most reasonable approaches to the problem were those of Rhee [305] and Tovbin [306]. Rhee proposed the relationship

$$\cos \theta = (a\sigma_{sl} + 1) - a\sigma_{la}, \tag{113}$$

which is the equation of a straight line and that is true when there are no sorption processes to accompany the contact interaction. According to Tovbin

$$\sigma_{sa}(1 - \cos \theta) = \sigma_{la}(1 + \cos \theta). \tag{114}$$

However, it is a rather rare case when the dependence is a valid one [307]. This is quite evident since at $\theta = 90°$ eq. (114) is reduced to the equality

$$\sigma_{sa} = \sigma_{la} \tag{115}$$

which, in the absence of adhesive interaction, follows from relationships (106) and (107). At $\theta = 0°$ the resultant equation

$$\sigma_{sl} - \sigma_{sa} = \sigma_{la} \tag{116}$$

contradicts the basic equation (97).

Some dependences of the Young's type relating σ_{la}, σ_{sa} and σ_{sl} are based on the analogy with the notorious equation of Berthelot [308] combining the constants of attractive interaction between two molecules

$$c_{12} = (c_{11}c_{22})^{1/2}. \tag{117}$$

In the simplest case, according to Fowkes [309],

$$\sigma_{sl} = \sigma_{sa} + \sigma_{la} - 2(\sigma_{sa}\sigma_{la})^{1/2}.^\dagger \tag{118}$$

This dependence, however, is inadequate to match the experimental data as demonstrated by Good [311] and Schwartz [312].

A more elaborate approach was developed by Good [313] who modified eq. (118) by introducing the constant α [314]

$$\sigma_{sl} = \frac{(\sigma_{sa}^{1/2} - \sigma_{la}^{1/2})^2}{1 - \alpha(\sigma_{sa} - \sigma_{la})^{1/2}}. \tag{120}$$

For the contact angle the equation is as follows[‡]

---

† To be precise, eq. (118) was first introduced, though in another form,

$$\sigma_{sl} = (\sigma_{sa}^{1/2} - \sigma_{la}^{1/2})^2, \tag{119}$$

by Rayleigh [310], who considered it to be of limited value.

‡ The empirical nature of eq. (121) is clear and was noted as soon as the equation was introduced [315, 316]. Solving eq. (121) with respect to σ_{sa}, Lipatov and Feinerman [317] obtained a family of parabolic curves. The straight lines corresponding to σ_{la} were shown to have a limited number of intersections with these curves. It was shown, using the example of mercury [318], that the region of real solutions extends on the positive side of the parabolic curve to $\cos \theta = 0.3$ only, whereas σ_{sa} values on the negative side do not have any real meaning.

$$\cos\theta = \frac{(\alpha\sigma_{sa} - 2)\sigma_{la}^{1/2}\sigma_{sa}^{1/2} + \sigma_{la}}{\sigma_{la}(\alpha\sigma_{la}^{1/2}\sigma_{sa}^{1/2} - 1)}. \tag{121}$$

Relying on computer calculations, he later introduced another constant β, the ultimate equations acquiring the form [319]

$$\sigma_{sl} = \frac{\sigma_{sa} + \sigma_{la} - 2\beta(\sigma_{sa}\sigma_{la})^{1/2}}{1 - 2\alpha(\sigma_{sa}\sigma_{la})^{1/2}} \tag{122}$$

and

$$\cos\theta = \frac{(2\alpha\sigma_{sa} - 2\beta)(\sigma_{sa}\sigma_{la})^{1/2} + \sigma_{la}}{\sigma_{la}[2\alpha(\sigma_{sa}\sigma_{la})^{1/2} - 1]}. \tag{123}$$

In eqs (120), (121) α is accepted as being constant (0.015), whereas in eqs (122) and (123) the values vary over a broad range depending on the nature of substrate. For polytetrafluoroethene and poly(tetrafluoroethene-co-perfluorolauric acid) these constants were found to be $\alpha = 0.00784$ and $\beta = 1.004$ [319]. Recently tables compiling the values of these parameters for different contact angles with liquids of known surface tension were published [320] which makes the application of equations (122) and (123) more useful.

Pittmann [321] attempting to clarify the distinct meaning of the variables in eq. (55), assumed that a liquid wetting a solid is always saturated with particles of the solid substrate; hence, σ_{la}^s denotes the surface tension of this liquid phase. Taking into account that the change of the specific free energy accompanying substitution of the solid–gas interface for the solid–liquid one is given as $(\sigma_{sa} - \sigma_{la})$, he modified eq. (97) to the form

$$W_{Ad} = \sigma_s^0 + \sigma_{la}^s - \sigma_{sl} \tag{124}$$

where σ_s^0 is the free energy at the solid–vacuo interface. Then

$$W_{Ad} = (\sigma_s^0 - \sigma_{sa}) + \sigma_{la}^s(1 + \cos\theta) \tag{125}$$

and, subsequently neglecting the decrease in free energy on transferring a solid from vacuum into air the following equation is obtained

$$W_{Ad} = \sigma_{la}^s(1 + \cos\theta). \tag{126}$$

According to Adam [322], eq. (126), in contrast to eqs (55) and (98) which have identical physical meaning, is the only relationship involving the work of adhesion that can be experimentally verified.

However, more fundamental and promising are, perhaps, approaches which tend to supplement the basic Young equation [323], rather than making it more explicit. Consider the equilibrium conditions in the system depicted in Fig. 2.23 [324]

$$\sigma_{la}\cos\theta \leqslant \sigma_{sa} - \sigma_{sl}, \tag{127}$$

$$\sigma_{la}\cos\beta \leqslant \sigma_{sl}' - \sigma_{sa}'. \tag{128}$$

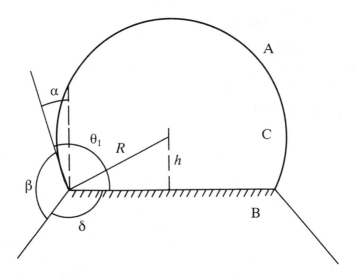

Fig. 2.23 — Schematic representation of a droplet on the ridge of a solid body.

The prime refers to the parameter applying to the tilted plane whereas the normal case is that of a horizontal plane.[†] At $\theta + \beta = 180°$ eqs (127), (128) are reduced to eq. (55). Noting the unequal values of the cosines of the contact angles for the horizontal and tilted planes eq. (55), one can show that the wetting perimeter is at equilibrium when

$$\theta \leqslant \theta_1 \leqslant 180° + \theta' - \delta \qquad (129)$$

where θ' is the contact angle on the tilted plane. If the wetting perimeter is not fixed (for instance, when phase B is growing through phase C), the lens of phase C can acquire the equilibrium form (shape) for which solution of eq. (45) (as complemented by eq. (46)) leads to the following relationship

$$\sigma_{sl} = \sigma_{la} \sin \alpha, \qquad (130)$$

where $\alpha = \theta_1 - 90°$. From geometric considerations (see Fig. 2.23) eq. (130) can be presented in the form

$$\sigma_{sl}/\sigma_{la} = h/R, \qquad (131)$$

corresponding to the known equation of Wulff [325], which is thus a supplement to the Young equation.

Hence, the system of eqs (46) and (130) can be constructed on the basis of two experiments, one of which provides for mechanical equilibrium of the system, according to Young, and the other, the equilibrium shape of the liquid lens, according to Wulff.[‡] This system of equations may be solved with respect to the interfacial tension of a solid

---

[†]　For isotropic bodies $\sigma'_{sa} = \sigma_{sa}$ and $\sigma'_{sl} = \sigma_{sl}$.

[‡]　Generally, Wulff's equation serves a premise allowing the calculation of surface energies of solid bodies, particularly of crystals and their distinct faces [146].

$$\sigma_{sa} = \sigma_{la} (\sin \alpha + \cos \theta) \tag{132}$$

at the solid–air interface, and

$$\sigma_{sl} = \sigma_{la} \sin \alpha \tag{133}$$

at the solid–liquid interface.

Analysis of these proposals demonstrates a rather limited applicability of wetting thermodynamics, in the most rigorous form, to a description of the processes of adhesive interaction of polymers. In practice this situation is usually avoided in at least two ways. The first one requires that the variables entering in the initial theoretical equations are substituted by semiempirical ones associated with the nature of high-molecular mass compounds, and this approach is discussed in section 2.2.1. The second involves the introduction of additional variables into basic thermodynamic relationships, these variables being characteristic of the attractive interactions between the contacting phases.

As such a variable Good suggested [326] a correction coefficient Φ, which is a function of the dipole moment μ, polarizability ϵ, ionization potential and volume V. The general appearance of this function is quite complex. However, Kirkwood [327] reduced it to a simple relationship

$$\Phi = (V_1 V_2)^{3/4}/(V_1^{2/3} + V_2^{2/3}). \tag{134}$$

Subsequently Davis [328] succeeded in finding analytical expressions for the parameter Φ at the polymer-(low molecular mass liquid) interface. Strictly speaking, as Φ is related to the parameters of the fundamental Lennard-Jones potential [330] it does have a certain physical meaning [195, 329]. Neither of these approaches, however, can provide for practical calculations in polymer adhesion. Therefore, to evaluate Φ Good, having involved the analogy with the above mentioned principle of Berthelot, expressed the work of adhesion as [331]

$$W_{Ad} = 2\Phi(\sigma_{sa}\sigma_{la})^{1/2}. \tag{135}$$

Then

$$\sigma_{sl} = \sigma_{sa} + \sigma_{la} - 2\Phi\sigma_{sa}^{1/2} \sigma_{la}^{1/2}. \tag{136}$$

Adamson obtained the same equation from a more general approach by considering the interactions of molecules with their nearest neighbours [332].

On the basis of eqs (55) and (136) one can readily derive the relationship for the contact angle as a function of the surface energies of the solid and liquid phases and of the quantity π [326]

$$\cos \theta = 2\Phi\sigma_{sa}^{1/2} \sigma_{la}^{1/2} - 1 - \pi/\sigma_{la}. \tag{137}$$

Ignoring surface pressure, Good [333] considers that the last term of eq. (137) might well be neglected. We believe that this may be the case for low-energy polymers only. By expanding eq. (137) in a Taylor series [330]

$$\cos \theta = 1 - \frac{\sigma_{la} - \Phi^2 \sigma_{sa}}{\Phi^2 \sigma_{sa}} + \frac{3}{4} \left(\frac{\sigma_{la} - \Phi^2 \sigma_{sa}}{\Phi^2 \sigma_{sa}} \right)^2$$

$$- \frac{5}{8} \left(\frac{\sigma_{la} - \Phi^2 \sigma_{sa}}{\Phi^2 \sigma_{sa}} \right)^3 + \dots \tag{138}$$

With the aid of eq. (137) it is possible to determine Φ from the data on wetting. In the approximation adopted Φ is then presented as the difference between the polarities of the phases in contact, this parameter being quite important in the thermodynamics of adhesive joint formation. When the difference is small, or the polymeric substrate is non-polar, $\Phi \approx 1$. For instance, for polyethene and polytetrafluoroethene the Φ values were found to be 1.021 and 0.98, respectively, despite the fact that the contacting liquids involved in the study were rather polar, their σ_{la} values reaching 49.7 mN/m [334] (see Table 2.1). For high energy surfaces this simplification does not hold true. In fact, as the surface energy of the substrate increases the contact angle with the adhesive diminishes regularly as is illustrated in Fig. 2.24 for the substrate polypropene [335]. In this case eq. (134) should also involve the polarizabilities of the contacting phases. According to de Gennes [206] it is only when

$$\epsilon_s > \sigma_l \tag{139}$$

that the adhesive interaction is developed on wetting. In other words, it is due rather to the fact that the polarizability of solids significantly exceeds that of liquids than to the high σ_{sa} values that high energy surfaces are wetted by molecular liquids. Hence, it is clear that generally the magnitude of Φ should differ reasonably from unity. Indeed, according to Girifalco and Good [331] it varies from 0.51 for hydrocarbons to 1.15 for alphatic alcohols.

Table 2.1 — The Φ values for non-polar polymers as a function of the nature of wetting liquid

Wetting liquid	σ_{la} (mN/m)	Φ	
		PTFE	Polyethene
Ethanol	23.7	1.038	
Hexadecane	27.5	1.009	
Hexylbenzene	30.0	1.006	
Di-(2-ethylhexyl)phthalate	31.2	0.923	1.001
Nitropropane	31.2	0.961	1.015
Acetic anhydride	34.7	1.050	0.916
Benzylphenylundecanoate	37.7	0.970	1.038
Tricresylphosphate	40.9	0.915	1.065
Bromonaphthalene	44.6	0.977	1.091
Tetrabromoethane	49.7	0.951	

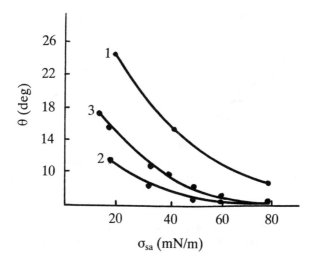

Fig. 2.24 – Contact angle versus surface energy of solid substrates. 1, atactic polypropene melt; 2, 3, 14% and 30% solutions of atactic polypropene in heptane.

Having analysed the problem under discussion, Good [311] has come to emphasize, as a matter of primary importance, that the intra- and interphase forces have to be fairly similar. When they are of the same origin, $\Phi \rightarrow 1$ resulting ultimately in $\Phi = 1$. It is only in this case that σ_{sl} values can be evaluated from measurements of contact angles θ [336]. However, for systems in which cohesive integrity is provided by dipolar, metallic or ionic forces, while their adhesive bonding is due to Van der Waals interaction, Φ is less than unity, $\Phi < 1$.

This approach is a deliberate simplification. The meaning of Φ_{12}, as defined by Good, is not confined merely to being the difference between the polarities of the contacting phases. By analogy with eq. (117) the general relationship is given as

$$\Phi_{12} = Y_{12}/(Y_1 Y_2)^{1/2} \tag{140}$$

where Y denotes an extensive thermodynamic property, e.g. entropy [328], total [311] or free [314] energy. Precisely, it is not the properties themselves but, according to Gibbs, their surface excesses $\Delta^s Y$ over the initially separate phases accepted as the standard state that should be taken into account. Then [306]

$$Y_{12} = \Delta^s Y_{12} - Y_1^s - Y_2^s. \tag{141}$$

As is seen from eq. (136), Φ and σ_{sl} are related via a linear relationship. Neumann *et al.* [319] presented experimental verification of this dependence by measuring contact angles on polytetrafluoroethene and poly(tetrafluoroethene-co-ethene) (50 : 50). Corresponding data are shown in Fig. 2.25. Consequently, knowing Φ it appears possible to evaluate σ_{sl}. Good claims [288] that the mean error in calculating σ_{sl} by means of eq. (136) does not exceed 1–2 mN/m. On the other hand, by presenting eq. (136) in the following form

$$\Phi = \sigma_{la}^{1/2} (1 + \cos \theta)/2\sigma_{sa}^{1/2} \tag{142}$$

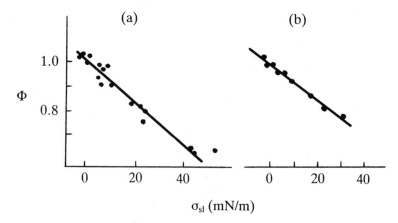

Fig. 2.25 – Parameter Φ of polytetrafluoroethene (a) and ethene–tetrafluoroethene = 1 : 1 copolymer (b) versus interfacial energy with different liquids.

one can easily see that Φ and $\sigma_{la}^{1/2}$ are interrelated also by a rectilinear dependence. Fig. 2.26 provides experimental confirmation of the dependence established as exemplified by the studies on the wetting of polystyrene [330]. Hence, (136) and (142) can be used for quantitative assessment of the effectiveness of adhesive interaction. In fact, Toyama [337] demonstrated that the adhesives based on natural rubber and (ethylacrylate-2-ethylhexylacrylate) copolymers exhibit maximum adhesive strength explicitly with those polymeric substrates (polystyrene and poly(methyl methacrylate)) that are characterized by maximum σ_{sl} values, these being close to the values of W_{Ad} calculated by eq. (98).

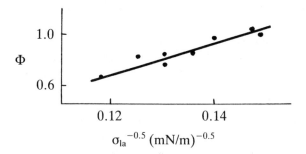

Fig. 2.26 – Parameter Φ of polystyrene versus surface energy of the wetting liquids.

Relying on computer processing of the data of Fig. 2.25, as well as of the results with other substrates [195], Good proposed a generalized empirical relationship [319]

$$\Phi = -\alpha\sigma_{sl} + \beta \qquad (143)$$

and determined its coefficients. Φ can thus be easily calculated by means of this relationship instead of the otherwise troublesome and laborious procedures involving eqs. (120)–(123). Making use of the constants involved in these equations it became possible, with the aid of eq. (122) [319], to express interfacial surface tension via the surface tension of the liquid phase and surface energy of the solid phase. This possibility is quite reasonable, as it is implied, when using eq. (136), that θ has to be determined with sufficient accuracy; the possible discrepancy in the ultimate result significantly increases due to the surface energies of different polymers being observed as close to each other (the background of this phenomenon is discussed at the end of section 2.2.1).

The major consequence of wetting in adhesive–substrate systems is the spreading of adhesive over the substrate enabling the formation of as complete an interfacial contact as possible. Similar approaches to the problem are of molecular–kinetic origin though some phenomenological aspects of the subject can be treated in terms of thermodynamics.

Spreading can be defined as the effect of spontaneous streaming of one condensed phase over the other resulting from the decrease of surface free energy of the system when it is not subjected to any external influence. On advancing the liquid phase the interfacial area between the phases increases, while on receding it decreases. The term characterizing this effect was first suggested in 1915 [338] as a response to the necessity to evaluate the effectiveness of the treatment of plant leaves with insecticides.[†] As regards the understanding of the physical meaning of this term, we are indebted to Harkins [340]. According to his classical definition, the spreading coefficient is given as the difference

$$\chi = \sigma_{sa} - (\sigma_{sl} + \sigma_{la}) \tag{144}$$

which is related to the Helmholtz energy, being its partial derivative

$$\chi = - (\partial F / \partial A)_{T, \mu_i}. \tag{145}$$

On the other hand,

$$\chi = W_{Ad} - W_{Coh}. \tag{146}$$

In a three-phase system, for instance, the spreading of phase 2 over the 1–3 interface can be characterized by the spreading coefficient χ expressed via the work of adhesion [341]

$$\chi_{2/13} = W_{Ad_{12}} + W_{Ad_{23}} - W_{Ad_{13}} - 2\sigma_{2a}. \tag{147}$$

In the particular though quite common case, when the open air is phase 3

$$\chi_{2/13} = W_{Ad_{12}} - 2\sigma_{2a}. \tag{148}$$

The physical implications of χ become clear when considering the situations where

$$\chi > 0 \tag{149}$$

and

$$\chi < 0. \tag{150}$$

---

† More than half a century later the problem of spreading over leaves arose once again. This time it was related to the need to retard transpiration — evaporation of moisture from leaves, stems and fruits of plants — an unfavourable process for plant physiology. A number of polymeric anti-transpirants was suggested, their choice having been governed by their adhesiveness to plant tissues [339]. This somewhat unexpected but practically important aspect of the problem under discussion provokes us to emphasize the principal importance of investigations concerning adhesion in biological systems noted in section 1.1.

The first corresponds to total (complete) wetting resulting in the formation of a liquid phase monolayer on a solid surface and the second to limited spreading characterized by a certain contact angle being ultimately established.

The appeal of the approach involving the concept of spreading is associated with its providing a simple correlation between the surface energies of a liquid (adhesive) and a solid (substrate) governing the development of adhesion. At $\sigma_{la} > \sigma_{sa}$ wetting is thermo-dynamically favourable and is liable to result in adhesive joint formation. Thus the efficiency of varying σ_{la} in manipulating adhesive interaction can be evaluated. Acknowledging the fact that for multicomponent systems σ_{la} is an integrated term, it can therefore be varied by introducing into the adhesive composition components with different values of σ^i. Indeed, according to Gibbs, the interphase boundary is enriched by the component which lowers the surface tension. This conclusion, resulting from the fundamental principle of energy minimization, can be described by the well-known equation

$$C_{if} = \frac{C_v}{KT} \frac{d\sigma_{la}}{dC_v} ,$$

(151)

where C_{if} is the concentration excess of the corresponding component at the interface and C_v is the mean concentration of this component within the phase.[†] Eq. (151) provides the means to predict the effect of various components added to a composite adhesive [344]. For instance, modification of polyvinylacetate with a small amount of decanoic acid produces a two-fold decrease in the strength of adhesive bonding of metals [345]; the opposite effect is achieved by adding oligomeric phenol-formaldehyde modifiers [110]. The number of examples can be easily extended and indeed the approach described provides the basis for the targeted development of adhesives.

The spreading coefficient appears to be a concept quite useful for practical purposes, as it enables a thermodynamic assessment of the tendency to adhesive joint formation to be made. Taking into account σ_{sl} as defined by eq. (118), the expression for the work of adhesion is given as follows

$$W_{Ad} = \sigma_{la} + \sigma_{sa} - \sigma_{sl} = 2(\sigma_{sa}\sigma_{la})^{1/2}.$$

(152)

For liquids

$$W_{Coh} = 2\sigma_{la}.$$

(153)

By substituting both in eq. (146), one has

$$\chi = 2[(\sigma_{sa}\sigma_{la})^{1/2} - \sigma_{la}].$$

(154)

With the aid of this relationship the values of χ can be calculated for various materials. For instance, assuming that the interaction of polyacrylate adhesive ($\sigma_{la} = 26$ mN/m)

---

† Concerning eq. (151), here again it is worth citing Gibbs, who wrote: 'When the film is extended, there will therefore not be enough of these substances ['those components which diminish the tensions' — authors] to keep up the same volume and surface densities as before, and the deficiency will cause a certain increase of tension' [129, p. 301]. Quantitative dependence of the elastic modulus of multicomponent liquid films on their thickness as well as on concentrations and activity coefficients of separate components was established by Rusanov and Krotov [343].

with the substrates is governed by dispersive forces only, Kaelble [346] calculated the values of χ for seven (polyacrylate adhesive/substrate) systems, mJ/m^2:

polycaproamide	16.4	polyvinylidenefluoride	4.0
polyvinylfluoride	15.4	poly(tetrafluoroethene-	
polytrichlorofluoroethene	10.2	co-hexafluorpropene)	−10.2
polystyrene	6.2	polytetrafluoroethene	−12.2

Furthermore, he concluded that all of these systems, except for the last two for which $\chi < 0$, would be fractured cohesively. Similarly, by means of eq. (154) Bragole [347] estimated the spreading of polyurethane adhesive over polyethene ($\sigma_{sa} = \sigma_{sa}^d = 35$ mN/m). As 15–20% of the secondary bonds in polyurethane are hydrogen bonds, the dispersion component to the overall $\sigma_{la} = 43$ mN/m is reduced to $\sigma_{la}^d = 34$ mN/m. Then $\chi = 2[(35 \times 34)^{1/2} - 43] = 17$ mJ/m^2 and, consequently, in conformity with the requirements of eq. (150) the spreading of polyurethane adhesive over polyethene is quite favourable. These calculations are easy to perform for any system comprising phases with known σ.

Modification of the substrate can affect the magnitude of χ. For instance, silicon plates with a 2 nm thick silica layer at the surface display a positive spreading coefficient, whereas when coated with octadecyltrichlorosilane they exhibit a spreading coefficient that is close to zero. It is these factors that determine the different spreading modes of polydimethylsiloxanes, including the kinetics of spreading and the structure of the films coating the substrates [348]. This example clarifies the role of either sorbed or grafted compounds in affecting χ and consequently, the liability to wetting by liquids [349, 350].

Analysis of the example with the silicon substrate (for which $\chi \to 0$) leads to an important conclusion essential with regard to the applied aspects. According to de Gennes [206], the spreading efficiency of a liquid film is a function of its density ρ_l and thickness d and, taking into account the density of the gaseous phase

$$\chi = \frac{4b}{3d^3} - \frac{gd^2(\rho_1 - \rho_a)}{2} \tag{155}$$

where b is a constant and g the acceleration due to gravity. Numerical calculations give $b/d^3 = 10^{-7}$ mN/m and $\chi \approx 0.1$ mJ/m^2. The term in eq. (155) related to gravitation can thus be neglected. Relating the surface energy of a liquid, to a reasonable range of the wetting (liquid) layer thickness Joanny and de Gennes [351] obtained the following relationship

$$d = k(3\sigma_{la}/2\chi)^{1/2}. \tag{156}$$

It appears then that the film thickness, when spreading is complete, is less the greater the χ value. In its turn this implies that the surface area covered with a liquid is increased. Hence, the liquids with high values of χ are liable to spread more readily and efficiently than those with low spreading coefficients. This effect was discovered by Cooper and Nuttal [338] and it serves as a guide to developing solution adhesives.

At the same time it should be borne in mind that the two phases may tend to become mutually saturated. When one of the phases is solid the approach proposed by Pittmann and summarized in eqs (124)–(126) comes into effect. In liquid–liquid systems mutual saturation may also take place. For instance, benzene rapidly spreads over the surface of pure water; however, after the phases have become mutually saturated, the benzene phase contracts to a lens, the surface tension of water being decreased to 62.2 mN/m because of the formation of the monomolecular benzene film equilibrated with saturated benzene vapour [112]. The final magnitude of the initially positive spreading coefficient χ (8.8 mJ/m^2 [342]) is −1.6 mJ/m^2. This is why the initial and equilibrium values of the spreading coefficient may significantly differ (for some hydrocarbons and alcohols spread over water the difference may be as large as to cause a change in the sign of χ). This effect is most noticeable when the polymeric substrate can be partially dissolved by the adhesive or the constituent solvent.

In spite of the fact that χ is a function of the nature of the contacting phases, the surface structure and the conditions under which the contact is established, the concept of spreading coefficients appears to be inadequate to give a description of the modes of spreading in adhesive systems and it is therefore used only in initial assessments. A more complete approach should obviously involve the spreading coefficient as one, but not the only, variable.

As one of the approaches we would suggest the theory, recently developed by Hocking [352, 353], elaborating on the spreading of small drops, that appears to account for the specific features of polymer spreading. Within the framework of this theory asymptotic expansions are constructed relating to the three regions of a drop, the 'inner' region near the rim of the drop, the 'outer' (central) region whose form is almost spherical, and the 'intermediate' region which links together the first two. These regions may be identified as the precursor film, the spherical cap, and the macroscopic foot in de Genne's description of a spreading droplet [206]. The shape of the central region is indeed very close to spherical. As the motion of the fluid is not hindered here the pressure in the system reaches equilibrium, indicating the constant pressure difference between the inside and outside of the drop. Then, according to the Young–Laplace equation, the curvature of the drop is constant, as stated above.

For such a system, neglecting gravitational effects, one can easily relate the magnitude of the contact angle with the drop thickness and radius as time-dependent functions [206]

$$d = 0.5 \, r\theta, \tag{157}$$

$$r = (V_d/0.5\pi d)^{1/2} = 0.7979 \, (V_d/d)^{1/2} \tag{158}$$

where V_d is the volume of the drop. With the assumption $V_d = \mathrm{const}$, along with $\cos\theta \ll 1$, eqs (157) and (158) are valid in the most common case relating to the complete wetting of the substrate by thin drops. This is the case that is most specific to the formation of adhesive joints under an external load. Taking into account that, as verified by Dussan [354], at the start of the spreading process $r(\tau)$ changes rapidly, subsequently slowing down as the spreading proceeds one then has

$$\pi r^2 (\tau) \approx V_d^a \tau^b, \tag{159}$$

a varying within a sufficiently narrow range [355]; however, it is markedly temperature dependent [356]. Usually, a varies from 0.20 [206] to 0.21 [357] and b from 0.60 [206] to 0.67 [356].

The cases of greatest interest, with respect to the applied aspects are those of complete spreading and incomplete wetting.

For the first of the cases in a one dimensional approximation the thickness of the film is a function of the magnitude of one coordinate x in the plane of the substrate surface. In the general expression for the free energy [206]

$$F = F_0 + \int_{x_{min}}^{x_{max}} dx \left[-\chi + 0.5\sigma_{la} \left(\frac{dd}{dx}\right)^2 + P(d) + G(d) \right] \tag{160}$$

the last term accounts for the hydrostatic and gravitational effects (which can be neglected at the microscopic level), the penultimate term deals with the long range forces and at $d \to \infty$ tends to zero

$$P(d \to \infty) = 0. \tag{161}$$

The term related to the surface energy is due to the expansion of the length element $dl^2 = dx^2 + dd^2$ on the assumption that

$$dd/dx \approx 0. \tag{162}$$

In conformity with the concepts discussed above, σ_{la} and χ retain their thermodynamical implications in the case of thick wetting films. The required conditions having been taken into account, the equilibrium condition for a drop can be obtained by minimization of eq. (160) with respect to d(x):

$$-\sigma_{la} \frac{d^2 d}{dx^2} + \frac{dP}{dd} + \frac{dG}{dd} = 0. \tag{163}$$

Integrating eq. (163)

$$0.5\sigma_{la} \left(\frac{dd}{dx}\right)^2 = P(d) + G(d) - \chi. \tag{164}$$

This relationship is most important as it involves parameters that can be controlled under the conditions of forced wetting of the substrates by adhesives.

In the case of incomplete wetting, which is no less important

$$\chi = -0.5\sigma_{la}\theta^2. \tag{165}$$

Substituting it into eq. (164) and neglecting the effect of gravitation and macroscopic pressure drop, the following equation is obtained for θ values different from zero ($\cos \theta \ll 1$)

$$\left(\frac{dd}{dx}\right)^2 - \theta^2 = 2 P(d)/\sigma_{la}. \tag{166}$$

Solution of eq. (166) results in a hyperbolic dependence [351]

$$d^2 = (x\theta)^2 + (k/\theta)^2, \tag{167}$$

which is true under conditions somewhat different from those defined by the condition eq. (162), namely when

$$dd/dx \ll 1. \tag{168}$$

However, at small contact angles the form of the drop is hyperbolic up to a thickness of not less than 10 nm, as evaluated by de Gennes [206]. Hence it is necessary to consider these ideas when examining the formation of adhesive joints.

At the same time, one should bear in mind that the treatment above does not allow one to distinguish between the rates of wetting and spreading, though both quantities are of great practical significance.

Evaluating the rate of wetting by the rate at which the contact angle changes

$$v_w = f \left(-\frac{d\theta}{d\tau} \right) \tag{169}$$

and the spreading rate by the time-dependent variation of the radius of the base of the wetting drop on a solid surface

$$v_{sp} = f \left[\tau \Big/ \left(\frac{dr}{d\tau} \right) \right], \tag{170}$$

one infers that $(dr/d\tau) \neq 0$ even at $(d\theta/d\tau) = 0$. However, as shown by Popel [358, 359] for a great variety of objects the patterns of the corresponding kinetic dependences are dissimilar, demonstrating the limitations of the approach described. Therefore, investigations into spreading are more common along other lines.

The change in the free energy of a system is assumed to be the 'driving force' of spreading. Cherry and Holmes [360], and Blake and Haunes [361] independently proposed the use of the theory of absolute reaction rates to examine the process. The first authors suggested that the rate of spreading is limited by the transfer of the molecules of a liquid from within the bulk of the phase, then the rate constant is determined by the activation energy of viscous flow E_η^*:

$$k_{sp} = \frac{2KT}{h} \, \text{sh} \left(\frac{\Delta F}{2KT} \right) \exp \left(-E_\eta^*/RT \right). \tag{171}$$

Analysis of this relationship is beyond the scope of thermodynamics, as it would involve rheological factors, viscosity in particular. On the other hand, Blake and Haunes [361] consider that in the case examined the processes of interfacial interaction are of primary importance; then, in a first order approximation

$$\Delta F = \sigma_{sl}(\cos \theta_\tau - \cos \theta_\infty) \tag{172}$$

where θ_τ and θ_∞ are the values of contact angle at some specified time τ and at equilibrium. Hence, the spreading rate is as follows

$$v_{sp} = 0.51 \, \text{sh} \, [\sigma_{la}(\cos \theta_\tau - \cos \theta_\infty)/N_s KT], \tag{173}$$

and eq. (171) transforms to

$$k_{sp} = (KTQ^*/hQ_o) \exp(-E_\eta^*/KT), \tag{174}$$

where N_s is the number of active sites on the substrate surface, Q^* and Q_o are the statistical sums of the transient and initial states of the molecules of liquid and h is a constant.

The validity of these approaches was verified experimentally in the studies of the spreading of metals at high temperatures [362] when

$$\sigma_{la}(\cos \theta_\tau - \cos \theta_\infty)/N_s \ll KT. \tag{175}$$

In the opposite case, i.e. when

$$\sigma_{la}(\cos \theta_\tau - \cos \theta_\infty)/N_s \gg KT \tag{176}$$

eq. (173) was found to be adequately valid [186], since the dependences $(\cos \theta_\tau - \cos \theta_\infty)$ vs. $\lg v_{sp}$ are linear. However, when v_{sp} increases an inflection appears on these dependences, which is attributed to the viscous effects in the spreading fluid [363]. A similar conclusion was obtained when the hyperbolic sine in eqs (171) and (173) was expanded into a series, the first member only of the series having been taken into account [359].

Obviously, these speculations hold true primarily at the final stages of the spreading process, when $\theta_\tau \to \theta_\infty$. For the initial stages [362] referring to eq. (170) one has

$$\frac{dr}{d\tau} = k_{sp}\sigma_{la}(\cos \theta_\tau - \cos \theta_\infty). \tag{177}$$

It is therefore necessary to distinguish between the kinetics of wetting proper and subsequent spreading, and of the transport of the molecules of liquid to the interfacial zone, which is of specific importance when dealing with highly viscous polymeric adhesives.

The effect of the kinetic resistance of the fluid on the overall spreading process must also not be neglected. Its magnitude is determined by the surface diffusion of the molecules of adhesive adsorbed at the substrate surface. This is probably the mechanism of fluid 'slippage' along a smooth surface as the contact line moves on [364, 365]. In fact, from the physical standpoint it appears more favourable (profitable) for shearing in the fluid to be concentrated within a narrow region near the interface than to be distributed in the volume of the phase. This effect was observed experimentally by Kraynik and Schowalter [366], and by Burton et al. [367]. In this case the description of the spreading process can be given in thermodynamical terms.

The effect of rheological factors is particularly important when treating the second aspect which is related to the transport of molecules of the liquid to the interface. In a reasonable approximation viscosity appears to be the one and only parameter that needs to be taken into account in this matter, as, according to the reptation model [368]

$$\eta = \eta_0 N^3 / N_c^2 \tag{178}$$

where N is the number of monomeric units constituting a macro-molecule (in other words — the D.P.) and N_c is the number of monomeric units between the two anchoring points. The thickness of the adhesive film required for the film to spread was calculated to be not less than 1 mm at $N = 10^4$. It is then clear that at high η values v_{sp} is a function of the equilibrium contact angle, diffusion coefficient, and the size of the spreading drop [369].

The concepts surveyed in this section, whether more or less closely related to fundamental thermodynamic principles (23)–(28), do not take into account the distinct nature of the interfacial interaction. This can be considered neither as a benefit nor as a draw-back of the thermodynamic approach as it is its inherent feature. Therefore, when applying formalized thermodynamic analysis to real adhesive bonds, it is obligatory that the nature of the forces acting at the interface between the condensed phases be additionally considered. However, the study of the energetics of adhesive interaction between polymers should, naturally, be preceded by a study of the surfaces of polymers proper with regard to their energy characteristics.

2.2 INTERACTION BETWEEN POLYMERS DURING ADHESIVE CONTACT

2.2.1 Energy characteristics of solid surfaces

Instrumental and mathematical procedures for the determination of surface parameters of condensed phases, solids in particular, have been extensively treated. This is illustrated by the vast literature compiled in a large number of the relevant reviews; some of them have been already referred to, though of the most recent we would suggest the following [370–375]. The list of these publications is quite impressive and illustrates the thorough elucidation of the problem. In this regard it is noteworthy to mention Halsey's comment concerning the calculation of surface energies: '. . . it is striking that even though some of the more difficult calculations are in danger of diverging, the real values of surface energies and associated surface tensions are notably unexciting. They do not vary much in value, and seem to increase with the melting or boiling point of the substances, in intuitively obvious fashion. If we take the heat of vaporization of a rare-gas atom at the absolute zero to be Λ, unrelaxed nearest-neighbour calculations of the crudest sort give a surface energy of. . . . Suitably modified for geometry, this equation should give enough *variety*, if not the *accuracy* [italicized by Halsey] of the calculated values so far available for comparison' [376, pp. 508–509]. We believe that in relation to polymers this conclusion is quite premature. In a number of cases the physical meaning of the distinct methods requires more accurate definition. Besides, the agreement between the results obtained by different authors canot always be recognized as satisfactory.

We will consider those methods for evaluating the surface energy of solids that are principally suitable in producing thermodynamic estimates for the effectiveness of adhesive interaction. The basic ones, having primarily a fundamental and not techno-logical background, are given in Table 2.2 together with their important references. By their physical characteristics these methods may be conventionally divided into five groups as follows:

1. *Destructive methods* involving the measure of σ in a destructive test of the material. For polymers, with the exception of brittle materials, these methods are infrequently used since, in general, it is difficult to identify the component relating to the pre-fracture deformation of the sample. This group comprises the methods of spalling, zero creep, flexing of a plate, electron microscopy, and that of smoothed grooves. There is no need to consider these methods in great detail as the principal disadvantage noted above makes them rather unsuitable for the investigation of high polymers.

2. *Solvolytic methods* based on the measurement of σ during failure of the material; the latter, however, being the result of the effect of a solvent and not of mechanical stress. In these methods the side effects of interaction between the low and high molecular mass

Table 2.2 – Major methods for determination of energy characteristics of solid surfaces

Method	Objects	Determined parameter	Physical basis	Merits	Demerits	Error (%)	Reference
1	2	3	4	5	6	7	8
Destructive methods							
1. Simple shear	Mica, crystals polymers	Work of surface formation	Measurements of wedge transfer force along artificial crack	Differentiation of cohesion and adhesion energies	Structural factors are difficult to regard for, inapplicability to plastic solids	10–20	[377]
2. Zero creep	Metals, glasses	Surface free energy	Measurements of fatigue load compensating surface shrinkage of the sample	No restrictions on the nature of objects	Internal stresses are difficult to account for[a]	10–15	[378]
3. Flexing of plate	Crystals, glasses	Surface free energy	Measurements of the flexure radius of a thin plate with polished surface	Simplicity and reproducibility	Significance of the quality of surface treatment		[380]
4. Autoelectron microscopy	Refractory metals	Surface free energy	Measure of direct voltage on emitter at constant current	Broad temperature interval	The effect of the surface is difficult to account for	10–20	[381]
5. Smoothed grooves	Metals	Surface free energy	Measurement of groove profile on polished surface of annealed sample	No restrictions regarding the nature of objects	Difficulty of measuring groove characteristics	10–15	[382]

Table 2.2 (continued)

Methods	Objects	Determined parameter	Physical basis	Merits	Demerits	Error (%)	Reference
1	2	3	4	5	6	7	8
Solvolytic methods							
6. Calorimetry	Metals, crystals	Specific surface entropy	Measurements of the difference between the heats of dissolution of the massive and the finely dispersed samples	No restrictions regarding the nature of objects	Significant effect of surface area	20–50	[383]
7. Solvolytic	Crystals	Surface free energy	Measurements of concentration of the sample in solution in equilibrium with the solid	No restrictions regarding the nature of objects	Significant effect of surface area	20–40	[384]
Methods of multi-phase equilibrium							
8. Trihedral pyramid	Metals polymers, glasses	Surface free energy	Measurement of three contact angles in a three-phase system	No restrictions regarding the nature of objects	Significant effect of surface area		[385]
9. Critical surface tension	Polymers	Conditional surface tension	Measurement of θ for homologous series of liquids; subsequent extrapolation to $\theta = 0$	Simplicity and reproducibility	Lack of unambiguous physical implications		[386]
10. Critical tension of wetting	Polymers	Conditional surface tension	Measurement of σ_{la} and θ for homologous series subsequent extrapolation to $\sigma_{la} \cos \theta = 0$	Simplicity and reproducibility	Lack of unambiguous physical implications		[387]
11. Wetting by binary mixtures	Polymers	Conditional surface tension	Measurement of σ_{la} and θ for mixtures of liquids; $\sigma_{sa} = 0.5\sigma_{la}(1 - \cos \theta)$	Simplicity and reproducibility	Arbitrary basic assumption $W_{coh} = W_{Ad}$		[388]

Table 2.2 (continued)

Method	Objects	Determined parameter	Physical basis	Merits	Demerits	Error (%)	Reference
1	2	3	4	5	6	7	8
Extrapolation methods							
12. Molecular mass extrapolation	Polymers	Surface tension	Measurement of contact angles for homologous series; extrapolation to infinite molecular mass	Simplicity and reproducibility	Lack of unambiguous physical implications	20–30	[389]
13. Concentrational	Polymers	Surface tension	Measurement of surface tension of polymer solutions; extrapolation to solid state	Simplicity	Lack of unambiguous physical implications		[390]
14. Temperature extrapolation	Polymers	Surface tension	Measurement of surface tension of polymer melts at different T with extrapolation to solid state	Simplicity	Secondary processes due to increased temperatures are difficult to isolate		[391]
Cohesive methods							
15. Parachor	Polymers	Surface tension	Correlation between surface tension and density of sample	No restrictions concerning the nature of objects	Inadequacy of the additive scheme	10–15	[392]
16. Cohesive	Polymers	Surface energy	Correlation between surface energy and solubility parameter	No restrictions concerning the nature of objects	Poor reproducibility		[393]
17. Polarization	Polymers	Surface energy	Correlation between surface energy and polarizability, and diamagnetic susceptibility	No restrictions concerning the nature of objects	Difficult ot determine the corrleated physical characteristics	10–30	[394]
18. Refractometry	Polymers	Surface energy	Correlation between surface energy and refractive index	No restrictions concerning the nature of objects	Polarity of the sample has to be evaluated	5–7	[395]

Symbols: σ surface tension; θ contact angle; W_{Ad} work of adhesion

a Compensation of internal stresses extends the applicability of the zero creep method [379] decreasing the error to 1–3%.

compounds are difficult to quantify. The improvements in instrumentation (e.g. calorimetry), however, account for a substantial decrease in experimental error.

3. *Methods of multi-phase equilibrium* are also based on the interaction between the low and high molecular mass compounds. The σ values however are introduced into the equations for the mechanical equilibrium of the system which rely on the thermodynamic concept of wetting. Corresponding to the numbers of liquid and solid phases the methods of trihedral pyramid (three solid phases and one liquid) and of binary mixtures (one solid phase and two liquid phases) are distinguished. Quite popular are the measurements of semi-empirical parameters, e.g. critical surface tension and critical wetting tension, whose meaning is close to that of σ.

4. *Extrapolation methods* are based on the measurements of the surface tension of liquids (solutions, melts, or oligomers) with subsequent extrapolation to the solid state relying on the dependences of σ upon molecular mass, concentration, or temperature. Because of the high level of experimental technique required for the determination of surface characteristics of liquids the methods of this group are most commonly used for the evaluation of σ of solid polymers although the limitations of the extrapolation principle are quite evident.

5. *Cohesion methods* are based on the relationship between intraphase characteristics of condensed phases and their surface energy. Interpretation of the results in this case is most complicated because cohesion parameters can be of a very different origin. However, this group of methods is regarded as the most physically sound, provided that the theoretical premises of the basic relationship are of sufficient depth. The common intraphase polymer characteristics are density, solubility parameter, dielectric constant, polarizability, diamagnetic susceptibility, refractive index etc. although this list does not restrict the potential of this approach.

We will give examples of the major methods used to evaluate energetic parameters of polymer surfaces, paying special attention to their application to solving the problems of adhesion and to questions that have not yet been discussed elsewhere in the survey.

One large group is concerned with methods involving the measurement of the surface tension of a liquid (either a solution or a melt) and its subsequent extrapolation to the solid state. The procedure holds true only where the quantity to be extrapolated obeys a law that is equally justified for both the liquid and the solid states. Concentration, obviously, does not meet this requirement as the sorption isotherms are different for the liquid and the solid states. For the very same reason the method of homologous series cannot be considered as sufficiently sound. This, in principle, is also the case with the temperature extrapolation procedure. Its tempting simplicity, however, has provoked certain semi-empirical assumptions that claimed as valid the feasibility of extrapolating the surface tension of polymer melts to their solidification temperatures.

For such cases Wu [294] suggested a general dependence interrelating the surface parameters of the initial and the infinitely large molecular masses, corresponding to the liquid and solid states, respectively

$$\sigma^{-1/\nu} = \sigma_\infty^{-1/\nu} + k_\sigma/M \qquad (179)$$

where k_σ is a proportionality coefficient, ν is the power index in MacLeod's equation [396] relating the surface tension of the sample to its density

$$\sigma = \sigma_0 \rho^\nu. \tag{180}$$

The value of ν is usually within the range of 3.0–4.4 (for instance, for non-associated liquids $\nu = 4.0$ [397]. Listed below are the values of ν for some polymers:

polyethene	3.2 [398]	poly(vinyl acetate)	3.4 [398]
polypropene	3.2 [398]	poly(methyl methacrylate)	4.2 [399]
polyisobutene	4.1 [398]	polychloroprene	4.2 [400]

Despite the apparent empiricism this quantity is related by a power dependence to the parachor [401]

$$\sigma_0 = (P/M)^\nu, \tag{181}$$

as well as to the thermal expansion coefficient c [294]

$$d\sigma/dT = -c\nu\sigma \tag{182}$$

(this latter expression illustrates the soundness of the analogy with the free volume concept treated below). Eq. (182) gives values for the temperature gradients of surface tension that satisfactorily agree with the experimental data (Table 2.3).

The ultimate expression describing the temperature dependence of surface tension may be written as follows [294]

$$\sigma_T = \sigma_{293} + \frac{d\sigma}{dT}(T - 293). \tag{183}$$

This form of the equation makes it possible to apply the temperature extrapolation procedure for the evaluation of σ of solid polymers† from the data on the properties of

Table 2.3 – Surface tension temperature gradients of polymers

Polymer	$-\dfrac{d\sigma}{dT}10^3$	T, K	Ref.
Polyethene	58	413–473	[402]
Polypropene	56	413–463	[398]
polyisobutene	64	323–453	[402]
Polystyrene	65	418–498	[403]
Poly(vinyl acetate)	66	373–473	[402]
Poly(methyl methacrylate)	76	403–453	[402]
Poly(n-butyl methacrylate)	59	373–453	[402]
Poly(isobutyl methacrylate)	60	393–453	[400]
Poly(tert.-butyl methacrylate)	59	373–453	[400]
Polychloroprene	86	323–423	[400]
Polydimethylsiloxane	48	298–473	[400]

† In relation to metals the theoretical background of such extrapolation procedure is rather conditional [404] and it leads to complex ultimate expressions [405] whose simplification brings in additional uncertainty [406].

polymer melts obtained by common methods (capillary rise, Wilhelmy slide etc.) [397]. One should bear in mind, however, that the temperature extrapolation principle is not strict even for polymer melts. This inadequacy, first emphasized by Boyer [407], originates from the assumption of the absence of secondary phase transitions and of entropy changes in a polymer undergoing a transition from viscous flow to the solid state. Apparently, the initial presumption is true only in a narrow interval around the glass transition temperature of a polymer.

At the same time, it should be borne in mind that, as is hinted at by eq. (182) the physical interpretation of the temperature gradient is not confined merely to that of the experimentally measured quantity. As is evident from the data of Table 2.3, $d\sigma/dT$ is not related to the polarity of the sample. In fact the $d\sigma/dT$ derivative is associated with thermophysical processes. Hence, taking into account the relationship between the melting points of the polymers and their internal chain rotation postulated somewhat *a priori* by Bunn [408], it would have been natural to relate $d\sigma/dT$ to the flexibility of the macromolecules. Defining chain flexibility as the number of repeat units in the segment s [409] we discovered that s and $d\sigma/dT$ are linearly dependent (Fig. 2.27), a fact that serves to validate further the temperature extrapolation procedure. The preceding treatment of the subject has noted that, on the whole, extrapolation procedures are of a semi-empirical nature. In fact, the initial eq. (179) may be made more precise by using the adjustment coefficients a and b [411]

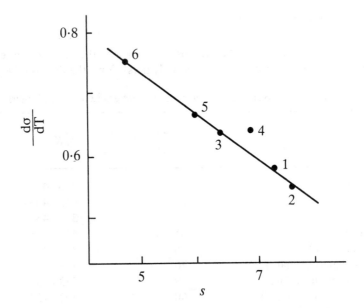

Fig. 2.27 – Temperature gradient of surface tension versus the number of repeating units in the segment of a macromolecule. 1, polyethene; 2, polypropene; 3, polyisobutene; 4, polystyrene; 5, poly(vinyl acetate); 6, poly(methyl methacrylate).

$$X = X_\infty + a/(b + l). \tag{184}$$

Here X is any cohesive parameter and l is the length of a macromolecular chain. Moreover, the dependences of the type of eqs (179) and (184) with no less success may be written not as a sum but as a difference. For this purpose one may use the Fox–Flory approach relating the length of a macromolecule with different X-quantities

$$X = X_\infty + k_X/M \tag{185}$$

which is valid, however, only in the case when X is a volume [412], and not a surface property [413, 414]. Analogous to the linear relationships between molecular mass and certain properties of high molecular mass compounds related to the lattice models of polymeric liquids [415], Legran and Janes [416], having retained the canonical form of eq. (185), however modified it by introducing into its denominator the power index 2/3. The resultant semi-empirical equation is as follows:

$$\sigma = \sigma_\infty - k_\sigma/M^{2/3} \tag{186}$$

where the linearity [417, 418] is assured by the proper choice of values for k_σ [419].

In spite of these limitations extrapolation procedures are undoubtedly helpful to the solution of applied problems. Principally however, and more important, is the fact that these procedures are bound to reveal the dependences of energy characteristics of the surfaces of solids on various cohesive properties. It would be appropriate to consider this point separately for two major types of solids, viz. for metals and, especially, for polymers.

The interrelationship between surface energy characteristics and cohesive properties is based on the possibility, following from Gibbs' thermodynamics, of expressing the so-called specific surface energy of a two-component system, determined from the dependence of the type eq. (183), as a function of the cohesive energy of the separate components, their volume difference and interfacial area [165]

$$\sigma^s = \sigma - T\frac{d\sigma}{dT} = (E_{coh}^{(1)} - E_{coh}^{(2)})\Delta V/A. \tag{187}$$

Direct application of eq. (187) to metals is rather difficult as in this case it is difficult to evaluate the cohesive energy. Therefore, instead of the latter indirect quantities are involved, the parachor being the most suitable, as confirmed by eq. (181). In accordance with eq. (181) the parachor is related to the surface properties of metals; on the other hand, it was found to correlate with sound propagation rate [420] which is characteristic of cohesive properties.[†] The interrelationship under discussion becomes even more explicit when using the rheochor R instead of the acoustic parameter. According to Friend [422] the rheochor is a function of the dynamic viscosity η and density ρ_1 of a fluid, and of the air density ρ_a

$$R = (10\eta)^{1/8}M/(\rho_1 + 2\rho_a). \tag{188}$$

Actually, for a large number of metals both parachor and rheochor are correlated strictly linearly [423] as is illustrated by the data in Fig. 2.28. It seems necessary to emphasize an important feature of this method, and that is the possibility of calculating the parachor by using the additive scheme, i.e. by summing the separate increments relating either to the composition or to the structure of a sample. On the other hand, the

---

[†] Ultrasonic waves propagation rate in polymers was shown to be related to the value of internal pressure [421], which in turn is very sensitive to the balance between interchain attraction and repulsion forces, i.e. to the purely cohesive parameters.

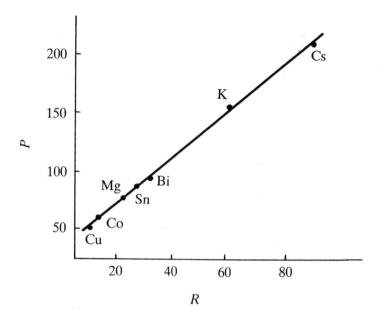

Fig. 2.28 – Correlation between parachor and rheochor for different metals.

rigorous calculation of the parachor for metals must take into account the contribution of valence electrons [424], i.e. eq. (181) appears to be over simplified to suit the task. It is quite clear that for a close-packed face (e.g. eq. 111) the electron density distribution would be different from that of the loosely packed face (e.g. eq. 110). As is illustrated by the scheme in Fig. 2.29, in the latter case the lattice of positively charged ions and the 'gas' of free electrons impose a greater mutually disturbing influence on each other [425], hence the greater surface energy corresponding to this face e.g. for copper the corresponding surface energy increase amounts to 17% [426].

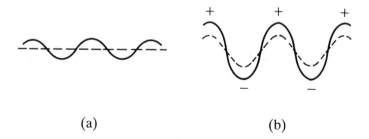

(a)   (b)

Fig. 2.29 – Pattern describing density distribution of ions (solid lines) and electrons (dashed lines) in more (a) and less (b) close-packed metallic surfaces.

The direct relevance of the surface energy of metals to their intraphase properties as characterized by the sublimation energy is illustrated by Fig. 2.30. Such dependences quantitatively relating surface energy to, for instance, the melting point [405, 427], heat of vaporization [428], energy [429], and the so-called cohesion pressure of metals (the ratio of the product of temperature and the isobaric thermal expansion coefficient to the

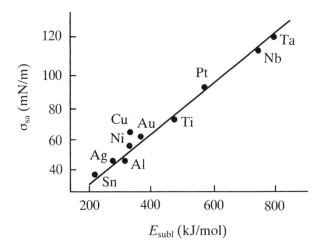

Fig. 2.30 – Surface energy of a number of metals versus their sublimation energy.

isothermal compressibility) [430] are very valuable in describing the behaviour patterns of both molten [431] and solid metals [370, 426], particularly in their adhesive inter-action with polymers [122]. It is important to emphasize that the magnitude of the surface energy of solids must be very intimately related to the strength of the ideal structure characterized, for instance, by the ultimate strength [432] and microhardness [433] of metals. Hence the physical implications of the known equation relating surface energy with the elasticity modulus of a sample become clear [434]

$$\sigma = (E/l_{1-2})(r_a/\pi)^2 \tag{189}$$

where r_a is the radius of interatomic attraction forces and l_{1-2} is the distance between the adjacent fracture planes. Tribological properties of solids appear also to be governed by their surface energy [435] (here again we are drawn to the fundamental characteristic of adhesion phenomenon discussed in section 1.1); the interrelationship between surface energy and the microhardness of metals correlates also with the characteristics of their adhesion in vacuum [436].

However, the most fundamental parameter of metals related to their surface energy is the work done on the escape of electrons φ. The relevant dependences are given in Table 2.4. It is clear that the variations in inter- and intraphase parameters are due to the variations in chemical nature (governed by the quantities A, z and ψ) and structure (accounted for by the quantities r_a, a, ρ_s and m_s) of metals. According to Zadumkin [444], this conclusion is also valid for alloys.

Having expressed cohesion characteristics of polymers in the form of eq. (185), Martin et al. [445] obtained the following equation

$$X_{coh} = X_{coh\infty} + k_{coh}/M \tag{190}$$

where X is any property; for example, the glass transition temperature. From eqs (185) and (190) there comes the relationship establishing a dependence between the properties of a material in the bulk and at the surface

$$\sigma_\infty^{-1/\nu} - \sigma^{-1/\nu} = k_\sigma(X_{coh\infty} - X_{coh})/k_{coh}. \tag{191}$$

Table 2.4 − Correlation between the work done on the escape of electrons and surface energy of metals

Cohesion characteristics	$\sigma_{sa}(\varphi)$	Ref.
Atomic radius r_A	$\sigma_{sa} = (444.5\varphi/r_A) - 110$	[437]
Lattice constant a	$\sigma_{sa} = 0.6\varphi/\pi N_A a^2$	[438]
Number of free electrons per atom $z^\dagger$ of the	$\sigma_{sa} = 1150\varphi(z\rho_s/A)^{5/6}$	[439]
metal,	$\sigma_{sa} = 0.104\, z\varphi/\pi r_A^2$	[440]
density ρ_s, and atomic mass A	$\sigma_{sa} = 530.5\, z\varphi/r_A^2$	[441]
Total potential threshold	$\sigma_{sa} = \alpha\psi\varphi^2/m_s$	[442]
of electrons $\psi^\ddagger$, coordination number in the surface layer m_s		
Tensile strength P_p	$\sigma_{sa} = 12.539\,(P_p\varphi)^{1/2}$	[443]

† Zadumkin gives z the meaning of valence [439].
‡ $\psi = 26.07\,(\rho_s z/A)^{2/3} + \varphi$.

A more detailed form of this equation involves the binary and unitary functions [447], hence, its direct solution encounters serious problems.

When estimating the surface energy of polymers from their characteristics in bulk one must bear in mind that the known approaches assume an *a priori* uniformity of the composition and properties of high molecular mass compounds over the entire volume of the sample right up to the interface. Undoubtedly, such approximation is not precise (cf. section 2.2.1), e.g. it does not take into account the differences in packing of the boundary layers and the bulk phase. The nature of this phenomenon is discussed below in section 3.1.1. Generally, apart from the energetics of interphase interaction, the calculations must take into account the existence of a density gradient in polymers within the contact zone. The authors of [448] also point out that at frontal solidification and cooling of polymers (contrary to the homogeneous pattern of the process) the surface energy is a function of the molecular structure of the samples, as well as of their pre-history, i.e. of such factors as sample formation (cooling rate, shrinkage) and sample geometry (size and shape).

Hence, it seems more reasonable to use the free volume of polymers as the X-parameter characteristic of their cohesive properties. This concept, however, was developed with sufficient rigour predominantly for polymer melts [449, 450]. In the case of solid-state polymers, when calculating surface energy, one would prefer to use the more common characteristics of high molecular mass compounds.

Among such characteristics the parachor, eq. (181), is one of the most thoroughly examined

$$\Pi = M\sigma^{1/4}/(\rho_1 - \rho_1').\qquad(192)$$

This concept was introduced [392] using Bachinskii's equation [453] which is related to McLeod's formula eq. (181) [451] (as originally proposed [452]).

$$\sigma = k_1(\rho_1 - \rho_1')\qquad(193)$$

where ρ_1' is the density of the saturated vapour of the liquid. The precise analysis of the unitary function representing the expansion of surface tension with respect to the correlation functions of the bulk phases [447] produces the value for the power index in eqs (192) and (193) equal to 3/11 and 11/3, respectively and regression analysis of the experimental data produced the values 0.2222–0.2857 and 3.5–4.5 [454]. In the general case the value of the power index is inversely proportional to the molecular mass of homo- and copolymers [455], being, in the latter case, almost independent of the co-polymer composition [456]. Taking into account that the parachor can be easily calculated from the contributions of separate atoms and functional groups this method for the estimation of σ has gained a certain popularity.

Selection of other parameters as a cohesion characteristic, e.g. solubility parameter [457] appears initially to be more justified. Michaels [458] expressed surface energy in terms of the energy of intraphase interactions and by minimizing the Lennard–Jones potential he derived the following equation

$$\sigma = e_{LJ}^{min}\,(1 - m_s/m_v)/a, \qquad (194)$$

where m_s and m_v are coordination numbers in the surface layer and in the bulk and a is the cross-sectional area of the molecule. Eq. (194) is fully consistent with the general equation of state for polymers [459]. Now it is not difficult to get to the equation first proposed by Hildebrand for chains of infinite length [393]

$$\delta = A(\sigma/V_m^{1/3})^\nu. \qquad (195)$$

Here $\nu = 0.43$ (Voyutskii *et al.* [460], having calculated σ from the parachor obtained $\nu = 0.545$) and A is equal to 1.1 [461], 2.4 [460], 3.0 [462], 4.0 [461], 4.1 [393], or 6.0 [463] depending on the geometric model of the sample structure. The validity of eq. (195) was verified by Toyama [464] who observed that the $\lg \delta$ vs. $\lg(\sigma/V_m^{1/3})$ dependence was linear and it is further supported by the fact that other groups of polymers closely related in terms of their chemical structure gave separate straight lines. In spite of the fact that the results of calculations using eq. (195) are in rather good agreement with experimental data [465] its wide application is restricted because of the difficulties in finding the solubility parameter δ (usually from the data on polymer swelling [140]), coefficient A, and the volume V_m.

Precise determination of this latter quantity V_m obviously involves the need to choose an element of the macromolecular chain to which V_m applies. For instance, eq. (195) is taken to include the total volume of a spherical molecule [393], which for polymers is evidently the volume of the repeat unit when $z = 1$ in the expression

$$\sigma = A\delta^2 (V_m/z)^{1/3}. \qquad (196)$$

However, besides the case where $z = 1$ [466] adequate agreement between the theoretical and experimental σ values is achieved when the number of atoms ($A = 0.2289$) [402] or similar groups ($A = 0.168$) [462] are taken for z, or even at an arbitrary chosen value for $A = 0.073$ [467]. Thus, the rigour of these relationships cannot be accepted as adequate.

To eliminate these drawbacks the values for δ are calculated on the basis of an additive scheme by the summation of separate contributions ΔF_i as introduced by Small [468], and these δ values are made use of within the framework of a semi-empirical approach. Wu [469] expressed the ratio m_s/m_v of eq. (194) via the number of atoms in a segment of the macromolecule m'_s and has developed an equation of the Hildebrand type with numerical A and v values lying close to the ones cited above. Then the equation for σ is

$$\sigma = 0.327 \, (m'_s/V_m)^{1.52} \left[\left(\sum_i \Delta F_i \right)^5 \Big/ m'_s \right]^{1.85}. \tag{197}$$

Deviation between the calculated and experimental values does not exceed 7–10 mN/m. To overcome the uncertainty in estimating m'_s Bonn and van Aartsen [461] suggested using a less ambiguous parameter, i.e. the sum of the interatomic distances in the main chain of the repeat unit r_a comprising the macromolecular backbone (the respective values corresponding to some functional groups have been compiled by Godfrey [470]. Then

$$\delta^2 = k_a \sigma (r_a/V_m)^{1/2}. \tag{198}$$

At $k_a = 6.6$ the dependence is linear [461].

Aside from their empiricism these approaches have the demerit of not taking into account the influence on δ and V_m of the actual polymer structure as reflected by its coefficient of molecular packing, moreover, the solubility parameter by its definition is secondary in respect of a more general concept of cohesive energy. We shall touch upon these problems in more detail.

Generally, the coefficient of molecular packing k_0 gives the relationship between the van der Waals volume and the free volume [471]

$$k_0 = \rho V_m/M. \tag{199}$$

The free volume, associated with the notion of 'unoccupied' volume (space), is very sensitive to the structural features of a material. For instance, for the systems of equivalent differently packed spheres the following values for k_0 were obtained [472]

Cubic system	
diamond-type	$\pi 3^{0.5}/16 = 0.340$
simple	$\pi/6 = 0.524$
body-centred	$\pi 3^{0.5}/8 = 0.680$
Tetragonal system	$2\pi/9 = 0.698$
closest-packing arrangement of equivalent spheres	$\pi 2^{0.5}/6 = 0.740$.

When determining the coefficients of the molecular packing of polymers their physical state has also to be considered [471, 473]:

At glass-transition temperature T_g	0.667
Amorphous and low-crystallinity states	0.681
Cast as a film	0.695
At $T = 6$ K	0.731

At the same time, one should distinguish between packing in the bulk phase and in the surface layers of high molecular mass compounds. The number of molecules per unit surface area depends on their radius [474] and equals $0.2887 \times 10^{16}/r^2$ ($[r] = \text{Å}$). Then for the ideal hexagonal structure the maximum possible occupied volume can be easily obtained by multiplying the latter quantity by $4\pi r^3/3$ ($1.2093 \times 10^{16} r$). In a final account, taking $2.10^{16} r$ for the total volume of the surface layer, one obtains the value for the packing coefficient as 0.605 [475] (this value is 18.2% less than that for the bulk). Hence, the fraction of molecules at the surface is $k_o^s/0.605 = 1.653 \ k_o^s$, where k_o^s is the real coefficient of molecular packing in the surface layer.

Hence, strictly speaking, cohesive energy appears to be a function of the packing coefficient, especially in the case of polymers. However, to account for the connection between the two quantities Askadskii introduced the concept of effective molar cohesive energy [476, 477]:

$$E_{coh}^x = k_o E_{coh}. \tag{200}$$

On the other hand, the solubility parameter, defined as [140]

$$\delta = E_{coh}^{1/2}, \tag{201}$$

is related to the cohesion characteristic of polymers, i.e. most unequivocally, to the van der Waals volume. Ultimately, the relationship between these quantities is determined in a simple way by the following equation [476]:

$$\sum_i \Delta E_i^x = N_A \delta^2 \sum_i \Delta V_i, \tag{202}$$

where N_A is the Avogadro number. Using eq. (202) it is not difficult to find solubility parameters for homopolymers relying on the known values of effective cohesive energy and the van der Waals' volume and which are calculated by the summations of ΔE_i and ΔV_i via the additive scheme [478].

Such calculation procedures are quite effective [478–481]; however, the a priori assumed additivity appears to be deficient since, strictly speaking, it is nothing but a first approximation in the search of the actual dependence. It should be noted that the deviations from additivity of one or other factor have a stimulating effect on the development of any concept [482]. Frequently, an additive scheme is artificially preserved by introducing supplementary parameters. Such attempts however always refer to specific definite cases [473, 479, 480]. For instance, the additive scheme is inadequate in the determination of the solubility parameters of ternary [483] and even of binary [294] copolymers δ_{ij} since they are not linear functions of δ_i and δ_j. Hence, the common formula [397] of the type

$$\delta_{ij} = \sum_i \sum_j c_i^m c_j^m \delta_i \delta_j \tag{203}$$

(c^m denote the molar fractions of ith and jth comonomers) give substantial deviations from the experimental data [484]. As it was demonstrated in [485], δ_{ij} can be evaluated avoiding the drawbacks of the additive approach by the following formula

$$\delta_{ij} = (c_i^m \delta_i)^2 + (c_j^m \delta_j)^2 + c_i^m c_j^m \delta_i \delta_j \tag{204}$$

which leads to a much better agreement with the experimental data [485, 486].

These ideas concerning the physical implications of cohesion parameters were made use of in calculations of the surface energy of polymers. For instance, using eq. (196) with the rather arbitrarily chosen constant $A = 0.073$ [467] and subsequently combining it with eqs (199) and (202) Beriketov et al. have obtained the following simple relationship [487]

$$\sigma = 0.093 \, N_A^{-1} \sum_i \Delta E_i^x \Big/ \sum_i \Delta V_i \qquad (205)$$

(the mean value 0.681 was taken for k_0). This formula produces an error ranging from 0.55% for poly(vinyl acetate) to 20.9% for polystyrene. One should not forget, however, that the basic postulates are rather tentative and the tempting simplicity of eq. (205) gives rise to doubts that all of the parameters related to the changes in the energetic properties of polymer surfaces are appropriately accounted for.

The approach of Askadsku [475], considering the globular structure of amorphous polymers, seems to be much more physically sound. For polymers of this type the fraction of the volume in which intermolecular interactions are feasible is $R_m^3 - (R_m - r)^3 / R_m^3$ (R_m is the effective radius of a macromolecule) as derived from purely geometric considerations. Instead of the effective cohesive energy and the van der Waals volume Askadsku suggested using the reduced quantities $\sum_i \Delta E^x / N$ and $\sum_i \Delta V_i / N$ (where N is the number of atoms in a repeating unit). Then [475, 488]

$$\sigma = \frac{A_f \sum_i \Delta E_i^x}{N^{1/3} \left(\sum_i \Delta V_i \right)^{2/3}} \qquad (206)$$

where A_f depends on the chemical nature of the functional groups of a polymer, the respective values being listed below [488]:

$-OH; -COOH$	0.0476
$-CN$	0.0600
$-C(O)O-; -C(O)NH-; -NO_2$	0.0751
$-H; -F; -O-$	0.1277

Our attempts to elucidate the relationship between σ and various fundamental cohesion parameters have led us to infer that certain correlations are observed only when Small's attraction constants [468, 489] and molar diamagnetic susceptibility [490] are used as such parameters, otherwise no correlations were observed. This is quite reasonable as one can hardly anticipate that the properties of such different types of functional groups would by any parameter give close correlation. Also one should recognize the purely experimental origin of the coefficient A_f and the fact that it has the meaning of an adjustment factor. Nevertheless, the results of calculations by eq. (206) are in no worse agreement than in the preceding case, though in contrast to eq. (205) the relationship (206) is much more justified. The fact that eqs (205) and (206) produce (as well as other even simpler formulae) results that fall within a relatively narrow range urges one to recall the remark of Halsey quoted at the beginning of this section. Hence, the inadequacy of cohesion parameters in giving a physically correct measurement of the surface energy of polymers.

When looking for new quantities to rely on in surface energy determinations one cannot ignore the dielectric permittivity e which, in physics, is a fundamental characteristic. Then, in principle, e can be involved as the characteristic of intraphase and of interfacial interactions (for instance, e and σ produce a similar effect on the polymer friction coefficient [491]). Therefore e should be related to various cohesion parameters. This conclusion is supported by the relationship between the solubility parameter and dielectric permittivity established for epoxy oligomers [492]

$$\delta = 8.005 + 0.206\,(e - 1). \tag{207}$$

Simple transformations result in the corresponding expression for the effective molar cohesive energy [493]

$$E^x = \frac{0.029\,M}{\rho}\,(e^2 + 75.719\,e + 1433.321). \tag{208}$$

The validity of eq. (208) was confirmed in the studies of cross-linked oligoepoxides, for which cohesive energy values obtained by eq. (208) were found to be in good agreement with those obtained in other ways [494]. Hence, this quantity is really important in that it governs the tensile properties of epoxy compositions [495, 496], which are the most commonly used adhesives. In fact, Fig. 2.31 illustrates this statement showing the direct relationship between cohesive energy of these materials and their failure stresses, as well as the impact viscosity. For the four systems under discussion an increase in E^x leads to a practically linear decrease in the strength parameters [496].

The successful application of this approach stimulates the use of other cohesion parameters for the estimation of polymer surface energy. Among these polarizability takes on a specific role insofar as, strictly speaking, the fundamental eq. (195) is based on the interrelationship between δ and ϵ. Quantitatively it is expressed by Fowler's equation, the constants of which are determined by the diamagnetic susceptibility of Kirkwood–Müller. Relying on these prerequisites, Davies suggested the following formula [394]

$$\sigma = 0.621\,k_\epsilon\,\epsilon r_0^4/V_m^2. \tag{209}$$

However, its application for practical calculations is rather difficult due to the need to estimate the equilibrium distance r_0 from the Lennard–Johnes potential which, according to [497], requires a consideration of the lattice model.

It is clear nevertheless, that application of polarizability as a cohesive parameter in the estimation of surface energy of solids is a feature of major significance. The fact that the efficiency of van der Waals interaction between the contacting materials is, at a first approximation, proportional to the product of their polarizabilities (a more precise calculation on the basis of Lifschitz' approach should consider the frequency dependence of ϵ [498]) has led de Gennes to conclude that the main prerequisite for the development of wetting processes is not a positive spreading coefficient χ, but rather the fulfilment of condition (139) [206]. Then one must look for an approach to the calculation of σ that would not just tend to exclude ϵ from the number of 'surface-sensitive' cohesion parameters but would seek for a more readily calculated or measured characteristic for which to substitute it.

Refractive index n seems to be a justified choice for such a parameter. Indeed, combining the two previous approaches based on the usage of ϵ and e one must rely, on the one hand, on the Clausius–Mosotti equation

$$\frac{\epsilon - 1}{\epsilon + 2} = \frac{4\pi N_A}{3}\,\frac{\epsilon\rho}{M}, \tag{210}$$

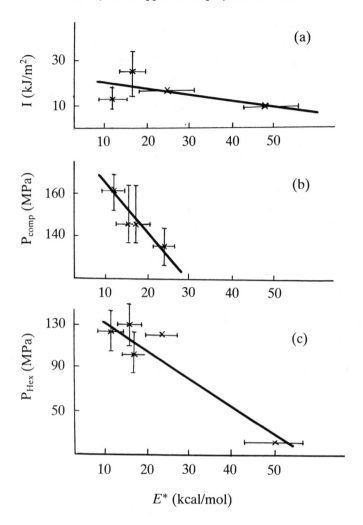

Fig. 2.31 – Impact strength (a), ultimate compression (b) and flexing (c) stresses of the epoxy-based adhesives cured with maleic anhydride at 343 K versus molar cohesion energy of the epoxy oligomers.

and, on the other hand, on Maxwell's equation

$$\epsilon = n^2. \tag{211}$$

Then from eqs (210) and (211) one has, according to Lorentz–Lorenz,

$$\frac{n^2 - 1}{n^2 + 2}\rho^{-1} = \frac{4\pi N_A}{3}\frac{\epsilon}{M} = r \tag{212}$$

where r is the specific refraction. Hence, refractive index is characteristic of the induction polarization of a medium occurring due to the induction of dipole moments in molecules via deformation of electron shells in a wave field (in the IR region displacement of nuclei also takes place). The obvious convenience of using n for a cohesive parameter rests in that it is only slightly dependent on molecular mass [499], concentration of end groups [500], and residual monomer content [501] in polymeric products.

On the other hand, the refractive index of a condensed phase is in fact always related to its surface characteristics. For alkanes and alkenes such a relationship involves refraction [502]. For non-polar liquids the general equation is as follows [503]

$$\sigma = 286 \left(\frac{n^2 - 1}{2n^2 + 2} \right) - 28.6. \tag{213}$$

The existence of such a dependence is confirmed by the fact that the surface energy and refractive index of a liquid are associated with the phenomenological characteristic, the optochor [504]

$$D_i = M_i \sigma_i^{x_i} / n_i. \tag{214}$$

Gambill [505] claims that of the known expressions it is eq. (213) that produces the most accurate assessments of σ. This relationship is equally true for high polymers [506], which in relation to their surface layers is discussed in section 3.1.1, while here it is illustrated by the data in Fig. 3.32.

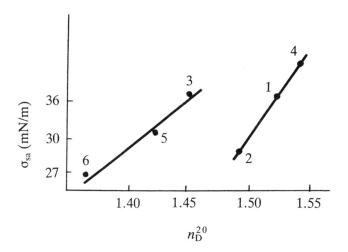

Fig. 2.32 – Surface energy versus refractive index of polyethene (1), polypropene (2), polyvinylfluoride (3), polyvinylchloride (4), polyvinylidenefluoride (5), polytetrafluoroethene (6).

These and similar consistencies provide for the possibility of establishing quantitative relationships between the refractive index and surface energy of high molecular mass compounds. Eq. (212) can be expressed in refractometric terms as follows [507]

$$\frac{n^2 - 1}{n^2 + 2} = \frac{k_0 \sum_i m_i a_i r_i}{N_A \sum_i \Delta V_i}, \tag{215}$$

where m, a and r are the number, atomic mass, and specific refraction of ith atoms in a repeat unit, respectively. Then, from eqs (202) and (215) one arrives at the following relationship between the refractive index and the cohesive parameters of polymers

$$\frac{n^2 - 1}{n^2 + 2} = \frac{k_0 \delta^2 \sum_i m_i a_i r_i}{\sum_i \Delta E_i^x} . \tag{216}$$

Substituting this expression into the basic Hildebrand's function (195) and performing simple transformations one obtains the final equation [395]

$$\sigma = k_0^{-7/6} A^{-7/3} \left(\sum_i \Delta E_i^x \right)^{7/6} \left[\frac{n^2 - 1}{(n^2 + 2) \sum_i m_i a_i r_i} \right]^{5/6} . \tag{217}$$

The only term whose meaning and magnitude have to be specified more precisely with regard to the above mentioned limitations of eq. (195) is the constant A.

According to its physical meaning A is a measure of the polarity of high molecular mass compounds. Therefore, since, according to [393], eq. (195) relates primarily to non-polar systems the value $A = 4.1$ suggested in [393] was taken as a first approximation in the calculation of σ. However, when the calculated data for 14 homo- and 5 copolymers were compared with the experimental ones no satisfactory agreement was observed; however, it was somewhat better for highly polar polyacrylonitrile and butadiene–acrylonitrile elastomers. In order to identify the effect of the chemical nature of polymers on the magnitude of the constant A its values were determined from eq. (217) using the most reliable of the published experimental values for σ. It was found that with respect to the magnitude of A the materials examined fall into three groups [395]. The first comprises low polarity ($A = 2.66$–3.19), the second polar ($A = 3.25$–3.39) and the third high-polarity polymers ($A = 3.52$–4.02).

Such differentiation has a physical explanation. To show this Fig. 2.33 depicts the dependence of calculated A values on the values for dipole moments derived from Debye's equation which relies on the data for ϵ and n_D [508]. It is seen that for each of the groups listed there is a rectilinear dependence (deviation is greatest for polystyrene, which is probably due to the effect of the phenyl radical). The fact that the magnitude of A associated with the type of intermolecular interaction is confirmed also by the fact that for polyvinylalcohol, for which hydrogen bonds are characteristic, the calculated value for A is 4.42, which is higher than that even for the highly polar butadiene–acrylonitrile elastomers.

Hence, for each of the groups listed there are certain average A values. For non-polar and low-polarity high polymers $A^{av} = 3.0$, for polar $A^{av} = 3.3$, while for high-polarity $A^{av} = 3.8$. This approach makes it possible to make a sound choice for the coefficients of Hildebrand's equation from the published data. Indeed, mainly low-polarity polymers were examined in [462] ($A = 3.0$), while in [461] ($A = 4.0$) and [393] ($A = 4.1$) the investigated polymers were highly polar. It should be emphasized that the A-coefficient in eq. (217) provides a quantitative measure for the as yet rather intuitive concept of the degree of polarity of polymers. It also has to be noted, in this connection, that the

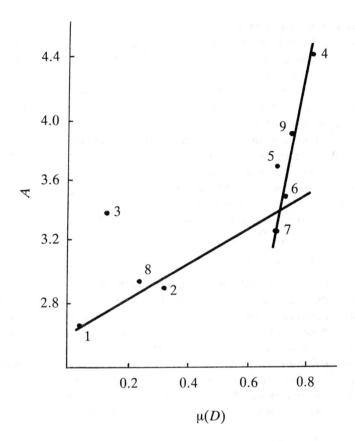

Fig. 2.33 – Constant A of the Hildebrand's equation as dependent on the dipole moments of polyethene (1), polytetrafluoroethene (2), polystyrene (3), poly(vinyl alcohol) (4), polyacrylonitrile (5), poly(vinyl acetate) (6), poly(methyl methacrylate) (7), polyiso-prene (8), poly(ethylene terephthalate) (9).

approach to evaluating the polymer characteristics being considered, first discussed in [395] and later developed in [475, 487], appears to be of general value. At the same time the attempts to endow A with a deeper meaning [475, 487] than initially implied in [395] can hardly be thought of as satisfactory, since these consider only the effect of either constitutional [475, 488], or structural factors [467], but not the whole complex of problems as reflected by the formula (217) involving cohesive energy, as well as the packing coefficient, and finally the refractive index.

Because of this surface energy values calculated by eq. (217) are those most accurately fitting the experimental data. The mean relative error is 8.2%, i.e. it does not exceed 3 mN/m, which is within the reliability range of the published data. For homopolymers the deviation between the calculated and experimental results is 8.6% and for copolymers it is 6.8%. The correlation coefficient between σ_{calc} and σ_{exper} is 0.926, which is sufficiently high to ensure adequate reliability of this procedure for the calculation of surface energy. It should be stressed that the latter seems to produce more reliable data than the experimental ones, which are obtained with certain assumptions made about the techniques used. This is shown convincingly when comparing the σ values obtained from eq. (217) with some fundamental polymer parameters for a large number of polymers. For instance, the dependences between σ and the published data for the van der Waals volume of a macromolecule [471], dielectric constant [490], molar cohesive energy [477] and glass transition temperature [509] are characterized by correlation coefficients of 0.741, 0.765, 0.856, and 0.905, respectively. Just as noteworthy is the relationship between σ and the friction coefficient [490]; the corresponding correlation coefficient being as high as 0.974.

Due to the continuing increase in the number of new high molecular mass compounds designed for adhesive purposes, evaluation of the surface energy of polymers based upon a knowledge of their chemical structure, i.e. even before they are synthesized, is of utmost importance. It should also be borne in mind that the magnitude of the refractive index of a polymer is determined to a certain extent by the pre-history of the sample and its purity. Similarly, there is a certain error in the calculated values of refraction r, as the more common form of eq. (212)

$$r = (n^2 - 1)/\rho(n^2 + 2) \qquad (218)$$

shows more explicitly the dependence of refraction on the density of real materials.

To eliminate these limitations it is preferable to use the refractometric characteristics of the initial monomers n_m and r_m instead of those relating to polymers n_p and r_p. For this purpose it is necessary to introduce the parameter M into the numerator of eq. (218), thus replacing the specific refraction by the molecular one. In the case of high polymers this parameter applies to the repeat unit of a macromolecule, for homopolymers it is equal to M_m (while for copolymers the more common content has to be taken into account). Then, writing down the functions eq. (212) for both the monomers and polymers and setting them equal[†] via M we obtain the following

$$\frac{n_p^2 - 1}{n_p^2 + 2} = \frac{\rho_p(n_m^2 - 1)\sum_i m_i^p a_i^p r_i^p}{\rho_m(n_m^2 + 2)\sum_i m_i^m a_i^m r_i^m}. \qquad (219)$$

Polymer density is easily eliminated from this expression by noting the fact that it determines the macromolecule's own volume $V' = M/\rho_p$, which according to eq. (199), is the ratio between van der Waals volume and the molar packing coefficient. On the other hand, the density of low molecular mass compounds is, as is known, related to n_m by means of the Eickman function [511]

$$\rho_m = (n_m^2 - 1)/k_E(n_m + 0.4), \qquad (220)$$

where $k_E = 0.6$ [512]. The benefit of this function is that it is actually independent of temperature. For 128 non-polar and polar hydrocarbons, carboxylic acids, halogen- and sulphur-containing compounds the maximum discrepancy between the experimental and calculated values in accordance with eq. (220) for ρ_m was less than 4.9% averaging 1.7%, i.e. is within the normal accuracy of measurement [513].

---

† Strictly speaking, for homopolymers r_p and r_m differ by the value of a double bond refraction. However, the latter value is small (it exceeds only that for the hydrogen atom [510] and hence can be neglected.

Finally, by substituting eqs (199) and (220) into eqs (219) and then into (217) we obtain the relationship [514]

$$\sigma = k_\sigma \left(\sum_i \Delta E_i^x \right)^{1/3} \left[\frac{M(n_m + 0.4) \left(\sum_i \Delta E_i^x \right)}{(n_m^2 + 2) \left(\sum_i \Delta V_i \right) \left(\sum_i m_i^m a_i^m r_i^m \right)} \right]^{5/6} , \qquad (221)$$

which involves those monomer parameters that are either determined experimentally (n_m) or can be derived from tabulated quantities (r_m [510]), as well as those polymer characteristics that are calculated by the summation of contributions ΔV_i and ΔE_i^x known for the materials of diverse chemical structure. The constant

$$k_\sigma = (k_E/N_A)^{5/6} (k_0 A^7)^{-1/3}$$

in eq. (221) is determined by the molar packing coefficients of polymers and Hildebrand's function (195) thus accounting for the physical state and polarity of the materials, respectively (Table 2.5).

Surface energies of polymers derived from refractive indexes of respective monomers [514] only deviate 1–2 mN/m from the results obtained in the basic refractometric method [395]. In certain cases the calculated results seem to be even more reliable than the published experimental data. For instance, for butadiene–acrylonitrile copolymers SKN-18, SKN-26, and SKN-40 (figures indicate acrylonitrile content) the σ_{sa} values obtained by the wetting method are 25, 31, and 34 mN/m, respectively [257]. The values are obviously low and close to the corresponding values for the non-polar poly-alkenes. When calculated by means of the suggested dependence eq. (221) and σ values obtained were 29.6, 34.5, and 45.2 mN/m, which are more representative in that they reflect the effect of the polar nitrile side-groups on the surface and adhesion properties of the elastomers considered.

Table 2.5 – The values for $k_\sigma \times 10^{20}$ constant of eq. (221) for various groups of polymers depending on their polarity and physical state, $k_\sigma \times 10^{20}$.

	Polarity		
	Low- and non-polar	Polar	Highly polar
Physical state	($A^{av} = 3.0$)	($A^{av} = 3.3$)	($A^{av} = 3.8$)
At 6 K ($k_0 = 0.731$)	0.0853	0.0683	0.0491
At T_g ($k_0 = 0.667$)	0.0881	0.0704	0.0506
Film-cast samples ($k_0 = 0.695$)	0.0867	0.0694	0.0500
Amorphous and low-crystalline bulk samples ($k_0 = 0.681$)	0.0873	0.0699	0.0503

Yet another convincing proof of the validity of this approach was presented by Pritykin et al. [84] who calculated σ for α-cyanoacrylate adhesives from their refractive indexes, recalling that these adhesives are monomers. For these adhesives of the general formula $H_2C = C(CN)–COOR$ surface energies were experimentally determined [515], which are as follows

–R	σ (mN/m)
$–CH_3$	37.4
$–C_2H_5$	35.1
$–CH(CH_3)_2$	33.3
$–C_4H_9$	31.5

These values can hardly be regarded as reliable even by comparison with the corresponding data for polyacrylates which are quite similar to the adhesives discussed. In fact, for poly(butyl methacrylate) $\sigma = 32.0$ mN/m, as compared to a higher value of 41.0 mN/m for poly(methyl methacrylate) [466]; substitution of the methyl group by the electron acceptor nitrile group should result in the increase of σ (for polyacrylonitrile such increase, as compared to polyethene, amounts to 40% [516]). Calculation by the formula (221) produces much more trustworthy results, which are in the range of 47.0–53.8 mN/m for cyanoacrylate adhesives. In the next section these values are correlated with those for the adhesive joint strength, these relationships being almost linear.

The approach developed here makes it possible to elaborate on a problem of intense theoretical and practical importance, i.e. the effect of molecular mass of a polymer upon its surface energy. In fact, relationships (179), (185) and (186) are rather implicit with regard to M and only hint at the inverse nature of the σ versus molecular mass dependence, which was actually observed by different authors (see for instance [22, 115, 318]). For example, the surface energy versus molecular mass dependence, as investigated with the help of narrow fractions of polyethyleneglycols at 333–413 K, was found to be almost linear within the precision of the experimental procedure [517] (Fig. 2.34). This fact was related to the stiffness g, contour length L and surface density ρ^s of polymer chains [517]:

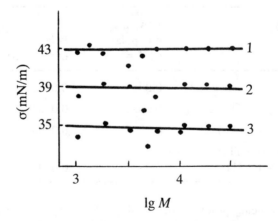

Fig. 2.34 – Surface energy versus molecular mass dependences for narrow polyethene glycol fractions. 1, 313 K; 2, 373 K; 3, 413 K.

$$\sigma = \rho^s \left[k_L - 2(g\rho)^{1/2} / sh\ L(\rho/g)^{1/2} \right] \qquad (222)$$

where k_L is a constant depending on the chemical nature of a macromolecule. However, these parameters are themselves implicit functions of molecular mass. Speculations regarding $\sigma(M)$ determined by an energy of intermolecular interaction (difficult to estimate) [518] are just as generalized and, hence, are quite useless for any practical purposes.

The problem under discussion seems to be most capable of resolution in refractometric terms and moreover, such an approach appears to have general value in the physical chemistry of polymer adhesion. By combining eqs (202), (215) and (217) one comes to expresss σ via both cohesive energy and the solubility parameter:

$$\sigma = \frac{\left(\sum_i \Delta E_i^x \right)^{3/2} \left(\dfrac{n^2 - 1}{n^2 + 2} \right)^{7/6}}{k_0^{3/2}\ A^{7/3}\ \delta^{2/3} \left(\sum_i m_i a_i r_i \right)^{7/6}} . \qquad (223)$$

Let us consider the variables of this dependence. The only experimentally measured term in it is the refractive index. To eliminate the molar packing coefficient k_0 we make use of eq. (199), in which M is the molecular mass of a repeat unit $M_{(1)}$. Then, from eqs (199) and (223) follows the equation relating surface energy to the two experimental parameters n and ρ:

$$\sigma = \frac{M_{(1)}^{3/2}\ \delta^{7/3} \left(\dfrac{n^2 - 1}{n^2 + 2} \right)^{7/6}}{A^{7/3}\ \rho^{3/2} \left(\sum_i m_i a_i r_i \right)^{7/6}} . \qquad (224)$$

In order to take into account the molecular mass dependences of n and σ let us recall eq. (185). Then, linear functions relating the refractive index and density of polymer-homologues to their molecular mass can be written as follows [519, 520]:

$$n = a_n/M + b_n \qquad (225)$$

$$\rho^{-1} = a_\rho/M + b_\rho \qquad (226)$$

where a and b are constants. Substituting these functions into eq. (224), after simple transformations the final dependence is [521]

$$\sigma = (\delta/A)^{7/3} (a_\rho + b_\rho M) M_{(1)} / M^{3/2}$$

$$\left(\frac{(a_n + b_n M)^2 - M^2}{[(a_n + b_n M)^2 + 2M^2] \sum_i m_i a_i r_i} \right)^{7/6} . \qquad (227)$$

For the molecular mass dependences to be calculated the constants in eq. (227) had to be determined. This was done by means of eqs (225) and (226) relying on the measure-

ments of refractive indexes and densities of a variety of diverse materials: oligoethenes
[522], oligoisobutenes [523], oligotrifluorochloroethenes [524], oligomethyl- [525],
and oligobutylmethacrylates [526], oligopropene- [527] and oligobutene-oxides
(oligotetrahydrofuranes) [528], oligoepichlorohydrins [528], and oligoethyleneimines
[529]. The respective constants are compiled in Table 2.6. For the materials listed the
surface energy versus molecular mass dependences were calculated and these are depicted
in Fig. 2.35. What attracts attention firstly is that all the dependences have the form of
descending hyperbolic curves, the only exceptions being, in full conformity with
anticipation, polypropene- and polybutene-oxides exhibiting surfactant properties. On
the other hand, the monotonic character of the calculated curves[†] and their form do not
contradict the data reported concerning the σ versus molecular mass relationship. What is
really important is that in all cases the plateaux of the curves in Fig. 2.35 correspond to
the values of σ determined in other ways [530].

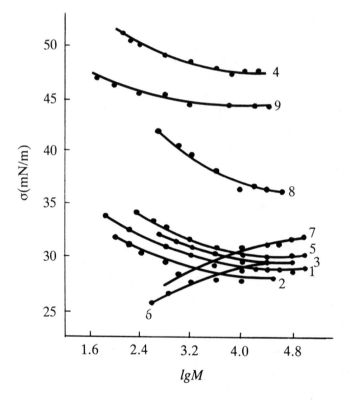

Fig. 2.35 — Surface energy versus molecular mass dependences for polyethene [(1), polyisobutene (2), polytrifluorochloroethene (3), poly(methyl methacrylate) (4), poly(butyl
methacrylate) (5), polypropene oxide (6), polybutene oxide (7), polyepichlorohydrin (8),
polyethyleneimine (9).

[†] Non-monotonic character of the σ(M) dependences can be considered as resulting from conformational transitions in macromolecules [318, 517] leading subsequently to variations in short-
range ordering.

Table 2.6 – Constants of eqs (225) and (226) for various polymers.

	Repeating unit	a_n	b_n	a_ρ	b_ρ
1.	$-CH_2-$	-8.1690	1.4698	29.20	1.1500
2.	$-CH_2-C-$ $\quad\;/\backslash$ $H_3C\;CH_3$	-12.8100	1.5067	35.65	1.0974
3.	$-CF_2-CFCl-$	-32.2600	1.4232	36.30	0.4800
4.	$-CH_2-C-$ $\quad\;/\backslash$ $H_3C\;COOCH_3$	-12.4600	1.4950	27.52	0.8254
5.	$-CH_2-C-$ $\quad\;/\backslash$ $H_3C\;COOC_4H_9$	-10.7900	1.4713	31.93	0.9670
6.	$-CH_2-CH-O-$ $\qquad\quad\|$ $\qquad\;\;CH_3$	-2.1000	1.4250	-3.60	0.9960
7.	$-(CH_2)_4-O-$	-2.7300	1.4541	-6.99	1.0388
8.	$-CH-CH_2-O-$ $\;\;\|$ CH_2Cl	-10.5600	1.5135	16.82	0.7324
9.	$-CH_2-CH_2-NH-$	-4.5200	1.5307	9.79	0.9509

These facts can be regarded as a proof of the validity of the approach taken. The position of the horizontal portions of the calculated curves makes it possible to predict the maximum threshold values of molecular mass after reaching which the surface energy remains constant.[†] The latter (constant surface energy) can be easily determined assuming that $a_n \ll b_n M$ and $a_\rho \ll b_\rho M$. Eq. (227) is then transformed to acquire the following form [521, 530]

$$\sigma = (\delta/A)^{7/3} (a_\rho M_{(1)})^{3/2} \left[\frac{b_n^2 - 1}{(b_n^2 + 2) \sum_i m_i a_i r_i} \right]^{7/6}. \qquad (228)$$

If the premises leading to the dependences shown in Table 2.4 are valid they can obviously be applied to polymers. We examined the feasibility of this latter assertion by comparing the magnitudes of the work of an escaping electron [531] and the surface energy of high molecular mass compounds [395, 514]. As is seen from Fig. 2.36 for five polymers with known σ values, the correlation found is almost linear. Other electrophysical characteristics of polymers might also be used to represent their cohesion

---

† The inflections on these curves are associated with the transitional regions from oligomers to polymers.

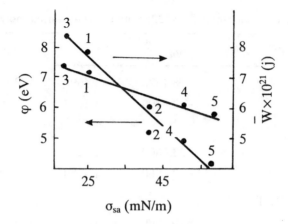

Fig. 2.36 – Surface energy versus the work on escaping of electon (φ) and versus the mean gas–kinetic energy ($\overline{W}$) of lead atoms revaporized from polymer films 5–15 μm thick in high vacuum at 90 K. 1, Polyethene; 2, poly(vinyl chloride); 3, polytetrafluoroethene; 4, poly(methyl methacrylate); 5, poly(ethylene terephthalate).

parameters. For instance, the heat of combustion of a polymer ΔH is directly related to the density of localized surface states ρ_s^s via the following expression [532]

$$\rho_s^s = -k_\rho \Delta H \tag{229}$$

where ΔH is in MJ/kg, ρ_s^s is in eV/cm^2 and $k_\rho = (0.0324-1.9200) \cdot 10^{-16}$. However, in the light of the previous discussion and of the data in Fig. 2.36 the dependence of σ versus the mean gas–kinetic energy $\overline{W}$ of lead atoms re-evaporated at 90 K in high vacuum from the surface of polymer films 5–15 μm thick is much more conclusive [533]. This relationship is also nearly linear.† Consequently, there must be a close correlation between the magnitude of surface energy σ, mean gas–kinetic energy $\overline{W}$, and the work of an escaping electron φ [535]. Indeed, this is actually the case and the corresponding dependences were found to be described by linear equations [536]

$$\sigma = 82.0 - 6.13 \cdot 10^{11} (\overline{W})^{1/2}, \tag{230}$$

$$\varphi = 1.48 + 0.64 \cdot 10^{11} (\overline{W})^{1/2}. \tag{231}$$

Then

$$\sigma = 96.2 - 9.6\varphi \tag{232}$$

which illustrates the role of parameter φ demonstrated in Table 2.4.

In practice, however, the most popular methods for the evaluation of polymer surface energy, just as they were traditionally, are the ones based upon consideration of the equilibrium state of a drop of liquid and measurement of contact angles. This is probably due to the level of instrumental technique and also to the fact that there is already a lot of experimental data available for comparison. Nevertheless, the majority of investigators

---

† It is appropriate to note that a series of polymers arranged by decreasing $\overline{W}$ coincides with the reported [534] tribo-electric set. This again is quite indicative of the fundamental role of adhesion in surface phenomena.

today have come to the conclusion that these methods provide the least objective and reliable results.

The equilibrium condition of a liquid drop on a solid surface [eq. (55)] is described, as was shown above, by an equation containing two unknown variables which does not allow a simple solution, and therefore the 'neutral drop' method shown schematically in Fig. 2.10 is more preferable. Besides the general relationship eq. (91), the equilibrium condition written down in the following form is added [537]

$$\sigma_{sa}/\sin{(\alpha + \beta)} = \sigma_{la}/\sin{\alpha} = \sigma_{sl}/\sin{\beta}. \tag{233}$$

Turning to the three-dimensional analogue of such a model it seems correct to place the liquid inside a trihedral curvilinear pyramid made of the material to be examined (Fig. 2.37). Then, the equilibrium condition is described by the system of three equations [385]

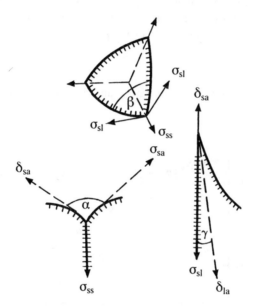

Fig. 2.37 – The resultant forces descriptive of the wetting of the solid surfaces of a trihedral pyramid by a liquid.

$$\sigma_{ss} = 2\sigma_{sa} \cos{(\alpha/2)},$$
$$\sigma_{ss} = 2\sigma_{sl} \cos{(\beta/2)}, \tag{234}$$
$$\sigma_{ss} = \sigma_{sl} + \sigma_{la} \cos{(\gamma/2)},$$

which simplifies the solution relative to σ_{sl} and σ_{ss}. Unfortunately, of the three methods, i.e. Young's, dihedral and trihedral angles, only the first one is in current use although it is the least analytically valid. Actually, the measurements are limited by one contact angle instead of two (Fig. 2.10) or three (Fig. 2.37) and all subsequent attempts to obtain a more precise solution are associated with the introduction of limiting assumptions [538].

One of the assumptions consists in equating the work of both adhesion and cohesion for the phase with minimum van der Waals interaction [388]. It then becomes possible to exclude σ_{sl} from eq. (55) and

$$\sigma_{sa} = 0.5\sigma_{la}(1 + \cos\theta). \tag{235}$$

It must be realized that satisfactory agreement between the calculated and experimental σ_{sa} values, observed for some systems, copolymers in particular [539], does not compensate for the fact that the initial assumption is valid only for one-component phases with an impermeable interface across which only intermolecular non-valence forces are able to act. A certain reduction in the relative error of eq. (235) is achieved if the substrates are wetted with binary solutions [540], rather than with pure liquids. Such experimental procedures, however, complicate the analysis of the behaviour of the system.

Extrapolation of $\sigma_{la}\cos\theta$ of eq. (55) to $\theta = 0$ is justified as, regardless of the nature of the wetting liquids, all experimental points, in coordinates of Fig. 2.38, lie on the one and the same curve. The parameter obtained from such an extrapolation procedure represents the so-called critical tension of wetting [387]:

$$\sigma_w = \lim_{\theta\to 0} \sigma_{la}\cos\theta. \tag{236}$$

It is clear that parameter σ_w is rather provisory and does not have the required physical meaning.

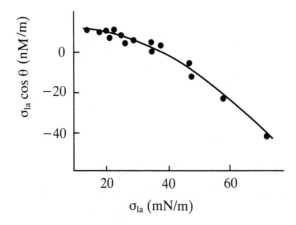

Fig. 2.38 – Graphical determination of the critical tension of the wetting of polymers.

Such extrapolations are more beneficial when using the compounds of one homologous series as wetting liquids. The dependence $\cos\theta$ versus σ_{la} is in this case linear [541], the parameter corresponding to the condition

$$\sigma_c = \lim_{\theta\to 0} \sigma_{la} \tag{237}$$

was described by Zisman, who was the first to suggest this method, as the critical surface tension. The value of this parameter is determined above all by the properties of a substrate, rather than by those of a liquid. This conclusion was supported by the results of numerous investigations [516]. σ_c can be estimated either graphically (Fig. 2.39) or by means of the linear equation [542]

$$\cos \theta = 1 + a(\sigma_c - \sigma_{la}) \tag{238}$$

describing the initial dependence.

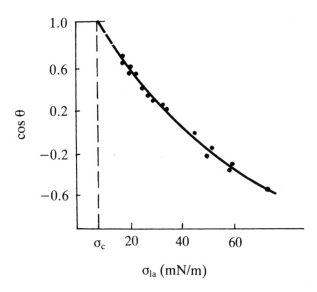

Fig. 2.39 – Graphical method for determination of the critical surface tension of polymers.

The latter expression coincides with eq. (113), thus supporting Rhee's attempts to simplify Young's equation. Combining eqs (55) and (238), Rhee obtained a simple formula relating σ_c to solid surface energy [543]:

$$\sigma_{sa} = (a\sigma_c + 1)^2/4a. \tag{239}$$

Hence, the parameter σ_c is endowed with the meaning of interfacial energy, although most investigators identify σ_c with σ_{sa}. The last dependence was confirmed in studies on wetting of crystalline penaerythritol trinitrate and cyclotrimethylene trinitrate [544]. Such results are not within the concept initially outlined by Zisman [545] who considered this approach to be true only for low-energy surfaces. Indeed, there is other evidence that measurement of σ_c by the wetting of solids of high surface energy is possible [546, 547]. Wu for instance, proposed a general equation of state which included σ_c [548]. Introducing Zisman's condition, eq. (237), into eq. (137) and expressing a more accurate value of the critical surface tension via parameter Φ, surface pressure and σ_{sa} he obtained the following relationship

$$\cos \theta = 2(\sigma_c'/\sigma_{la})^{1/2} - 1 \tag{240}$$

where $\sigma_c' = \Phi^2 \sigma_{sa} - \pi$. By applying this equation for a number of polymers the results obtained [548–550] were better correlated with the experimental data, than those calculated by the normal method of Zisman (strictly speaking, $\sigma_c < \sigma_c^\pi$ [550]).

At the same time the deviations of experimental $\cos \theta$ versus σ_{la} dependences from linearity were also observed. Attempts to interpret this fact within the framework of Zisman's concept involve the combining of eqs (98) and (238) leading to a parabolic equation [341]:

$$W_{Ad} = (2 + a\sigma_c) \sigma_{la} - a\sigma_{la}^2. \tag{241}$$

Neuman and co-workers [319] proposed the quadratic equation

$$\sigma_{la} \cos \theta = x\sigma_{la}^2 + y\sigma_{la} + z \tag{242}$$

which generally reflects the non-linearity of experimental dependences more precisely. However, eq. (241) appears to give a better description of the wetting of high-energy polymer surfaces [341].

On the other hand, regarding the ability to deform polymeric substrates discussed in section 2.1.2, one should take into account the anisotropy of critical surface tension. Indeed, stretching of elastomers inevitably leads to a substantial change in σ_c value (Fig. 2.40). The nature of this effect is easy to understand if one recalls eq. (82) involving linear tension κ. Introducing it into eq. (55), one has [551]

$$\sigma_{sa} = \sigma_{sl} + \sigma_{la} \cos \theta + \kappa/r_1. \tag{243}$$

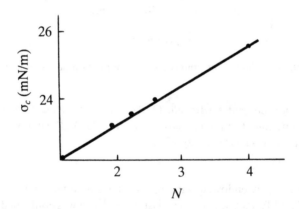

Fig. 2.40 – Critical surface tension of vulcanized isoprene elastomer SKI as dependent on the uni-axial elongation.

Then the considered quantity must be different in the cases of longitudinal ($\parallel$) and lateral ($\perp$) orientations of macromolecular chains. At $\pi = 0$ the anisotropy of critical surface tension is given by the following condition

$$\Delta\sigma_c = \sigma_c^{\parallel} - \sigma_c^{\perp} = -\sigma_{sl} + \Delta(\kappa/r_1) \neq 0. \tag{244}$$

Hence, σ_c is determined, like other substrate characteristics, by Young's modulus and Poisson's coefficient in accordance with eq. (78).

The generality of Zisman's method, as illustrated by the above discussion, coupled with its simplicity and ease of interpretation of experimental data ensured that of all the parameters characterizing a polymer surface σ_c has come to be the one most widely used.

At the same time the semi-empirical character of σ_c is beyond doubt and this must be borne in mind when treating adhesion phenomena. Having examined the wettability of low-energy liquid surfaces Johnson and Dettre [552] conclude: 'Any analysis based on equating surface tensions with critical surface tensions must be incorrect ... Since this [equating — authors] is not true for amorphous liquids, it certainly is not true for the more complex polymers.' This especially evident when making an assessment of σ_c relying on deliberately simplified considerations. For instance, the time during which a strip of fluid with a given σ_{la} retained its integrity at the substrate surface was once made use of instead of the critical surface tension derived from contact angle measurements [553]; however, the approach is obviously erroneous [554].

Let us now consider the evaluation of individual and separate contributing components to surface parameters and their relation to the rules of formation of polymer adhesive joints.

Generally, any energetic parameter of the surface comprises polar and non-polar components inasmuch as the constant A of the Lennard–Jones potential can be represented, according to Good [330], as increased by London, Debye, and Keesom forces. Similarly, the parameter Φ can be expressed as a corresponding summation also [341].

The degree of polarity x_i^p can be represented as the ratio between the polar component of the surface energy and its integral value

$$x_i^p = \sigma_i^p / \sigma_i. \tag{245}$$

The degree of 'non-polarity' is obviously a contribution of dispersion forces

$$x_i^d = \sigma_i^d / \sigma_i. \tag{246}$$

According to the definition

$$\sigma_i = \sigma_i^d + \sigma_i^p. \tag{247}$$

Here and in eq. (245) polar interactions have to be differentiated from valence and hydrogen bonds and the latter, in turn, include donor and acceptor components. This 'non-polar' contribution is however not obtained by a summation of the donor and acceptor components but by their product similar to the 'related-to-hydrogen-bonds' fraction of the solubility parameter [555]

$$(\delta^h)^2 = 2\delta_d \delta_a. \tag{248}$$

Hence, adhesive interaction assisted by hydrogen bond formation is nothing but a specific case of adhesive bonding owing to donor-acceptor interaction.

Additionally, in certain cases it is necessary to consider the contribution of acid-base interactions. These are obviously essential in adhesive joints involving carboxyl-containing polymers. These are also important in adhesive interactions between liquids and finely dispersed substrates (fillers and pigments) [556] and, generally, between Lewis acids and bases [557]. Fowkes [558] suggested the following relationship

$$W_{Ad} = 2(\sigma_1^d \sigma_2^d)^{1/2} + m\,h(-\Delta H_{12}) + W_{Ad}^p \tag{249}$$

where m is the number of moles of acid-base pairs per unit area, ΔH is the enthalpy change accompanying the formation of a mole of such pairs and h is a constant.

H can be estimated by means of the Drago equation [559]

$$- \Delta H_{12} = c_1 c_2 + e_1 e_2 \qquad (250)$$

where c and e are parameters related to the contribution of covalent and electrostatic bonding [560]. The usefulness of this approach is assured in that it can be associated with the fundamental ideas of Hammet and, in general, with the concepts of the quantitative theory of organic reactions within the framework of which the chemical nature of the reacting materials can be given an adequate description.

Depending on the ratio between σ_i^p and σ_i^d, low and high energy polymers are conventionally distinguished. For the former, relying on the principle of mean-harmonic geometric congruence, Wu [561] determined interfacial surface energy as

$$\sigma_{sl} = \sigma_{sa} + \sigma_{la} - 2(\sigma_{sa}^d \sigma_{la}^d)^{1/2} - 4\sigma_{sa}^p \sigma_{la}^p/(\sigma_{sa}^p + \sigma_{la}^p) \qquad (251)$$

whereas for the latter, relying on the principle of arithmetic congruence, as

$$\sigma_{sl} = \sigma_{sa} + \sigma_{la} - 4 \frac{\sigma_{sa}^d \sigma_{la}^d}{\sigma_{sa}^d + \sigma_{la}^d} - 4 \frac{\sigma_{sa}^p \sigma_{la}^p}{\sigma_{sa}^p + \sigma_{la}^p}. \qquad (252)$$

Relationships for parameter Φ are accordingly changed

$$\Phi^d = 2 \left(\frac{x_1^d x_2^d}{b_1 x_1^d + b_2 x_2^d} + \frac{x_1^p x_2^p}{b_1 x_1^p + b_2 x_2^p} \right) \qquad (253)$$

$$\Phi^p = [2x_1^p x_2^p/(b_1 x_1^p + b_2 x_2^p)] + (x_1^d x_2^d)^{1/2} \qquad (254)$$

where $b_1 = \sigma_1/\sigma_2$ and $b_2 = \sigma_2/\sigma_1$. It was shown [548] that the least divergence between σ_{sa} and σ_c was ensured by eq. (251). Listed below are the dimensionless characteristics of polarity x determined with the aid of this equation for some polymers [397]:

Poly(dimethyl siloxane)	0.04
Polychloroprene	0.11
Poly(isobutyl methacrylate)	0.14
Poly(n-butyl methacrylate)	0.16
Polystyrene	0.17
Poly(methyl methacrylate)	0.28
Polyoxyethene	0.28
Poly(vinyl acetate)	0.33

This approach allows one to obtain, in terms of polarity, the thermodynamic condition for adhesive spreading over the surface of a low-energy substrate. In the first approximation x can be regarded as the criterion governing the process [562]. However, strictly speaking, in accordance with the definition eq. (154), for the adhesive as a higher energy phase we obtain:

$$\chi_1 = 2 \left(\frac{2\sigma_{sa}^d \sigma_{la}^d}{\sigma_{la}^d + \sigma_{sa}^d} + \frac{2\sigma_{sa}^p \sigma_{la}^p}{\sigma_{la}^p + \sigma_{sa}^p} - \sigma_{la} \right). \qquad (255)$$

Introducing x_i^p into this formula after simple rearrangements we come to:

$$\chi' = \frac{\chi_1 + 2\sigma_{1a}}{4k_s\sigma_{sa}} = \frac{(1-x_1^P)(1-x_2^P)}{k_s(1-x_1^P)-x_2^P+1} + \frac{x_1^Px_2^P}{k_sx_2^P+x_2^P}. \tag{256}$$

Having taken into account the condition for optimum spreading

$$(\partial\chi'/\partial x)_{k_s}, \quad x_2^P = 0 \tag{257}$$

Wu obtained the quadratic equation [561]

$$k_s(2x_2^P-1)(x_1^P)^2 - 2x_1^Px_2^P\,[x_2^P(k_s-1)+1] + [(k_s+1)$$

$$-(2x_2^P-1)]\,(x_2^P)^2 = 0 \tag{258}$$

whose solution gives a meaningful selection provided that the polarities of adhesive and substrate are equal.[†]

This conclusion accounts for the well-known DeBruyne's rule, according to which polar and non-polar polymers are not liable to form strong adhesive joints with each other [566]. However, according to eq. (252), the dependence eq. (258) becomes meaningless when the difference between x_1^P and x_2^P is large. As a matter of fact, water ($\sigma_{1a}^d = 21.8$ mN/m), according to eq. (154), should not spread on the surface with $\sigma_{sa} < 243$ mN/m, the contact angles, for instance with silver, platinum, palladium, uranium and gold lying within $50-85°$ [567]. However, due to the existence of oxide films on the substrate surface, spreading actually takes place by the development at the interface of dipole–dipole interaction and hydrogen bonds, rather than by dispersion forces.[‡] Hence, adhesive interaction between components with different polarities can be rather effective [569] and in bonding practice cases are known when DeBruyne's rule was invalid [570].

Such discrepancies are sometimes determined not only by inadequate interpretation of the experimental data (as in [571]) but also, and this is more important, by incorrect theoretical treatment of the relative contribution of the polar and non-polar components to the polymer surface energy in accordance with eq. (247). To illustrate this statement let us consider the following dependence [572]:

$$\cos\theta = 2(\sigma_s^d\sigma_1^d)^{1/2}/\sigma_1 + 2(\sigma_s^P\sigma_1^P)^{1/2}/\sigma_1 - 1. \tag{259}$$

It has two consequences: $\cos\theta$ is linearly related to $(\sigma_1^d)^{1/2}/\sigma_1$ when both the polar and non-polar components are involved; at $\cos\theta = 1$ the contribution of the polar component can be neglected in comparison with that of the non-polar one. To examine these suggestions the wettability of polyethene, poly(methyl methacrylate), poly(ethylene terephthalate) and cellulose acetate, i.e. of polymers of different polarities, by water–propanol solutions was examined. Analysis of the data obtained demonstrates that neither the initial dependence eq. (259) nor the consequences are experimentally valid [573].

[†] The absence of a dispersion component in expression (258) can be seen as common generality. For instance, according to Spelt et al. [563], the magnitude of θ is determined only by σ_{1a} but not σ^d. This inference seems deliberately limited as its authors have finally come to the somewhat archaic conclusion that σ_{sa}^d and σ_{sa}^P are nothing but adjustment parameters [564]. Moreover, even for such thoroughly examined adhesives as polymer melts [449] σ_{1a} values derived within the framework of the concept discussed [563, 564] and the experimental ones disagree by orders of magnitude.

[‡] 'It is perhaps fortunate that most metals oxidise in air . . .', claims Brewis [568].

The difference in polymer polarities can be evaluated by parameter Φ. Usually, its values are very close to but not equaly to unity (see Table 2.1), and are determined by the energetics of interfacial bonding. Neglecting the latter fact, Fowkes [574] assumed that for dispersion interactions $\Phi = 1$. Then, eq. (136) can be reduced to

$$\sigma_{sl}^d = \sigma_{sa} + \sigma_{la} - 2(\sigma_{sa}^d \sigma_{la}^d)^{1/2}. \tag{260}$$

Thus σ_i^d is introduced as one of the variables in the basic relationships of wetting thermodynamics. From expressions eq. (55) and eq. (260) there follows a new form of Young's equation that accounts for the dispersion forces

$$1 + \cos \theta = \frac{2(\sigma_{sa}^d \sigma_{la}^d)^{1/2}}{\sigma_{la}}. \tag{261}$$

It seems natural to supplement this equation with the polar component $2(\sigma_{sa}^p \sigma_{la}^p)^{1/2}/\sigma_{la}$. If, however, we neglect this component and confine ourselves to dispersion interactions the corresponding component of solid surface energy would be as follows

$$\sigma_{sa}^d = \sigma_{la}^2 (\cos \theta + 1)^2 / 4\sigma_{la}^d. \tag{262}$$

Eq. (262) can be easily resolved graphically as with the coordinates of Fig. 2.41 this dependence is linear[†] for wetting liquids with known σ_{la}.

Such an approach makes it possible to evaluate experimentally the dispersion and polar components of the surface energy. For this purpose Kaelble [575] proposed the following equation

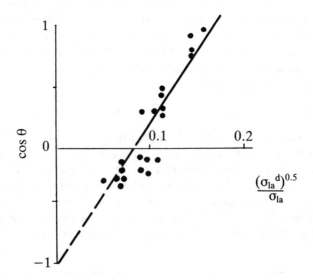

Fig. 2.41 – Contact angle with polypropene versus surface energies of the low-molecular mass wetting liquids.

[†] In general, for explicitly dispersion interactions when $(\sigma_{la}^d)^{1/2}/\sigma_{la} \to 0$ eq. (262) reduces to $\cos \theta = -1$.

$$W_{Ad}/2(\sigma_{la}^d)^{1/2} = (\sigma_{sl}^d)^{1/2} + (\sigma_{sl}^p \sigma_{la}^p/\sigma_{la}^d)^{1/2}, \tag{263}$$

which can be represented graphically as $W_{Ad}/(\sigma_{la}^d)$ vs. $(\sigma_{la}^p/\sigma_{la}^d)^{1/2}$. Then $(\sigma_{sl}^p)^{1/2}$ is determined as the slope angle, while $(\sigma_{sl}^d)^{1/2}$ corresponds to the intersection of the straight line with the ordinate axis and from which W_{Ad} can be easily calculated using eq. (98).

This method was checked for a number of materials (carbon fibres were the first of these [576]) for which the contact angle values were found to be a function of the cross-section perimeter l and of the force P required for releasing the filament from the binder [577]

$$\cos \theta = Pg/l\sigma_{la}. \tag{264}$$

Polar ($\sigma_{la}^p = 0$–51.8 mN/m) and non-polar ($\sigma_{la}^d = 18.4$–72.8 mN/m) compounds were used as wetting liquids. The procedure can be easily computerized by presenting the respective dependences in matrix form [578].

The approach described was also extended to solvent mixtures [579]. In this case the procedure involves preliminary calibration of a polymer surface by a standard liquid for which

$$W_{Ad} = \sigma_1(\cos \theta + 1) = 2[(\sigma_1^d \sigma_s^d)^{1/2} + (\sigma_1^p \sigma_s^p)^{1/2}], \tag{265}$$

while for the first liquid to be contacted

$$W_{Ad}/2(\sigma_s^d)^{1/2} = (\sigma_1^d)^{1/2} + (\sigma_1^p \sigma_s^p/\sigma_s^d)^{1/2}. \tag{266}$$

Then it is not difficult to obtain the dependences for dispersion and polar components of the surface energy of multi-component liquid systems. This method makes it possible to derive with considerable accuracy the surface properties of typical non-polar (polyethene) and polar (polyvinylchloride, polyimides) polymers from the studies on wetting by mixtures of ethyleneglycol monomethylether with methylethylketone and of methylethylketone with N-methyl-2-pyrrolidone. Similarly, Busscher et al. [580] used water–propanol mixtures and in plotting the variables of eq. (266) the corresponding dependences were represented by straight lines. As is seen from Fig. 2.42, the horizontal or the negative slope dependence corresponds to non-polar polymers, while the positive slope plot refers to the polar ones. This fact may serve to illustrate the sensitivity of the procedure and its efficiency when aiming at a quantitative evaluation of the individual components to surface energy of solid substrates.

Another approach to calculation of surface energy was proposed by El-Shimi and Goddard [581] who, by analogy with Rhee's method discussed above, combined eqs (55) and (237). For the wetting of low-energy polymers of two liquids, water (w) and methylene iodide (j),[†] with the contact angles θ' and θ'', respectively, the equilibrium condition can be written as a two-equation system:

---

† Strictly speaking, in this case one should take into account polymer swelling and dissolution, accompanying the wetting, as well as the aggregation of wetting liquids [582].

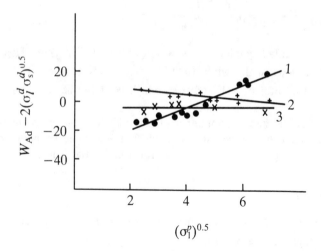

Fig. 2.42 – Interrelationship between individual components of the surface energy for: poly(methyl methacrylate) (1); polytetrafluoroethene (2); polyethene (3).

$$\sigma_{sa}^{d}\sigma_{sa}^{p}[\sigma_{lw}^{d} + \sigma_{lw}^{p} - 0.25\sigma_{lw}(\cos\theta' + 1)]$$

$$+ \sigma_{sa}^{d}\sigma_{lw}^{p}[\sigma_{lw}^{d} - 0.25\sigma_{lw}(\cos\theta' + 1)]$$

$$+ \sigma_{sa}^{p}\sigma_{lw}^{d}[\sigma_{lw}^{p} - 0.25\sigma_{lw}(\cos\theta' + 1)]$$

$$- 0.25\sigma_{lw}\sigma_{lw}^{d}\sigma_{lw}^{p}(\cos\theta' + 1) = 0.$$

$$\sigma_{sa}^{d}\sigma_{sa}^{p}[\sigma_{lj}^{d} + \sigma_{lj}^{p} - 0.25\sigma_{lj}(\cos\theta'' + 1)]$$

$$+ \sigma_{sa}^{d}\sigma_{lj}^{p}[\sigma_{lj}^{d} - 0.25\sigma_{lj}(\cos\theta'' + 1)]$$

$$+ \sigma_{sa}^{p}\sigma_{lj}^{d}[\sigma_{lj}^{p} - 0.25\sigma_{lj}(\cos\theta'' + 1)]$$

$$- 0.25\sigma_{lj}\sigma_{lj}^{d}\sigma_{lj}^{p}(\cos\theta'' + 1) = 0. \qquad (267)$$

This system is solved with respect to σ_{sa}^{d} and σ_{sa}^{p}. However, its practical use is quite complicated and calls for simplification. Instead of the system of eq. (267), Jaunczuk and Chibowski [583] proposed a single relationship

$$(\sigma_{s}^{d})^{1/2} = \frac{\sigma_{w}\cos\theta'' - \sigma_{wH}\cos\theta' + 2\sigma_{w} - \sigma_{H}}{2[2(\sigma_{w}^{d})^{1/2} - (\sigma_{H}^{d})^{1/2}]} \qquad (268)$$

(H represents hydrocarbon) that produces satisfactory (for all practical purposes) agreement between the calculated and experimental data obtained for polytetrafluoroethene. Yet another simplification, though a more rigorous one, was suggested by Zorll [584] who used one wetting liquid; the ultimately derived quadratic equations are as follows

$$(\sigma_{sa}^d)^2 \, (A_2 B_1 - A_1 B_2) + \sigma_{sa}^d (C_2 B_1 - C_1 B_2 - D_1 A_2 + D_2 A_1)$$
$$+ (D_2 C_1 - D_1 C_2) = 0,$$

$$(\sigma_{sa}^p)^2 \, (A_2 C_1 - A_1 C_2) + \sigma_{sa}^p (B_2 C_1 - B_1 C_2 - D_1 A_2 + D_2 A_1)$$
$$+ (D_2 B_1 - D_1 B_2) = 0, \tag{269}$$

which include as coefficients the parameters of the liquid phase only:

$$A = 0.25\sigma_{la}(3 - \cos\theta), \tag{270}$$

$$B = \sigma_{la}^p[\sigma_{la}^d - 0.25\sigma_{la}(\cos\theta + 1)], \tag{271}$$

$$C = \sigma_{la}^d[\sigma_{la}^p - 0.25\sigma_{la}(\cos\theta + 1)], \tag{272}$$

$$D = 0.25\sigma_{la}\sigma_{la}^d\sigma_{la}^p(\cos\theta + 1). \tag{273}$$

Contrary to eqs (267) and (268), the system of eqs (269) can also be used to investigate polar polymers.

An important consequence follows from eq. (261), when Zisman's condition, eq. (237), applies

$$\sigma_c = (\sigma_{sa}^d \sigma_{la}^d)^{1/2}. \tag{274}$$

In other words, if the surface energy of a wetting liquid is provided by the dispersion forces alone (as is often the case in practice [585]), the critical surface tension σ_c acquires the meaning of the dispersion component of the surface energy of a solid.[†] On wetting of the latter with polar liquids $\sigma_c < \sigma_{sa}^d$. In the opposite case, occasionally occurring as a result of incorrect calculations of the critical surface tension by summing up its individual components according to Fowkes, $\sigma_{sl} < 0$. This, evidently, bears no real meaning. As claimed by Kloubek [586], inclusion of the polar term restrains, in general, the possibility of the Fowkes' treatment. The concepts discussed make it possible to clarify the limitations of Zisman's method for evaluation of the energy parameters of a polymer surface.

Strictly speaking, in the general case eq. (261) must include surface pressure, as in eq. (137)

$$\cos\theta + 1 = (\sigma_{sa}^d \sigma_{la}^d)^{1/2}/\sigma_{la} - \pi/\sigma_{la}. \tag{275}$$

However, the last term is normally ignored which has proved to be correct for non-polar polymers only. Generally this term accounts for the probability of sorption of the molecules of adhesive at specific active sites on the substrate surface. Bantysh et al. [587] examined the case of slowly penetrating low-molecular mass compounds with different contributions to the total surface energy of the dispersion, polar and hydrogen bond components used as wetting liquids. By means of the iterative procedure for the

---

† Such an approach makes it possible to interrelate the regularities of the wetting hysteresis of polymers with the variation of σ_{sa}^d. It was found, for instance, that advancing and retreating contact angles are linear functions with regard to $(\sigma_{sa}^d)^{1/2}/\sigma_{la}$ [251].

contact angle up to its coincidence with the experimental value they have come to the following relationship:

$$\cos\theta = \frac{4}{\sigma_1}\left(\frac{\sigma_1^d\sigma_s^d}{\sigma_1^d + \sigma_s^d} + \frac{\sigma_1^p\sigma_s^p}{\sigma_1^p + \sigma_s^p} + k_hA_h\frac{\sigma_1^h\sigma_s^h}{\sigma_1^h + \sigma_s^h}\right) - 1. \tag{276}$$

Here k_h accounts for the change of they hydrogen bond strength due to the replacement of the solid–solid contact by the solid–liquid one and A_h is a steric factor reflecting the process of formation of more than one interfacial hydrogen bond between the adsorbed molecule and the active site on the substrate surface.

On the other hand, besides parameter π there is another factor specific for systems with a broad spectrum of bond energy that has to be considered. This is most easily seen when using Gibbs' theory discussed in section 2.1.1. Indeed, the general expression [588]

$$\sigma_s^d = U_s^d + T[(d\sigma_s^d/dT) - (dU_s^d/dT)] \tag{277}$$

includes the dispersion energy excess of the polymer surface layer per unit area. The second term of this equation was estimated to contribute not more than 1.5 MJ/m². The first term can be derived from atom–atom and group interactions using dipole–dipole, dipole–quadrupole and repulsion potentials. When evaluating σ_s^p it is also necessary to include the change of the free energy relevant to chain orientation. In spite of the need for additional verification eq. (277) produces values that are in reasonable agreement with those obtained experimentally. However, it is more likely that it is neglecting the contribution of the polar component to surface energy that causes the σ_{sl} values to be commonly over-estimated (when using eq. (260), in particular).

As a necessary correction factor Dahlquist [589] suggested the energy of polar stabilization I_{sl}^p which is determined by the nature of a wetting liquid. For the wetting of polytetrafluoroethene by water (w) and octane (o) Hamilton [590] established the dependence

$$I_{sl}^p = 50.8\cos\theta + \sigma_w - \sigma_o \tag{278}$$

and proved the necessity to include this parameter, whereas neglecting it makes analysis more complicated [591]. This unified approach allows it to be expanded to cover the dispersion interactions as well. Fowkes, who was the first to introduce this term (I^d), had however, no opportunity to estimate its value because the parameters of eq. (260) were obtained when investigating polymer wetting by one liquid (a hydrocarbon) only [592]. In order to reduce the calculation to a form permitting simple solution by decreasing the number of variables, Tamai et al. [593] proposed to wet the solids with two immiscible liquids, for instance, by a hydrocarbon and water. Then for each of the liquids (w and H) one has

$$\sigma_{sH} = \sigma_{sa} + \sigma_{Ha} - 2(\sigma_{sa}^d\sigma_{Ha}^d)^{1/2} - I_{sH}, \tag{279}$$

$$\sigma_{sw} = \sigma_{sa} + \sigma_{wa} - 2(\sigma_{sa}^d\sigma_{wa}^d)^{1/2} - I_{sw}. \tag{280}$$

Substituting eqns (279) and (280) into eq. (55) one obtains the following relationship

$$\sigma_{Ha} - 2(\sigma_{sa}^d\sigma_{Ha}^d)^{1/2} = \sigma_{wa} - 2(\sigma_{sa}^d\sigma_{wa}^d)^{1/2} + \sigma_{Hw}\cos\theta - I_{sw} \tag{281}$$

which can be verified experimentally. For such a three-phase system the condition for selective wetting in terms of free surface energy is given by the inequality

$$\sigma_{sl_i} - \sigma_{sl_j} - \sigma_{sl} < 0. \tag{282}$$

When

$$|\sigma_{sl_j} - \sigma_{sl_i}| < \sigma_{sl} \tag{283}$$

wetting is reversible and, according to Shanahan *et al.* [594], each liquid of the pair (l_i or l_j) can replace one another from the substrate surface. This fact should be taken into account from a practical aspect as well, for instance, when an adhesive is applied on to the substrate from a mixture of solvents or when bonding is accompanied by the replacement of some products from the interface.

Matsunaga and Ikada [595] made an attempt to combine this approach with eqns (55) and (136). For the case of simultaneous wetting of a polymer surface by two liquids (one of which is applied on to the other)

$$\sigma_w - \sigma_H + \sigma_{wH} \cos \theta = 2[(\sigma_w^d)^{1/2} - (\sigma_H)^{1/2}](\sigma_s^d)^{1/2} + I_{sw}^p + \pi. \tag{284}$$

It is assumed that $\sigma_H^d = \sigma_H$, whereas, according to eq. (247)

$$\sigma_w = \sigma_w^d + \sigma_w^p. \tag{285}$$

Neglecting the surface pressure it is easy to find $(\sigma_s^d)^{1/2}$ and I_{sw}^p from the dependence $(\sigma_w - \sigma_H + \sigma_{wH} \cos \theta_{sw})$ versus $2[(\sigma_w^d)^{1/2} - (\sigma_H)^{1/2}]$ as the slope and the intersection with the ordinate axis, respectively. Using this method the wetting of poly(methyl methacrylate), cellulose, poly(vinyl alcohol) [595], and of the copolymers of vinyl alcohol with ethene and butylvinyl ether [596] by the mixtures of *n*-alkanes with formamide, ethylene glycol, glycerol and water was investigated. It was shown that σ_s^d does not depend on the nature of the polar liquid, hence the validity of the approach described was demonstrated.

At the same time, there might be traces of a hydrocarbon (film) left on the substrate surface after the first wetting liquid had been replaced for the second one (water). This is quite probably the case and has to be considered in the thermodynamic description of a system, especially if a hydrocarbon film is sufficiently thick. Then, eq. (284) can be simplified [597]

$$\sigma_s^d = (\sigma_w - \sigma_H + \sigma_{wH} \cos \theta)/(\sigma_w^d)^{1/2}. \tag{286}$$

Nevertheless, even with regard to this fact the procedure described reveals the inadequacy of eq. (136) for the purpose of calculating W_{Ad} via σ_s, σ_w, and I_{sw}^p as the I_{sw}^p versus $(\sigma_w^p)^{1/2}$ is non-linear. More reliable data can be obtained by using eq. (249) which takes into account the acid-base interactions [556].

Tamai succeeded in an attempt to slightly simplify this approach by introducing another factor (θ'') and using three liquids (two hydrocarbons) which then leads to the following relationships [598]

$$\sigma_{sa}^d = \frac{(\sigma_{Ha}' - \sigma_{Ha}'') - (\sigma_{Hw}' \cos \theta' - \sigma_{Hw}'' \cos \theta'')}{2[(\sigma_{Ha}')^{1/2} - (\sigma_{Ha}'')^{1/2}]}, \tag{287}$$

$$I_{sw}^p = \frac{1}{(\sigma'_{Ha})^{1/2} - (\sigma''_{Ha})^{1/2}} \left\{ [(\sigma''_{Ha})^{1/2} - (\sigma^d_{wa})^{1/2}] (\sigma'_{Ha} - \sigma_{wa} - \sigma'_{Hw} \cos\theta') \right.$$

$$\left. - [(\sigma'_{Ha})^{1/2} - (\sigma^d_{wa})^{1/2}] (\sigma''_{Ha} - \sigma_{wa} - \sigma''_{Hw} \cos\theta'') \right\}. \tag{288}$$

The three-liquid method provides more reliable values for the dispersion components of the polymer surface energy along with the simultaneous determination of the value for I (Table 2.7). This latter quantity, called by Tamai [598] the work of adhesion (this leads however, to terminological ambiguity with regard to the implications of Dupré's equation), can be calculated from the one-liquid method relying on equation similar to eq. (278) and, according to Fowkes [592], from the relationship in eq. (277) by considering surface pressure. The results obtained [601] are in good agreement with those of Tamai [593]. This author believes that since quantity I accounts for the non-dispersion interactions then it can be used for the evaluation of interaction across the surface of both non-polar and polar solids [598], including metal oxides [602].

Table 2.7 — Non-dispersion components to the interaction between individual polymers and water, mJ/m².

Polymer	One-liquid method		Three-liquid method
	[589]	[599]	
Polytetrafluoroethene	3–5	3.8	0.05–0.1 [598]
Poly(methyl methacrylate)	30–32	32.3	27.0 –27.8 [598]
Poly(vinyl chloride)	18–20	21.9	
Poly(ethylene terephthalate)	–	–	26.4 –27.0 [600]
Poly(hexamethylene adipamide)	–	–	47.5 –48.1 [600]

Matsunaga [600] simplified the calculation by introducing increments ΔI_{sw}^p relating to x-groups and the parameter $y(x)$ determined by the number of these groups in the repeat unit m_x, the density and molecular mass M of a polymer:

$$y(x) = m_x (M/\rho N_A)^{-2/3}. \tag{289}$$

Then

$$I_{sw}^p = \sum_i \Delta I_{sw}^p(x)\, y(x). \tag{290}$$

The discrepancy between the values of I_{sw}^p calculated by this method and those found experimentally is 0.3–0.8%, increasing from non-polar to polar polymers.

The advantages of the method involving several liquids are clear. Besides increasing the accuracy of equilibrium conditions in multi-phase systems (which makes it possible to obtain more precise solutions of Young's equation) in terms of polymer adhesion, wetting of a substrate by two or three immiscible liquids results in less 'contamination' (as defined by Cazeneuve et al. [603]) of the surface under investigation, and so eliminating adsorption-related effects. However, as is the case in other procedures requiring application of eq. (55), substantial experimental difficulties are caused by the need to measure contact angles [538]. An attempt to eliminate this difficulty was examined by Schultz and co-workers [604]. They reported experiments in which carbon filaments were immersed vertically into two immiscible liquids, e.g. formamide with a hydrocarbon (hexane, hexadecane, decalin) layered on top. By analogy with the above expressions the relationship for interfacial interaction can be written as follows

$$\sigma_F - \sigma_H + \sigma_{HF} \cos \theta = 2(\sigma_s^d)^{1/2}[(\sigma_F^d)^{1/2} - (\sigma_H)^{1/2}] + I_{sF}^p. \tag{291}$$

Measurement of θ was avoided by weighing the filaments separately in a hydrocarbon (H) and in a two-phase hydrocarbon–formamide mixture, and subsequently by weighing the substrates while moving the filled vessel along its axis. Then with the aid of the relationship $\sigma_F - \sigma_H = F_{HF}\sigma_H/F_H$ versus $(\sigma_F^d)^{1/2} - (\sigma_H)^{1/2}$ (where F is the so-called 'adhesion force' [604]) one finds σ_s^d and I_{sF}^p, and from the definition of I_{sF}^p as of $2(\sigma_s^p \sigma_F^p)$ then σ_s^p is obtained. The relationship proposed was shown to be linear only for low-modulus fibres ($T = 300$); moreover, the calculated values of the dispersion and polar components of surface energy are in line with those of carbon-chain polymers used as coupling agents for the given substrates. For high-modulus carbon fibres (M-40) the dependence is non-linear, which can be attributed to surface inhomogeneity. The results obtained support the objectivity of the method and its high sensitivity.

As may be seen from the above discussion, Fowkes' approach has become quite common in use. At the same time, it must be remembered that this approach is based on an *a priori* assumption ($\Phi = 1$) though additivity of the different types of van der Waals forces is perceived rather by intuition than on the basis of strong arguments.[†] Only quite recently an attempt has been made to provide a statistical-mechanics background for this approach.

This relies on the treatment of a canonical partition function representing the thermodynamic properties of a two-phase fluid system comprised of N molecules, and having the total volume V and the interphase area A_{if}:

$$Q(T, V, A_{if}, N) = \cap Z(T, V, A_{if}, N), \tag{292}$$

where $\cap$ represents the kinetic contribution which is independent of A_{if}; the configurational partition function Z is

$$Z(T, V, A_{if}, N) = \frac{1}{N!} \int \cdots \int d(r) \, [\exp - U(r)/KT]. \tag{293}$$

According to definition eq. (24), the function Q of eq. (292) is associated with the Helmholtz free energy

---

† It would be improper however to confine the procedure discussed involving increments, to the semi-empirical or to compensating implications. Having taken advantage of Feinman's ideas [605] Askadsky demonstrated that within the framework of the additive scheme the behaviour of the repeat unit of a polymer can be described by means of mathematical techniques of thermodynamics relating to a set of non-harmonic oscillators [478, p. 29], i.e. on a sufficiently sound basis. This should be borne in mind when appraising the background and reliability of the calculative schemes either reported elsewhere [481] or those developed here.

$$F = -KT \ln Q,$$ (294)

which is known to be related to the surface energy via eq. (13). Then

$$\sigma = -\frac{KT}{Z} \left(\frac{\partial Z}{\partial A_{if}} \right)_{T,V,N}.$$ (295)

Using Green's technique to change the variables

$$r = (x, y, z) = \left(A_{if}^{1/2} x', A_{if}^{1/2} y', \frac{V}{A_{if}} z' \right).$$ (296)

Z of eq. (295) can be represented in explicit form

$$Z = \frac{V^N}{N!} \int_0^1 \cdots \int_0^1 d(r) \exp\left[-U(r', A_{if})/KT \right].$$ (297)

Introducing eq. (297) into eq. (295), Navascues [606] has come to the final relationship, which is as follows

$$\sigma = \frac{1}{ZN!} \int \cdots \int d(r) \left(\frac{\partial U}{\partial A_{if}} \right)_{T,V,N} \exp\left[-U(r)/KT \right].$$ (298)

Here the potential U obeys the additivity pattern and can be represented as a sum of contributions due to repulsive, London, polar, induction and other forces. Strictly speaking, these concepts were developed for fluid systems and they demonstrate that at least for polymer solutions and melts Fowkes' approach is also valid when interfacial bonds other than dispersion, polar, and hydrogen types are involved in adhesive joint formation.

Such a uniformity in background of a large number of apparently different theoretical concepts leads to calculations of the energy characteristics of polymer surfaces, relying on relevant equations, that only slighly deviate from one another. Undoubtedly, the degree of validity and rigour of some of the approaches discussed is quite different; however, within the limits of experimental accuracy one obtains essentially the same data for the initial parameters. This inference could have been readily proved by investigating one material using the different procedures in order to avoid subjective factors. Such a possibility has recently been considered (however, for other and purely applied purposes) for polypropene [607]. It was shown that the values for surface energy varied within the range of 0.4 mN/m (33.6 mN/m when determined according to Fowkes [574], and 34.0 mN/m according to Wu [294]), while for the interfacial energy with water it was within 1.5 mN/m (51.3 mN/m when determined according to Fowkes [574] and Kaelble [116], and 52.8 mN/m according to Wu [294].

However, what we believe is really worthy of discussion is that the values for surface energy obtained by different methods (and not only those mentioned above) are quite close and lie within a very narrow interval of σ_{sa} values regardless of the chemical nature of the high molecular mass compounds examined. This consistency is quite objective and, apparently, not incidental. Nevertheless, in spite of the major importance of this fact the reasons for it have not, to our knowledge, yet been discussed.

We believe that the reasons for this phenomenon are the specific difference in properties of the bulk volume and of the surface layers adjacent to the geometric boundary of the polymer sample and the cause of this difference lies, principally, in the character of the macromolecular packing. The influence of this effect, associated with macromolecular packing, on the magnitude of the polymer surface energy is quite significant and we believe that it can prevail over that of the chemical nature of polymers. It seems reasonable to assume that in the surface layers reorientation of functional groups is taking place in such a manner that they are turned away from the surface layers and inwards towards the bulk phase, so that the measured surface energy is predominantly due to the polymer backbone. In other words, the nature of functional groups should not decisively affect polymer surface energy.

To prove this assumption let us consider the results of recent studies of polymer surfaces performed by a variety of methods including some most novel ones.

Precision measurements of contact angles on polystyrene films cast on various substrates demonstrated that their values were regularly smaller than those of self-supported films contacting air only, though a similar but an even more pronounced effect was observed for poly(butyl methacrylate) [608]. These data can be interpreted as resulting from the orientation of the functional groups (either the low-polarity phenyl groups of polystyrene or the polar oxygen-containing groups of polybutyl methacrylate) towards the polymeric bulk phase and away from the substrate.

This conclusion has been recently confirmed in studies with poly(butyl methacrylate) and poly(2-ethyl hexylacrylate) [609]. A convenient model for studying the matter is provided by polymeric gels. It is known that for hydrogels contact angles are rather small, although the mobility of functional groups in the surface layers seems to be adequately high. The mobility of these groups can be regulated by corona discharge treatment. Yet this procedure, contrary to expectations, for polypropene, gelatin and agar, did not affect in the slightest the θ values, which remained just as low as they were [610]. One can thus conclude that the increased mobility of the side chains results in their orientation inwards towards the bulk phase, and surface energy measured by the wetting method is indicative of the backbone of a macromolecule rather than of a given polymer as a whole.

This conclusion was confirmed directly in the ESCA measurements of the concentrations of individual types of atoms in the surface layers of polymers. For polyethylene terephthalate these concentrations were measured to be 72% for carbon and 28% for oxygen [611]. The calculated carbon and oxygen content is 62.5% and 33.3%, respectively. Hence, carbon atoms are obviously in excess at the polymer surface. More impressive are the data provided by the method based on the measurement of the energy loss of elastic scattering of noble gas ions as, unlike ESCA, this method makes it possible to obtain information about the atoms located at the very outer boundary of the surface layer. Large numbers of polymers of various chemical nature were investigated by this method, e.g. polyethene, poly(methyl methacrylate), poly(vinyl alcohol), polycarbonate, polyoxymethene and polybutadiene acrylate. For all of the polymers examined atomic intensities on the surface did not correlate with the stoichiometric ratio of oxygen and carbon atoms in the macromolecule [612] the surface layers of polymers being 'enriched' with the latter. For instance, for polyoxymethene (theoretical ratio $O:C =$

1.0) the value measured was only 0.15. In other words, the surface of the phase is comprised predominantly of carbon atoms, hence, its energy, regardless of the general nature of the material, varies only within very narrow limits.

The above considerations and relevant experimental data (incomplete here since some of the results are discussed in other sections) provide, we believe, the most sound explanation of the narrow range of surface energy values of high molecular mass compounds of very varied nature and of the only slight dependence of surface energy on polymer composition.[†] The validity of such an approach, first discussed in the Russian edition of this monograph (1984), was independently supported in the following two years by the work of Andrade and Chen [614] who stated that reorientation of macromolecular fragments in the boundary layers of polymers resulted in a regular decrease of the interfacial free energy.

2.2.2 Energetics of adhesive interaction between polymers

A thorough analysis of the energetics of adhesive bond formation should ensure that all types of forces acting across the interface be considered. In principle the problem is very simple, the type of force exhibited being determined by the nature of the contacting phases. For instance, polyvinyl alcohols and cellulose that contain many hydroxy groups are liable to form hydrogen bonds, the chemically interacting polymers form covalent bonds, semicor.ductors tend to form ionic bonds etc. However, as discussed earlier, dispersion interactions are, in all cases, the most probable and should be considered first of all.

At the present time the direct calculation of the interaction energy between the adhesive and the substrate is probably not feasible. Even for simple systems comprised of several simple molecules the application of current knowledge requires a large number of assumptions. That is why in practice one is usually compelled to rely on *a priori* considerations regarding the nature of the bonds which are presumably associated with adhesive interaction. Obviously, the cognitive value of such results is almost non-existent. Sometimes, however, knowledge of the chemical nature of the contacting polymers suffices to permit an approximate evaluation of the adhesive interaction.

Conclusions concerning the subject are more reliable when they are dependent on studies of the behaviour of adhesive joints under destructive conditions. Qualitative ones can easily be derived from the data on solvolytic stability of such systems. For instance, liberation of the adhesive from the substrate indicates that the interaction between these elements is a low-energetic one, whereas stability of the joint in sufficiently active solvents leads one to assume that a network of interfacial chemical bonds exists. In principle, the introduction of solubility parameters could have provided a quantitative description [615]. However, the necessity arises to consider a number of effects which are not directly related to adhesion but are associated with diffusion factors, sorption, specificity of the interaction of a polymer with a solvent etc.

---

† We believe it a somewhat simplified explanation of the similarity of σ values of various polymers that an 'atmospheric layer' is formed at the polymer surface in air [613]. Such an interpretation, based in essence on the well-known views of Bikerman [275], is probably valid only for metals whose surfaces, being of higher energy (at least by an order of magnitude) than of polymers, are always covered with compounds adsorbed from the atmosphere.

The approach based on the fracture studies of adhesive joints is free from these drawbacks. Methodologically it is not a perfect approach as it implies that, on fracture of the systems, the same amount of energy is liberated as was required for its formation. This premise is in apparent contradiction with the fracture mechanics of polymers. In addition, one has to bear in mind that the strength of adhesive joints is determined by a large number of other factors, e.g. mechanical effects, thickness of adhesive film, internal stresses, deformation properties of the adhesive and the substrate etc. However, at present there is no alternative to the approach described as it provides a semi-quantitative assessment of the efficiency of adhesive interaction.

We believe that at the present state of the art the best way for the evaluation of the energy of adhesive interaction in real polymeric systems should involve all three approaches and applied in sequence. As a first step the order of magnitude is evaluated on the basis of the potential ability of the major adhesive and substrate components to interact with each other. This preliminary information is then utilized in studies of the stability of the adhesive joint with regard to the action of inert and active solvents with specified cohesive properties. And, ultimately, the strength of the adhesive joint is measured under conditions which prevent a possible cohesive fracture of the joint. The data obtained are used to establish correlations with the basic thermodynamic parameters of adhesive interaction.

Solvolytic methods enable a qualitative assessment of the energy of adhesive interaction between polymers and are based on studies of the stability of corresponding systems in liquid media. In a majority of cases this action results in the decrease of the adhesive joint strength, as is illustrated by the data of Table 2.8. There are two reasons for this effect; firstly the decrease in cohesive characteristics of the transition layers of polymers on account of either the consequence of adsorption—active media [619] or the plasticizing effect of diffusing molecules [620], and secondly the weakening of interfacial bonds due to the liquid layer acting as wedge and the establishment of a adsorption—desorption equilibrium at the interface [621]. The first group of factors (i.e. relating to the cohesive properties of transition layers) reflects the influence bestowed upon adhesive joints by organic liquids [620, 622]; while the second, related to phenomena at the interface proper, refers to the effect produced by water [621]. It is easy to show that both of these result from the decrease of interfacial energy. Malkin and co-workers suggested the relationship between interfacial energy σ_{sl}, polymer life-time t and tensile strength P_p [623, 624]

$$t = ce^{k\sigma_{sl}} P_p^{-(a+b\sigma_{sl})}. \tag{299}$$

They also determined corresponding constants a, b, c and k for systems comprising polybutadiene and 1,2-alkanols.

Interaction of adhesive joints with water [625] is the most thoroughly examined due to its practical importance. As a result of such interaction the adhesive is almost completely separated from the substrate [625—628] even when other solvents do not affect the strength of systems involving interfacial chemical bonds. The ultimate effect depends on the duration of the exposure to water; the graph is represented by a descending curve asymptotically approaching the abscissa axis. Analytically the corresponding dependence can be described by a logarithmic equation [629]. The magnitude of the observed effect is largely governed by the nature of the substrate, e.g. poly-ϵ-caproamide is more susceptible to the action produced by water than is polyethene

Table 2.8 — Relative strength of adhesive joints (%) exposed to various solvents.

Elements of the joint		Active medium							Ref.
Adhesive	Substrate	H_2O	C_6H_6	CH_3COCH_3	CH_3OH	C_2H_5OH	C_3H_7OH	C_4H_9OH	
Mica[a]		21.1	56.5	—	—	31.3	—	—	[616]
Natural rubber[b]	—	—	—	30.0	—	7.6	—	—	[109]
Gutta-percha[b]	—	—	—	75.5	—	10.5	—	—	[109]
Epoxy	Steel	—	—	—	—	49.9	—	—	[617]
Paraffin	Aluminium	103.3	—	—	21.9	49.6	81.0	102.5	[618]
PMMA	Aluminium	85.4	—	—	28.8	—	—	59.4	[618]

a   Cleavage of the crystals of muscovite.
b   Autohesive bonding.

[630]; aluminium, titanium, and steel are more susceptible than copper [628]. As regards polymers, such patterns of behaviour are related to the reversible changes in supermolecular structure of a substrate accompanying water imbibition and drying [622].

Consequently, after water has been removed the strength of adhesive joint could be restored up to the initial level. This is the case in systems with an interfacial network of dispersion forces, e.g. those involving poly-ε-caproamide, polyethene [631], and (butyl methacrylate)-(methacrylic acid) copolymer [627]. In the case of polyethene variations in joint strength are reversible if the joints are exposed to benzene, xylene and also acetone. The same kind of reversible response has also been discovered for polystyrene joints exposed to water–methanol mixtures [632]. Moreover, this effect is also observed in systems with an interfacial network of hydrogen bonds, such as, for example, the metal–epoxy adhesive joint [633], or the epoxy-based carbon-fibre reinforced composite [634]. By examining autohesive joints of polyethene pretreated with corona discharge Owens [635] provided a direct confirmation of this reversible response. The system recovered its initial strength after it had been exposed to water and then subsequently dried. At the same time, the bonding was found to be somewhat weakened by typical proton-donating (ethanol) and proton-accepting (acetone, tetra-hydrofuran) compounds, which verified with existence of hydrogen bonds in the system examined.

Clearly, the existence of a network of interfacial valency bonds eliminates, in the majority of cases, the possibility of the adhesive joint strength being decreased due to the effect of active media. Table 2.9 compiles the data of the studies of elastomer bonding by the adhesive modified with resorcinol–phenolformaldehyde oligomer to ensure covalent bonding across the interface [636]. These data are quite illustrative. Similar results were obtained in systems involving oxidized polymeric substrates [630, 631, 637] or acid-treated metals [626, 638, 639]. In both cases oxygen-containing functional groups produced during the pretreatment procedures enable the formation of interfacial covalent bonds. For instance, the treatment of titanium with various inorganic acids was shown to produce two kinds of surfaces differing in the magnitude of surface energy (below 50 and above 72.8 mN/m, respectively). Correspondingly, adhesive joints comprising each of the two types of surface differed essentially in water resistance [633]. As is demonstrated in Fig. 2.43, the relatively low-energy surface provides for a joint strength in water that is at least half that of the joint with the high-energy surface. Similar results were obtained when examining the solvolytic stability of adhesive joints with magnesium [640] and aluminium [61] this last case being most representative. The treatment of the substrate (aluminium) with either sulfuric acid or a mixture of phosphoric and chromic acids results in surfaces with σ_{sa} 54.5 and 151.5 mN/m, respectively. Interfacial interaction of sulfuric acid-treated aluminium with butadiene–styrene and butadiene–acrylonitrile copolymers is due to physical bonds only (the joints lose their integrity in liquid media) whereas the situation with the phosphatized substrate is essentially different. When there is no chemical interaction between an adhesive and a substrate the strength of the adhesive joint as a function of time is determined only by the initial and boundary conditions for the sorption of a liquid and by its diffusion coefficient [642].

The strength of adhesive joints can be increased by liquids possessing surface active properties. The effect is attributed to the fact that the diphylic surfactant molecules are oriented at the substrate surface so as to favour the energetics of interfacial interaction.

Table 2.9 – Relative strength of adhesive joints (%) exposed to benzene[a]

Adhesive Substrate	Ethene–propene– dicyclopentadiene elastomer SKEPT	Butadiene elastomer SKB-X
Ethene–propene–dicyclopentadiene elastomer SKEPT	45.0/35.0	0/50.0
Polyisoprene elastomer SKI-3	0/33.3	31.6/40.0
Butadiene–α–methylstyrene elastomer SKMS-30ARKM	0/20.0	40.0/34.6

a   The value in the numerator refers to pure elastomer; the value in the denominator refers to the
   adhesive compound containing 6% of resorcinol–phenolformaldehyde modifier FE-12.

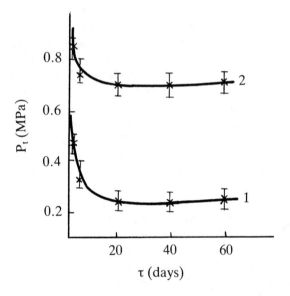

Fig. 2.43 – The strength of titanium adhesive joints produced with epoxy adhesive versus
exposure-to-water time. 1, $\sigma_{sa} \approx 50$ mN/m; 2, $\sigma_{sa} > 72.8$ mN/m.

The rise in adhesive joint strength is most pronounced in the case of aqueous solutions of
surfactants [639] and a similar pattern was observed for aliphatic alcohols (see Table 2.8)
whose surface activity is directly related to the length of hydrocarbon radical in its
molecule.

   Certain attempts were undertaken to give a quantitative assessment of the decrease of
adhesive interaction taking place in liquid media. These attempts considered, for example,
the dielectric permittivity of liquid media [616], or the solubility parameter of a polymer
[643] but they failed, however, to give an unequivocal interpretation of the results
obtained [615]. Nevertheless, increased polarity of the active medium does result in an
easier separation of the adhesive from the substrate [644]. Indeed, due to higher polarity

the rate of liquid penetration along the interface increases as compared to its diffusion rate in the adhesive [645]. In its turn diffusion processes lead to an accumulation of internal stresses and, consequently, to different fracture mechanics. On the other hand, it should be borne in mind that the liquid attacks the adhesive joint via an interaction that takes place in the region of compressive forces. This factor determines the S-shaped pattern of the rates of penetration of low molecular mass substances into the interphase zone [646], and commonly described as anomalous sorption. This may be attributed specifically to the hampered relaxational processes within the boundary layers of polymers.

In general, three phenomena are involved in the polymer–liquid medium interaction, namely adsorption (resulting in a decrease of interfacial energy), absorption (increasing the free volume of a two-phase system and providing for weaker intermolecular inter-action) and, ultimately, destruction of the interfacial bonds. Depending on the nature of a polymer, the contribution of each differs. For instance, the life-time and the relative variation in strength of polybutadiene samples immersed in fatty alcohols and water are related linearly to the interfacial energy [623]; the creep in water of the phenolphthalein-terephthalic acid polyarylate is governed by the absorption effect [647], while that of polyethyleneterephthalate by the adsorption effect which is due to the presence of surfactants [648]. Note that straining a polymer in an adsorption-active medium is somewhat different from straining it in air, the specific feature being that orientation of the polymer proceeds without a neck being formed [649] and low molecular mass substances can also become entrained [650, 651]. Increasing the deformation rate can drastically reduce the effect of the medium on the mechanical properties of polymers [652], the opposite effect being attributed to the changes in supermolecular structure of a polymer in the zone where it suffers maximum deformation [653].

Thus, these factors can lead to a decrease as well as an increase of adhesive joint strength in the presence of water. In a number of cases the molecules of water are bound to the functional groups of a polymer via strong valency and hydrogen bonds, therefore here the effect should be taken into account with regard to the influence it exerts not only on the interphase interactions, but the intraphase as well. Clearly, hydroscopic polymers are most liable to be affected by water; the strength of their adhesive joints may be anticipated to be a function of water content. Indeed, Gul' [654] demonstrated that the specific work of ply-separation of systems involving poly-ϵ-caproamide and ϵ-caprolactam–hexamethyleneadipate(hexamethylenesebacate) copolymers as polymeric substrates increased up to 1% water content and subsequently fell to zero. The duality of the effect produced by water with regard to the strength of the system was related by Comyn [655] to the diffusion modes of water in adhesive joints.

When discussing the diffusion modes of water it is often suggested that it penetrates into the system from the edges of the joint and subsequently diffuses along the interface. Such an apparently obvious explanation relies on the fact that the diffusion coefficients of water determined from the kinetic dependences of adhesive joint strength are several orders higher than those determined in sorption experiments [656]. Having examined the performance of the joints between titanium–aluminium alloy and a vapour deposited gold layer in a humid atmosphere Baun [657] has come to the same conclusion. By means of scanning electron microscopy and mass-spectrometry he detected that dis-integration of the system was due to the migration of water along the interface. The predominant penetration of water into the adhesive joints of polymers with metals in the lateral direction was independently observed for polyethene [645] and epoxy [658] adhesives.

However, we believe that this conclusion is true only when applied to the final stages of the process; it cannot be extended to cover the modes of interaction between adhesive joints and water in general. In fact, according to Brewis *et al.* [659], in a damp atmosphere the strength of aluminium constructions bonded with film epoxy adhesive is governed by the amount of water diffusing through the bulk of the joint, as is the case with adhesive joints affected by organic liquids (in this case the values for diffusion coefficients determined in sorption studies are quite close to those derived from the data on adhesive joint strength [660]. In an isolated film of epoxy adhesive exposed to water for the time period of 1000 h water clusters were detected [661]. The liquid penetration into the adhesive phase was so rapid that in some cases one of the substrates (glass) was torn away, while the thickness of the epoxy adhesive film increased by up to 12% [662]. These results correlate with the data derived from the calculated diffusion coefficients of tritium labelled water through epoxy adhesive joining aluminium [663]. Their validity was independently confirmed in the studies of the distribution patterns of moisture in carbon fibre-based composites with epoxy binder [664].

The data presented demonstrate that the most probable mechanism by which water affects the adhesive joint involves migration of water into the layer of adhesive and its subsequent distribution within the said layer. Hence, adhesive plasticization by moisture has to be taken into account, e.g. with such polymers as polystyrene [665], and also the rather rigid epoxy adhesives [666]. On the other hand, small amounts of water can be sorbed in the amorphous regions of an adhesive loosening its structure and intensifying interfacial interaction [654] as is the case with polyamides. Ultimately, the presence of water can lead to boundary layers with decreased cohesive strength being formed that unfavourably affect the strength of the system as a whole. The opposite effect (i.e. strengthening of the joint) is probable only for some distinct types of metal–polymer adhesive joints in which water intensifies interfacial interaction, e.g. polyamide is rather strongly bonded to aluminium due to the hydroxy functions being formed on the substrate surface [667].

Hence, liquid media can affect adhesive joints in either a favourable way, resulting in their being strengthened, or an unfavourable one, leading to their being weakened. Nevertheless, the solvolytic method is far from being superfluous since it depends upon the non-contradictory premise which relates to the balance of the adhesive and cohesive forces, which, as determined by the strength of the system, is affected by the active medium. In fact, water resistance of adhesive joints shows a satisfactory correlation with weathering data under the natural conditions of combined environmental factors [668], this approach being considered as more informative than the more common stress–strain tests of dry samples [669–671]. It is important to note when examining the adsorption of polymers that the solvent directly affects the interaction energy with the adsorbent and, in adhesive joints, this effect is much more pronounced because of statistical factors [672], hence the connection between this approach and thermodynamics.

Having analysed eq. (97) Schonhorn [673] suggested the following expression relating the thermodynamic work of adhesion between the substrate and the sorbed liquid W_{sl}^l and the work of adhesion at the three interfaces

$$W_{sl}^l = 2\sigma_{la} + W_{sl} - W_{sa}^l - W_{la}^l. \tag{300}$$

If $W_{sl}^l \approx W_{sl}$, the system is not subject to spontaneous separation into elements as the formation of the new interface is energetically unfavourable. When $W_{sl}^l \leqslant 0$ the effect of the liquid is that the adhesive separates from the substrate.

To conclude this discussion of the problem we may note that an investigation of the rules and features of the interactions between adhesive joints and liquid media provides the basis of modern developments in adhesives to be cured in the presence, for instance, of water [674] or oils [675] as well as the basis for working out effective substrate pre-bonding treatment procedures [626, 676]. These aspects are discussed in sections 4.2.1 and 4.2.2.

A much more common approach to investigation of the energetics of adhesive inter-action is to consider the strength of adhesive joints as related to surface characteristics of the bond elements. These characteristics may be either the overall parameters or may refer to individual components. While the second approach has been developed quite recently, the first one has been already adequately reviewed in the literature (see for example [329, 372, 677]), and hence, several examples will suffice to illustrate it. One should bear in mind, however, that there are three conditions (all three can be described thermodynamically) that provide for the direct relationship between the surface energy of a polymer and the strength of its adhesive joint [678]. There should be an interface separating the substrate and the adhesive; the wetting in the system should be incomplete and there is a third phase between the elements of the system.

Linearity between the adhesive joint strength and the surface energy is most impressive in the case of metallic substrates. Fig. 2.44 presents the corresponding relationships for seven metals bonded together with polyvinylbutyral (rigorously, the metallic surfaces should have been freshly prepared; however, in the examples being considered the presence of oxides and of other adsorbed impurities on the surface does not alter

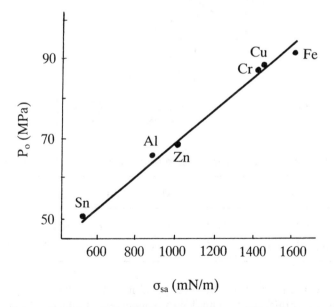

Fig. 2.44 — Conventional strength (reduced to zero thickness of the adhesive layer, which is polyvinylbutyral KA) of the adhesive joints of various metals versus substrate surface energy.

essentially the character of the dependence). To eliminate the effect of technological factors [122] a conventional parameter, reduced to zero thickness of the adhesive layer, was used to characterize the joint strength. For other metals such as Sr, Cd, Al, Zn, Au, Ag, Ir, Cu, Cr, Fe, and Ni Miedema and Nieuwenhys [679] obtained the dependence entirely similar to that in Fig. 2.44. The situation is the same in the case of an 'inverted' system in which a thin film (0.15 μm thick) of metal (aluminium) is coated by metalliz-ation in vacuum on to a polymeric substrate (polyethene, polytetrafluoroethene, poly(methyl methacrylate), polyethyleneterephthalate) [680].

The mechanism responsible for the surface energy–adhesive joint strength correlations involves the direct effect of a metallic surface on the properties of the contacting layers of a polymer, these properties being their energy (thermodynamic aspect) and their structural characteristics (molecular–kinetic aspect). Indeed, having measured σ_{sa} and the degree of crystallinity of polyethene coatings on various metals Schonhorn [681] demonstrated that these were determined by the nature of a metal-substrate (see Table 2.10). Obviously, these premises underly the influence directed by energetic factors on adhesive joint strength. Hence, the adhesive strength of polymeric coatings on metal surfaces is linearly related to the surface energy of the latter (Fig. 2.45).

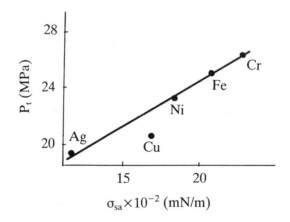

Fig. 2.45 – Tensile strength of tetrafluoroethene–hexafluoropropene copolymer films coated on to different metals versus substrate surface energy.

Table 2.10 – The effect of substrate on the properties of polyethene coating.

Substrate	Cu	Ni	Sn	Al	Cr	Au
σ_{sa} (mN/m)	37.4	51.3	53.8	54.8	56.1	69.6
Degree of crystallinity (%)	5.1	53.3	60.1	63.2	66.2	93.6

It should be borne in mind, however, that in real systems such correlations are suppressed by the influence of technological parameters of adhesive joint formation, of the geometry of the joint, of the adhesive layer thickness etc. that are not included in the majority of cases. In real systems the dependences discussed are not strictly linear;

however, their similarity and monotonic characteristics confirm the general conclusion about the strength of adhesive joints of polymers and their (polymer) surface energy being directly related to each other. Fig. 2.46 shows, as an example, the corresponding dependence for joints comprising a number of polymeric materials bonded with an epoxy adhesive [680].

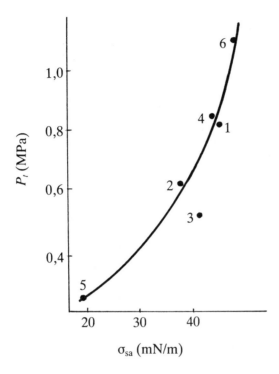

Fig. 2.46 – The strength of polymer adhesive joints produced with the epoxy adhesive versus surface energy of polymeric substrates. 1, polystyrene; 2, polyvinylfluoride; 3, poly-vinylchloride; 4, polyvinylidenechloride; 5, polytetrafluoroethene; 6, poly(ethylene terephthalate).

An important technological factor that tends to overshadow the influence of thermo-dynamic parameters on the adhesive joint strength versus surface energy dependences is that of the internal stresses arising within the adhesive on curing. The role of this factor is often overestimated, the correlation between surface energies of polymers and the strength of their adhesive joints even being claimed as impossible to establish. Neverthe-less, its influence can be limited by summing the stresses within the glue line in the region where they (stresses) are of maximum concentration with the resistance to failure of the system on the whole. Then, for a metallic substrate it appears possible to trace the dependence between the joint strength and the difference between the surface energies of a polymer at the interface with the substrate and with air. This dependence, obtained for a steel substrate with epoxy, polyurethane, polyester and polyacrylate adhesives [683], is depicted in Fig. 2.47. It is, as anticipated, close to linear and serves to illustrate the validity of concepts being considered.

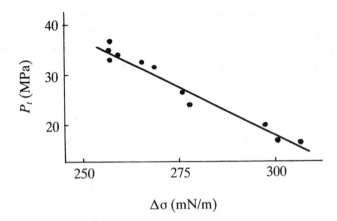

Fig. 2.47 – The strength of adhesive joints between different polymeric adhesives and steel versus the difference between surface energies of these adhesives at the interface with air and with the substrate.

The effect of technological factors can also be eliminated by expressing the strength of adhesive joints in units of tack. As tack can be regarded as instant adhesion, the corresponding assessment is independent of kinetic considerations relating to bond formation. Comparison of the values for the peeling force of various adhesive tapes from a number of polymeric and metallic substrates with the values for adhesives' (tacky layers) surface energies σ_Σ demonstrates that in the region of minimum difference between surface energies of substrates and tacky layers (adhesives) the ply-separation resistance versus $\Delta\sigma$ dependence is rectilinear (Fig. 2.48). σ_Σ is determined by summation of the corresponding characteristics of the separate components σ_i that are proportional to concentrations C_i of the components [684]

$$\sigma_\Sigma = \sum_i C_i \sigma_i. \tag{301}$$

Toyama and co-workers [685] and Ito and Kitazaki [686] demonstrated that the strength of adhesive tapes' attachment to various substrates was a function of the critical surface tension of the latter the position of the maximum corresponding to the equality between the energetic characteristics of the elements comprising a system [685, 686] even with variable failure conditions [687]. The maximum strength of such adhesive joints was found to correspond to $\Delta\sigma = 10$ mN/m [688]. It is therefore natural to consider that this effect is of general significance as it is seen in materials of very different nature. This effect is cited as an example of the fundamental principle of minimization, i.e. maximum adhesive joint strength corresponds to the minimum energy threshold at the adhesive–substrate interface provided that the influence of the factors of molecular–kinetic origin is eliminated. This problem is elaborated on in section 3.1.2; here we would like to stress that the function of parameter $\Delta\sigma$ [684] revealed here illustrates at its best the role of the thermodynamics of interphase interactions in adhesive joint formation.

At the same time, an unjustified exaggeration of the formalism of the thermodynamic approach can lead to contradictory interpretations of experimental data. For instance, regardless of whether a distinct polymer is used as the adhesive or as the substrate the strength of its adhesive joints ought to be the same. However, as observed in

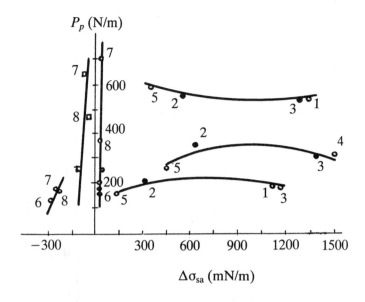

Fig. 2.48 – Peeling strength of the adhesive joints between various substrates and adhesive tapes based on polyvinylchloride (○), cellophane (□), and type 70 (●) versus the difference between surface energies of the tacky layers and substrates. 1, stainless steel; 2, steel, grade 08 KP; 3, aluminium alloy D-16; 4, brass; 5, copper; 6, polytetrafluoroethene; 7, poly-(methyl methacrylate); 8, unplasticized polyvinylchloride; 9, silicate glass; 10, polyethene; 11, plasticized polyvinylchloride; 12, paper.

polystyrene–poly(vinyl alcohol) systems, the peeling strength of the system involving polystyrene as the adhesive is no less than seven times greater than when polystyrene is the substrate [689]. This effect is obviously an illustration of factors of molecular–kinetic origin, and in particular it is due to the different intensities of the diffusion processes developing between the phases. Which of the elements of the PS-PVA pair is in the solid state and which is in the liquid state interacting with the substrate determines the relevant intensities.

The effect of technological factors on the character of adhesive joint strength versus surface energy of polymers dependences can be eliminated almost completely if surface energies of polymers are calculated on the basis of the molecular characteristics of the constituent monomers, as described above (see p. 98). We have tested this for α-cyanoacrylates, which today are most effective adhesives. For these the values of σ_{sa} were calculated according to eq. (221), which were then compared with the reported data on the strength of aluminium alloy adhesive joints [690, 691]. The strict linearity of the obtained graph (see Fig. 2.49, plot II) verifies the validity of this proposition [692]. On the other hand, as stated above, the adhesive ability of the samples must also be related to the cohesive characteristics of adhesives and as such, the elastic compression modulus E_{com} is a most reasonable property to select. According to Cheung et al. [693], it is more sensitive to the chemical nature of α-cyanoacrylates than even the glass transition temperatures of the corresponding polymers. In fact, the results depicted in Fig. 2.49, plot I presents direct evidence of the linear correlation between the calculated values of σ_{sa} and the measured values of E_{com} [691, 694]. The data compiled [84, 692,

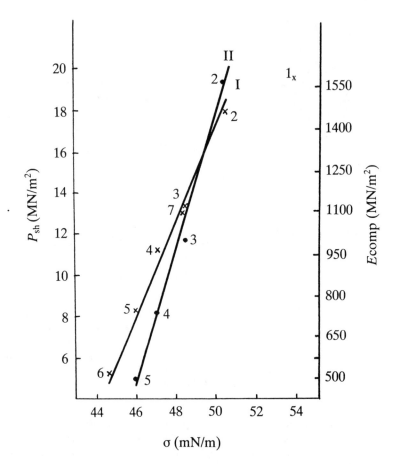

Fig. 2.49 – Calculated surface energies of polymerized cyanoacrylate adhesives versus compression modulus (I) and shear strength, of their adhesive joints with aluminium alloy D-16 (II). 1, methyl-; 2, ethyl-; 3, propyl-; 4, butyl-; 5, amyl-; 6, heptyl-; 7, allyl-α-cyanoacrylates.

694] are, probably today the most unequivocal illustration of the guiding role of the surface energy of polymers in their adhesive joint formation processes.

In practice, however, the strength of adhesive joints of polymers is commonly compared with the critical surface tension. In spite of the previously mentioned conventionality of σ_c, other corresponding dependences are also close to linear. Since first observed by Zisman [545], this fact has had adequate verification in the literature. This conclusion holds true for compatible pairs of polymers, as well as for substrates that are liable to form adhesive joints via mechanisms other than the diffusional one [695]. Hence, the dependences of adhesive strengths of corresponding systems on the critical tensions of their components [680] are close to but not strictly linear (Fig. 2.50). There is another example illustrating the objective character of such relationships. The critical surface tensions and adhesive joint strengths with steel of the ethyene—vinyl acetate copolymers display similar forms of the dependences on copolymer composition [696]. It is, however, significant that dependences of the type shown in Fig. 2.51 are valid only

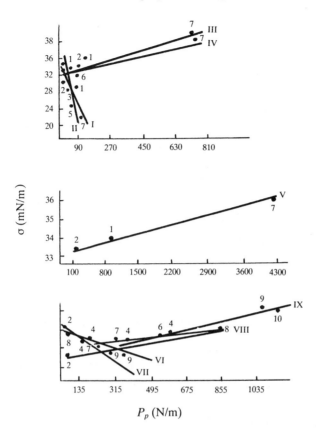

Fig. 2.50 – Peeling strength of various adhesive joints versus surface energy of polymeric substrates. Substrates: I, polyethene; II, polypropene; III, poly(hexamethylene adipamide); IV, cellophane; V, butadiene–acrylonitrile elastomer SKN-40; VI, aluminium, VII, zinc; VIII, copper; IX, steel. Adhesives: 1, SKN-40; 2, butadiene-styrene copolymer SKS-30; 3, polyisoprene elastomer SKI-3; 4, polyisobutene P-118; 5, polybutadiene elastomer SKB; 6, SKN-26; 7, SKN-18; 8, polychloroprene; 9, natural rubber; 10, chlorosulfopolyethene.

within a certain interval of copolymer compositions (in the general case these dependences are extremal as the strength of adhesive joints is affected both by the surface characteristics and by chain flexibility, i.e. by factors of a molecular–kinetic origin). Similar dependences were observed for composites based on modified epoxy oligomers reinforced with boron or pyrogenized carbon fibres, σ_c was shown to correlate with the product of the intralayer shear of the system and the extent of interphase contact [697], the regions of low and increased critical surface tension of the substrates being described by the dependences similar to those referred to in the studies of the water resistances of adhesive joints with metals. Such effects are of general significance [698], particularly for the purposes of predicting the strength of composites which is essentially governed by the σ_c of reinforcing fibres [699]. At the same time, when considering the role of this parameter, one should always keep in mind its limited physical implications, as emphasized in the previous section.

Hence, the energy characteristics of solid surfaces are directly related to the efficiency of adhesive interaction, particularly to the strength of corresponding adhesive joints.

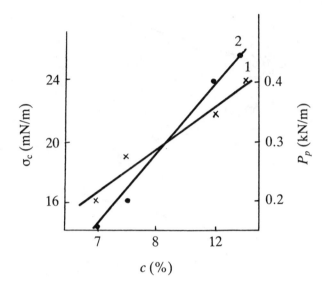

Fig. 2.51 – Critical surface tension of ethene–vinyl acetate copolymers (1) and peeling strength of their adhesive joints with steel (2) versus vinyl acetate content in the adhesives.

Quite obviously, surface energy appears to be the basic one of these characteristics. In no sense can the attempts to establish empirical dependences between for instance σ_{sa} and peeling strength be observed as useful. For instance, the suggested relationship between these parameters [700]

$$P_p = 40\sigma_{sa} - 1160 \tag{302}$$

is valid only for polyethene in the particular case of $P_p \leqslant 1450$ N/m. A more general approach involves derivation of the $P(\sigma)$ relationships, where P is the peeling (P_p) or shear (P_s) strength. To carry out the calculations most reliable data concerning the values of P_p and P_s were used. The values of the peeling strength were measured for the following systems: polyethene bonded with polyisobutene and butadiene elastomers [701], polypropene bonded with polyisobutene and butadiene–acrylonitrile elastomers [702], polyisobutene [703] and poly(hexamethene adipamide) [701] with natural rubber, butadiene–styrene, and butadiene–acrylonitrile elastomers, polychloroprene [704], poly(butyl methacrylate) [705], and butadiene–acrylonitrile elastomer SKN-40 (butadiene–co–acrylonitrile 60/40) [704] with polyalkenes poly(vinyl alcohol), poly-(methyl methacrylate), and poly(ethylene terephthalate), poly(ethylene terephthalate) with polyethene, poly(vinyl acetate), and butadiene elastomers [706]. The values of shear strength P_s were determined for systems comprising six elastomers butadiene, chloroprene, butadiene–styrene (SKS-30), butadiene–acrylonitrile (SKN-26), butyl rubbers and natural rubber in various combinations [707]. Each of the substrates listed was tested with at least four adhesives. In this way the majority of significant combinations between polymers varying in nature, polarity and relaxational spectra were used. The values of the surface energies of the adhesives and substrates were calculated with the aid of eq. (217).

By a computerized procedure a polynomial function involving groups ranging from the second to the fifth order was constructed to represent the $P = f(\sigma)$ function for the listed adhesive systems of 14 substrates. The coefficients of the independent variables were chosen so as to minimize the mean square deviation of the calculated values for P from the experimental ones provided for the best approximation.

Resulting from the operation of this procedure was a conclusion having a major significance — in all of the cases examined the functions that best fitted the P versus σ dependences were approximated by polynomials of second order

$$P = a - b\sigma + c\sigma^2. \tag{303}$$

It seems that polynomials of higher orders would have provided for a better coincidence of the $P_{\text{calc.}}$ and $P_{\text{exp.}}$; however, it is eq. (303) that gives the least sum of the squares of corresponding deviations. Eq. (303) is represented by a parabolic graph whose form corresponds to the most frequently observed changes of the adhesive joint strength as related to variations in different polymer characteristics.

To determine how the polynomial coefficients are related to the chemical nature of the materials it is reasonable to examine the reduced form of eq. (303)

$$P^x = 1 - b'\sigma + c'\sigma^2 \tag{304}$$

where $P^x = P/a$, $b' = b/a$, and $c' = c/a$. The 14 polynomials thus obtained are compiled in Table 2.11. What attracts attention is that the values of b' as well as those of c' do not differ much in the tests according to each of the two types of loading. This may be taken as an indication of the informal character of the computational procedure, which gives the form of the polynomials and their coeffients, that reflect the specific features of the structure of the materials and their adhesive interaction [709].

In the peeling tests the values of both coefficients vary within a relatively narrow range and are independent of the polarity of the polymeric substrates. This enables averaging of coefficients to be performed and resulting in the generalized relationship

$$P_{\text{p}}^x = 1 - 0.0583\sigma + 0.0007\sigma^2. \tag{305}$$

The situation is different with the shear measurements on the adhesive joints. The range within which the polynomial coefficients vary is almost as narrow as in the previous case; however, by separating the substrates into two groups, the low-polarity (polybutadiene, natural rubber, butyl rubber, SKS-30 elastomer) and the high-polarity (polychloroprene, SKN-26 elastomer) the range can be split into two narrower groups. These are $b'_{\text{lp}} = 0.06443$ to 0.07990 and $c'_{\text{lp}} = 0.00105$ to 0.00163 in the first case, and $b'_{\text{hp}} = 0.0526$ to 0.05414 and $c'_{\text{hp}} = 0.00070$ to 0.00098 in the second. The generalized relationships are then as follows

$$P_{\text{sh}}^{\text{lp}x} = 1 - 0.0687\sigma + 0.0012\sigma^2, \tag{306}$$

$$P_{\text{sh}}^{\text{hp}x} = 1 - 0.0534\sigma + 0.0008\sigma^2. \tag{307}$$

We believe that these features are related to the fact that the fracture of adhesive joints under shear is more susceptible to the extent of deformation of a substrate than when they are under a peeling load.[†] Indeed, in contrast to the peeling stress shear implies that

---

† This fact is in line with the concepts of mechanics, according to which peeling is related to the work of fracture, while the shear is related to the force or stress.

Table 2.11 – Adhesive joint strengths of various polymers expressed as polynomial functions of their surface energies

Substrate	Fracture type	$P^x(\sigma)$	a (303)
Polyethene	Peeling	$1 - 0.04222\sigma + 0.00047\sigma^2$	652.5
Polypropene	Peeling	$1 - 0.06482\sigma + 0.00099\sigma^2$	−49460.0
Polyisobutene	Peeling	$1 - 0.05298\sigma + 0.00094\sigma^2$	−3923.0
Poly(butyl methacrylate)	Peeling	$1 - 0.05532\sigma + 0.00052\sigma^2$	−667.4
Poly(ethylene tere-phthalate)	Peeling	$1 - 0.07701\sigma + 0.00150\sigma^2$	1135.0
Poly(hexamethylene adipamide)	Peeling	$1 - 0.05663\sigma + 0.00081\sigma^2$	5004.0
Polychloroprene	Peeling	$1 - 0.06204\sigma + 0.00058\sigma^2$	−912.7
Butadiene–acrylonitrile elastomer SKN-40	Peeling	$1 - 0.05568\sigma + 0.00004\sigma^2$	−211.4
Polybutadiene	Shear	$1 - 0.06519\sigma + 0.00107\sigma^2$	35060.0
Natural rubber	Shear	$1 - 0.06539\sigma + 0.00108\sigma^2$	11750.0
Butyl rubber	Shear	$1 - 0.07990\sigma + 0.00163\sigma^2$	13640.0
Butadiene–styrene elastomer SKS-30	Shear	$1 - 0.06443\sigma + 0.00105\sigma^2$	22090.0
Polychloroprene	Shear	$1 - 0.05414\sigma + 0.00098\sigma^2$	3025.0
Butadiene–acrylonitrile elastomer SKN-26	Shear	$1 - 0.05268\sigma + 0.00070\sigma^2$	8553.0

the substrate is first deformed and then separated from the adhesive. The increase in polarity of the substrate promotes adhesive interaction and consequently, reduces the difference between the two types of loading bringing eqs (305) and (307) into coincidence, thus supporting the objectivity of the computational procedure.

One can see that the values of coefficients c' in eqs (305)–(307) are rather small indicating, as we see it, that these equations can be considered to be generalizations of the commonly reported linear functions $P(\sigma)$ (see for example eq. (302)). Since the exponential term is neglected, these (linear functions) are limited both in their physical implications and in their practical significance. Apparently, the linear P versus σ relation for polymers is only an initial approximation. The introduction of an additional non-linear term (minute but still different from zero) complicates the task and calls for a more sophisticated, accurate and precise procedure to establish the relationship under discussion. In this regard the approach described is a natural step towards a quantitative treatment in energetic terms of the regularities of adhesive interactions.

Proving that the physical meaning of coefficient a is related to the chemical nature and structure of the substrate and to the type of loading presents a task of particular interest. It might be suggested that this parameter reflects the interfacial term of the adhesive joint strength as being dependent on the testing procedure (for polychloroprene, for instance, $a_{sh} \neq a_p$). If this assumption is true then coefficient a should clearly be related to the

Table 2.12 —Correlation coefficients between a and some physical characteristics of polymers

Substrate	Fracture type	V	T_g	K	A
All polymers	Peeling	0.751	0.485	0.565	0.521
Non-polar polymers	Peeling	−0.944	−0.874	0.931	−0.993
Polar polymers	Peeling	0.835	0.557	0.788	0.544
All polymers	Shear	0.511	−0.371	−0.195	−0.587
Low-polar polymers	Shear	0.814	−0.041	−0.168	−0.528

fundamental cohesion parameters of polymers. We have taken as such the van der Waals volume V [471], the glass-transition temperature T_g [509], and the constants K of the Mark–Houwink [490] and A of the Hildebrand [395] equations. The extent of interrelation between a and the parameters listed was quantified by means of correlation coefficients.

The data of Table 2.12 confirm that there is a certain interrelationship existing between a and the cohesion characteristics of polymers. In accordance with the considerations referred to above this interrelationship is closer when the joints undergo fracture by peeling. According to calculations the peeling test appears to be more informative than the shear one — this is not in contradiction to existing ideas and indicates that the approach suggested is soundly based. In this regard the high values of correlation coefficients in the case of peeling in systems with non-polar substrates are quite illustrative.

The results presented thus confirm the correctness of the approach described. Within the framework of the computational scheme developed [708] it is easy to determine the resistance of adhesive joints to peeling and shear relying on the values of a and σ. Application of the procedure may be illustrated by way of the following example involving systems other than those listed in Table 2.11. For the adhesive joints of polypropene ($\sigma_{sa} = 33.1$ mN/m) and of poly(ethylene terephthalate) ($\sigma_{sa} = 57.2$ mN/m) [395] bonded with polyisobutene ($a = -3923$) the relative resistance to peeling, as determined according to eq. (305), is

$$P_p^{PPx} = 1 - 0.0583 \times 33.1 + 0.0007 \times 33.1^2 = -0.1628$$

and

$$P_p^{PETx} = 1 - 0.0583 \times 57.2 + 0.0007 \times 57.2^2 = -0.0445.$$

Then

$$P_p^{PIB-PP} = -0.1628 \times (-3923) = 0.64 \text{ kN/m}$$

and

$$P_p^{PIB-PETF} = -0.0445 \times (-3923) = 0.17 \text{ kN/m}.$$

Taking into account the limited precision of P_p measurements (the error is not less than 0.01 kN/m) on the one hand and that the values of coefficients of the polynomial eq. (305) resulted from the averaging procedure on the other, there is actually close agreement between the calculated values and those reported [704]: 0.61 kN/m and 0.19 kN/m, respectively.

On the whole the range of the energy spectrum of adhesive interactions is rather broad and interfacial bonds of different types can be involved. For instance, the adhesive interaction between ethene-(acrylic acid) and (vinyl acetate)-(dimethylaminoethyl methacrylate) copolymers [709], or polyimides, or even copper [710], as well as between poly(vinyl chloride) and other metals [712] was shown to be due to ionic bonds (in the latter case ionic bonding occurs via formation of oxides and hydroxides on the substrate surface, as revealed by Mössbauer emission spectroscopy using ^{57}Co isotope [711]. However, in metal–polymer adhesive joints other types of high-energy interfacial bonds are also feasible. In the case of polyacrylates, bond formation is assumed to be due to the carbonyl functions [713], while in the case of polybutadienes, Chang and Gent [714] claim it to be due to double bonds, based on the fact that the work of peeling is directly proportional to the concentration of double bonds.[†] In systems involving epoxy adhesives and copper or tin, interfacial interaction is governed by donor-acceptor bonds involving unpaired electrons on the p- and d-sublevels of the atoms of metals [716]. This is also true for the adhesive joints of copper and its alloys with urethanealkydepoxy adhesive [717]. Formation of hydrogen bonds — and these are different from purely dipole interaction (contrary to what is asserted in [717]) — presents a particular example of bonds of this type [555].

Even in systems involving non-polar polyhydrocarbons in which the existence of only dispersion interfacial bonds could have been anticipated, adhesive interaction can be due to higher-energy forces.[‡] For instance, the peeling strength of autohesive joints of polybutadiene and of ethene–propene–diene copolymer (SKEPT) was shown to be linearly related to the fraction of the interfacial chemical bonds[§] evaluated by means of the Mooney–Rivlin equation [714]. Introduction of acrylic acid, as a third comonomer, into the ethene–butyl acrylate copolymer to the amount of only 1.5 mole % increases the adhesive adsorption to α–FeOOH 1.7-fold and to α–Fe$_2$O$_3$ 2.1-fold [720].

Hence, interfacial chemical bonds do appear to be quite common. Regarding the last of the examples listed it seems necessary to recall the concept of acid-base interaction treated in section 2.1.1. Reactions between a metallic substrate and an acidic adhesive can be described by the two schemes [721]:

$$-MOH + HXR \longrightarrow -MO(H)\ldots HXR \rightleftharpoons -MOH^+ \ldots {}^-XR, \qquad (308)$$

---

† Such conclusions can be inferred from a purely mechanical basis. When factors which affect adhesive joint failure are reduced by raising the temperature and reducing the peeling rate, Gent demonstrated the existence of chemical bonds across the polybutadiene–glass interface [715].

‡ This fact is associated with the formation of precursor films discussed in section 2.1.2. Adamson claims [718] that near the pressure of its saturated vapour water is sorbed on to the surfaces of low-energy polymers (polyethene, polytetrafluoroethene); hence the wetting proceeds over the new surface of the ice-type structure.

§ Such bonds are probably formed by co-vulcanization of polymers, via a diffusion mechanism. As a result, high-energy interfacial bonds are formed even in non-polar polydimethylsiloxane [719].

$$-MOH + XR \longrightarrow -MOH \ldots XR \leftrightharpoons -MO^- \ldots {}^+HXR, \qquad (309)$$

where M denotes the atoms of metal, X the atoms of oxygen, sulfur, nitrogen and, in eq. (309), of halogens as well. Equilibrium parameters of these reactions Δ are determined by the values of dissociation constants (expressed as pK) and isoelectric points I_e [722]:

$$\Delta_{(308)} = I_e - pK_{(308)} = \log \frac{[MOH^+] \, [^-XR]}{[MOH] \, [HXR]}, \qquad (310)$$

$$\Delta_{(309)} = pK_{(309)} - I_e = \log \frac{[MO^-] \, [^+HXR]}{[MOH] \, [XR]}. \qquad (311)$$

At $\Delta_{(308),(309)} \gg 0$ interfacial interaction is so intensive that it can even result in metal corrosion. However, when Δ is negative the adhesive and the substrate are kept together predominantly by dispersion forces. This is the case with polyimides and manganese, copper, aluminium and even silicon oxides [722]. Fowkes [557] examined the silicon oxide–poly(methyl methacrylate) system by using eqs (249) and (250) and the interfacial energy due to acid-base interaction was shown to amount to 87% of the total work of adhesion. The contribution of dispersion bonding can be increased by varying the chemical nature of the contacting materials.

Dispersion forces were shown to add to the overall adhesive interaction even in systems specifically designed to ensure the formation of electrostatic bonds. These are the powdered xerographic materials based on epoxy oligomers, poly(vinyl chloride) and polyamides [723]. At the same time, in the adhesive bonds between elastomers for which, in a majority of cases, adhesive interaction is mainly due to dispersion forces (provided the elastomers lack reactive functional groups) the contribution of electrostatic forces reaches 15% [724]. Today quantization of the contributions from both types of forces in the overall efficiency of the adhesive interaction can be carried out only for the simplest cases.[†]

One of such simple cases providing for a quantitative description involves adhesion of finely dispersed polymeric powders to solid substrates. Treating an adhesive joint similarly to a condenser (capacitor) for which the electrostatic forces is defined as

$$F_e = 2\pi \rho_e^2 A \qquad (312)$$

(ρ_e is the charge density, A is the contact area), Derjaguin et al. [726] found that for a spherical particle of radius r the electrostatic force is given as

$$F_e = 2\pi^2 \rho_e^2 r A. \qquad (313)$$

On the other hand, the contribution of van der Waals forces is

$$F_m = -Hr/6r_{\min} \qquad (314)$$

---

† The general solution of the problem concerning the determination of the fractions of inter-molecular and chemical bonds from fracture studies of adhesive joints involves the concepts of fracture mechanics which are not always unequivocally interpreted [725].

where H is Hamaker's constant. Then the ratio between the contributions of both components to the adhesive interaction can be expressed via Young.'s (E) and Poisson's (ν) moduli as

$$F_e/F_m = 18.654\,(1 - \nu^2)^{2/3}\rho_e^2(r \times r_{\min}^2/E^2 H). \tag{315}$$

Numerical calculation shows that for spherical particles of polystyrene ($r = 10\ \mu m$) with charge density $\rho_e = 3.34 \times 10^{-2}\ C/m^2$ the contribution of van der Waals forces to overall adhesive interaction is only around 0.01%. However, for ionic crystals the contributions of both components are comparable [727].

Hence, the van der Waals forces, particularly the dispersion ones, are an obligatory component of the energy spectrum of adhesive interaction. Therefore, their effect and contribution must be treated with the utmost care and attention.

Direct data that were free of the influence of the technological factors of adhesive joint formation were obtained for metal-coated polymers. What is important is that in these systems adhesive interaction involves the entire range of interfacial contact (resulting in what may be called full interfacial contact), an aspect discussed in detail in section 3.2.3. By means of vacuum vaporization Yoshito and co-workers [728] coated a number of polymeric substrates with aluminium; the strength of the systems evaluated by the scratch technique was found to be due to dispersion forces. Studies with systems involving polymers galvanically coated with other metals such as copper, gold, and silver verify this conclusion [729]. The results of other procedures used to produce adhesive joints with these metals are similar. For instance, interfacial interaction between poly-urethane elastomer and glass coated glass microspheres was shown to be governed by dispersion forces [730]. There were also no interfacial chemical bonds detected in the high-strength [38 MPa] selenium-coated polycarbonate system, the experimental methods having been those of impact debonding and X-ray desorption from the interface [731].

The contribution of dispersion forces to the surface energy of various polymers reaches 90–95% even in those special cases initially designed for bonds of higher energy to be established. This is just as true for polymer blends [312]. There are certain exceptions however, and these are the sufficiently polar polymers, e.g. poly(vinyl acetate), poly(methyl methacrylate), and cellulose triacetate for which the contribution of dispersion forces is, in agreement with anticipations, appreciably lower [561], and also the copolymers of non-polar and highly-polar monomers [578]. In the case of graft-copolymerization the surface energy is greatly affected by the thickness of the grafted (outer) layer. For instance, when acrylonitrile and styrene are grafted together on to a polytetrafluoroethene surface the σ_{sa} versus thickness of the grafted layer dependence is most pronounced in the range of 60–80 nm [732]. The effect is probably due to factors of molecular–kinetic origin.

Summing up the data presented we may claim it possible to control the energetics of adhesive interaction[†] by varying the chemical nature of polymers.

[†] Within the confines of the monograph we do not examine here other approaches which lead to the same objective. Among the 'other' approaches those based on loading mechanics are worth looking at. For instance, in adhesive joints of polyimide films mutual compensation of the dispersion and polar components was observed at stress values around $5 \times 10^7\ N/m^2$ [160]. The effect may be attributed to the partial deconjugation of the cyclic system via sp^2-sp^3 transitions and is not associated with the rearrangement of crystalline microphases within the adhesive.

Varying the content of polar groups appears to provide the most obvious means of affecting the values of distinct components to the surface energy of high-molecular mass compounds. For instance, Saito and Yabe [733] determined as 35% the acetylation degree of the formylated polyvinylformal corresponding to the minimal σ_{sa}^p value, i.e. to the maximum value of σ_{sa}^d and the data were similar for acetylcellulose and cellophane [734]. In ethene–vinylacetate copolymers the polar component to the surface energy σ_{sa}^p was shown to be linearly related to the vinyl acetate content up to 30% [734] (at 35% vinyl acetate content the increased stiffness of the macromolecule reduces the strength of adhesive joints to a minimum [735]. For melts, due to the specific features of their spreading over the substrate's surface, the limiting value is as low as 14% (see Fig. 2.51) [696]. The surface energy of polystyrene is actually determined by the dispersion forces alone [736]; however, its sulfonation to the sulfonic ions content of 2.24×10^8 ions/m^2 (as determined by radioactive isotopes) results in that their (dispersion forces) contribution is reduced by 10%, while simultaneously σ_{sa}^p rises from 0.5 to 27.8 mN/m [737]. Similar results were obtained for copolymers of styrene with p-styrenesulfonate [738].

The decrease in concentration of polar groups results in the opposite effect as confirmed by the data of Ko, Ratner and Hoffman [739] who radiation grafted a mixture of ethyl- and 2-hydroxyethyl methacrylates on to polyethene and, relying on the data on their wetting with methylene iodide, evaluated the dispersion and polar components of the surface energy by way of Wu's scheme. In agreement with expectation the dependence of the polar component to the surface energy on the content of the more polar comonomer shows a discontinuity at 20% content of 2-hydroxyethylmethacrylate σ_{sa}^p rising sharply to the higher level (Fig. 2.52). Similarly, thermal treatment of the Hercules HT-S carbon fibres for an hour at 730 K in a hydrogen atmosphere results in a two-fold decrease of σ_{sl}^p, while in vacuum (under residual pressure of 0.133 MPa) the

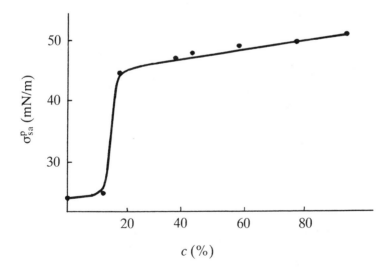

Fig. 2.52 – Increment of the polar component to the surface energy of ethyl- and 2-hydroxyethyl methacrylate copolymers radiation grafted on to polyethene versus the polar comonomer content.

decrease is three-fold, at the same time σ_{sl}^d rises by 60 and 50%, respectively [576]. Another way to solve the problem involves the blocking of polar groups by substrate modification. The treatment of Thornell 400 fibre with hexachlorobutadiene leads to a four-fold decrease of σ_{sl}^p and a 40% rise in σ_{sl}^d [576].

The prevalent effect of dispersion forces in adhesive interaction even for a highly polar adhesive is convincingly demonstrated by a simple calculation relating to the system comprising polyethene and epoxy polyamide composition. The contact angle characteristic of the system is 35.4° [741]. According to Barbarisi [740], the surface energy of the adhesive composition σ_{sa} is 41.7 mN/m. If one neglects the effect of surface pressure, eq. (261) can then be used instead of eq. (275) to carry out the calculation. The product then is $\sigma_{sa}^d = 34.3$ mN/m, which is in good agreement with the experimental data [741].

It appears that the suggestion of the predominant contribution of dispersion forces to overall adhesive interaction must be incorrect for such polar substrates as, for example, cellulose (for which $\sigma_{sa}^d = 9.2$ mN/m [742]. This indeed is the case though, only for adhesive joints at the equilibrium state. Detailed investigation of the modes of wetting of paper and of parchment revealed that at least at the start of the process and until the plateau on the wetting kinetic curves had been reached, i.e. within the first 20 s of the wetting process, adhesive interaction was due to dispersion bonds alone [743]. Beyond the 20 s time-span the dependence deviates from linearity (Fig. 2.53).

The results discussed are not confined merely to illustrating the various aspects of adhesive interaction. They are of general importance since a comparison of the strengths of adhesive joints with the magnitudes of the distinct components of the surface energy of adhesives provides the means of distinguishing the type of interfacial interaction which provides the bonding. For instance, a knowledge of σ_{sa} and σ_{sa}^i for polytetrafluoro-

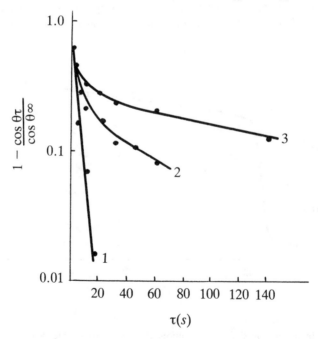

Fig. 2.53 – Time dependent variation of the contact angle with parchment at $\cos \theta_\infty$ values of 0.5 (1), 0.6 (2), and 0.7 (3).

ethene, polyvinylchloride, polyvinylidenechloride, polyvinylfluoride, polystyrene and polyethyleneterephthalate on the one hand and the strength of their adhesive joints on the other was sufficient to draw a conclusion concerning the energetics of adhesive joint formation in these systems [695]. Moreover, polymers can be distinguished with regard to their σ_{sa}^i values from a knowledge of their refractive indexes [506]. This is demonstrated in Fig. 2.54 showing rectilinear dependences between surface energy components and the refractive indexes of various polymers.

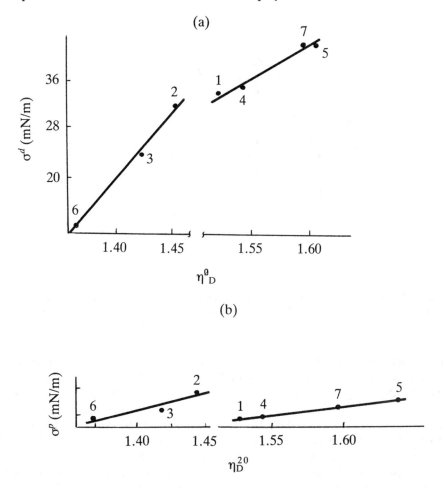

Fig. 2.54 – Dispersion (a) and polar (b) components of the surface energy versus refractive index. 1, Polyethene; 2, polyvinylfluoride; 3, polyvinylidenefluoride; 4, polyvinylchloride; 5, polyvinylidene-chloride; 6, polytetrafluoroethene; 7, polystyrene.

The approach described, used for the first time to its full extent by Raevsky and Pritykin [695], was further developed in a number of subsequent studies. Arslanov and Ogaryov made us of the graphical method of Kaelble to determine the values of σ_{sa}^d, σ_{sa}^p and σ_{sa} for aluminium subjected to alkaline etching, acidic oxidation and anodic oxidation (anodization) [744]. The values determined were related to the strength of adhesive joints produced with the aid of (butyl methacrylate)/(methyl methacrylate) (95/5) copolymer as adhesive. Comparison of the energetic and mechanical characteristics shows that the contribution of dispersion forces in the overall adhesive interaction is constant regardless of the substrate pretreatment procedures. However, $P(\sigma_{sa})$ and $P(\sigma_{sa}^p)$ are parallel to each other (Fig. 2.55). Consequently, in these cases polar interfacial forces predominate. The intersection point of the σ_{sa}^p graph with the ordinate axis gives the contribution of dispersive interaction to the strength of the system (2.5 MPa while

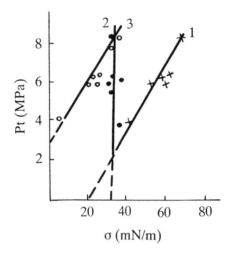

Fig. 2.55 – The strength of adhesive joints of aluminium with AKF-113F acrylic adhesive versus surface energy (1) of aluminium and its disperson (2) and non-dispersion (3) components varied by means of different pretreatment procedures.

the maximum value for the peeling strength P_p is 8.2 MPa). The data presented demonstrate that the peeling strength of adhesive joints is, within a broad range of values, governed by the energetic characteristics of the surface of aluminium [745], actually determined by the surface treatment procedure. Anodizing was found to be most effective in this respect [744]. Having chosen the film thickness d as a measure of the efficiency of the treatment procedure Arslanov and Ogaryov [744] succeeded in demonstrating that d is related linearly to the magnitude of the dispersion component of the surface energy of the substrate, while σ_{sa} versus d and σ_{sa}^p versus d dependences exhibit maxima corresponding to the maximum electrial resistivity (Fig. 2.56). Consequently, in spite of the influence of electrostatic effects, the efficiency of the pre-bonding treatment of an aluminium surface is determined directly by the contribution of dispersion forces. However, when acidic oxidation is the pretreatment procedure polar interfacial bonds have the major effect on the strength of adhesive joints.

This is one of a number of most informative correlations which help in understanding the nature of interfacial interactions. The examples surveyed are merely a fraction of the data supporting and proving its validity and usefulness. We believe that at the present stage of the art in the physical chemistry of adhesion this route is the most promising and simple to use.

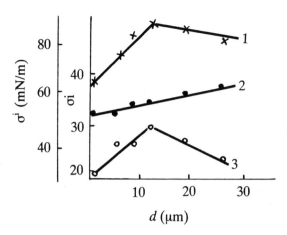

Fig. 2.56 — Surface energy (1) of aluminium, its dispersion (2) and non-dispersion (3) components versus thickness of oxide film produced in different pretreatment procedures.

Thus, these methods enable a sufficiently reliable assessment of the energy spectrum of adhesive interaction between adhesives and substrates of different nature to be made. Corresponding data are extremely valuable for the thermodynamic analysis of the features of adhesion. In fact, thermodynamics of interfacial processes acquires implications relating to reality only when one has to consider certain values for the energies of adhesive joint formation. Then it (thermodynamics) will become a useful tool in solving a variety of problems, particularly applied ones. Indeed, the targeted development of adhesive compositions and the substrate pre-treatment procedures whose application is likely to provide most favourable energetics of adhesive interaction are tasks of considerable practical urgency. These aspects will be discussed in section 4.2.

Summing up the treatment of polymer adhesion on the basis of thermodynamics it should be emphasized that, in spite of the fact that physically correct ideas (as applied to real systems) are difficult to accommodate, corresponding analysis based on a limited number of sufficiently reliable data makes it possible to understand, at least qualitatively, the general picture of the phenomenon of adhesion. At the same time, within the framework of this approach some factors related to rheological effects, the existence and specific properties of boundary layers in polymers, the effect of interfacial contact area etc. cannot be accounted for. Consequently, the general analysis of the theoretical modes of adhesion naturally implies the necessity to supplement thermodynamic concepts with molecular–kinetic ones.

3

Molecular–kinetic approach to polymer adhesion

Within the framework of the molecular–kinetic approach it is possible to consider the specific features concerning the nature of the contacting polymers (their composition and structure) and the interfacial zone in regard to the fundamentals of the formation and behaviour of adhesive joints. In this respect the molecular–kinetic approach appears to be a necessary supplement to thermodynamic analysis of the adhesion phenomenon. At the same time, its separate contribution is to provide the possibility of elucidating and specifying the role of technological factors (processing parameters) in the adhesive interactions of macromolecular compounds.

3.1 THE CONCEPT OF THE INTERFACE IN POLYMER ADHESIVE JOINTS

3.1.1 Boundary and transition layers in polymers

Within the bounds of the molecular–kinetic approach concepts of the surface of a condensed phase and of the interface between the condensed phases acquire real physical meaning, whereas Gibbs' approach discussed in section 2.1.1 is based, strictly speaking, on formal considerations. Generally, the surface may be considered as the limit of the spatial continuity of some phase state defined by its property.[†] Hence, the zone between the condensed phases, polymeric in particular, represents a region displaying a complex of intrinsic physical characteristics. In accordance with the terminology accepted in molecular physics [22, 747], the region between the bulk of a condensed phase and its geometric surface should be termed 'the transition layer', while the portion of it directly adjacent to this surface is designated 'the boundary layer', the latter being adsorptive

---

† The extreme expression of this position is manifested in the thesis that the concept of a surface, in the strict meaning, cannot be applied to describe the boundary of a real continuum of matter [746].

in nature. Such differentiation enables a more definite understanding of the current term 'surface region', distinguishing within it the zones that are different in their physical origin and, hence, in their properties.

Indeed, van der Waals regarded the interface, even in the most simple liquid–gas system, as a layer of finite thickness whose density decreases from ρ_l at the liquid side to ρ_g at the geometric interface. Rigorously, such a conclusion is true only for the temperature region near the critical point (as was noted by Frenkel [166]), however, it can also be derived from the lattice model of a liquid. The analysis of molecular distribution functions shows the density change of a condensed phase in the transition layer to have a step-wise oscillatory character [748], the oscillations being gradually attenuated as the liquid phase is penetrated, with the period close to the mean inter-molecular distance [749]. Such discreteness is confirmed by the results of optical measurements [750]. In fact, the reflected light is plane-polarized only in the case where the refractive index changes discontinuously from 1 to n, while it is elliptically polarized if the density change is gradual and smooth. Such effects are related to the actual thick-ness of the real zones between the contacting phases.[†] In the general case the thickness must be finite, increasing with the rise of temperature up to infinity at the critical point [753]. Hence, the reasons for the complex character of the dielectric profile in liquids near the interface become clear [754].

Such effects are more noticeable in polymeric materials. In this case the gradient in optical and dielectric properties of the phase over its thickness (as moving away from the geometric boundary) can be detected in refractometric studies. For instance, Soushkov et al. [755, 756] recorded the refractive index profile in PET films using the highly sensitive waveguide technique [755] and the method of White and Heidrich [757]. The results unequivocally demonstrated the existence of transition layers (Fig. 3.1)

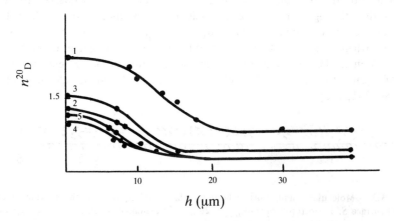

Fig. 3.1 – Refractive index profile in polyethylene terephthalate (PET) film after treatment with p-chlorophenol in dioxane (1), acetone, (2), methanol (3), methylene chloride (4) and dichloroethane (5).

[†] For example, initiation by even the slight temperature gradient of surface chemical reactions [752] is a consequence of the actual existence in ionic crystals of a surface layer in which deviations from stoichiometry and unbalanced spatial electric charge are observed [751].

irrespective of external effects, e.g. different chemical solvents. Taking into account the constant constitution of the polymer sample, the increase in refractive index quite obviously indicates an increase in the density of these layers. Then, the strict linearity of the $n_D(\rho)$ function (Fig. 3.2), observed in ellipsometric studies [758], confirms the validity of this conclusion.

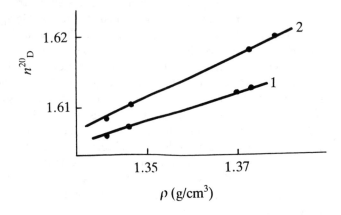

Fig. 3.2 – Refractive index of PET films versus density. Refractive index determined by ellipsometry (1); by internal reflection spectroscopy at 943 cm^{-1} (2).

In accord with expectations, boundary and transition layers are most evident in systems which contact solid substrates. Possible types of molecular orientation in boundary layers in liquids are shown in Fig. 3.3. Each type of orientation is characterized by a light absorbance coefficient of its own (α). Moreover, direct measurements performed with the 0.15 μm thick layers of nitrobenzene on a quartz surface [759] disclosed that the light absorbance coefficients corresponding to the three types of orientation were interrelated as follows: $\alpha_h < \alpha_i < \alpha_p$. Similar effects can be even more impressive when the contacting phases exhibit liquid–crystalline properties [760]. However, even in the case of simple liquids the substrate surface extends its effect to a depth of 3 μm into the liquid. This effect is due to the long-range character of surface forces [84, 761, 762].

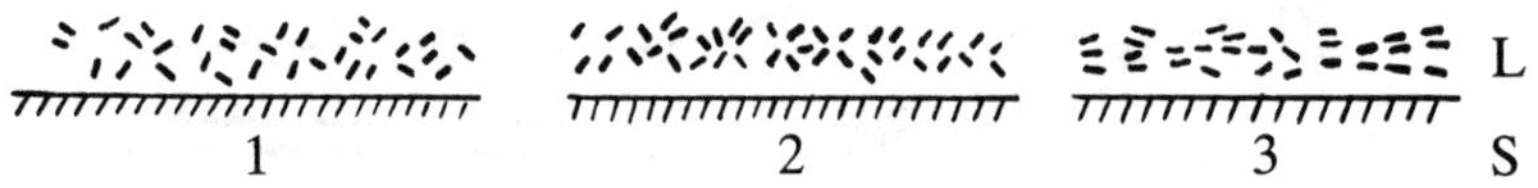

Fig. 3.3 – Molecular orientations in boundary layers of liquids L at the interface with a solid surface S. 1, Isotropic, $\alpha_i = (\alpha_{max} + 2\alpha_{min})/3$; 2, homeotropic, $\alpha_h = \alpha_{min}$; 3, planar, $\alpha_p = (\alpha_{max} + \alpha_{min})/2$.

For instance, close to a polished quartz surface a significant decrease of the viscosity of benzene [763] and dibutylphthalate [764] was observed, the thickness of the boundary layer in dibutylphthalate being as high as 0.5 μm, in the nitrobenzene–glass system the specific heat capacity of the liquid was found to be an extremal function of the distance from the solid substrate [765]. Clearly, the effects imposed by the solid surface are greater the higher the molecular mass of the contacting liquid. Indeed, for the homologous series of alkoxyalkanes of the general structures

$$H_3C-\left[-CH-CH_2-\atop |\atop OC_2H_5\right]_x \begin{array}{c}-CH-CH_2-CH-OC_2H_5\\ |\qquad\qquad|\\ OC_2H_5\qquad OC_2H_5\end{array}$$

and

$$H_3C-\left[-CH-CH_2-\atop |\atop OC_4H_9\right]_y \begin{array}{c}-CH-OC_4H_9\\ |\\ OC_4H_9\end{array}$$

a regular rise of viscosity in the boundary layer η_1 as compared to that in the bulk η_2 was observed [766]

	x or y	$\Delta\eta = \eta_1 - \eta_2$
x	1	0.80
	2	1.25
	3	2.50
y	1	1.10
	2	1.50

Then, from general considerations it becomes clear that in polymers, for which x and y are much greater, one should anticipate much more substantial differences in the properties of the bulk phase and of the boundary and transition layers. In this case viscosity anomalies were detected at distances up to $500\,\mu m$ from the solid surface [767]. Development of instrumentation, e.g. from centrifugation [768] to the blowing-off technique [769], made it possible to distinguish even more precise details. For example, a drastic decrease of viscosity in layers within $0.2-0.3\,\mu m$ of the solid surface was observed for liquid polydimethylsiloxanes of narrow molecular mass distribution $(1.2-1.5)$ and with viscosity of $(0.05-20.0) \times 10^{-4}\,m^2/s$, when they were layered on to a polished steel substrate. In layers of up to $1.5-2.0\,\mu m$ thick the viscosity exceeds its value in the bulk by 30–40% [770].

Such features are believed to be related to the rigidity of macromolecules and to the intermolecular interactions between them; at least three flow rate profiles being distinguished for a polymeric liquid of a given composition, depending on the cohesion energy of the liquid [771]. It is thus clear that the effect of a solid substrate depends on the molecular mass of a polymer. Cohen and Reich [772] have recently estimated the role of this factor using a birefringence technique. For low-molecular mass polystyrene (MM = 800) the ordering effect induced by a glass surface extends for not greater than $1\,\mu m$, while for the polymers with molecular mass of up to 10^5 this distance increases by no less than an order of magnitude. The studies of viscosity anomalies in polymer solutions flowing through porous media have disclosed that the effect of the solid surface extends to a distance which is less than the characteristic linear dimension of the layer [773].

The features described verify the validity of the concepts concerning the role of the surface layers as applied to polymer adhesion. Fig. 3.3 illustrates the conventional classification of these layers. On the other hand, the non-linear character of these dependences reflects the fact that the structure of transition zones is quite complex,[†] especially of those involved in adhesive contact, where the transition layers transfer the external stress applied to the system from the adhesive to the substrate. Thus the reasons for the different performances of the boundary and transition layers of high molecular mass compounds should be made more precise. While the latter can be described by a consideration of the complex of intrinsic characteristics, nevertheless, for the mechanical properties of adhesive systems, it is the specific features of the boundary states of polymers which govern the efficiency of interfacial interaction with the surface of the other phase. Taking into account that a molecule at the boundary is subjected to the effect of only three forces (instead of the 'usual' four) it seems natural to suggest that a macromolecule within a boundary layer is held less tightly than within a transition layer and, especially, in the bulk of the phase. Otherwise, the adhesive would have been inert, in a molecular–kinetic sense, with respect to the substrate and the adhesive interaction must be limited only by the balance between the surface energies of the contacting phases. To overcome this 'inertness' adhesives are used in the form of solutions or melts. However, we believe that there should be some fundamental grounds for the molecular–kinetic activity of the boundary layers in polymers, these being free of the influence of technological factors.

To prove the statement concerning the less tight bonding and, hence, the more loose packing of the macromolecules in boundary layers let us examine the data obtained by high-resolution electron energy-loss spectroscopy. This method was first introduced for the investigation of the surface of polyethene films under non-destructive conditions [775]. It was found that at a distance of up to 0.1 μm, i.e. at the depth of a monomolecular layer, the state of the polymer can be described as that of 'condensed gas'. Another proof of this premise was obtained in the studies of polymers brought into contact with various metals. X-ray photoelectron spectroscopy was used to evaluate the contact area, while the amount of polymer having adhered to the substrate during contact with the bulk polymer sample was detected by mass-spectrometry [776]. It was demonstrated that complete coverage of nickel and, especially, of tantalum with poly(methyl methacrylate) was achieved at a layer thickness that was close to the size of the macromolecular coil (Fig. 3.4). Earlier mass-spectrometric data indicated that only 3% of the adhered macromolecules sustained chain scission [777], i.e. the transfer of macromolecules to the substrate during contact could not be attributed to destruction processes (these aspects will be thoroughly examined at the end of this section). Hence, the data of Fig. 3.4 support the existence of the loosely bound macromolecules (with the neighbouring chains from the inside as well as the lateral neighbours) in boundary layers.

When the boundary layers of an adhesive interact with a substrate a certain ordering of the corresponding macromolecules occurs, and this appears to be the most fundamental result of the interaction. Then, regardless of the composition of the phases in contact and of the contact conditions the efficiency of this interaction can be evaluated in terms of entropy. Generally, the entropy gain which accompanies the transfer of

---

† Chalykh [774] suggested that the transition layer in polymers be regarded as a superposition of the three zones, diffuse, interfacial and structurally-inhomogeneous. The latter two are characteristic for incompatible systems or for those at phase equilibrium; in all other cases the first of the trio must be taken into account.

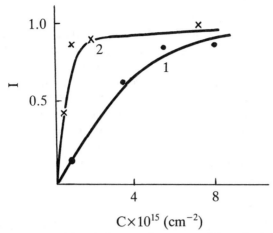

Fig. 3.4 – Relative intensities relating to carbon atoms of the main chain observed in photo-electron spectra of PMMA ($\overline{M}_n = 2.6 \times 10^6$) having adhered to the surface of nickel (1) and tantalum (2) versus amount of adhered PMMA expressed as monomeric unit concentration per unit area.

molecules from the bulk to the vicinity of the geometrical surface within which its orientating effect increases, is determined, according to Frenkel [166], by the ratio of the mean frequencies of thermal vibrations:

$$\Delta S = N_s \ln (\overline{\nu}_v / \overline{\nu}_s) \tag{316}$$

where N_s is the number of molecules in the transition and boundary layers, and subscripts v and s refer to the bulk of the phase and its surface, respectively. For experimental evaluation of the entropy change ΔS Lipatov [778] proposed an ultrasonic technique, having related ΔS to the amplitude of high-frequency vibrations. The latter were expressed previously in the form of the probability integral by calculating the inter-mediate functions [779]. Hence, entropy decrease was established as accompanying the macromolecules' transfer from the bulk of the phase into transition and boundary layers, e.g. for poly(methyl methacrylate) $S_v = 0.107$ and $S_s = 0.093$ W s/(cm^3 deg) [778].

These results demonstrate the orientating effect of a surface, giving rise to certain ordering of macromolecules (segments) when they acquire a less probable non-equilibrium state. The resultant difference in characteristics of the bulk and of transition layers of polymers is of prime importance for a molecular–kinetic interpretation of any surface phenomenon in polymers and adhesion in particular. It is therefore reasonable to discuss the mechanism accounting for this difference in the properties of the bulk and the transition layers in polymers.

The thermodynamic and molecular–kinetic distinctions of the boundary and transition layers from the bulk of the phase demonstrated above should obviously result in substantially different free volumes for the corresponding regions. Curro, Lagasse and Simha [780] considered the 'hole', as the most common (and obvious) type of single n-defect, to be the diffusible carrier of the free volume. However, due to reasons of energetic origin such defects are liable to form conglomerates of 10–20 elements. Hence, such clustered systems are assumed to be the carriers of excess (free) volume [781, 782]. It is the mode of movement (propagation) and the life-time τ_{cl} of these clusters that control the diffusion mechanism of volume relaxation in amorphous polymers. Diffusion of such clusters, characterized by the fact that the τ_{cl} values are limited, leads to the formation of a more compact transition layer of macroscopic thickness (of the order of magnitude stipulated by $\sqrt{D\tau_{cl}}$, where D is the diffusion coefficient), as, during the specified time τ_{cl}, the layers of the bulk phase are unable to get rid of the excess volume thus reducing the free volume in the regions adjacent to the interface [450].

The resulting differences can be demonstrated by various methods [783]; the optical one, as mentioned above, providing the most direct data. For instance, the different free volumes of the bulk and of the transition layers result in a phase difference Δf_d of birefringence. The relationship between this parameter and the angle of incidence β is given by the simple expression [784]

$$\Delta f_d = b(1 - \cos 2\beta). \tag{317}$$

Its validity is supported by the linear plots in Fig. 3.5 [785]. Hence, it is clear that it is the coefficient b of eq. (317) that is susceptible to the thickness d of transition layers, this thickness being determined by the 'saturation level' of b (Fig. 3.6). The thickness values obtained by this method are sufficiently correct, being markedly different, in agreement with the expectations, for structurally different polymers, namely linear, cyclic-chain and network types. Listed below are the corresponding thickness values in μm:

Polystyrene	1.2^a; 25 [784]; 45^b [786]; 90^c [787]
Poly(α-methyl styrene)	4.4^a [784]
Poly(methyl methacrylate)	0.9^a [784]
Poly(vinyl butyrate)	73^b [786]; 160^c [787]
Polyimide	20^a [784]
Epoxy systems based on oligomer ED-20 cured with m-phenylenediamine	500–600 [785]

a   Minimal thickness of anisotropic transition layer with planar chain arrangement at zero value of the polar angle. In [450, pp. 159–160] d is treated as an overall thickness; the values of this parameter in SI units are erroneously increased by an order of magnitude.

b   These values were derived from the graphs in references (Fig. 3 [784] and Fig. 4 [786]), according to [785] as the half-thickness of the film sample. The necessity for this manipulation follows from the fact that the orientation-induced ordered chain arrangement extends explicitly to the reduced d value [784].

c   With the account of the value for d reported in [786], the value presented relates to the overall thickness of the film sample; it has to be reduced two-fold to obtain the characteristic of transition layer.

For silicate glass the approach described produces a value of $d = 0$ [784] which appears to be as expected.

The principles are generally valid, irrespective of the arrangement (orientation) of macromolecular chains in boundary and transition layers with regard to the geometric interface, be it parallel or perpendicular (the function of the polar angle $0.5 (3 \overline{\cos^2 \nu} - 1)$ is negative in the former case, and positive in the latter) [784]. These concepts are just as valid for linear polymers as for network types [788], including interpenetrating polymer networks [789]. The physical correctness of these ideas, as well as the fact that they rely on diffusion phenomena and can, therefore, be described in

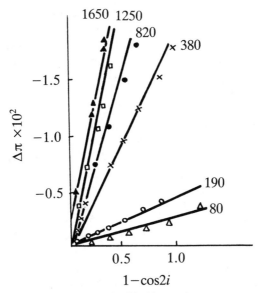

Fig. 3.5 – Phase difference between the birefringent beams versus angle of incidence for epoxy-based film adhesives; the figures indicate film thickness (μm).

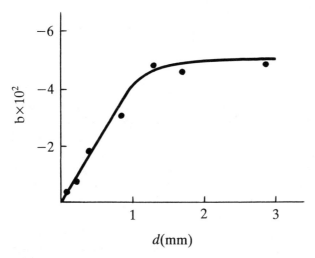

Fig. 3.6 – Coefficient b of eq. (317) for epoxy-based film adhesives versus film thickness.

terms of the variations of diffusion coefficients D, lead one to infer two obvious consequences. First, macromolecules in the boundary and transition layers as well as in the bulk should be different in their conformational states. Second, and most importantly for adhesive joints, the properties of polymers in zones adjacent to the interface must be very susceptible to the substrate.

The effect of a solid surface on the transiton layers in polymers is exhibited, according to Malinsky [790], via two interrelated aspects, spatial and energetic.[†] Geometrically the

† This approach originates from Balandin's multiplet theory of catalysis [791].

substrate imposes limitations on the material in contact at the three levels of its structural organization, segmental, macromolecular and supermolecular. Energetically, the role of the substrate is confined to changing the character of interphase interaction as a function of distance from the surface. Consequently, mobility of the structural elements in transition layers is markedly hindered as compared to that in the bulk of the material due to the loss in the number of feasible conformations for a macromolecule in contact with a solid surface. These predictions were verified in a large number of studies (the most thorough are probably those by Lipatov and coworkers [318, 792–796]) concerning the behaviour of various types of adhesive systems ranging from glue bonds and filled systems [797] to the composites [798, 799]. Let us examine some of the major findings.

The boundary layer of a polymer, being the one in intimate contact with a solid surface, is adsorptive in origin and is several monolayers thick. It is natural to assume that on moving away from the geometric interface the properties of these layers and of those underlying them change continuously, but not discretely. Besides, as a rule, it is not isolated macromolecules that are adsorbed at the interface, but their aggregates, due to the aggregative character of adsorption [800], for multicomponent systems, selective adsorption can be observed. Polymer adsorption in systems in which there is no diluent (solvent) results in the conformational spectrum of the macromolecules in the transition zone being drastically altered, while the zone itself is expanded to an extent determined by the number and volume of the segments [801]. Hence, the profile of chain unit density distribution in the adsorbed flat layer below the θ-point must comprise, according to de Gennes [802], three regions. A rapid decrease of the density near the geometrical interface is observed in the first one; here

$$\rho \sim (1 + r_x/d)^{-1} \tag{318}$$

where r_x is the distance from the interface. In the second region levelling off to a plateau is to be anticipated. Finally, in the third region an exponential reduction of ρ to the value characteristic of the bulk should take place [803]. More complex functions were suggested to describe the density gradient in transition layers [804–807].[†] It is natural to interpret such features in terms of the above mentioned free volume concept (as was done for epoxy adhesives whose density [808] and packing coefficient [809] were shown to rise regularly on approaching the substrate surface; the results were correlated with free volume relaxation).

On the other hand, one cannot exclude the case when certain components of the adhesive (e.g. curing agents) are selectively adsorbed by the substrate resulting in a less crosslinked layer being formed [810], and reducing the strength properties of the system [674, 811]. It should also be taken into account that, according to present day views [812], the adsorbed chains are multi-point attached to the substrate surface (contrary to the single chain-end attachment assumed earlier by the anchoring model [813]) and the thickness of transition layers should be determined predominantly by the terminal portions of the macromolecules [801]. Solution of the corresponding problem is substantially complicated when analysing the behaviour of chains placed between two solid surfaces either adsorbing, or non-adsorbing. Four models can be distinguished, the 'loops' (both chain ends are attached to the single substrate), the 'bridges' (the ends of the chain are anchored at different substrates), 'lashes'/'cilia' (only one chain end is

---

† Simplified description produces a smooth hyperbolic dependence $\rho(r_x)$ [796].

anchored), 'floating chains' (no anchoring points) [814]. However, no precise analytical expressions could be derived for the chain distribution functions within the framework of these models.

Because of the effect of the factors mentioned, major physical properties of polymers must be represented by a function of the distance-to-the-surface.[†] Such concepts are supported by studies of stress–strain and relaxational properties of polymers in the bulk and in thin layers over a broad temperature range [816, 817], as well as by investigations with the molecular probe [818] and dielectric relaxation methods [819]. In these studies the inhomogeneity of the transition layers was established, the layers having been shown to be comprised of relatively loose as well as more compact regions (this problem is treated below). The orientating effect of the substrate provides for the restrained molecular mobility in such regions and hindered relaxation processes [790, 820] in composites. It also provides for the change in molecular-mass distribution of macro-molecules in transition layers [821] and for the change in visoelastic properties of these layers *per se* [822]. It accounts for the anomalous internal stress concentration in transition layers of the samples subjected to a heating–cooling cycle [823], the decrease in heats of fusion in composites [824] (for instance, for polyethene filled by calcium carbonate particles of 1 μm in diameter the decrease is 120 J/g from the initial value of 225 J/g [825]) etc. These effects originate from the fact that the glass-transition temperatures of the polymer in the bulk and in transition layers are different (in poly-alkenes, for instance, it provides for the multiplicity of the α-relaxational transitions [826]).

On the other hand, in IR-dichroism studies it was shown that drawing of polyethene [827], polypropene [827, 828], polybutadiene, polyisoprene [829], poly-ϵ-caproamide [830], polypyromellitamide [827, 831], and various polyheteroarylene fibres [832], which are typical of the flexible and rigid-chain polymers, results in the orientation of macromolecules in transition layers being intrinsically higher than in the bulk of the phase.[‡] A detailed examination of this feature was performed with polytetrafluoroethene [834]. Ordering parameters relating to transition layers k_R^s and to the bulk k_R^v of the PTFE films ($d = 50$ μm) were determined by the IR-dichroism technique [835]. As is evident from Fig. 3.7, the k_R^s versus uniaxial elongation dependence is much sharper than that of the k_R^v. Hence, the transition layers of PTFE appear to be much more susceptible than the bulk of the phase with regard to orientation induced by the drawing stress. This conclusion is valid for any PTFE films thicker than 25 μm [836] and can be related to the fact that it is in this (i.e. transition) layer that the fluoro-containing functional groups tend to become involved [837]. Quantitative consideration of the problem makes it possible to express the thickness of transition layer via the parameters discussed [834].

$$d_s = 0.5\ k_R^v\ d/k_R^s. \tag{319}$$

---

† This conclusion is obviously of general importance. For instance, by means of photoelectron spectroscopy it was established that as the layer thickness of a metal coated in vacuo on to a polymer surface was increased, the energy of electron binding decreases almost to the value characteristic of the metal in the bulk [815].

‡ A similar effect was observed in silica-filled polystyrene the macromolecules of polystyrene in transition layers were shown to have a more straightened conformation than in the bulk of the phase [833].

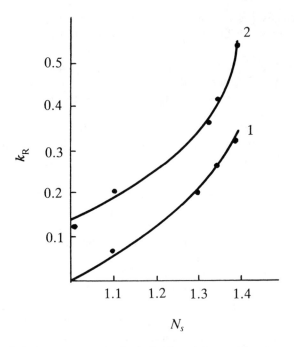

Fig. 3.7 – Ordering parameter in the bulk (1) and in transition layers (2) of PTFE film
(50 μm thick) versus uniaxial elongation.

Analysis of the data of Fig. 3.8 makes it possible to distinguish three stages of the
drawing process of a PTFE film. During the first stage, at elongation ratio $N_s = 1.0$–1.5,
a layer-by-layer orientation of macromolecules takes place, proceeding into the bulk of
the phase to quite a reasonable depth (up to 12 μm); during the second (corresponding to
the plateau in Fig. 3.8) ordering within the oriented transition layer occurs, while during
the third stage the transition layer deteriorates, resulting in that d_s is somewhat decreased
to a value of 11.3 μm.

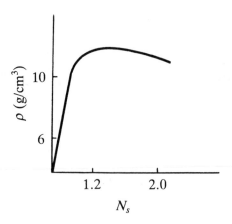

Fig. 3.8 – Thickness of transition layers in PTFE films (50 μm thick) versus uniaxial
elongation.

The reasons for these structural differences between transition layers and the bulk of polymers were investigated for poly(ethylene terephthalate) films as an example. By ATR spectroscopy it was established that within the transition layers the aromatic rings were oriented predominantly parallel to the interface, while the ester bonds were disposed normal to it [838]. As follows from the data in Figs 3.1 and 3.2 such differences cannot avoid affecting the refractive indexes in polyethyleneterephthalate. Indeed, the refractive indexes (determined by ellipsometry and ATR techniques [758]) were shown to increase regularly with increase of ordering and, hence, the density of the transition layers, as is illustrated in Fig. 3.2.

These results lead to at least two major inferences. Firstly, following the ideas discussed in section 2.2.1, the linear increase of the refractive index implies a similar growth of the polymer surface energy. Hence, the transition and, particularly, the boundary layers are indeed characterized by an energy excess which provides a thermodynamic prerequisite for adhesive interaction. Then orientational processes in polymers can be considered as a powerful delayed method of increasing the effectiveness of the interaction, and this is supported by the data presented above. Secondly, thermal conditions for adhesive joint formation have to be chosen so that the thickness of the transition layers will not be decreased due to annealing of the polymer. For instance, when polyethyleneterephthalate films were heated for less than 60 s the transition layer thickness was found to be 4 μm and the refractive index 1.58, while on longer thermal treatment d_s decreased and the refractive index n approached the value characteristic of the bulk of the phase [839].

When discussing the effect of solid substrates on the boundary and transition layers of polymers with regard to the molecular mechanism of this effect it is reasonable to recall the basic concepts of polymer diffusion. Utilizing these concepts is justified at least by the fact that they make it possible to elucidate the relationship between the diffusion of active sites in polymers and the distance to the substrate surface, imposing the orientating effect, in terms of the thermodynamics of irreversible processes [840]. It was demonstrated by numerical methods that diffusion was at a maximum just within the range of the transition layer of the adhesive. Hence, the nature of the polarity gradient in the layers of polyalkenes adjacent to the steel substrate, observed in mechanoluminescence studies [841], becomes clear. Generally, taking into account the concept discussed on p. 153, one may claim as the reason for the properties gradient in polymers [450] the superposition of diffusional (specific for the more densely packed transition layers, which are, hence, in this structural respect more like crystalline materials) and the local (specific for amorphous materials) mechanisms of free volume relaxation.

The last conclusion accounts for the fact that experimentally measured transition layer thickness can be as high as few hundreds of micrometers which corresponds to the macroscopic level. This fact cannot be attributed to the effect of the solid substrate surface only as the data in Fig. 3.9, depicting the film density versus film thickness dependences for epoxy-based coatings, unequivocally indicate the existence of a certain effect due to scaling [842]. The diffusional mechanism of volume relaxation is supported by the generalized relationship between the coating density and the reduced diffusional characteristic τ_T/d^2 (where τ_T is the annealing time) presented for the same epoxy-based adhesives in Fig. 3.10. The increase of d orientation of the internodal chains of the network is also observed [843]. Consequently, the free surface of a film can be considered

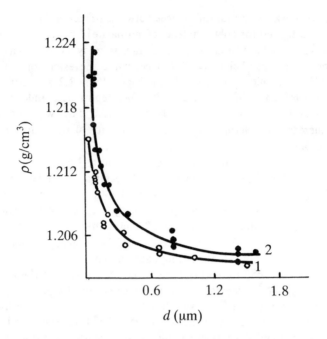

Fig. 3.9 – Density versus thickness of epoxy-based coatings (m-phenyl diamine as curing agent) formed on the following substrates: aluminium (1), (dimethyl dichlorosilane)-modified glass (2).

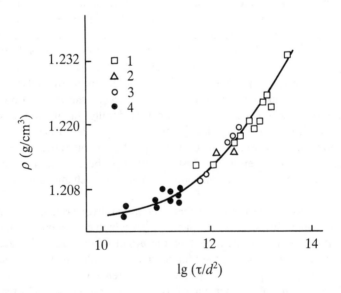

Fig. 3.10 – Density of annealed epoxy-based films, cured with m-phenyl diamine, on dimethyl dichlorosilane-modified glass versus reduced diffusional characteristic τ_T/d^2 (where τ_T is the annealing time). 1', Film thickness 70 μm; 2, 110 μm; 3, 200 μm; 4, 770 μm.

as an active boundary with respect to the diffusing free microvolumes. When there is no adhesive interaction the boundary is liable to be the 'absorbing' one, i.e. it tends to accumulate the free volume at the interface, whereas in the opposite case it is liable to be the 'reflecting', i.e. the one accumulating microvoids in transition layers, thus ensuring the non-uniform density distribution in polymers normal to the interface (boundary) [842]. Mechanical properties of the regions under consideration are, probably, altered in a similar manner insofar as the density of thin films was found to be linearly related to the elasticity modulus, as shown in Fig. 3.11 for epoxy—resol adhesives [844]. This is exactly what accounts for the effect produced by strong interfacial interaction on the cohesive characteristics of transition and boundary layers of polymeric adhesives.

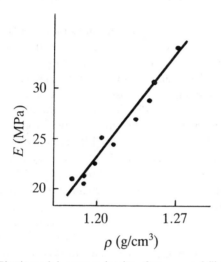

Fig. 3.11 — Elastic modulus versus density of epoxy-resol films 10 μm thick.

When dealing with the adhesion phenomenon, most important are the data concerning the changes of macromolecular mobility in surface layers. Qualitatively, mobility is evaluated by monitoring ESR-spectra of a stable radical e.g. iminooxide, introduced into the polymer [845]. The energy state and segmental motion of the macromolecules at the interface can thus be interrelated [846, 847]. Quantitative analysis relies on the application of NMR-spectroscopy. It was found that the changes of macromolecular flexibility in the transition layers, resulting in the corresponding changes of relaxation times, are due to the bending of segments with respect to the geometric interface and to their 'twisting' [848].

A detailed investigation of the problem was carried out by means of IR-spectroscopy. It was discovered [846] that the boundary layers and the bulk of the films of polyvinyl-chloride and polystyrene, having been formed on a glass surface, differ in the relative contents of trans- and gauche-isomers. For poly(methyl methacrylate) the conformational set of a macromolecule in these layers is low due to the decrease of the number of flat trans-conformations [850]. Corresponding conformational transitions, taking place in free films of these particular polymers, proceed via the structures presented in Table 3.1. The data of Table 3.1 indicate that the boundary layers are rich in those conformers that enable the efficient spatial separation of the polar and non-polar functions in a macro-

Table 3.1 — Physical characteristics of polymers and of their boundary layers

Polymer	d_s (μm)	C_{synd} (%)	Syndiotactic favourable chain conformations	Conformational transtion near the interface
Polyvinylchloride	1.5	70	$tttt,\ ttg^+g^+$	$ttg^+g^+ \Rightarrow tttt$
Poly(methyl methacrylate)	5.5	80	$tt,\ tg^+$	$tt \Rightarrow tg^+$
Polystyrene	6.0	60	$tttt,\ ttg^+g^+$	$ttg^+g^+ \Rightarrow tttt$

molecule. It was demonstrated [851] that, depending on the polarity of the surrounding medium, either hydrophilic or hydrophobic macromolecular fragments are selectively localized at the surface, orientational effects providing for the minimization of cuncompensated intermolecular interactions within the transition layer and thus reducing the interfacial energy of a system. In this case the surface of a polymer in contact with the polar substrate is more polar than that in contact with air or a non-polar substrate. Indeed, the surface of a polyvinylchloride film adjacent to glass contains 17.3% polar groups more than that which is contact with air [851]. Predominant localization at the surface of the non-polar structural fragments [850] serves as an additional proof of the validity of the concepts discussed in the final part of section 2.2.1 (p. 120).

The combined effect of all of these factors produces the equilibrium thickness of the boundary and transition layers.

On the other hand, it is not only the nature of polymers and of the 'near-the-substrate-surface' interaction (which, in the most general case is, apparently, of the van der Waals type [852]) which govern the packing and thickness of transition layers [807, 853]. They are determined also by the procedure for contact formation. The most perfect packing is achieved when the phase is formed from a dilute solution [854], in this case the adhesive film can be regulated to have a thickness from 0.1 μm to 1–2 μm [855] (note for comparison that cooling of isotactic polypropene melt on the substrate does not affect the mechanism of helix formation in macromolecules [839], whereas for (vinyl chloride)–(vinyl acetate) copolymers the loops in the chains are changed for the statistical coil [856]). Qualitatively, this effect is in accord with the ideas concerning the properties of transition layers in polymer solutions [857] supported recently in studies of laser light-scattering from the capillary waves on the surface of aqueous poly(vinyl acetate) solution [858]. As the result of the inevitable uncoiling of macromolecules [859], i.e. of the conformational changes in macromolecules, the boundary layers in solutions should become depleted in the basic component, i.e. polymer content. In fact, with the increase of the incidence angle of the exciting light up to the critical value, corresponding to total internal reflection, the fluorescence intensity of 9-methacryloyl-oxymethylanthracene-tagged polystyrene in ethylacetate solution is smoothly decreased [860].

These effects, eventually, are capable of changing the chemical composition in the boundary layers as compared to the bulk of a polymer. In polyamides, for instance, the boundary layer is excessively rich in carbon atoms [861] (which serves to confirm the suggestion, declared in section 2.2.1, concerning the nature of the nearly equal surface energies of various high-molecular mass compounds). Relying on the studies of methyl methacrylate—glycidyl methacrylate copolymers treated with gaseous trifluoroacetic anhydride, Hammond [862] has come to a similar conclusion for fluorine atoms. By means of X-ray photoelectron spectroscopy Thomas and O'Malley [863] have shown that 5 nm thick transition layers of binary and ternary copolymers of ethene oxide with styrene are enriched in the latter component. A similar result was observed when polystyrene and polyetheneoxide films of nanometric thickness were applied one after another from chloroform solution on to an aluminium substrate [864]. These facts are explained by the higher (as compared to polyetheneoxide) surface energy of polystyrene. The validity of this conclusion was demonstrated for hexamethylene sebacate—demethyl siloxane copolymers, which are described by minimal σ_{sa} values [863]. As in earlier examples, such differences in composition of the transition layers and of the bulk are related to the sample prehistory. Formation of the styrene-dimethyl siloxane copolymer from cyclohexane results in the 1.3 μm thick boundary layer enriched with styrene, while from a solution in styrene a layer of poly(dimethyl siloxane) at least 4 nm thick is produced [865]. Obviously, the solvent should affect the conformational spectrum of a macromolecule. The σ_c values of 35—45 μm thick polymer films applied from solutions on to a steel substrate do not depend on the thermodynamic quality of a solvent when the polymer is polystyrene in contrast to poly(methyl methacrylate) and styrene—butylacrylate—methacrylic acid copolymer [866, 867].

The effect of all of the above factors is increased when there is a strong interfacial interaction, which provides for a marked decrease of the thickness of transition layers and that cannot be attributed to the orientating effect of the interface alone. For low-polarity polymers this results in the formation of more compact (dense) boundary layers, while for high-polarity polymers it results in the decreased mobility of macromolecules in these regions [868]. Such an effect, i.e. diminishing of boundary layer thickness with the increase of interfacial interaction, is also observed in the sliding friction of polymers. For instance, strong interfacial interaction of polyethene, polypropene, and poly-ϵ-caproamide with steel at sliding friction reduces the normally high (for the free samples) thickness of surface layers down to 100—500 μm [869]. This quantity is reduced to an even greater extent for epoxy adhesives, e.g. to 7—8 μm [870], and for polyethene, and ethene—(vinyl acetate) copolymer is further lowered to 2—6 μm [871], all with steel as substrate. The relationships observed between the intensity of interfacial interaction and the thickness of transition layers are, undoubtedly, liable to affect the strength of adhesive joints. Such a relationship is clearly illustrated for polyimide coatings formed on 25 μm thick copper foil by stepwise heating of pyromellitic dianhydride and oxidianylinepolyamine acid in N-methyl pyrrolidone [710]. As is seen from Fig. 3.12, the peeling strength of adhesive joints versus thickness of the adhesive layer dependence is a monotonic almost linear descending one.

However, the relationship between the thickness of the transition layers and interfacial interaction intensity is most markedly revealed in filled polymers.[†] For instance, in epoxy compositions filled with finely dispersed aluminium or iron powders transition layers degenerate into boundary ones [873]. Direct evidence of this phenomenon was

---

† It may be suggested that poor adhesive interaction, for instance, due to incompatibility of the polymers, must result in the opposite effect. This clarifies the reasons for the loose packing in transition layers of the filled polyethene/polyurethane blends [872].

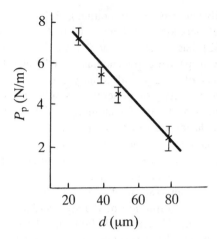

Fig. 3.12 – Peeling strength of polyimide–copper joints versus adhesive film thickness.

obtained in birefringence studies of the interaction efficiency of polystyrene and poly(vinyl butyral) with powdered graphite and keratin [874]. For both polymeric matrices the thickness of boundary layers substantially decreased on filling, for polystyrene from 45 μm down to 22 and 19 μm with graphite and keratin, respectively and for poly(vinyl butyral) from 73 μm down to 25 and 28 μm. Note the different effect of one and the same filler on different polymers. This is probably due to the differences in polymer polarity which result in independent mechanisms of interfacial interaction being realized. If this assumption is true then filling polystyrene and poly(vinyl butyral) should result in different orientations of the macromolecules at the solid phase surface. Indeed, for the more polar adhesive the addition of either graphite or keratin is accompanied by a decrease in the factor of orientational ordering (see p. 155) from −0.07 (polymer matrix proper) [787] to −0.18 and −0.09 [874], respectively for the two fillers. In the case of the less polar matrix, polystyrene, the intrinsic value of this factor −0.28 [787] is increased to −0.06 on introduction of the inert filler (graphite), while the reinforcing filler (keratin) decreases it to −0.35 [874].

Thus, the boundary layers of polymers involved in adhesive interaction are subjected to a rather strong 'compressing' effect imposed by the substrates. It is quite clear, since, as the energy of the segment-to-the-substrate-surface interaction is increased, the thickness of the boundary (adsorptive) layer decreases as may be demonstrated by computer simulation [875]. Only a slight change of the segment density in the boundary layer followed by its subsequent slow decrease corresponds to the interaction energy of 0.5 KT, while at 0.9 KT compact packing of the adsorbed macromolecule is achieved. Analysis of the density distribution in a multi-link chain assuming that each of its links interacts both with the neighbouring links and with the solid surface of a substrate demonstrates that when the interaction is strong the chain is adsorbed as an unstructured layer whose thickness only slightly exceeds that of a segment. When interfacial interaction is weaker then loops and terminally anchored chains etc. are formed.

At the same time, it should be borne in mind that the concepts discussed above are valid for boundary layers only. With respect to transition layers, located between the boundary layers and the bulk of the phase, the enhancement of adhesive interaction should imply the 'expansion' of the zone sustaining the effect of a substrate. In other words, the response of the boundary and the transition layers to the variation in the efficiency of adhesive interaction is quite different. For the former the enhancement of adhesive interaction results in the formation of more compact (dense) layers of decreased thickness, while for the second it leads to more loose layers of increased thickness. Let us examine how this thesis corresponds with the experimental data.

On introducing various fillers into a polyethene matrix the thickness of transition layers in polyethene was found to vary similarly to the surface energy of a filler σ_s : 2.5 nm for silica, 27.6 nm for kaolin and 377 nm for finely dispersed quartz [876]. Variation of the free volume in polystyrene and poly(methyl methacrylate) was shown to follow a similar pattern when fillers with different surface energies i.e. glass, calcium oxide, and sodium chloride, were introduced [877]. Naturally, such effects are more easily observed at the macroscopic interface, i.e. in traditional adhesive joints. Thus, by the molecular probe technique it was demonstrated that the thickness of transition layers in polystyrene and poly(methyl methacrylate) bonded to high-surface-energy quartz was 30 and 60 μm, respectively, whereas it was only 3–4 μm when the substrate was a low-surface-energy polytetrafluoroethene [818].

Hence, the existence of strong adhesive interaction imposes a different effect on the boundary and transition layers in polymers, endowing real meaning to the above classification of these layers. At the same time it should be borne in mind that these differences originate from the fact that the change in the density of the transition layer, as compared to the bulk, cannot be accounted for by thermodynamic factors,[†] but is due to molecular–kinetic ones, since it is the flexibility of macromolecular chains s, and not surface energy σ, that determines the behaviour of a polymer in the transition layer. In fact, for a given polymer σ is a constant and it therefore demonstrates no functional dependence on the distance from the surface, whereas parameter s is directly related to the orientating effect of the interface.[‡] Therefore these variations of the density and thickness of different layers can be correlated with the mobility of macromolecules in these layers, the higher the mobility, the more effective is the adhesive interaction.

As a result of the effect of these factors various macroscopic characteristics of boundary and transition layers of polymeric adhesives, mechanical in the first place, may be changed. For instance, the elastic modulus of an epoxy-resol adhesive film in contact with the inactive Wood's alloy is a monotonic function of film thickness, while it is a more complex one when the adhesive film is in contact with aluminium (Fig. 3.13). The decisive role of adhesive interaction in this instance is proved by the fact that prolonged annealing of aluminium adhesive joints changes the dependence, converting it to a monotonic one [844]. The Young's modulus of a film of poly(methyl methacrylate) is 32.4% higher than of the bulk sample [880]. For epoxy–resol adhesives [844] the

---

† In this connection worth mentioning is the remark by Rusanov [878], who emphasized that though thermodynamics cannot indicate the thickness of surface layers, it, nevertheless, makes it possible to determine its lower margin, at least in regard to liquid phases.

‡ A quite peculiar demonstration of this effect was disclosed in investigations of the solvent mobility profiles in polymer solutions, i.e. of the change of solvent mobility with the distance from a segment surface. Elmgren [879] demonstrated that the influence of the macromolecules of hydroxyethylcellulose and of ribonuclease on the water mobility changed as the inverse distance to the power of 1.6 and 2.2, respectively; at the distance of 4 nm from the cellulose segment, or from the surface of ribonuclease molecule, the mobility was reduced by 10%, from that in bulk solvent.

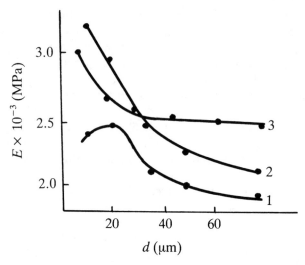

Fig. 3.13 – Elastic modulus versus thickness of epoxy–resol films formed on aluminium.
1, Starting film; 2, film annealed for 30 hours, 3, Wood's alloy.

difference between a bulk sample and a 16 μm thick film amounts to 36.6% for Young's modulus, 23.5% for tensile strength, and 57.2% for ultimate tensile strain. Strong interfacial interaction between a fibrous substrate and a polymeric matrix in composites decreases the compliance of a polymer by 65%, whereas for systems with weak adhesion this decrease is as low as 8% [881].

Such effects are particularly related to the character of relaxation processes [882] which are normally described by the glass-transition temperature T_g of a polymer. For instance, in a fibrous composite T_g is 40–50° lower than for the bulk sample of the binder, the most drastic decrease being observed in surface layers of the matrix [883]. This effect is one of general importance. This is verified by the identical glass-transition temperature versus film thickness graphs for polyethene (Fig. 3.14) and polystyrene (Fig. 3.15) films in contact with steel [884] and glass [849]. More detailed data concerning the effect of adhesive interaction on relaxation processes in polymers were obtained by radiothermoluminescence. Using this technique, Bartenev and coworkers discovered structural inhomogeneities in the boundary layers of polyethene in contact

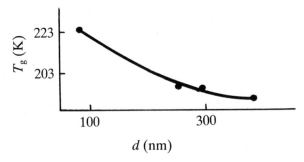

Fig. 3.14 – Glass–transition temperature versus film thickness of polyethene cast on a steel surface.

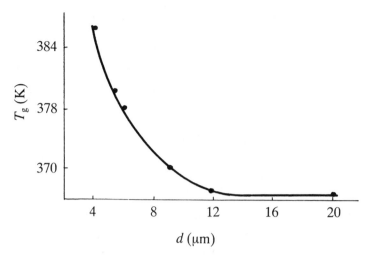

Fig. 3.15 – Glass-transition temperature versus film thickness of polystyrene cast on a glass surface.

with steel [885]. In the range of d_s values from 1 to 4 μm the α-maximum of the luminescence versus temperature curves is split into two peaks ($T_g' = 227.7$ K and $T_g'' = 258$ K), at $d_s = 4$–20 μm the bimodal α-band is retained but $T_g' > 227.7$ K, while at $d_s > 20$ μm the splitting vanishes and T_g' continues to increase up to a constant value of 245 K at $d_s > 500$ μm [886, 887]. It was also shown that boundary layers of increased density were formed in polyethene contacting a steel substrate [887, 888]. Similar conclusions follow from the electron microscopic studies of polyethene in contact with titanium [889, 890], and from investigations of composites [891].

As regards the more loose transition layers, sandwiched between these regions and the bulk phase of polymers, their actual existence is indicated by the indirect data discussed above. This problem was investigated by ellipsometry for adhesive joints involving elastomers. This is a non-destructive method that provides the most reliable information concerning the processes taking place at the interface between the condensed phases. Natural rubber (NR) and butadieneacrylonitrile elastomers (SKN-18, -26 and -40, where figures denote the percentage acrylonitrile content) were used as adhesives, which were layered on to the glass substrate (prism of ellipsometer) from solution. The properties of these systems in the bulk and at the interface with air were also examined.

It follows from the measurements of thickness d and refractive index n of the elastomers [892] that the boundary and transition layers are formed at the interface, their properties being different from that of the bulk of a polymer (Table 3.2). The boundary layer of the elastomer, adjacent to the glass surface, is thinner than that of the free surface of elastomer, as it is subjected to a stronger force field. The latter conclusion is in agreement with the data of section 2.2.1 establishing $n \sim \sigma_{sa}$. Therefore, the increase of n is evidence of a more restrained macromolecular mobility in the former case than in the latter. In fact, strong interaction across an elastomer–glass interface results in that, irrespective of the nature of the elastomers involved, the d and n values of corresponding layers are constant. In the case of elastomers in contact with air the effect of interfacial interaction is insignificant (the values of refractive index are rather low and constant),

Table 3.2 – Equilibrium values of thickness and refractive index of the boundary layers in elastomers

	Bulk	Elastomer–air		Elastomer–glass	
Elastomer	n_D^{18}	d (nm)	n_D^{18}	d (nm)	n_D^{18}
Natural rubber	1.5228	19 ± 1	1.037	10 ± 1	1.537
SKN-18	1.5224	22 ± 1	1.036	10 ± 1	1.537
SKN-26	1.5220	26 ± 1.5	1.036	10 ± 1	1.536
SKN-40	1.5215	38 ± 1.5	1.036	10 ± 1	1.536

consequently, the thickness of the transition layer is determined by the nature of an adhesive. Such data are related to the difference in elastomer density in the bulk and in the transition layers. At the same time, it was established [892] that adjacent to the dense boundary layer, directly contacting the substrate, there lies a less dense[†] transition layer, its thickness being almost two orders of magnitude greater than that of the boundary layer. This provides the first direct evidence of the differentiation of surface strata of polymers into transition and boundary layers.

Relying on ellipsometric studies a schematic pattern of the arrangement of surface layers of different densities was proposed [892] for elastomers contacting a high energy surface (Fig. 3.16). With the aid of the diagram in Fig. 3.16 it is not difficult to show that the reason for the discrepancies in the published data (e.g. by Lipatov and coworkers [816]) concerning the assessment of the effect of the substrate on the density and other

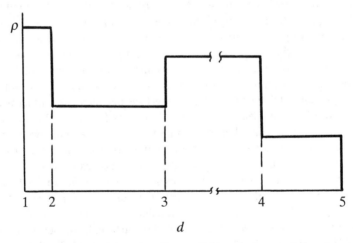

Fig. 3.16 – Elastomer density versus distance d from the geometric interface with a high-energy substrate. 1–2, Boundary layer at the interface with glass; 2–3, transition layer at the substrate side; 3–4, bulk of elastomer; 4–5, boundary layer at the interface with air.

---

† This conclusion was confirmed by small-angle X-ray scattering for poly(methyl methacrylate) and polystyrene as examples [893].

characteristics of surface regions of polymers is that two features are neglected, namely that there exists a thin outer (boundary) layer and that the mechanisms governing the effect of a solid surface on the boundary and transition layers in polymers are different. Hence, the lack of correspondence between the data of different studies is essentially due to the fact that these data describe the overall properties of surface layers including both the boundary and transition layers.

Thus, adhesive interaction provides for the unequivocal differentiation of the surface strata in polymers into transition and boundary layers. Strictly speaking, this differentiation is particularly valid for adhesive joints since in the case of condensed phases in contact with a gaseous medium or vacuum the absence of interfacial interaction merely smooths out the distinctions between these layers. Hence, not only do the theoretical concepts concerning the properties of the boundary and transition layers in polymers acquire a deeper meaning, but also the specific significance for the purposes of describing the processes of formation and the modes of behaviour of various adhesive joints, from traditional glue bonds and coatings to vulcanized rubbers[†] and modern composites.

However, consistent application (and subsequent development) of these ideas is substantially restricted by the lack of specific physical characteristics pertaining to boundary and transition layers (although there is a great deal of indirect data). The most rigorous method presumably likely to fill the gap is the method of X-ray photoelectron spectroscopy. Using this technique, Lukas and Jesek [899] demonstrated that the density of various polymers was linearly correlated with the inelastic run of photoelectrons l_λ. Taking into account that the penetration depth of electrons into the phase is rather small (of the order of magnitude of 1 μm [900, 901]) this effect can be attributed only to the influence of the boundary layers. To provide a quantitative assessment Grigorov [902] proposed combining this method with secondary-ion mass-spectrometry (SIMS). By means of the SIMS technique the surface produced in layer-by-layer etching of the sample by the beam of high energy particles is analysed in order to obtain the distribution function $f(d)$ of thickness d over the surface area A_d:

$$f(d) = A_\Sigma^{-1} \frac{\mathrm{d}A_d}{\mathrm{d}d}. \tag{320}$$

The time-dependent variation of this quantity is a function of the rate of destruction processes v_d within the boundary layers, destruction time τ_d and the intensity of the secondary-ion signal from the boundary layers I_p and from the original substrate I_s. Corresponding expressions were derived by Grigorov [902] for a plane substrate (traditional adhesive joints like common glue bonds and coatings):

$$f(v_d \tau_d) = v_d^{-1} \frac{\mathrm{d}}{\mathrm{d}\tau_d} \left[\frac{I_p}{I_s} - \frac{l_\lambda}{v_d} \frac{\mathrm{d}(I_p/I_s)}{\mathrm{d}\tau_d} \right] \tag{321}$$

and a highly disperse substrate (as adhesive joints in filled-polymer systems):

[†] In this connection quite significant is the development of an independent adhesion theory of elastomer reinforcement [894, 895], which is valid also in essence for plastomers [896]. This fact stimulated its recent development [897, 898].

$$f(v_d \tau_d) = (2v_d)^{-1} \left[3 \frac{d(I_p/I_s)}{d\tau_d} + \left(\tau - \frac{4l_\lambda}{v_d} \right) \frac{d^2(I_p/I_s)}{d\tau_d^2} \right.$$

$$\left. - \frac{l_\lambda \tau_d}{v_d} \frac{d^3(I_p/I_s)}{d\tau_d^3} \right]. \tag{322}$$

Calculations using eqs (321) and (322) showed that at an etching speed of 10^{-4} μm/cm (which is common in SIMS technique) the method provides data relating to polymer boundary layers as thins as 10^{-3} μm.

However, these methods are obviously too complicated for the everyday studies of adhesive joints, at least at the present state of the art. Therefore, taking into account the prerequisites and advantages of the refractometric method discussed in section 2.2.1, we find it appropriate to make use of the latter to determine the physical characteristics of boundary and transition layers in polymers.

Assuming that the van der Waals' volume of a macromolecule in the bulk and in the boundary layer are equal, the molar packing coefficient can be calculated by substituting the values for refractive indexes (see Table 3.2) and specific refraction [395] into eq. (215), while from eq (216) the effective molar energy of cohesion can be obtained. Then the density of a boundary layer can be determined from expression (199). The results of calculations for the materials listed in Table 3.2 are compiled in Table 3.3.

Table 3.3 – Physical characteristics of elastomers in the bulk and in boundary layers.[a]

Elastomer	Bulk			Elastomer–air			Elastomer–glass		
	k_0	ρ	E^*	k_0	ρ	E^*	k_0	ρ	E^*
Natural rubber	0.661	0.913	11.783	0.053	0.073	11.819	0.676	0.934	11.939
SKN-18	0.649	0.995	11.924	0.054	0.077	12.029	0.710	1.018	12.034
SKN-26	0.696	1.009	13.301	0.054	0.078	13.084	0.711	1.031	13.338
SKN-40	0.699	1.037	16.092	0.055	0.082	16.210	0.715	1.061	16.085

a $\rho = [\text{g/cm}^3]$; $E^* = [\text{kJ/mole}]$.

When surveying Table 3.3, the feature to note in the first place is the regular increase of the density[†] and the packing coefficient of the boundary layer at the interface with glass as compared to characteristics of the bulk phase. The observed 2.3% increase is in direct agreement with the above assumption about a denser packing of high molecular compounds in boundary layers. This conclusion is confirmed by experimental data on packing of polyethene and polyethyleneglycolterephthalate copolymers (60 : 40) cast on to cellophane [905]. As the thickness of the transition layer decreases to that character-

---

†   Reliability of the calculated ρ values is confirmed by their close agreement with the published data: for natural rubber $\rho = 0.91$–0.92 g/cm^3, for SKN-18 $\rho = 0.945$ g/cm^3, for SKN-26 $\rho = 0.962$ g/cm^3, for SKN-40 $\rho = 0.986$ g/cm^3 [904].

istic of the boundary layer, the k_0 value increases in a regular manner which is independent of the temperature of system formation:

T (K)	$k_0{}^a$
343	0.6914/0.6883
383	0.6893/0.6882
403	0.6885/0.6877

a numerator refers to $d = 10 \mu m$/
denominator $d = 40 \mu m$.

This fact, consistent with the concepts discussed, should also be noted when considering a low-energy substrate (polytetrafluoroethene); the decreased efficiency of adhesive inter-action leads to substantially increased values of k_0 which reaches 0.6992 at temperatures of 343 K and 383 K.

As a consequence, the thickness of boundary layers is directly related to the structure of elastomers at the interface with air, the density versus thickness graph being linear for a series of closely chemically related butadieneacrylonitrile copolymers (Fig. 3.17). On the other hand, elimination of the orientating effect of a solid results in the boundary layers acquiring a substantially looser packing as is demonstrated by more than an order of magnitude decrease of the density and packing coefficient (see Table 3.3). As the values of effective molar cohesive energy (as well as the calculated values of σ) are actually constant in the bulk and at the interface [906] one may infer that factors of molecular–kinetic origin, predominantly macromolecular mobility, provide for different properties of the polymer bulk and of the transition and boundary layers to a much greater extent than those of thermodynamic origin.

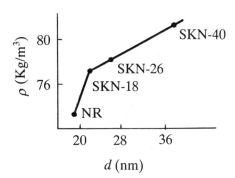

Fig. 3.17 – Density of natural and butadiene–acrylonitrile rubbers at the interface with air versus thickness of boundary layer.

To verify the validity of this conclusion the number of repeat units in the segment s, which is a common measure of the mobility of macromolecular chains, could be used. However, its assessment by existing techniques (mostly by the electro-optical method [907] is a quite sophisticated task. For this reason there are no published data concerning the evaluation of s for polymers, butadieneacrylonitrile copolymers in particular (for natural rubber $s = 2.0$ was reported [908]). A special procedure for calculation of s was developed to overcome this obstacle.

Relying on general considerations one may define s as the ratio between the molar volume of a segment V_m^s and that of the repeat unit V_m [909]:

$$s = V_m^s / V_m. \tag{323}$$

From eq. (202) it is easy to derive the equation involving effective molar cohesive energy which would make use of the procedures reported for the evaluation of V_m [478]. By analogy, the concept of molar cohesive energy of a segment was defined as follows [909, 910]:

$$\sum_i \Delta E_i^s = N_A \delta^2 s \sum_i \Delta V_i = \delta^2 V_m^s. \tag{324}$$

Then the mobility of macromolecular chains can be characterized by the ratio between the cohesive energy of a segment and that of the repeat unit. ΔE_i^s sums the contributions from the different atoms in the main chain and in the polymer side chains as well as from the different types of interactions between them. Summations were obtained by computer processing of 37 equations with 19 unknown variables [910]. The method developed provides for the reliable evaluation of the flexibility of macromolecules of different structure, from polyalkenes [910] to polyphenylquinoxalines and polypeptides [911] and, taking into account the $\delta_{12}(\delta_1, \delta_2)$ function [485, 486] described by eq. (204) this methods could be applied to copolymers also [486, 912]. With the aid of this procedure parameter s for butadieneacrylonitrile copolymers was evaluated to be 3.9 for SKN-18, 4.6 for SKN-26 and 6.0 for SKN-40. Reliability of the data obtained is confirmed by the fact that s values are linearly correlated [912] with the values of the braking coefficient of internal rotation [913] and other related physical parameters [7985] for acrylonitrile elastomers.

As is seen from Fig. 3.18, s on the one hand, and d, ρ, and k_0 on the other, are interrelated, the observed interrelationship being strictly linear for acrylonitrile elastomers [903]. The dependences of macromolecular mobility on the experimental values of the density in the bulk (Fig. 3.18, bIV) and in the boundary layer at the interface with air (Fig. 3.18, bII) follow a linear pattern for all the materials considered. Implications of the latter situation are of major significance. The linearity of the plot in Fig. 3.18, bII demonstrates that the density of the boundary layers of elastomer films at the interface with air is governed predominantly by the flexibility of macromolecules. When the films are formed in contact with glass the dependence, as illustrated in Fig. 3.18, bI, deviates somewhat from linearity, which is due to the restricted mobility of macromolecular chains at the interface with glass.

The predominant effect of macromolecular flexibility on the properties of boundary layers in elastomers can also be confirmed by using the braking coefficient of internal rotation ω in macromolecular chains as a measure of flexibility. Plotting the published ω values [913] versus the calculated values for ρ and k_0 reveals the strictly linear relationship between these quantities relating to the boundary layers of elastomers at the interface with air (Fig. 3.19), the case when the structure of boundary layers is not determined by the orientating effect of a solid substrate. The relationship between ω and packing coefficients of elastomers in the bulk and in the boundary layers at the interface with glass follows a similar pattern (Fig. 3.19b). It is not difficult to see that the results

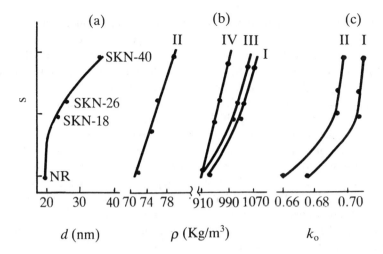

Fig. 3.18 — Mobility of macromolecular chains of natural and butadiene–acrylonitrile rubbers versus thickness of boundary layer at the interface with air (a) versus density of boundary layers at the interface with glass I, air II, and in the bulk (III calculated, IV experimental values) (b) versus packing coefficient of boundary layers at the interface with glass I and air II (c).

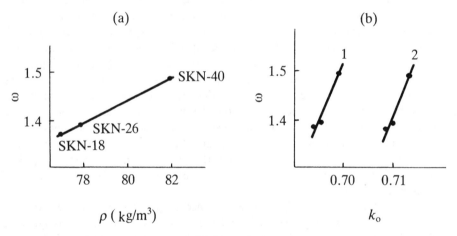

Fig. 3.19 — Braking coefficient of internal rotation in chains of butadiene–acrylonitrile rubbers versus density of boundary layers at the interface with air (a) versus packing co-efficient in the bulk 1 and in boundary layers at the interace with glass 2 (b).

presented do not contradict the qualitative inferences of the original work by Vakula and coworkers [892], which were later made more precise by Gorshkov *et al.* [914] using the dynamic method of Mandelstamm–Khaikin.

Hence, the mobility of macromolecular chains appears to be, as might be expected, the parameter that determines the properties of the boundary layers as well as of the closely associated transition layers in polymers. This fact is of crucial importance within the framework of the molecular–kinetic approach to surface phenomena and adhesion in particular. In spite of the apparent self evidence of this statement it was necessary to conduct a great deal of laborious experimentation to provide strict proof [22]. However, it is necessary to note that, to the present, the reported attempts to predict the adhesive ability of high polymers relying solely on this conclusion are of a qualitative character only (see, for example, [915]). Thus, there is every reason to believe that the approach described was, to our knowledge, the first to enable at least a semi quantitative treatment of the properties of surface layers of high molecular mass compounds as reflecting the mobility of their chains. In fact, by evaluating the strength of an adhesive joint by its instantaneous value on contact (so called 'instant adhesion' neglecting the kinetics of adhesive joint formation) one can observe, as we have found, a very close correlation between the parameter s for polyethene, polypropene, polystyrene, poly(methyl methacrylate), and polyethyleneterephthalate and their tack to poly(butyl methacrylate) [916], the correlation coefficient being as high as 0.990. Apparently, the generalized molecular–kinetic theory of the formation and performance of polymer adhesive joints must involve the parameter s as one of the basic variables.

As regards substrates of any other type, not only polymeric ones, this conclusion is of much broader implication. It may be demonstrated that the molecular–kinetic generalizations of adhesive joint formation are specified by both energetic and by structural factors since, for the realization of interfacial interaction, the mutual arrangement of topological fragments of the contacting phases becomes a matter of prime importance. In polymers these fragments are segments, macromolecules or super-molecular structures, for metals coordination numbers at the surface, i.e. specific concentrations of atoms at the interface. By means of the X-ray method Buckley [917] determined specific concentrations for the three crystallographic faces of copper (111), (100), and (110) with coordination numbers 9, 8 and 7, respectively, and measured the peeling strength of gold foil pressed to a copper substrate under 20 mg load for 10 s. Graphical presentation of Buckley's results (Fig. 3.20) demonstrates the rectilinear dependence between the specific concentration of atoms at the surface of a metal and its adhesive ability. It is essential that due to the short formation time of the adhesive joint between Cu and Au, metals which exhibit low mobility of surface atoms, the effect of purely kinetic factors can be neglected. Nevertheless, the straight line in Fig. 3.20 supports the thesis of the prevailing effect of the structure of materials on the modes of their adhesive interaction.

In addition to the factors already discussed it is necessary to take into account that the surface energy of the boundary layers of condensed phases is the highest as compared to that of the layers beneath and this factor can produce an unfavourable effect on the strength of adhesive joints. The reasons for this unfavourable effect lie in the enhanced sorption by the polymer boundary layers from the environment as well as from the bulk of the phase itself.

Factors comprising the first group were considered by Bikerman]6] who demonstrated that sorption of the impurities notably affects the wetting modes due to changes in the energetic spectrum of the surface, on the one hand, and to the decrease in the real contact area between the two phases, on the other. Incomplete wetting is one of the major technological reasons for adhesive bond strength not reaching the optimal level. A large number of procedures aiming to reduce the negative effect of these factors is used in practice. The whole complex of these forms the basis of the technology of substrate pretreatment prior to bonding.

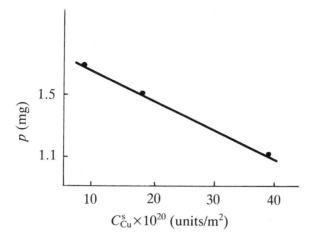

Fig. 3.20 – Copper–gold adhesive joint strength versus specific concentration of copper in surface layers of the substrate.

Besides the formation at the substrate surface of boundary layers with decreased cohesive strength (so-called weak boundary layers, classified in [6, 918, 919]), the adsorptive interaction at the interface may give rise to structural inhomogeneities at the molecular and supermolecular levels [920] which are decisive for the mechanics of deformation and fracture [921]. For instance, investigation of the crystallization kinetics of oligoetheneglycoladipate melt filled with glass and highly-disperse silicon dioxide revealed the structural inhomogeneity of thin polymer layers depending on the distance from the solid substrate [921]. Similar results were obtained in ATR–IR studies of polyurethane–oligourethaneacrylic systems cast on a glass substrate [923].

While the effect of the environment may, in principle, be almost eliminated by common technological methods aimed at the protection of the substrate surface, it is a qualitatively more difficult task to prevent the migration to the substrate surface of low molecular mass impurities from the bulk of the phase. This process, from the viewpoint of thermodynamics, is governed by the balance between surface energies of the starting polymer and of its boundary layer. The existence of such energy gradients leads to the boundary layers becoming enriched, in accordance with eq. (151) by compounds with the higher surface energy. Indeed, the data on wetting of various polyurethane films demonstrate that due to the enrichment of their surface layers by the more flexible ester fragments their surface energy is actually independent of polymer composition [924]. More definitely the migration of low molecular mass additives from the bulk of the phase to the surface and thus affecting the σ_{sa} value was proved by X-ray photoelectron spectroscopy. According to Gaines [925] the surface of a block copolymer is enriched, as a rule, by the low-energy component, while Schmitt et al. [926] measured its concentration in the interphase region of a blend of polymers with deliberately different σ_{sa} values (polycarbonate 34 mN/m, polydimethylsiloxane 24 mN/m) to be as high as 85% within the 0.5 μm thick layer.

Such a distribution of the components in transition layers of polymeric multiphase systems according to the values of their surface energy provides for the corresponding fractionation by molecular masses as well. Such an effect was observed in composites

[927] and filled polymers [928]. It was discovered that on adsorption of polystyrene from the melt on to a glass surface the fractions of low molecular mass are accumulated at the interface [929]. It is clear that such effects, directly following from the concepts discussed above (see pp. 100–103), should be taken into account when examining adhesive interaction involving high molecular mass compounds.

Real adhesives and substrates contain low molecular mass impurities as unreacted starting compounds for the synthesis (monomers, stabilizers, special additives) or as products of conversion (e.g. of breakdown) of individual components of the system. Their concentrations can reach rather high levels [6]. This factor can evidently affect the strength of adhesive joints. The validity of this assumption was verified by the method of radioactive tracers where it was shown that the amount of plasticizer migrating to the interface from out of the bulk of the volume is linearly related [930] to autohesive peeling strength (Fig. 3.21). Quite satisfactory data in support of the assumption were obtained when comparing the monomer (caprolactam) concentration, migrating from the bulk of polycaproamide to the interface with elastomers, with the strength of corresponding adhesive joints [931].

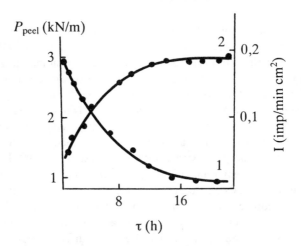

Fig. 3.21 – Autohesive peeling strength of elastomers (1) and radioactive tracer intensity (2) versus migration time of plasticizer from the bulk of elastomer.

Hence, elimination of weak boundary layers is an important aspect of increasing adhesive joint strength. For instance, elimination from a polyethene surface of the ever present low molecular mass paraffins should, apparently, provide an increase in surface energy of the substrate; this was done by annealing polyethene with a degree of crystallinity of 80–90% at a temperature below the temperature of equilibrium melting [932]. As is seen from the data in Fig. 3.22, thermal treatment of polyethene for 1 h and for 200 h produces a linear rise in σ_{sa}. Bikerman and Marshall [933] attained the same result by means of multiple re-extraction of paraffin impurities by heptane. As a result, the polymer strength exceeded the initial value (93 N/m instead of the initial 91 N/m), while adhesive joints of the so-treated polyethene substrate fractured cohesively.

These results demonstrate the obvious benefits of controlling the diffusion of low molecular mass compounds into polymer surface layers, since elimination of these weak

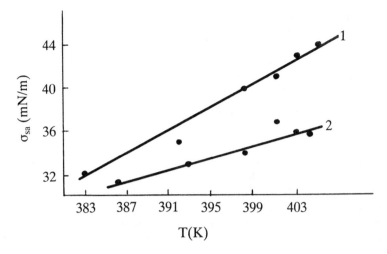

Fig. 3.22 – Surface energy of polyethene versus annealing temperature. 1, Annealing for 1 h; 2, annealing for 200 h.

regions is the most effective, but not the only way to increase the strength of adhesive joints. In a similar way, where rubbers are concerned, it was suggested that the surface layer be made more elastic than the bulk phase [938].[†] This can be attained by varying the concentration of certain accelerators as a function of the distance from the interface [939]. At the same time it is possible to toughen the weak boundary layers by using reactive rubber modifiers that tend to diffuse (migrate) from the bulk of the phase to its surface.

Owing to the influence of the factors discussed above, concentration of the end groups in polymer surface layers is an order of magnitude higher than in the bulk phase of the free polymers [940]. This fact is associated with thermodynamic condition (2) and appears to determine the decrease in intermolecular interaction in the phase with increasing distance from the geometric surface [941]. This conclusion is confirmed by the data of IR-spectroscopy studies of submicron layers of polymers [942]; it is in accord with the data on packing coefficients of the boundary layers in elastomers listed in Table 3.3. Moreover, in about 1 μm thick surface layers interatomic bonds are stretched, via the fluctuation mechanism, 1.5–20 times more than in the bulk of the sample [943], thus increasing the local excessive stresses [944]. The stresses accumulated in surface layers can be as high as 1.5 MPa, as for polyvinylidenefluoride on a steel substrate [948]. As a result, the concentration of ruptured bonds and submicron cracks in the surface of polymers at the polymer/substrate interface is almost an order of magnitude higher than in the bulk [945–947].

---

† Taking into account the definition of an adhesive joint given in section 1.2, the latter condition obviously acquires general significance. In fact, Gul' *et al.* disclosed the relationship between relaxation properties of transition layers in polymers and the strength of their adhesive joints [934]. They discovered that the intensity coefficient of relaxation processes was correlated with the coefficient of combination strengthening [935]. This is the basic effect providing for the strengthening of multilayer systems involving plastics [936] and elastomers [937].

In other words, stressing of a polymer results in the stress being accumulated primarily within the transition and boundary layers of a polymer. This is proved by the asymmetric broadening of certain bands in IR-spectra of polymers subjected to mechanical loading [949], which is related (see [950] for a semiquantitative treatment) to the non-uniform distribution of the stress over individual portions of macromolecules. Hence, the most 'dangerous' zone in a solid body is located at its surface. It is in the surface layers that the destruction, particularly fracture, of the sample starts, the rate of related processes in these layers being much greater that in the bulk [830, 944–946, 951]. For instance, loading of a quartz crystal produces a three-times greater frequency shift of the lattice vibrations associated with the surface than of those in the bulk [951]. Drawing of poly-ε-caproamide results in increased molecular orientation of its transition layers as compared to the bulk [830]. A quantitative treatment is based on the studies of the kinetics of end group accumulation due to mechanical stress-induced destruction of macromolecules in transition and boundary layers of polymers. In the studies with polyethene it was shown [952] that accumulation kinetics is described by a rate equation of the first order:

$$C_\tau = C_\infty (1 - e^{-k_s \tau}) \tag{325}$$

where the accumulation rate constant k_s is exponentially related to the load and temperature.

These effects produce a notable influence on the performance of heterophase systems in force fields. In fact loading, for instance, of fibrous composites should affect the number of overloaded bonds in the fibres. Indeed, it was demonstrated by IR-spectroscopy [953] that under identical external loads the largest fraction of overloaded bonds was observed in the polypropene filler (8%). As is seen from Fig. 3.23, even under conditions remote from those of fracture (e.g. in the case of polyethene or phenol-

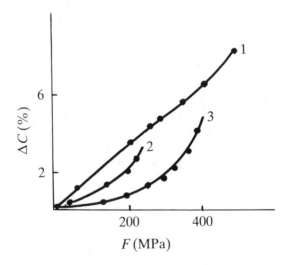

Fig. 3.23 – Concentration of over-stressed bonds in polypropene versus applied load. 1, Oriented film 0.02 mm thick; 2, ditto with a polyethene coating on both sides 0.03 mm thick; 3, ditto with a coating of phenolformaldehyde adhesive BF-6 on both sides 1.5–2.0 mm thick.

formaldehyde-polyvinylbutyral matrix with $P_{fracture} = 650$ MPa) this fraction amounted to 4%. Further increase of the load results in an additional rise in the number of overloaded bonds in these systems (see Fig. 3.24) [954]. Similar effects may arise under the influence of oxidation [955] or other factors.

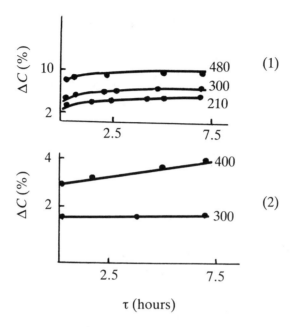

Fig. 3.24 — Concentration of over-stressed bonds in polypropene versus loading duration. 1, Oriented film 0.02 mm thick; 2, ditto with a coating of phenolformaldehyde adhesive BF-6 on both sides 1.5–2.0 mm thick (figures denote the load in μPa).

These generalizations are of special importance when attempting to forecast the performance of polymer adhesive joints. For purely statistical reasons the fracture of the latter cannot occur at once along the entire interphase boundary, especially when dealing with a substrate in the elastic state. Yamomoto et al. [956], having calculated the propagation of deformation modes in the zone of interaction between the boundary layer of adhesive and the flexible substrate at a deformation rate of 10^{-3} m/s and 1 s duration, demonstrated that propagation of the geometric interface was rather large and purely kinetic in nature (Fig. 3.25). Therefore, the number of the mode (the distance from the interface) is a complex function of interfacial forces even at contact time of only 1 ms (Fig. 3.26).

Hence, when calculating the strength of adhesive joints one cannot ignore the stressed transition layers even for the two-dimensional model (stretching of composite rods, complicated by flexing), which does not take into account the residual stresses. Assigning these layers the thickness d^* and the shear modulus G^* for a lap joint shown in Fig. 3.27, Tourusov and Wuba [957] derived the following expression to describe the mean shear strength of the adhesive joint:

$$P_{sh} = \frac{P_0(E_a d_a + 2E_s d_s) + 2E_s d_s E_a d_a \, \Delta\epsilon_T(x/l) \, thx}{E_a d_a x \, tgx + (E_a d_a + 2E_s d_s) \, y \, cthy} \tag{326}$$

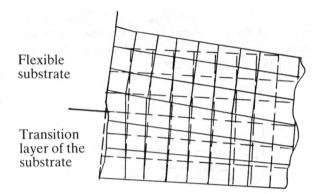

Fig. 3.25 – Diagram of the propagation of deformation modes in the interfacial zone of an adhesive joint at deformation with a rate of 1 mm/s for 1 s (solid lines, deformed modes; dashed lines, non-deformed modes).

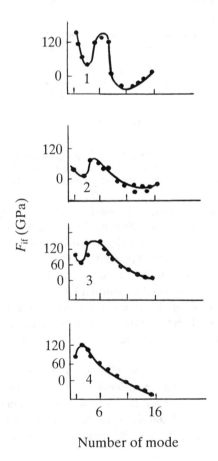

Fig. 3.26 – Distribution of interfacial forces by the deformation modes of adhesive joints versus adhesive–substrate contact time. 1, 10^{-3} s; 2, 10^{-2} s; 3, 10^{-1} s; 4, 1 s.

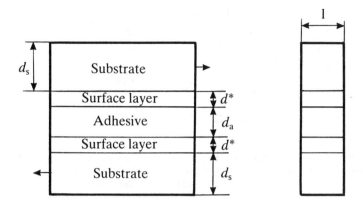

Fig. 3.27 – A model for the calculation of the strength of a lap adhesive joint.

where $x^2 = l^2G^*[(1/E_sd_s) + (2/E_ad_a)]/4d^*$ and $y^2 = l^2G^*/4E_sd_sd^*$, and the sub-indices a and s refer to the adhesive and substrate, respectively. Applying eq. (326), it is found that the dependence of P_{sh} on the geometry of the system and testing temperature is in good agreement with experimental data [957]. This case is quite complicated (in spite of the simplified model involved in the analysis), as both the substrate and the adhesive are under the effect of the active force field. The situation is somewhat simpler with composites; in this case the matrix can cause certain load relief within the substrate boundary layers [953, 954] and hence decreasing the concentration of overloaded bonds thus enabling an increase in the strength of the system. We believe that the correct approach to describing adhesion processes that would take into account both the formalism of the elasticity theory and interphase interaction must also include factors representing transition and boundary layers of the polymers constituting the adhesive joint.

One of such approaches is based on the studies of the fracture of adhesive joints under conditions when the effect of the deformation characteristics of the bulk of the contacting species may be neglected. In the procedure suggested by Regel *et al.* [958] thin polymeric films (0.1–10.0 nm) were cast from dilute polymer solutions on to the substrate to form adhesive joints which were subsequently subjected to thermal destruction by rapid heating (flash procedure) in the vacuum chamber of a mass-spectro-meter analyser, fitted with a special accessory (Fig. 3.28). A typical kinetic curve for the yield intensity of the products of thermal destruction in the system, e.g. stereoregular poly(methyl methacrylate)/nickel [959] is depicted in Fig. 3.29. Its pattern indicates that there were no side-factors affecting thermal destruction. From the results of the experiment kinetic parameters of the process can be derived. Fig. 3.30 presents the example of such data processing. Essential information concerning the removal of a polymer from a metal substrate is thus obtained [960].

As is evident from a comparison of the dependences of the thermal destruction of polymers in the bulk and in thin layers (Fig. 3.31), the values for activation energy coincide, the coincidence being retained, e.g. for polystyrene, down to a layer thickness of 1 nm. In thinner layers the mechanism of thermal destruction is probably different [961] as the activation energy rises to the values characteristic of the carbon–carbon

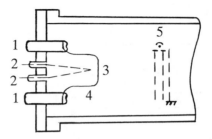

Fig. 3.28 – A schematic drawing of the accessory to the mass-spectrometer for investigation of thermal destruction in ultra thin polymer films. 1, Inlets; 2, thermocouple; 3, polymer sample; 4, adsorbent-heater; 5, ions source.

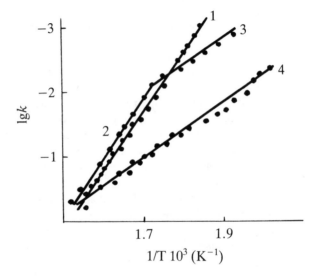

Fig. 3.29 – Kinetic curve for the yield intensity of thermodestruction in the system PMMA-nickel at polymer layer thickness of 20 nm.

bond scission. This is indicative that the metal substrate produces no catalytic effect, and that at $d < 1$ nm the thickness of polymer layer decreases to a value that is less than of a monomolecular one so that the macromolecules within this layer are isolated from one another[†] [776, 962]. A different pattern is observed in the case of thermal destruction of poly(methyl methacrylate) (see Fig. 3.29). It is supposed to be due to the interaction of carbonyl groups of the polymer with the active sites of the substrate. At a layer thickness of 1–7 nm activation energy of poly(methyl methacrylate) destruction is less than in the bulk though one can deduce that the total number of adhesive–substrate contacts is rather small. Steel was shown to produce enhanced interfacial interaction [963] owing to the fact that the water sorbed on the surface causes partial hydrolysis of poly(methyl methacrylate) to the polymeric methacrylate acid.

---

† This inference supports the views concerning the structure of boundary layers in polymers discussed on p. 152.

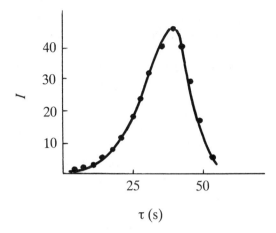

Fig. 3.30 – Thermodestruction rate constant for stereoregular (1, 4) and atactic (2, 3) PMMA on tantalum versus temperature. $k_1 = 10^{15} \exp (45 \times 10^3/RT); k_3 = 10^5 \exp (20 \times 10^3/RT); k_2 = 10^{14} \exp(44 \times 10^3/RT); k_4 = 10^6 \exp (20 \times 10^3/RT);$ PMMA layer thickness, 20 nm (1); 0.2 nm (2, 3); 15 nm (4).

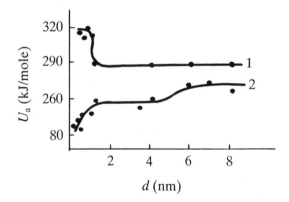

Fig. 3.31 – Activation energy of thermodestruction of the films of polystyrene (1) and PMMA (2) on tantalum substrate versus film thickness.

Hence, this method makes it possible to investigate at one and the same time the effects of the adhesive, substrate and the nature of the interfacial interaction, [964] on the performance of adhesive joints and that, combined with the layer-by-layer etching technique, it provides the additional possibility of exploring into the boundary layers of the contacting phases [965]. The efficiency of interfacial interaction is quite obviously determined by whether there are active sites existing at a substrate surface. In fact, the kinetic dependences of thermal destruction of thin polymer films are essentially different for ionic crystals, carbon fibres and metals as substrates [959]. In the latter case the active sites were detected as anionic vacancies within the oxide layers of the substrate. Exoelectron emission [966] and photodestruction [967] were the experimental techniques involved in these studies. Similar analysis can be carried out for systems more adequately modelling common adhesive joints. A polymer and a metal, after being compressed in vacuum (1.33–13.3 nPa) at a given temperature and load, were separated and examined in a mass-spectrometer by the flash-technique as described above. The results are depicted in Fig. 3.28. It was thus found [777] that when the elements of the adhesive bond display reasonable chemical interaction a thin polymer layer is left at the surface of the metal after the bond has been ruptured (for poly(methyl methacrylate)–tantalum this layer is only 0.5 nm thick [961, 968] and contains a reasonable number of ruptured macromolecules, the resultant radicals being retained under high vacuum. A more detailed analysis provides information concerning the position and orientation of functional groups of the adhesive at the substrate surface. For instance, electron-stimulated desorption of poly(methyl methacrylate) from tantalum and nickel involves three steps, differing in the cross-section of the process, desorption of H_3C^+ ions liberated from the ester groups disposed normal to the surface, parallel to it, and oriented into the bulk of the polymer phase [969].

Summing up the discussion of this section, once again we would like to emphasize that while from the viewpoint of thermodynamics the adhesive–substrate interaction is predominantly determined by the energetics of the phases in contact, it is the mobility of macromolecules in the transition and boundary layers in polymers that is decisive within the framework of the molecular–kinetic approach. Interrelationship of the basic ideas of both concepts appears to build up a foundation for a generalized theoretical treatment of polymer adhesion.

3.1.2 Interfacial interaction between polymers during adhesive contact

The adhesive–substrate interfacial interaction starts as a sorption process. While the analogy between adhesion and adsorption is almost obvious in terms of thermodynamics, complete identification of both phenomena in terms of molecular–kinetic ideas is unjustified. The adsorption contact of a macromolecule with a substrate is determined by the conformational set of a macromolecule, adsorption isotherms being complicated by the competitive effect of a solvent, developed substrate surface etc. Rigorous analysis demonstrates that during polymer adsorption the interaction energy varies continuously, indicating that it (polymer adsorption) is a phase transition of the second order [970–972]. On the other hand, transfer of a macromolecule from the volume of the solution into the confined space of a microdefect at the substrate surface is an energetically discrete process and takes place at a certain threshold value for the contact energy between the repeat unit and a substrate and represents a first order phase transition with an enthalpy effect proportional to the size of a microdefect [971, 973]. For instance, in computer simulation studies of the conformational characteristics of aliphatic hydro-carbon chains, involving the Monte-Carlo procedure in Metropolis' approximation, it was established that interaction with the substrate was essentially a function of the surface curvature of a substrate, the free energy minimum corresponding to the 'liquid' adsorbate at small values of the curvature radius, while at larger values to the phase transition to the more ordered 'crystalline' state [974].

At the same time, in some cases and relying on the analysis of interfacial interaction on the grounds of adsorption, it is possible to establish certain generalities valid for both adsorption and adhesion and to uncover the factors responsible for their manifestation. Generally, the free energy of a macromolecular chain in a single conformation attached to the substrate by isolated chemical bonds is determined by the free energy of the isolated chain F_1, those of the isolated local attachment F_2 and of the fraction of the chain between the attachment points F_3:

$$F = xF_1 + N_z F_2 + F_3 - \text{RT} \ln W_z. \tag{327}$$

F_2 is related to the energy associated with the change of stiffness in the local region, F_3 depends on the locus of the chain-substrate attachments, x denotes the degree of polymerization and N_z is the average number of attachment points. The last member in eq. (327) accounts for the probability pertaining to the different modes of interfacial contact. Then, the free energy change accompanying adsorption of macromolecules from solution, in the simplest case, is given as follows [975]:

$$\Delta F = -\Delta F_1 + \Delta F_2 + \Delta F_3. \tag{328}$$

In a more general solution the conformational spectra of the initial and adsorbed macromolecules are taken into account [976, 977].

When discussing the features of adhesion in energy terms one cannot avoid mentioning the approach based on Griffith's concept of brittle fracture [978] that has become common in the mechanics of composite systems. The analysis of this approach is beyond the scope of physical chemistry. Here it is appropriate to note that Griffith's fracture energy U_G, determining the critical stress for the propagation of a crack with the length l_G via the simple relationship

$$(2EU_G/\pi l_G)^{0.5} \tag{329}$$

can be expressed in terms of surface energy. For instance, Smith [979] established the relationship between Griffith's fracture energy and both the dispersion and polar components to the surface energies of adhesive and substrate

$$U_G = [\sqrt{\sigma_{sa}^d} - 0.5\,(\sqrt{\sigma_{sa}^d} + \sqrt{\sigma_{sa}^d})]^2 + [\sqrt{\sigma_{sa}^p} + 0.5\,(\sqrt{\sigma_{la}^p} + \sqrt{\sigma_{sa}^p})]^2$$
$$- 0.25\,[(\sqrt{\sigma_{la}^d} - \sqrt{\sigma_{sa}^d})^2 + (\sqrt{\sigma_{la}^p} + \sqrt{\sigma_{sa}^p})^2] \tag{330}$$

that demonstrated satisfactory correlation with the dependence of the peeling strength of adhesive tapes on the nature of various substrates [979] though less successful attempts to establish such a relationship had been undertaken previously. It can be assumed that provided that the initial prerequisites and the usage of thermodynamic parameters are adequately correct these attempts may serve as the basis for a unified approach to the assessment of the efficiency of interfacial interaction, combining both the physicochemical and the mechanical approaches. The validity of this statement is illustrated by Fig. 2.49, I, depicting the dependence between the compression modulus and the surface energy of polycyanoacrylates, which is rectilinear [692].

A different approach to the investigation of the adhesive–substrate interfacial interaction relies on the analysis of the data on adhesive joint fracture in terms of the kinetic concept of strength [980]. In this case the stress γ and the durability (life-time) t are interrelated via the exponential function

$$t = t_0 \exp{(U_a - \lambda\gamma/KT)} \tag{331}$$

where U_a is the activation energy of the process and λ is the structural coefficient. In heterogeneous systems internal stresses γ_0 arising from sample prehistory and joint formation procedure have to be taken into account

$$t = t_0 \exp \{[U_a - \lambda(\gamma \pm \gamma_0)]/KT \}. \tag{332}$$

The parameters of eqs (331) and (332) can be determined from the temperature–load dependences of the durability of the system.

Note that the time-variation of γ_0 is also described by an exponential function. Indeed, according to the definition [981]

$$d\gamma_0 = E_a \, d\xi_a \tag{333}$$

where E_a is the elasticity modulus of the adhesive, and ξ_a is the relative contraction of the adhesive due to crosslinking after gelation. Taking into account that E_a and ξ_a are related to the kinetic parameters describing the growth of adhesive joint strength [982, 983], then the internal stresses should be dependent on the apparent activation energy of crosslinking U_a^* and to τ_τ and τ_g which relate to the starting time of the crosslinking process and to the time interval required for the adhesive to form a crosslinked gel [984]:

$$\gamma_0 = 0.5 \, E_\infty \xi_\infty \left[1 - \exp \left(- \frac{\tau_\tau - \tau_g}{\tau_g \exp (U_a^*/RT)} \right) \right]^2 \tag{334}$$

(where subscript ∞ denotes the equilibrium state). The validity of eq. (334) was verified [984] for epoxy adhesives as an example and the effect of internal stresses on the strength of adhesive joints was described by a complex expression [985, 986, p. 22].

The activation energy of the fracture process is determined predominantly by the nature of the substrate. This conclusion follows from the data on the fracture of systems, particularly fibrous composites, for which the activation energies of fracture of the system on the whole and of the matrix *per se* are equal [987, 988]. In the case considered, the volume fraction of the fibre (i.e. of the substrate) is important. For instance, the tensile strength of composites is linearly related to the filler content (for short fibres only), the slope of the line being determined by the nature of the matrix [989]. It is quite illustrative that discrete variations of U_a from values corresponding to fracture of the adhesive up to those relating to the fracture of the substrate is character-istic of such dependences (Fig. 3.32) [777]. Detailed analysis [990] demonstrated that discrete values of the activation energy of the fracture of glass-reinforced plastics are multiples of the activation energy of the process of chemical destruction of the substrate U_s, i.e.

$$U_a/U_s = N_i \tag{335}$$

where N_i denotes whole numbers associated with the number of the bonds ruptured in a stressed composite during thermal fluctuations. A certain variation in the values of activation energy can be attributed to the presence at the substrate surface of various impurities, e.g. of water sorbed on the glass fibres [991, 992], as was the case with thermal destruction of thin polymer layers (*vide supra*, section 3.1.1).

Consequently, the activation energy of adhesive joint fracture cannot be accepted as an unambiguous criterion of the efficiency of adhesive interaction. This conclusion is confirmed by the data compiled in Tables 3.3 and 3.4 listing the activation energies of fracture of autohesive (Table 3.3) and adhesive (Table 3.4) joints, the reported activation energies having been determined from thermal dependences described by the Arrhenius equations [22]. Recent calculations of U_a for the adhesive joints of polystyrene with

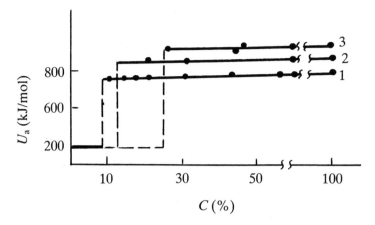

Fig. 3.32 – Activation energy of the failure process in composites based on aluminium alloy
D-16 versus volume fraction of tantalum (1), molybdenum (2) and boron (3) fibres.

polyamide and with polyethyleneterephthalate [1004] unequivocally indicate that the
data obtained relate to the viscous flow of the materials rather than to the fracture
process.

The pre-exponential factor t_0 of eqs (331) and (332) whose magnitude is close to
0.1 ps [777] is also valid for the purpose. Hence the structural coefficient λ [1005]
appears to be the only parameter related to the efficiency of interfacial interaction.
According to its physical meaning [980], λ is determined by the structural features of
an adhesive and a substrate as well as by the structure and mechanical characteristics of
the boundary and transition layers, by the geometry of the system etc. This is confirmed
by the data compiled in Table 3.5, listing the values of the parameters in eq. (331).

A similar conclusion follows from the data illustrating the force dependence of
durability for epoxy-based composites filled with polyvinyl alcohol fibres (Fig. 3.33)

Table 3.3 – Activation energies of the fracture of autohesive joints

Polymer	T (K)	U_a (kJ/mol)	Ref.
Low density polyethene	354–378	251.2	[993]
High density polyethene	371–405	268.0	[993]
Polyisobutene	293	11.7	[994, 995]
Polystyrene	353–468	209.3	[996]
Poly(butadiene–co–acrylonitrile) (88.3 : 11.7)	293	6.7	[997, 998]
Poly(butadiene–co–acrylonitrile) (80.4 : 19.6)	293	20.1	[997, 997]
Poly(butadiene–co–acrylonitrile) (72 : 28)	293	36.4	[997, 998]
Poly(butadiene–co–acrylonitrile) (63.1 : 36.9)	293	69.5	[997, 998]

Table 3.4 – Activation energies of the fracture of adhesive joints kJ/mol

Adhesive	Substrate				
	Steel	Aluminium	Copper	Glass	Cellophane
Polyethene	–	––	–	–	18.9 [999]
Polyisobutene	–	11.3 [1000]	16.6 [1000]	–	–
Poly(vinyl chloride)	52.4 [1001]	–	–	–	–
Polybutadiene	–	21.8 [1000]	22.2 [1000]	–	–
Poly(vinyl chloride–co–vinyl acetate)	–	–	–	–	52.4 [1002]
Poly(butadiene–co–styrene) (70 : 30)	–	–	–	52.4 [1003]	–
Poly(butadiene–co–acrylonitrile) (82 : 18)	–	25.6 [1000]	51.5 [1000]	–	–

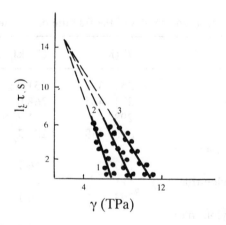

Fig. 3.33 – Durability of composites reinforced with polyvinyl alcohol fibres versus stress (at 293 K). 1, The fibres *per se*; 2, composite with an epoxy-based matrix; 3, ditto except that the fibres were modified by methylene diisocyanate.

Table 3.5 — Energetic parameters of the fracture process of various polymers and of their adhesive joints[a]

Adhesive			Substrate			Adhesive joint		
Name	U_a	λ	Name	U_a	λ	U_a	λ	Ref.
Polypropene	126	—	Polyvinyl alcohol	147	0.108	147	0.077	[1006]
Polypropene	126	—	Polyvinyl alcohol[b]	147	0.108	147	0.065	[1006]
Polyvinyl alcohol	130	0.602	Viscose	176	0.172	184	0.172	[992]
Epoxy oligomer	147	2.666	Glass fibre	335	0.086	293	0.043	[992]
Epoxy oligomer	147	2.666	Carbon fibre	1275	0.473	1257	0.602	[992]
Unsaturated polyester	189	—	Glass fibre	356	—	159	—	[1007]

a U_a [kJ/mol]; λ [kJ m² (mol μN)$^{-1}$].
b The surface was modified by diisocyanate to generate interfacial chemical bonds.

[1006, 1008]. Durability of the system notably varies when passing from the individual substrate (Fig. 3.33, 1) to its adhesive joint in which either the van der Waals' (Fig. 3.33, 2) or chemical (Fig. 3.33, 3) (due to coupling with diisocyanate) bonds are realized. However, the three lines depicting these three cases, on being extrapolated to the low-stress region, intersect in a single point. This fact serves to show that the efficiency of interfacial interaction is related to the structural coefficient, which is given as the slope of the dependences, while the activation energy of the process stays constant. As λ increases the strength of adhesive joints decreases regularly due to stress accumulation in the local sites of fracture [980]. On the other hand, coating the substrate surface with the coupling agent leads to an increase in its surface energy resulting in enhanced adhesive interaction with the matrix, as was demonstrated for polyethylene terephthalate mono-filaments [1009].

The distinct mechanism of interfacial interaction leading to the formation of macro-scopic contacts in a system comprising a polymer is determined by the specific features of polymer structure. Among these features the prime one is the mobility of macro-molecular chains, which is liable to change during the adhesive interaction. According to the fundamental minimization principle the driving force for adhesive interaction is provided by the free energy decrease at the interface. Thus, the activated mobility of macromolecules and their individual fragments or segments can lead eventually, via chain diffusion, to mutual intertwining and entanglements forming interfacial bonds. Even such stiff chains as those of epoxy-oligomers can diffuse into polyheteroarylene fibres, the resulting being the strengthening of corresponding composite structures at a rate proportional to the penetration depth of epoxy-oligomers [1010]. Thus, a transition layer of diffusion origin is formed e.g. for polystyrene—poly(methyl methacrylate) adhesive joints its thickness is $2-8\,\mu m$ [1011, 1012]. In authohesive joints of natural rubber diffusion of the vulcanizing agent across the interface into the polymer bulk can be as deep as 2 cm [1013]. The existence of such layers in various polymeric systems was reliably detected by electron microscopy [1014], X-ray diffraction studies, mechanical and dielectric loss studies [1015], hydrodynamics, attenuated total internal reflection spectroscopy, and ellipsometry [1016]. These investigations provide evidence that the formation and properties of these layers are governed, in the final analysis, by the chemical nature and mobility of contacting macromolecules.

Such an approach to treating the problem of interfacial interaction of high molecular mass compounds is sufficiently consistent and non-contradictory. In fact, it is an integrated approach combining thermodynamic concepts relating to surface phenomena with the molecular–kinetic considerations regarding the behaviour of polymers at the interface. It is therefore not accidental that the equation proposed by Saito [1017] to calculate the work of adhesion at peeling and including the effect of the diffuse layer comprises two terms, the first of which is identical to the thermodynamic dependence (eq. 98), while the second accounts for the molecular–kinetic dissipation of the energy in the system. When one neglects the effect of the interfacial layers, the calculated thermodynamic parameters of the interaction in binary mixtures of polymers are in notable disagreement with both the original as well as with the novel Flory theory [1018]. Obviously, one should not overestimate the influence of diffusion factors on the modes of adhesive joint formation [22, 1019], especially when dealing with incompatible systems or with those at phase equilibrium [774]. However, such considerations create a physically non-contradictory basis when seeking to examine interfacial interaction of high molecular mass compounds from uniform principles.

The mechanisms providing for the mobility of macromolecular chains at the interface are determined by the nature of the phases in contact, particularly, by their polarities. For instance, for equally polar polymers interfacial interaction is most probably due to diffusion across the interface to a small depth, probably, not deeper than the size of a segment. Interdiffusion to a larger depth is quite unfavourable, if at all possible from both thermodynamic and structural viewpoints. In the case of adhesive interaction between polymers of different polarities Voyutskii and Vakula [1020–1023] suggested a mechanism of local diffusion mindful of microheterogeneity of the polar materials and the possibility that only certain structural units may cross the interface. In a system comprising both polar and non-polar fragments microseparation (as in blends of incompatible polymers) due to segregation of the fragments with different overall dipole moments should take place. Such a model makes it possible to grasp the basic notions concerning the compatibility and diffusion of polymers. However, one should not identify the conventional regions with the real structures like domains or with the *a priori* introduced [1024, 1025] 'smallest elements of supermolecular structures'.

Local diffusion is most probable in systems of relatively low-polarity polymers comprising flexible and mobile segments as diffusing elements [1004]. This can account for the crack-healing of notched [1026] or even broken [1027] samples of poly(methyl methacrylate) or poly(methyl methacrylate–*co*–ethyl methacrylate) during thermal treatment as well as for the decreased strength of adhesive joints in which the adhesive is a non-polar elastomer and the substrate is a polar plastomer (polar elastomers are actually incapable of interfacial interaction with non-polar plastomers [1021, 1028]). During the adhesion of amorphous and crystalline polymers, clearly, dissolution of the former in amorphous regions of the latter takes place. In both cases, according to Flory [1029] and Eyring and Kauzmann [1030], diffusion should involve segments as the diffusing units, hence the activation energy of the process is actually independent of the size of the diffusing macromolecules [1031].

Indeed, in contrast to the diffusion of low-molecular mass compounds into high polymers when the diffusing molecules are moving independently of each other [1032], in the case of interdiffusion of polymers the diffusing macromolecular segments have to overcome strong inter- and intramolecular interactions as well as structural impediments such as network nodes, entanglements, loops etc. This may result in the potentially mobile segments exhibiting a significantly lower number of degrees of freedom than predicted in the idealized model [1030] due to internal stresses arising at the molecular level and the cooperative character of the diffusion process. As a result, while the peeling strength of autohesive joints of polydimethylsiloxane elastomers is 15–25 J/m^2, it falls to 60–97 mJ/m^2 after even very moderate preliminary crosslinking [1034]. As a matter of fact effects of this kind govern the rheological properties of polymers. In this case however they are associated with a certain gradient of the applied external stress favouring, contrary to diffusion, the forced straightening and disentanglement of the chains. In fact, for polyethene–polytetrafluoroethene [1034] and polyamide–steel [1035] adhesive joints it was discovered that the activation energy of wetting is higher than that of the viscous flow of adhesive, thus influencing the deformation mode and, consequently, the fracture mechanism. Indeed, at low peeling rates the 'pulling apart' of the diffused macromolecules takes place, while at higher rates, when the role of diffusion is actually negligible, the macromolecules *per se* are ruptured [1036].

Diffusion of macromolecular chains is accompanied by rearrangement of polymer transition layers. In the case where the polymers are compatible this factor can affect the formation of supermolecular structures in amorphous and crystalline polymers in close proximity to the boundary layers [1038]. Therefore, the entire span of the region with changed structure may be as large as 10–100 nm. It should also be taken into account that diffusion can proceed not only in the normal direction with respect to the interface but laterally [22, 1037] as well. In the case of adhesive joints lateral (surface) diffusion leads to the situation that during the formation of the joint the functional groups at the surfaces of the phases migrate so as to occupy that position which ensures enhanced interfacial interaction.

The rate of diffusion characterized by the frontal migration of the ends of diffusing elements increases constantly up to a certain limit and then slows to a stop. Naturally, even when diffusion has ceased the system is still far from its equilibrium state corresponding to the entropy maximum. It is therefore reasonable to infer that in real polymeric systems diffusion phenomena take place only in quite definite confined regions. As regards the size of these regions, most modern theories of diffusion assume that they are of the size of individual macromolecular segments [1039, 1040].

Within the framework of di Benedetto's model [1041] the diffusing molecule is represented as a three-dimensional oscillator entrapped inside the cell comprised of four segments. Diffusion starts when any of these segments is ruptured. By making the model more exact Pace and Datyner [1042] succeeded in deriving the diffusion equation in which the only mass parameter of the pre-exponential term was that of a segment.

On the other hand, it is only the diffusion of segments that determines the effective change of conformations of the macromolecules in boundary layers [1043]. In fact, according to de Gennes [368], the characteristic time τ_s required for a segment to be displaced to the distance l_s, which is of the same order of magnitude as the size of the segment itself, can be taken as the measure of local conformational rearrangements. This quantity was shown to govern [1040, p. 43] both the number of neighbouring segments with which the motion of any distinct segment correlates at the time τ_x

$$N(\tau_x) \simeq (\tau_x/\tau_s)^{0.5} \qquad (336)$$

and the mean-square displacement of a macromolecular segment in the course of diffusion

$$<R^2(\tau_x)> \simeq 6D(\tau_x \tau_s)^{0.5}. \qquad (337)$$

In its turn the diffusion coefficient is dependent on the size of a statistical segment

$$D \simeq l_s^2/\tau_s. \qquad (338)$$

Note the consistency of this approach with the data in Fig. 2.27 depicting the relationship between the temperature gradient of surface energy in polymers and the number of repeat units in their segments [410].

Obviously, though the conformational effects and interdiffusion cannot be the only causes for the transition regions in polymers this does not imply that the concept of local diffusion is incorrect, as might have been suggested when relying on the theory of the interface developed on the basis of statistical physics. According to Helfand [1004, 1046] the structure of the transition layer is determined by the energy of contacts between the repeating units of macromolecules and by the conformational entropy gain, if there are no such contacts. The Flory–Huggins interaction parameter, which is rather small for incompatible polymers, was taken into account in assessing the entropy factor [1044]. The results obtained led the researchers to conclude that compressibility of the systems of non-compatible polymers is limited [1045] and the thickness of transition layers in these systems is almost negligible [1046]. However, these deductions are contrary to the experimental data presented in section 3.1.1, hence the concept of segmental solubility appears to be true. The discrepancy between the calculated and experimental data is due to the limitations of the physical implications supporting Helfand's concept, since measurements of the interaction parameter in polybutadiene–polydimethylsiloxane performed recently by Anastasiadis et al. [1047] were in poor agreement with the predictions of the theory [1048, 1049] concerning the effect of molecular mass.

The best and the most illustrative proof that diffusion is an inherent feature of polymer adhesion is supplied by electron microscopy studies [1014, 1039, 1050]. In these it was demonstrated that interphase contact between polymers is accompanied either by interdiffusion resulting in the formation of the transition layer characterized by a certain concentration gradient of the starting materials, or by the unidirectional diffusion resulting in the formation of a boundary layer with a more compact packing of macromolecules [1051–1054]. The latter case is supported by the data of Table 3.2. Moreover, it was demonstrated that even in systems of incompatible polymers the size of diffusing macromolecular fragments is comparable to that of the segments. In semi-compatible systems, e.g. in those of poly(methyl methacrylate) with chlorinated polyethene and polyvinylchloride, the thickness of the transition layers and of the interdiffusion regions is the same, as demonstrated by X-ray microanalysis and SEM [1055].

This conclusion is in reasonable accord with the data of electron microscopy studies of the diffusion fluxes as well as of the structure and concentration profiles in interfacial regions in poly(methyl methacrylate)–polyvinylidenefluoride [1056]. For less compatible polymer pairs, like poly(methyl methacrylate)-chlorinated polyethene (chlorination degree 66%), this conclusion is true only after the system has been thermally treated at 420–454 K to stimulate interfacial transfer processes [1057]. It is therefore clear that in systems comprised of compatible polymers (authohesion is the extreme case) the rate and the depth of diffusion should be much higher than in previous cases. This inference is supported by the data of studies with radioactive tracers [1058, 1059] and conventional light-transmitting microscopy [1060]. Electron microscopy was unable to detect the termination of diffusion process in autohesive joints of polystyrene and butadiene-acrylonitrile copolymer SKN-18 (the elements of the joint differed in molecular mass) [1053].

Hence, taking into account the kinetic character of diffusion, the growth of adhesive joint strength P during the formation period can be attributed to the effect of time-dependent factors. Thus at the initial stages of adhesive contact the high tensile properties cannot be expected. For the reptation model [1061] Kim and Wool [1062] demonstrated that $P \propto \tau^{1/4}$ (however, applicability of this model is limited as real polymeric chains are branched as shown in measurements of the self-diffusion coefficient of polybutadiene by small-angle neutron scattering [1063]). This, however, does not justify the doubts expressed by Tchalykh [1039], concerning the universality of the diffusion mechanism of adhesive joint formation, provided that there is molecular contact between the phases. In this respect the results of simultaneous measurements of adhesive joint strength and of the diffusion coefficient in polyamide–steel adhesive joints at contact times less than 30 ms are quite illustrative. Ueno et al. [1035], who performed the study, demonstrated that the joint strength was due to the diffusion of the segments of the adhesive. The studies of durability of adhesive joints in liquids [1064] and an analysis of the strength versus contact duration dependences for autohesive joints of elastomers [1065] have led to similar conclusions. There are grounds to believe that neither thermo-dynamic nor molecular–kinetic features impose any principal limitations on the manifestation of such a mechanism of interfacial interaction.

While examining the interrelationship between diffusion and compatibility phenomena, as demonstrated in polymers, it is appropriate to note that the features characteristic of the latter are in agreement [1038, 1066, 1067] with the considerations discussed above. While interfacial surface tension σ_{if} at the interface between solutions of incompatible polymers is close to zero [1068], and therefore, cannot be taken as a measure of compatibility, in the case of adhesion contact between solid polymers, even with similar polarities, σ_{if} acquires sufficiently high values, e.g. mN/m:

			(Ref.)
poly(methyl methacrylate)/ poly(butyl methacrylate)	(293 K)	3.4	[400]
polyethene/polydimethyl–siloxane	(423 K)	5.1	[1069]
polyhexamethyleneadipamide/ polystyrene	(543 K)	5.2	[1070]
poly(methyl methacrylate)/ polyethene	(293 K)	11.4	[400]

Then, as a first approximation σ_{if} may be assumed to be characteristic of the height of the energy barrier between the contacting polymers and, therefore, must have been correlated with the strength of the adhesive joints. This thesis is confirmed by the data in Fig. 2.48 depicting the peeling strength of various adhesive joints versus the difference between the surface energies of the contacting phases. This approach [684] has gained a certain popularity (e.g. see [688]). Michel [569], for instance, discovered that the strength of joints with polyacrylate as adhesive was the highest with those polymers whose surface energies were very close to those of the adhesive (Fig. 3.34).

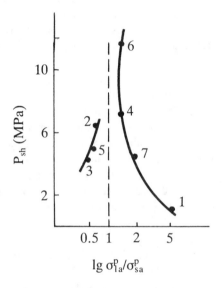

Fig. 3.34 – Shear strength of adhesive joints versus ratio between polar components to surface energies of adhesive and substrate. Bonded species: polypropene (1), poly(vinyl-chloride) (2), poly(methyl methacrylate) (3), polychloroprene (4), polyhexamethylene adipamide (5), polystyrene (6), styrene–acrylonitrile copolymer (7). Adhesive polyacrylate formulation.

Hence, thermodynamic compatibility in polymeric blends should be closely related to the excess of the Gibbs' potential. Actually, however, there is no correlation between ΔG values [1071] and stability of two-phase systems [1067, p. 24]. This is quite understandable as the value of ΔG is determined both by the Flory–Huggins interaction parameter χ and by the thickness of the transition layer. Having assumed that at a thickness of this layer less than 0.5 nm the interphase boundary can be considered as sharply defined, Krause [1072] demonstrated that at $\Delta G = 0.008$–0.05 for two- and and three-block copolymers $\Delta G/RT < 0$, the tendency being such that for systems with a less diffuse transition layer the value of the group was more negative. At $\Delta G = 0.07$ $\Delta G/RT$ is positive when the interphase boundary is sharply defined and is negative when it is diffuse.

Thus, when using these characteristics of compatibility to assess adhesive interaction one should consider the sign of the group rather than its absolute value, despite the fact that the mixing energy excess in two-component systems was once [1067, p. 24; 1073] claimed to be identical to the free surface energy. For instance, in the system comprising polystyrene and poly(methyl methacrylate), for which $\chi > 0$, interdiffusion is due to the increase of the interphase contact area resulting from wetting of the substrate with the adhesive [1074]. However, for autohesive joints with $\chi = 0$, and, moreover, for adhesive joints with $\chi < 0$ (e.g. between PMMA and styrene-acrylonitrile copolymer with 9.5–33% acrylonitrile content) diffusion is highly favourable and hence the strength of the joints increases with contact time [1074] being greater the more negative the χ value.

The intensity of the diffusion process is determined, in the final analysis, by the mechanism of interfacial interaction. For instance, according to Krause [1075], in 75% of the 282 pairs of compatible polymers interaction is due to specific bonds, e.g. hydrogen bonds, between the phases. Compatibility (miscibility) of the blends of cis-1,4-polyisoprene with atactic (97%) polybutadiene-1,2, which are homogeneous at the segment scale, arises owing to the almost equal polarizabilities of the components. Therefore, dispersion forces, which were shown to dominate in the blends, remain largely unchanged in magnitude upon blending [1076]. This case is an exception to the condition requiring a negative Flory interaction parameter for high polymers to be miscible. Due to the invariance of the intermolecular potential, mixing is not accompanied by changes in the conformational state of the polymer chains. This results in a density of topological constraints in the blends that varies monotonically between those found in the pure components.

These views are in agreement with both the experimental obervations [1077] and with the concepts of the theory of polymer adsorption. According to the statistical thermodynamics approaches of Helfand [1044, 1049] and Roe [1078], minimization of the surface energy leads to monolayer adsorption [1079] and hence it is at the segmental level that interdiffusion takes place (according to Kuleznev [1067], even with incompatible pairs, though this is unacceptable to Chalykh [1039, p. 253]). This is confirmed by studies on polymer blends using ESR [1080] and electron microscopy [1052]. The analogy with the features reported concerning the formation of interfacial layers in block-copolymers [1081] supports the above thesis also. Silberberg [1082] claims that the properties of macromolecules adsorbed at solid surfaces are governed by the energetics of the interactions at the segment level. For instance, Northolt [1083] demonstrated for a number of polymers that the activation energy of segmental motion is directly related to Young's modulus and to the glass-transition temperature:

$$\Delta H_s^* \propto E^{1.4 \pm 0.5} \propto T_g^{2.3 \pm 0.3}. \tag{339}$$

It is clear that the effectiveness of blending processes and, consequently, the efficiency of adhesive interaction [1084] are associated with the flexibility of macromolecular chains. An increase in their stiffness results in cooperative binding of the repeat units with the substrate surface [1085]. Recalling that the number of repeat units in a segment may serve as the characteristic of adhesive properties of polymers (see section 3.1.1) one may find the above speculations quite useful.

On the other hand, it is known that the thickness of the transition layer does not always correlate with the segmental size as well as the time-dependence of the strength of adhesive joints with segmental solubility rate. Hence, one may suggest that segments are not the only structural elements of transition layers [1086]. Perhaps, thermodynamic factors govern compatibility only at the initial stages of the process, the thickness of the resultant transition layers being not greater than the size of a segment. Subsequently, owing to the accumulated stresses and overall non-equilibrium character of the system, macromolecules on the whole and, as in adsorption, their aggregates, can become involved in mass transfer.

The data presented demonstrate that polymer compatibility is governed predominantly by structural factors, these being manifested, as in diffusion, at the segmental level [1038, 1039, 1067]. Indeed, the critical molecular mass of polystyrene at which it is entirely compatible with polyisoprene and poly(methyl methacrylate) is comparable to the molecular mass of its kinetic segment [1087]. In systems comprised of non-polar elastomers, like polybutadiene SKD or poly(butadiene-co-styrene) DSSK-18, and butadiene–acrylonitrile copolymers this effect is observed at a molecular mass, which is less than that of a segment [1088]. Hence, the physical implications of segmental solubility become clear [1068] (as well as the reasons for the proposal to use the alternative term 'forced segmental compatibility' [796, p. 89]). In agreement with these views, the transition layer between polymers of different chemical type is diffuse, its thickness being inversely proportional to the difference in polarities of the individual components. For instance, in systems of gutta percha with butadiene–styrene or butadiene–acrylonitrile elastomers, or of polybutadiene with chlorinated poly(vinyl-chloride) the diffuse interphase boundary is 1 μm thick (as measured by luminescent microscopy), whereas in blends of polybutadiene with natural rubber, whose polarities are very close, the interface actually ceases to exist [109]. As a result, the complex of physico-mechanical characteristics of blended systems is changed, those blends with a thinner segmental layer exhibiting lower brittleness and fatigue resistance [1088]. The condition that the sizes of the segments of polymers brought into contact are comparable is of general validity and therefore, the theory of segmental solubility is assumed to be universal in describing the structure and properties of polymer blends in the glassy, elastic, and viscous-flow states [1088].

Hence, the diffusion mechanism of interfacial interaction is supported by the generalities guiding the compatibility of high molecular mass compounds. The common features of both processes (enabling particularly, an assessment in terms of diffusion the compatibility of polymers with ester plasticizers [1088] and oligomers [1089, 1090] to be made) are due to the fact that they are both acting at a segmental level. This inference appears to be more sound than the hypothesis [1091, 1092] which assumes the possibility of infinite diffusion in an adhesive joint, since migration of macromolecules on the whole is scarcely feasible, whereas migration of their segments is provided for by their molecular masses corresponding to the region of solubility. Even when compatibility between the components is low this type of dissolution can take place via conformational changes in boundary layers resulting in increased surface entropy.

The analogy discussed makes it possible to elaborate on the principal question of how and to what extent structural factors affect the mechanisms of polymer compatibility and diffusion. The effect of structural factors is associated with the concept of segmental mobility in macromolecular chains. According to current ideas, the latter is related either to conformational entropy (thermodynamic concept) [1093], or to the free volume (molecular kinetic concept) [1094]. Clearly, the former cannot, in principle consider the real nature of high molecular mass compounds, and is confined to giving a phenomenological description showing the direction of the processes, this is well illustrated by the analysis of the initial model theories [1095, 1096]. On the other hand, the free volume concept appears to be hopeful in that it enables one to assess the effect of structural factors on inter- and intramolecular interactions. The addition of one polymer to the other affects the local chain mobility; however, these changes, according to present ideas [1076], are related to the fact that the free volume depends on the composition of a system. In this regard Chalykh claims [1039, p. 115], 'The rate of the processes of structural rearrangement is explicitly associated with segmental mobility, which, in its turn, is determined by the average fraction of the free volume of the diffusion medium.'

The free volume of a macromolecule can be represented as the total of the geometric V_g and physical V_v components† interrelated by Eyring's equation [1097]:

$$V_g = - V_v \ln V_v. \tag{340}$$

V_g is usually derived from the thermophysical characteristics of polymers [1098] with the aid of the Simha–Boyer equation [1099]

$$V_g = (c_1 - c_2)T_g, \tag{341}$$

in which c_1 and c_2 are the thermal expansion coefficients at temperatures below and above the glass–transition temperature T_g, respectively. A fact of peculiar interest when speaking of adhesive joints is that at constant T_g ($V_{g(T_g)} = 0.025 \pm 0.003$ [1098]) the free volume decreases with the increase of the filler content [1100, 1101]. This is evidently related to the fact that the packing coefficient is actually constant at the glass transition temperature. Hence, the decisive contribution to the decrease of the free volume at adhesive interaction is produced by Δc (for instance, heat transmittance of polyvinyl butyral filled with cured phenol formaldehyde oligomer is linearly related to the thickness of the interphase layer [1102]), which is, in the final account, determined by the mobility of macromolecular chains.‡ In fact, eq. (341) is identical to the relationship combining the glass transition temperature with the internal rotation braking coefficient [1106].

The magnitude of V_g is quite sensitive to the polymer chemical structure. In fact, for the sufficiently flexible macromolecules of polybutadiene, polyisoprene, natural rubber, and butadiene-styrene (with the styrene content not exceeding 30%) elastomers $V_g = 0.097 \pm 0.008$, while for the stiff-chain butadiene–acrylonitrile SKN-40 and fluoro-containing SKF-26 rubbers V_g is consistently higher i.e. $V_g = 0.113 \pm 0.11$ [1101].

Campion [1107] has made an attempt to clarify the reasons which determine the variations of the geometrical component of the free volume in relation to the features of polymer autohesion. In Stewart's model of a macromolecule he distinguished the voids, and the intermolecular (due to packing of the chains) and intramolecular (due to the voids in the repeating units) free volumes. The latter is determined by the molecular mass M_1 and the length l_1 of a repeat unit, and by the structural coefficient k_g, which, depending on the coordination number (varying discretely from 3 to 6), acquires the value of 2.0, 1.0, 0.67 and 0.5, via the following expression

---

† Strictly speaking, the third component, which is due to anharmonic aspect of the atomic vibrations, has to be accounted for as well.

‡ For network structures the crosslink density has to be taken into account. Correlations between the crosslink density and T_g [1103], and between T_g and V_v [1104], established for epoxy oligomers, serve as an indirect proof of the statement, whereas the linear relationship between the packing coefficient and the concentration of hydroxy functions [1105] in these oligomers (which pre-determines the crosslink density) provides the direct evidence.

$$V'_g = V_g M_1 / N_A k_g l_1.$$ (342)

The calculated contribution of the intramolecular component to the geometric free volume [1107] amounted to 41.6% for natural rubber, 19.2–22.7% for butadiene–styrene elastomers (depending on the styrene content), 26.5% for *cis*-polybutadiene, 16.5% for butyl rubber and 4.4% for ethene–propene copolymer. As can be seen, the intramolecular free volume decreases in stiff-chain macromolecules and tends to zero in crystalline polymers. This pattern is clearly due to purely steric reasons. Actually, the order described provides the prerequisites which enable one to relate the parameters of diffusion processes to the structural parameters of the objects. In fact, in polymers incorporating repeat units which contain double bonds the intramolecular free volume is greater than in the ethyene–propene copolymer and correspondingly, the autohesive ability of this copolymer is moderately lower.

These features are assumed to be closely associated with the concepts of diffusion. This thesis was first noted by Frenkel [166]; however, the problem was treated quantitatively only in terms of the diffusion model of volume relaxation. This model relies on two initial assumptions (a) the free volume diffusion coefficient is a function of local concentration and (b) the entire variation of the free volume is due to its diffusion from the bulk phase to the surface and from the surface into the bulk on cooling and heating, respectively [1108, 1109]. For the problem under discussion most important is assumption (b). Analysis of the consequences of this model showed [450, p. 147] that the free volume is associated with clusters of *n*-defects. This fact makes it possible to relate the concepts under consideration to the behaviour of polymer surface layers surveyed in section 3.1.1. When examining the other aspects of the problem one has to take into account the relationship of the free volume concept to the thermal fluctuation ideas on polymer strength. As demonstrated above, of all the parameters of eq. (331) the structural coefficient λ is most susceptible to the variations in efficiency of the adhesive interaction. From the standpoint of mechanics λ is indicative of the stress concentration in a loaded body [980] however, Regel and Slutsker [1110] assume that in a general treatment λ is equivalent to the volume of a macromolecular chain with the diameter equal to the length of the carbon–carbon bond. Taking the fluctuation volume as a characteristic of λ clarifies the interrelationship between λ and the work of adhesion observed in systems of incompatible polymers [1111].

However, despite the objective character of the standpoints discussed the approach based on the free volume concept has not yet been systematically developed in the physical chemistry of polymer adhesive joints.

The efficiency of interfacial interaction between polymers providing, in the final analysis, for strong and durable adhesive joints is associated with two conditions. Firstly, the system has to be chosen so that mass transfer across the interface is thermo-dynamically favourable. Secondly, macromolecular chains of the contacting phases have to be sufficiently mobile to ensure maximum contact area.

Let us examine how these concepts coincide with experimental data.

The first one stems from the fundamental principle of interfacial energy minimization determining the drop in the height of the energy threshold at the interface. More or less explicitly this principle is involved in the treatment of various aspects of the phenomenon surveyed. The validity of the minimization principle is supported by the data in Fig. 2.48, which shows that close values of the surface energy of an adhesive and a substrate provide for the highest tack of various compositions [684]. Results obtained elsewhere [336, 668] also lend suppert to this assertion. In the current section this conclusion was further extended to involve the shear strength of adhesive joints (Fig. 3.34). It should be emphasized that this principle is a requirement of monolayer adsorption [1079] and is related to the compatibility maximum in systems comprised of components with minimal differences in polarities [109].

The validity of this approach was supported in a number of recent studies. Toyama [1112] verified the uniformities depicted in Figs 2.48 and 3.34 for systems involving three adhesives (derived from natural rubber, polyacrylate, and polyvinyl ester) and eight distinct substrates ranging from the high-energy poly(hexamethylene adipamide) to the low-energy polytetrafluoroethene. Wu [1113] calculated that the interfacial energy at the interface between poly(hexamethylene adipamide) and poly(ethene$-co-$propene) (63 : 37) is 9.7 mN/m, which is quite a large value to ensure adequate adhesive inter-action. Introduction into poly(ethene$-co-$propene) of carboxy functions to the amount of 1% decreases σ_{if} to 0.25 mN/m [1113] producing a much better adhesive interaction. Similar results were obtained in studies with other materials [1114], the maximum adhesive strength being achieved when the magnitudes of the surface energy of adhesive and substrate were close to each other. Having taken these considerations into account, one may recognize the criterion of haemocompatibility suggested recently by Ruckenstein and Gourisankar [1115], according to which interfacial energy should not exceed 1–3 mN/m.[†]

What are the mechanisms providing for the minimization of interfacial energy? Apparently, the only plausible molecular mechanism of this process should involve structural rearrangement in the boundary layers and clearly, the properties of these layers will be correspondingly altered. This conclusion is supported by the data produced by a diverse variety of methods, mechanical and dielectric relaxation, inverse-phase gas chromatography, and differential scanning calorimetry [1118]. In fact the rod-like mesogenic molecules at the interface with a nematic liquid-crystalline phase are oriented normal to its surface, while when the phase is isotropic these molecules are oriented lateral to the interface[‡] [1119]. As compared to the free films, i.e. those in contact with air, the contribution of the polar component of surface energy of a PMMA film in contact with glass is increased to 1.1 mN/m (from the initial value of 0.1 mN/m), whereas for polystyrene film contacting glass the fraction of the surface occupied by the polar groups is increased from 38 to 61% [1120]. Similarly, the surface concentration of the polar component in a styrene–butyl acrylate copolymer is increased during exposure of the copolymer to water [1121] leading to a decrease of interfacial energy in the system. With segmented polyurethanes comprised of 4,4'-diphenylmethanediisocyanates, ethylene-diamine and polyethers the surface energy at the interface with air tends to a minimum due to accumulation of the ether fragments at the interface [1122] (accordingly, in agreement with the criterion stated above, these materials acquire the initial haemo-

---

[†] Raising an objection against such an interpretation of this criterion, Sharma [1116] claims that the effect of selective sorption by the substrate of individual blood components, for example of proteins, has to be taken into account. However, it is because of the fundamental character of the basic condition that the suggested criterion is of general meaning, the different stages of the process of adhesive interaction being 'automatically' reckoned for [1117].

[‡] This effect is attributed to the change of excluded volume [1119] driving one back to the structure–volume concept, including the diffusion aspect, considered above.

compatibility due to minimal platelet adhesion). Finally, we would like to note that all the described correlations are also valid for polypeptides [1123]; this topic, however, is beyond the scope of this monograph.

Thus, the first of the prerequisites providing for the efficient interfacial interaction between polymers is also of general physical validity. It is easy to see that, strictly speaking, this condition also determines the second one, which requires that the molecular kinetic aspects giving rise to polymer adhesion be taken into account. Hence, as was emphasized in the Introduction, it is physically incorrect to treat the effects of each of the approaches separately.

At the same time, one cannot completely ignore cases where the factors of molecular–kinetic origin prevail over the thermodynamic ones. Particularly, this is the case with butadiene–acrylonitrile elastomers. Due to the high cohesive energy of these elastomers the content of nitrile groups in the copolymer would not actually affect the strength of adhesive joints produced at room temperature. This is illustrated by the data of Fig. 3.35 with regard to the fractionated samples.[†] For commercial products characterized by a broad molecular mass distribution and the presence of impurities the growth in the concentration of polar groups results in a more regular decrease of the adhesive joint peeling strength (Fig. 3.36). These results demonstrate that the dependences of the type depicted in Fig. 3.36(1) and observed in most practical cases are due to the different purity of the materials under investigation.

On the other hand, increasing the contact temperature up to 423 K leads to the patterns of the corresponding dependences exhibiting maxima. The rise of the adhesive joint strength produced by thermal treatment is greater the higher the polarity of the adhesive. The latter fact is attributed to the rise of the mobility of elastomer chains with

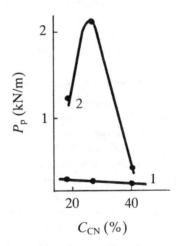

Fig. 3.35 – Peeling strength of adhesive joints of poly-εcaproamide with fractionated butadiene–acrylonitrile copolymers (the fraction with $M = 320 \times 10^5$) versus concentration of nitrile groups in macromolecules of adhesives. 1, 293 K, 2, 423 K.

[†] Here and elsewhere the dependences applying to butadiene–acrylonitrile copolymers were confined to a limited set of the corresponding elastomers. Therefore, these results hould be regarded as a qualitative illustration of the validity of the features discussed.

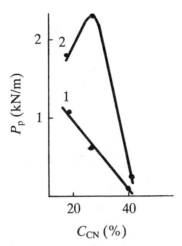

Fig. 3.36 – Peeling strength of adhesive joints of poly-ε-caproamide with unfractionated butadiene–acrylonitrile copolymers versus concentration of nitrile groups in macromolecules of adhesives. 1, 293 K; 2, 423 K.

temperature owing to the loosening of intermolecular interaction in transition layers, e.g. thermal treatment of tetrafluoroethene-perfluoropropylvinyl ester copolymer at 533 K decreases the surface energy by 8.5% [1124].

However, polarity of elastomers is not the only factor affecting the efficiency of interfacial interaction. Since nitrile adhesives differ in their dispersion, one may presume that it is this factor that imposes the decisive effect on the intensity of diffusion processes across the interface. In fact, the fraction with a molecular mass less than 25 000 amounts to 91.5% in SKN-18, 14.6% in SKN-26 and 9% in SKN-40 [1125]. Obviously, according to Fick's second law, the diffusion-controlled mass transfer is intensified as the molecular mass of the diffusant is decreased, hence, SKN-18 copolymer should be the one of the three to produce the maximum adhesive joint strength. However, the effect of molecular mass is significant predominantly at elevated temperatures thus enabling enhanced chain mobility. As a result graph (1) in Fig. 3.35 is almost parallel to the abscissa. The data of Fig. 3.37 obtained at room temperature add proof to the above statement. As the temperature is increased, the latter dependence, in agreement with prediction, acquires a continuously ascending character (Fig. 3.38). For the elastomer with the maximum nitrile group content this effect is much less pronounced (Fig. 3.37(3)).

The consistencies of diffusion processes, as relating to the intensity of adhesive interaction, are more explicitly demonstrated in autohesive polymer joints. Here the height of the energy barrier at the interface is minimal and the ultimate strength of the system is, to a large extent, determined by the interpenetration depth of macromolecular segments. Two cases are considered. If the interpenetration depth is relatively small, the work of peeling is predominantly due to overcoming the intermolecular interaction [1126]

$$W_p = \pi^{1/2} k_s^{1/2} (2N_A)^{1/3} F_m (\rho/M)^{1/3} \tau^{0.5\,(1-k_D)}/l_{C-C} \cos(90 - \varphi/2) \qquad (343)$$

where F_m is the intermolecular bond rupture force, l_{C-C} the carbon–carbon bond length, φ the valence angle and k_s and k_D the constants pertaining to the mobility of

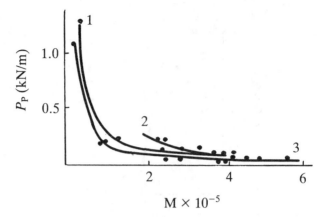

Fig. 3.37 – Peeling strength of adhesive joints of poly-ε-caproamide with butadiene–acrylonitrile copolymers versus molecular mass of the adhesive (bonding temperature 293 K) 1, SKN-18; 2, SKN-26; 3, SKN-40.

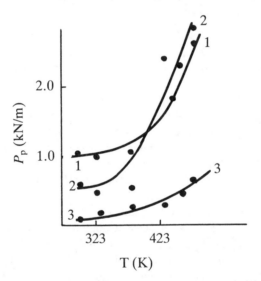

Fig. 3.38 – Peeling strength of adhesive joints of poly-ε-caproamide with unfractionated butadiene–acrylonitrile copolymers versus bonding temperature. 1, SKN-18; 2, SKN-26; 3, SKN-40.

macromolecular chains and to the diffusion coefficient as a function of time, respectively. On the other hand, if the segment penetration depth is sufficiently large the peeling (separation) of the system is necessarily accompanied by rupture of the macromolecules. Then Young's modulus of a single carbon–carbon bond E_{C-C} has to be considered, while the parameter F_m has to be substituted for F_{C-C} describing the scission of a covalent bond. Finally, one comes to

$$W_p = \pi^{1/2} k_s^{1/2} (2N_A)^{2/3} F_{C-C} (\rho/M)^{2/3} \tau^{0.5\,(1-k_D)} / 2E_{C-C} l_{C-C}$$

$$\cos (90 - \varphi/2). \tag{344}$$

The joint strength dependences, corresponding to relationships (343) and (344), are represented in terms of diffusion coefficients D, molecular masses M and densities ρ of the separate components 1 and 2 [1039, p. 253] via the following expressions: for autohesion

$$P_{Aut} \propto D \tau^{0.5\,(1+k_D)}, \tag{345}$$

for adhesion

$$P_{Ad} \propto [D_1^{1/2}(2\rho_1/M_1)^{2/3} + D_2^{1/2}(2\rho_2/M_2)^{2/3}]\tau^{0.5\,(1-k_D)}. \tag{346}$$

A more precise analysis is based on the macromolecular reptation theory, which considers the mean length and the average number of interfacial 'bridges', the mean length and the average number of the chains crossing the interface, the mean penetration depth and the average number of the bonds being ruptured [1127]. It is clear that even in the case of the autohesion joint corresponding expressions are rather complex, while introduction of simplifications leads to dependences of limited physical implications. The expression for the number of chains crossing the interface in an autohesive joint of mono-disperse amorphous polymers can serve as an example [1128]:

$$N(\tau) \propto N_A \rho A_n M^{-3/4} \tau^{1/4} \tag{347}$$

where A_n is the geometric interfacial contact area.

The two fracture mechanisms of adhesive joints described by expressions (343) and (344) can be easily distinguished by analysing the work on peeling versus molecular mass of adhesive dependences. For instance, for the quasiequilibrium systems formed at constant and rather large contact times [1125, 1129] it was found that

$$W_p \propto M^{-2/3}, \tag{348}$$

which is in accord with eq. (344).

The strength of polymer adhesive joints as a function of their molecular mass can also be examined within the framework of the thermodynamic approach. In this case surface energy is the only fundamental property of the adhesive involved and there is no need to use the hard-to-obtain parameters of single bonds; however, and this is most important, this approach is not confined to considering the diffusion mechanism of adhesive joint formation alone. The problem is solved with the aid of eq. (227). Making use of the data in Table 2.6 we have calculated the surface energy of unfractionated polyisobutene samples with M in the range $(9.3-242.0) \times 10^4$ and compared it with the data on peeling strength of its autohesive joints [997] and of its adhesive joints with iron and copper [1130], and aluminium and zinc [1131].

As is seen from Fig. 3.39, 1, for autohesive joints between unfractionated polyiso-butene samples there is observed a linear dependence between P_p and σ. For fractionated samples of polyisobutene whose autohesive joint strength was described by the work on peeling [1125] the corresponding dependence is also strictly linear (Fig. 3.39, 2) and that is in good agreement with the initial concept. Note that contrary to the known approach developed by Vasenin [1132, 1133] the data obtained apply to systems tending to equilibrium and whose formation times are several hours rather than minutes. Fig. 3.40 depicts the metal–polyisobutene adhesive joint strength as a function of surface energy of the adhesive. The general character of this interrelationship [1134] is yet another

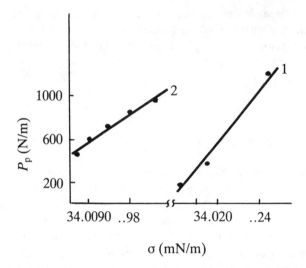

Fig. 3.39 – Peeling strength of autohesive joints of polyisobutene versus surface energy. 1, Unfractionated samples, bonding time 72 h; 2, fractionated samples, bonding time 1 h.

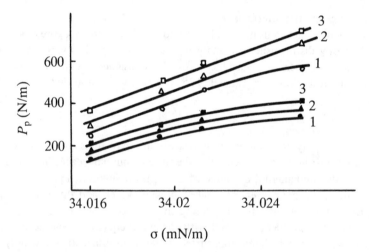

Fig. 3.40 – Peeling strength of the adhesive joints of steel (open circles) and copper (solid circles) versus surface energy of polyisobutene adhesive. 1, Untreated joint; 2, joint, thermally treated at 423 K; 3, joint, thermally treated at 448 K.

proof of validity of this approach. As is seen from Fig. 3.40, for copper, whose σ is relatively small, there is only a slight deviation from linearity of the P_p versus σ dependences due to the slight effect of the substrate on the internal stresses within the adhesive and relaxation of these stresses during thermal treatment linearizes the $P_p(\sigma)$ function. For steel substrates the observed consistencies are even more pronounced.

In the cases considered the absolute variation of σ in the range of high molecular masses is very small and its precise measurement is beyond the capabilities of present day experimental technique. However, the relative variation follows a regular pattern and is quite significant in stipulating the corresponding variation of the efficiency of adhesive interaction [1134]. This inference is supported, for instance, by the measurements of the energy of intermolecular interaction in polyethyleneterephthalate fibres formed from the polymer with different molecular masses [1135]. Taking into account the ideas discussed above it is probably more rigorous to use interfacial rather than surface energy. In this case, in accordance with expectations, for incompatible polymers the $\sigma_{sa}(M)$ function is given by eq. (348) as exemplified in the studies of Anastasiadis and Koberstein [1136] involving polydimethylsiloxane with polybutadiene-1,4 and polystyrene with hydrogenized polybutadiene-1,2 systems. For poly(propylene glycols) and their copolymers with poly(ethylene glycols) in contact with polyethene and poly(vinyl acetate) a linear relationship holds true [1137, 1138]

$$\sigma_{if} \propto M^{-1}. \tag{349}$$

Nevertheless, σ_{sa} is much more available than σ_{if} insofar as it can be easily calculated on the basis of a physically sound approach and using experimentally measured polymer characteristics [521, 530].

This trend was supported in the studies of adhesive joints between vulcanized butadiene–styrene rubbers and steel bonded by polychloroprene adhesive. To ensure the high strength of these joints primers based on chlorinated natural rubber (pergut) were used. As demonstrated by calculations using eq. (227), in a broad range of molecular masses $(2.5-15.7) \times 10^4$ a strictly linear function of the kind of eq. (349) is valid [1139]:

$$\sigma \propto M^{-1}. \tag{350}$$

Computer processing of the data (see Fig. 3.41) [1139] demonstrated that regardless of the duration of adhesive joint formation the observed dependences are approximated by first-order equations (see Table 3.6). The calculated values of Student's criterion $T_{calc.}$ are larger than the tabulated ones ($T_{tabul.} = 2.0146$), thus supporting the unequivocal reliability of the approximation. It is quite significant that, at longer times of adhesive interaction, paired correlation coefficients r start to decrease. This fact can be considered as one of the first pieces of direct physico-chemical evidence that P and M obey an inverse relationship, a fact that has been previously accounted for in terms of mechanics [22, 1126]. Then, the coefficients of eq. $P_p = AM + B$ should be directly determined by the

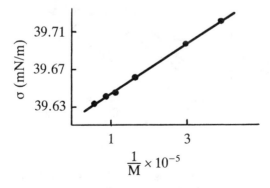

Fig. 3.41 — Surface energy of pergut as a function of its molecular mass.

Table 3.6 — The effect of molecular mass of the primer on the strength of vulcanized rubber–steel adhesive joints

τ (h)	Coefficients of eq. $P_p = AM + B$		r	$T_{calc.}$
	$A \times 10^5$	B		
2	0.57578	0.23214	0.915	15.3434
6	1.05697	0.57673	0.939	18.5862
24	1.42139	1.03984	0.934	17.7250
72	1.62838	1.25590	0.962	23.6547
168	1.51967	1.85599	0.962	23.6871
240	1.66550	2.22460	0.848	13.6480
312	1.52243	2.53483	0.835	10.2896

variation of adhesive joint strength. In fact, as is illustrated in Fig. 3.42, the A versus P_p dependences tend to linearity at large M, while those of B at lower values of M [1139].

At the same time, investigation of the kinetic dependences of A and B during adhesive interaction reveals an inflection point at 72 h exposure time prior to debonding the system (Fig. 3.43). The same threshold value was discovered when examining the time-dependent variation of adhesive joint strength as a function of the molecular mass of the primer. The latter fact reflects the effect of the two mechanisms of adhesive joint

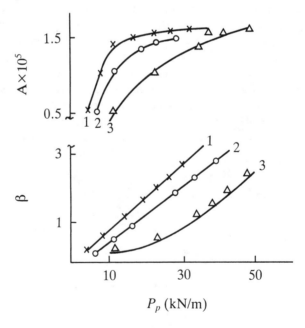

Fig. 3.42 – Coefficients of the first-order equation in Table 3.6 versus peeling strength of adhesive joints involving pergut primers of molecular mass 2.5×10^4 (1), 6.2×10^4 (2), and 15.7×10^4 (3).

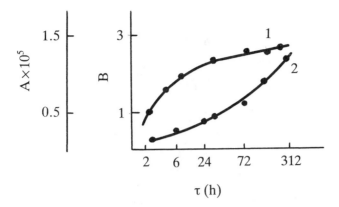

Fig. 3.43 — Coefficients of the first-order of equation (Table 3.6) versus exposure (bonding time) of adhesive joints prior to debonding.

strength (fracture) considered above, namely of inter- and intraphase origin. Apparently, during the first 3 days (72 hours) adhesive interaction is governed by the surface energy of the primer, which is a function of its molecular mass. Beyond this time-span diffusion processes prevail thus levelling off the effect of thermodynamic factors.

The extent of such processes, as demonstrated above, is governed by the mobility of the macromolecular chains of an adhesive. This is a quite generally valid conclusion and it holds true for both high and low contact temperatures. This is illustrated by the data of Fig. 3.44, where mobility is assessed by the value of s derived from eq. (323). It is quite clear that the dependence is almost linear for polydisperse samples for which the more subtle effects either become constant or are compensated for. However, for fractionated polymer samples the P_p versus s dependence is still a monotonic one with no maxima, as can be seen in Figs 3.35 and 3.36.

The strength of autohesive joints of ethyene-propene, butadiene, chloroprene, and butadiene-acrylonitrile elastomers follow similar patterns [1140]. However, most outstanding are the results of investigations with cyanoacrylate adhesives [84], which display unambiguous and reproducible characteristics. Fig. 3.45, which is similar to Fig. 2.49, shows the shear strength, while Fig. 3.46 shows the tensile strength of the joints produced with α-cyanoacrylate adhesives in both instances as a function of s of the corresponding macromolecular chains. Note the strict linearity of the graphs obtained [692]. Moreover, analysis of the variation of adhesive properties of the materials with regard to the values of parameter s makes it possible to reveal differences in the mechanisms of flexibility of the adhesive chains. As can be seen from Fig. 3.46, compounds with side chains containing either ether groupings (polyalkoxyalkyl cyanoacrylates) or ester groupings (polycarbalkoxyalkyl cyanoacrylates) are described by individual graphs. Taking into account the unambiguous nature of the interrelationships [694] it is natural to suggest the number of repeat units in a segment as being the molecular—kinetic equivalent of the thermodynamic characteristic (i.e. surface energy) of the adhesive properties of polymers. Then there should obviously be a direct relationship between these parameters and in fact, this is just the case, as illustrated by Fig. 3.47 [84].

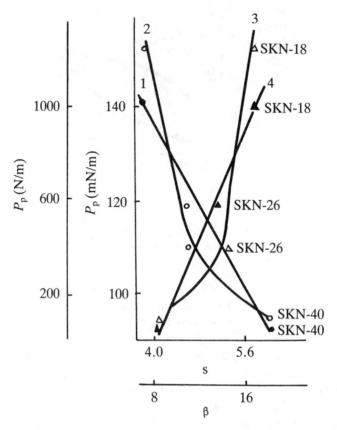

Fig. 3.44 – Peeling strength of the adhesive joints of poly-ε-caproamide with fractionated (open symbols) and unfractionated (solid symbols) butadiene–acrylonitrile copolymers produced at bonding temperature 293 K versus mobility of the macromolecules of adhesive (1, 2), and parameter β of their compatibility with the substrate (3, 4). Left ordinate is for fractionated samples; right ordinate is for unfractionated samples.

The fundamental role of parameter s necessitates the need to examine its physical meaning as related to the processes of polymer fracture. Firstly, the compression modulus of an adhesive is determined directly by the value of s; the corresponding relationship depicted in Fig. 3.45 is strictly linear. This is a fact of general importance, which is confirmed by a number of constant features of polymer mechancs examined in terms of cohesive energy E^s and segment volume V_{m_s} (s is defined [909] as the ratio of these two parameters). For instance, in the low temperature range [1141], when certain features of the behaviour patterns of the materials subjected to the effect of force fields have become constant, the impact viscosity of polymers and the size of the region of slow stable crack propagation are inversely proportional to V_{m_s}. The latter fact demonstrates the need to consider Giffith's ideas discussed earlier.

On the other hand, deformation, viscous flow, diffusion, and, consequently, the adhesive properties of polymers are determined, as is known, by α- and β-relaxation processes. The first are displayed near the glass-transition temperature, whereas the latter are seen below it. Their interrelationship follows from a simple expression describing the ratio between the activation energies of the corresponding transitions [1142]:

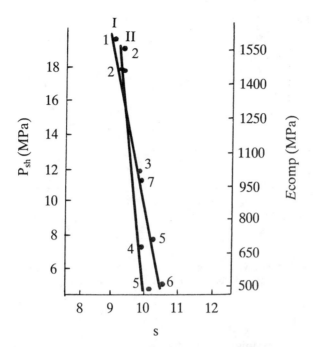

Fig. 3.45 – Shear strength of adhesive joints of aluminium alloy D-16 (I) produced with methyl- (1), ethyl- (2), propyl- (3), butyl- (4), amyl- (5), heptyl- (6), and allyl-α-cyano-acrylates (7); and compression modulus of polymeric adhesives (II) versus calculated equilibrium flexibility of macromolecules of adhesives.

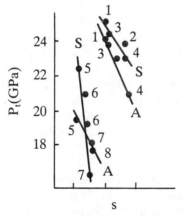

Fig. 3.46 – Tensile strength of the adhesive joints of steel H18N9T (S) and aluminium alloy D-16 (A) produced with polyalkoxyalkyl- $-CH_2-C(CN)-COOR$ (1–4) and poly-carbalkoxyalkyl-α-cyanoacrylates $-CH_2-C(CN)-COOCH_2COOR'$ (5–8) versus calculated equilibrium flexibility of macromolecules of adhesives. R = $-CH_2CH_2OCH_3$ (1); $-CH(CH_3)CH_2OCH_3$ (2); $-CH_2CH_2OC_2H_5$ (3); $-CH(CH_3)CH_2OC_2H_5$ (4); R′ = $-CH_3$ (5); $-CH_2H_5$ (6); $-C_3H_8$ (7); $-C_4H_9$ (8).

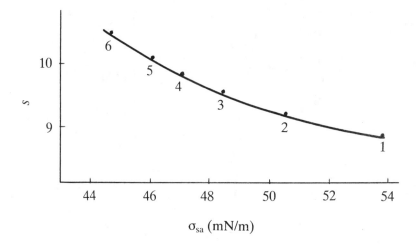

Fig. 3.47 – Surface energy versus equilibrium flexibility of macromolecules of methyl- (1), ethyl- (2), propyl- (3), butyl- (4), amyl- (5), and heptyl-α-cyanocrylates (6).

$$Q_\alpha/Q_\beta \approx 4 \pm 1. \tag{351}$$

For the purposes of examining the general rules of adhesion the development of β-relaxation processes is most important. To discuss the effects of these processes one can make use of the molecular model of local motion in chain molecules proposed by Skolnick and Helfand [1143]. According to the theory for the rate of conformational transitions (trans–gauche) of bonds in chain molecules (such as alkanes and polymers), developed on the basis of this model, the rotational motion of the transforming bond is accompanied by motion in neighbouring bonds, but the extent of this motion diminishes with distance. Berstein and Egorov [1144] demonstrated that this motion ceases at a distance from the rotational axis that is commensurate with the size of the equilibrium (Kuhn's) segment. Then, it becomes quite clear why the effective activation volumes of relaxation in the β-relaxation region actually coincide with the values for V_{m_s}. This was the case observed for polyethene, poly(vinyl chloride), poly(methyl methacrylate), polystyrene, and polycarbonate [1145, 1146]. In the course of conformational transitions the energetic parameters, namely the activation energy Q_β and cohesion energy of a segment E^s, must be changed in a similar way. In fact, the corresponding interrelationship was recently reported for 24 polymers of very different chemical nature (Fig. 3.48) [1144]. The equation describing this relationship is as follows [1147]:

$$Q_\beta \approx (0.30 \pm 0.05)E^s + 15 \text{ [kJ/mol]} \tag{352}$$

where 15 kJ/mol is the minimal increment of the β-transition to the potential barrier.

As far as the modes of adhesive interaction of butadiene–acrylonitrile copolymers are concerned, one should expect that with the account of the general principles discussed above the role of structural factors, which substantially affect the efficiency of interphase diffusion processes, is increased with a rise in temperature. Then the nature of the adhesive can be assessed by means of the parameter A of eq. (195) as it is susceptible to the structure of the transition layers in polymers as well as to their energetic characteristics [395]. The respective dependence is shown in Fig. 3.49. For fractionated

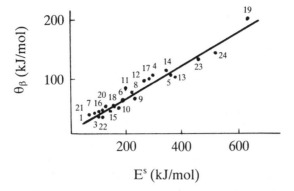

Fig. 3.48 – Activation energy of β-transition versus molar cohesion energy of segments of polyethene (1), polypropene (2), polyisoprene (3), polystyrene (4), poly-α-methyl styrene (5), polyvinyl fluoride (6), polyvinyl chloride (7), polytetrafluoroethene (8), polyvinyl alcohol (9), polyvinyl acetate (10), polymethyl acrylate (11), polymethyl- (12), poly-butyl- (13), and polycyclohexyl methacrylates (14), polyacrylonitrile (15), polycarbonate (16), polyoxy- (17) and polyphenyloxyphenylenes (18), polyethyleneterephthalate (19), poly-m-phenylene isophthalamide (20), polyhexamethylene adipamide (21), polydimethyl- (22) and polydiethyl siloxanes (23), polyimide and 3.3′, 4,4′-diphenyloxide tetracarbon acid (24).

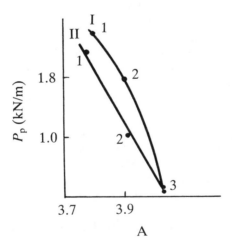

Fig. 3.49 – Peeling strength of the adhesive joints of poly-ε-caproamide with unfractionated (I) and fractionated (II) butadiene–acrylonitrile copolymers SKN-18 (1), SKN-26 (2), and SKN-40 produced at 423 K versus the value of the constant of Hildebrand's equation.

elastomers at 423 K the dependence is rectilinear. At the same time, no explicit relation-ship was found between P_p and A at 293 K, or between P_p and s at 423 K.

In order to determine which of the two factors, kinetic or compatibility, produces the predominant effect on the intensity of interfacial interaction between polymers one has to rely on a specific parameter that describes compatibility. The so-called compatibility parameter β characterizing the energy barrier at the interface can be used in this way. It is expressed as the difference between solubility parameters of the contacting phases

$$\beta = (\Delta\delta)^2. \tag{353}$$

Relying on the values of effective molar cohesive energy and the van der Waals' volumes of repeat units [476, 477] with the aid of eq. (202) we calculated the δ values for three butadiene–acrylonitrile elastomers and poly-ϵ-caproamide. The β values derived from eq. (353) for the three systems comprising each of these elastomers and poly-ϵ-caproamide as the common substrate were 16.8, 13.7, and 8.4, respectively. For the non-fractionated samples of elastomers at 293 K the P_p versus β dependence is linear (see Fig. 3.44) and, moreover, it is entirely symmetrical to that of P_p versus s. For monodisperse adhesives the dependence, as may be seen, deviates substantially from linearity. At the same time, in accordance with prediction the P_p versus β dependence becomes degenerate as the temperature increases, thus indicating that the role of compatibility has become constant. Such effects are seen when there are no significant differences in the residual stresses arising during formation of the adhesive joint, or when these stresses are counteracted by reducing the strength of a system, the difference between the glass-transition temperature of the adhesive and the peeling strength test temperature being the reducing parameter [1148].

These data serve as direct evidence of the different mechanisms of interfacial interaction between elastomers and poly-ϵ-caproamide at low and elevated temperatures. In both cases these mechanisms are governed by molecular–kinetic factors (mobility of macromolecular chains); however, it is firstly the effect of the translational flexibility (s) that dominates and secondly that of structural origin (A). As the temperature is increased the structure of transition layers undergoes most significant changes (reflected in A) promoting the transfer of the lowest molecular mass fractions of the elastomer and, consequently, stimulating the effective interpenetration of the structural units of adhesive across the interface. The fact that these units can be identified with segments is due to the physical implications of parameter s [909]. Investigation of adhesive interaction between butadiene–styrene elastomers containing carboxy functions and poly-ϵ-caproamide as substrate provokes similar conclusions. In this case formation of hydrogen bonds or of donor-acceptor complexes [1149] might be assumed to be intensifying factors.

In general, three types of adhesive systems should be distinguished, elastomer–elastomer, elastomer–plastomer and plastomer–plastomer. The latter two, as mentioned above, are quite specific as the increased rigidity of plastics (compared to elastomers) hampers the diffusion processes across the interface. The behaviour of elastomer–elastomer systems should, however, follow the described patterns and, moreover, more subtle effects related to specific features of molecular motion in adhesives might be revealed in this case.

In this regard investigation of the generalities of elastomer autohesion (i.e. there is no need to consider the effect of factors related to an activation barrier at the interface) appears to be quite informative. We have compared the shear strength of systems involving elastomers of differing polarities [1150] with the flexibilities of their chains [910–912] and with the constant A of eq. (195) [395]. Corresponding graphs are depicted in Fig. 3.50 and, as is seen, these are linear. In both cases, however, the data pertaining to butadiene elastomer SKB do not fit the graphs. This is probably due to ignoring the nature of the end groups and the structural features of the product obtained under specific conditions (polymerization in the presence of sodium). Note that, with the exception of the joints between butadiene–acrylonitrile copolymer and poly-ϵ-

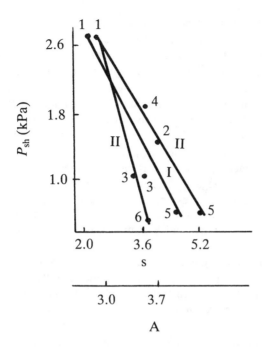

Fig. 3.50 – Shear strength of autohesive joints of elastomers versus mobility of macro-molecular chains (I) and the value of the constant of Hildebrand's equation (II). 1, natural rubber; 2, polybutadiene elastomer SKB; 3, butadiene–styrene elastomer SKS-30; 4, poly-chloroprene elastomer; 5, butadiene–acrylonitrile elastomer SKN-26; 6, butyl rubber.

caproamide the strength of autohesive joints is related to the structural coefficient A even at low temperatures. The relationship established makes it possible to differentiate the elastomers under investigation according to their polarity. Indeed, the dependences pertaining to low-polarity and to polar polymers are depicted as individual graphs. There is a point common to both, however, that corresponds to natural rubber. This is probably due to the fact that although, according to its chemical nature *per se*, natural rubber is a low-polarity elastomer its dipole moment is significantly larger than that of the closely related polymers because of the unavoidable low-molecular mass and proteinaceous impurities. For instance, the refractive index of natural rubber, being directly character-istic of its polarity, is 6.8% higher than that of polyisoprene. Along with the actual linearity of these dependences, this feature is indicative of the high sensitivity of the proposed parameters towards the modes of interfacial interaction between polymers and, consequently, the strength of their adhesive joints [410, 1151].

When extending these relationships to various combinations of elastomers, one should note that these dependences only tend to linearity whilst not being strictly linear. With the increase of s and A the shear strength of adhesive joints between elastomers regularly increases in the case of polar substrates (chloroprene and butadiene–acrylonitrile rubbers) and decreases in the case of low-polarity ones (butadiene, natural, butadiene–styrene, and butyl rubbers). As in the above case (see Fig. 3.50) the point corresponding to natural rubber is common for both graphs.

Differentiation of materials by their polarities is indicative of the different mechanisms of interfacial interaction involved. This can be due to bulk as well as to surface effects. Indeed, for some polymers the adhesive joint strength versus surface energy dependences can be distinguished for polar and low-polarity materials according to the slope of the graphs (see section 2.2.2). Quantitatively this case can be examined with the aid of a parameter a [708] as the assessment criterion. This parameter is the free term of the computed polynomials $P(\sigma)$ or, alternatively, the ratio P/P^* of eqs (305)–(307). It is related to the properties of the substrate and the loading mode of the system (either peeling or shear) only and its values are compiled in Table 2.11. According to the data of Table 2.12, as an interfacial parameter a is closely related to the characteristics of the bulk of a polymer. Note the regular increase of the correlation coefficient for the sets of polar and low-polarity substrates, as compared to their common set (in peeling). While the overall correlation coefficient lies within the range of 0.485–0.751, it is actually at the same level for polar materials and is significantly higher for the low-polarity ones, reaching 0.874–0.993.

These results are in good agreement with the conclusion drawn above claiming that interfacial interaction of low-polarity polymers is predominantly developed via a blending mechanism, being governed by the surface properties of high molecular mass compounds. For polar polymers the more pronounced effect of volume (structural) factors is characteristic.

With regard to these ideas it would be desirable to have an overall parameter that would reflect the fundamental polymer characteristics governing the modes of adhesive interaction. Such a parameter should take into account the effect of the properties of both the bulk and the surface of a polymer. It is quite usual to estimate the bulk properties by the effective molar cohesive energy, which is dependent on the packing coefficient and van der Waals volume, and the surface properties by the surface energy.[†] Hence, the parameter α was introduced [708] and defined as follows

$$\alpha = E^*/\sigma. \tag{354}$$

Maintaining the continuity of the approaches discussed above (sections 2.2.1 and 3.1.1) it is logical to express eq. (354) in refractometric terms. For this purpose by substituting E^* from eq. (216) into eq. (217) one can express σ as a function of the solubility parameter:

$$\sigma = k_0^{-3/2} A^{-7/3} \left(\sum_i \Delta E_i^* \right)^{3/2} \left(\frac{n^2 - 1}{n^2 + 2} \right)^{7/6} \delta^{-2/3} \left(\sum_i m_i a_i r_i \right)^{-7/6}. \tag{355}$$

Having performed rearrangements with eq. (355) taking into account of eqs (216), (217) and (354), one arrives at an expression for α relating it to refractometric characteristics, coefficients of Hildebrand's equation, molar packing coefficient and the solubility parameter [708, 1153]:

---

† Using similar reasoning Akopyan et al. [1151] proposed a much more complicated approach, according to which the cohesion properties of polymers are described by the conventionally equilibrium stress, while σ is determined for the strained sample. However, the relationship between these parameters reveals no dependable criterion of the type described and analysis involving two variables appears to be necessary in each case.

$$\alpha = k_0 A^{7/3} \left(\sum_i m_i a_i r_i \right)^{2/3} \delta^{-1/3} \left(\frac{n^2 - 1}{n^2 + 2} \right)^{-2/3}. \tag{356}$$

When analysing this expression the feature which initially attracts attention is that the ratio between the refraction of a repeat unit and the Lorenz–Lorentz function of the refractive index can be identified, according to eq. (215), with the volume of a macro-molecule. This fact is of fundamental importance for the physical chemistry of adhesion phenomena in so far as, on the basis of the simplest premises, it enables an unequivocal interpretation of the earlier conclusion stating that the molecular–kinetic features of adhesive joint formation between polymers are guided by structural factors. The latter are undoubtedly stipulated by the differences in the mutual arrangement of the topological units (these can be the main and side chains, entire macromolecules or supermolecular structures) identified with definite volumes. It is quite significant that this deduction was made in terms of thermodynamics, thus confirming the inherent uniformity of the thermodynamic and molecular–kinetic approaches to adhesion. Finally, note that the values for the power index of the basic variable in eqs (344) and (356) are the same. The molecular mass of a polymer appears to be such a variable in the first of the equations, while it is the volume in the second. Hence, the experimentally observed dependences of adhesive joint strength on $M^{2/3}$ [1125, 1136] get the physical support as, strictly speaking, the molecular mass and its distribution are associated with volume related effects.

Listed below are the values for parameter α, calculated according to eq. (356) using the reported values for refractive indexes [509], solubility parameters, molar packing coefficients [478], refractions and coefficients of Hildebrand's equation [395]:

Polyalkenes and their derivatives

polyethene	20.7	(1.0)
polypropene	66.6	(3.2)
polyisobutene	75.3	(3.6)
polyacrylonitrile	78.2	(3.8)
polyvinylchloride	65.4	(3.2)
polytetrafluoroethene	33.7	(1.6)
polystyrene	122.6	(5.9)

Oxygen-containing polymers

poly(methyl methacrylate)	99.4	(4.8)
poly(vinyl acetate)	103.6	(5.0)
polyethyleneterephthalate	192.4	(9.3)

Diene elastomers

polybutadiene	56.8	(2.7)
natural rubber	74.3	(3.6)
polyisoprene	73.0	(3.5)
polychloroprene	89.3	(4.3)
butyl rubber BK-2045	90.5	(4.4)
butadiene–styrene copolymer SKS-30	85.9	(4.2)

butadiene–acrylonitrile copolymer SKN-18   113.2   (5.5)
butadiene–acrylonitrile copolymer SKN-26   101.7   (4.9)
butadiene–acrylonitrile copolymer SKN-40   112.2   (5.4)

In accordance with expectations, polyethene is characterized by the minimal α value. Taking this polymer as the standard with $\alpha = 1$, one can then calculate the relative values for the parameter α and these are listed in parentheses. One should bear in mind that the data used in calculations were extracted from different references and hence there may be included effects due to slightly different structures and purities of the starting materials. Therefore, the parameters derived from eq. (356) are somewhat controversial, although major deviations from the true values would not be expected.

A series of polymers arranged according to their α values is fully consistent with today's ideas concerning the relationship between adhesive properties of polymers and their chemical nature, particularly with the known [109] donor-acceptor series, polyethene/polybutadiene/polypropene/natural rubber/polyisobutene/elastomer SKS-30/polychloroprene/butyl rubber/SKN-26/SKN-40/SKN-18/polyethyleneterephthalate. In fact, among the closely related polymers, e.g. of polyalkenes, polyethene displays the least adhesive properties and polyisobutene the maximum. Among polydienes the extremes are represented by polybutadiene and polychloroprene respectively (note that in agreement with previous data natural rubber and polyisoprene are different in their adhesive properties) and among butadiene–acrylonitrile copolymers the relative order of SKN-26 and SKN-28 seems to be reversed. The last pair may seem somewhat unexpected however, it is entirely consistent with the data in Fig. 3.38, showing that at elevated temperature, when the polymers are more liable to effective interfacial interaction, SKN-26 provides for a greater strength of the adhesive joints with poly-ϵ-caproamide compared to SKN-18.

Note that according to eq. (354) α is not dimensionless. By making use of the methods of the theory of congruence one can avoid this defect:

$$\alpha' = \delta^4 \left(\sum_i \Delta E_i^* \right) / N_A \sigma^3. \tag{357}$$

After substituting the constants according to SI requirements

$$\alpha' = 0.122 \, \delta^4 \left(\sum_i \Delta E_i^* \right) / \sigma^3 \tag{358}$$

However, the essential advantage in using α' instead of α from eq. (356) is that it does not affect the relative changes of the parameter. Hence, the approach of using this parameter to assess the merits of the adhesive properties of polymers appears to be quite consistent, and yet it is even more important to emphasize the physical soundness of this concept. Indeed, in Young's approximation σ can be considered as characteristic of the ability of a polymer to wet the substrate, while E as that of the opposite tendency to retain the initial shape of the spreading sample. Therefore, there should be an inverse relationship between α and the efficiency of interfacial interaction provided that the diffusion mechanism dominates, i.e. there is no activation barrier at the interface during adhesive contact [1153].

The validity of this conclusion is supported by the data on the peeling strength of the adhesive joints of three SKN elastomers (the fraction with the minimal molecular mass) with polyisobutene, 35 N/m for SKN-18, 75 N/m for SKN-26 and 70 N/m for SKN-40 [22]. In other words these elastomers are arranged in a set corresponding to the above theoretical series. Analysis of the relationship between α values of various polymers and the reported P_p values of their adhesive joints with polyethene, polyisobutene and polyethylene terephthalate leads to a similar conclusion. For the adhesive joints of polyethene (with the adhesives polybutadiene, polyisoprene, butadiene–styrene elastomer SKS-30, and butadiene–acrylonitrile elastomer SKN-40 [1019–1021, 1031]), of polyethylene terephthalate (with adhesives polyethene, polyvinyl acetate, polyiso- prene, elastomers SKS-30 and SKN-40 [706]), and of polyisobutene (with adhesives natural rubber, elastomers SKS-30, SKN-18, SKN-26, and SKN-40 [702]) the paired correlation coefficients are −0.842, −0.846 and −0.957, respectively, being indicative of the actual linearity of the dependences. The data of Fig. 3.51, depicting the strength of autohesive joints of elastomers (from Fig. 3.50) versus α provide additional evidence of the inverse relationship between α and P. One can easily see that the graphs in Figs 3.50 and 3.51 are similar coinciding even in that both reflect the differentiation of the sub- strates into two groups. The similarity of the P versus s (Fig. 3.50) and P versus α (Fig. 3.51) dependences indicates that the role of parameters s and α in interfacial interaction between polymers is actually the same [1151]. This conclusion is confirmed by the fact that the correlation coefficient characterizing the interrelationship between the α values calculated for 24 polymers of different nature and the reported [478] values of their van der Waals volumes was found to be as high as 0.858.

Hence, the parameter α does have a particular physical meaning, reflecting the independent variation of the bulk and the surface properties of polymers. The benefits of using α instead of other known physico-chemical criteria[†] arises from the fact that it is a complex parameter accounting for both the cohesive energy as well as that of the surface. In fact, when using the solubility parameter or that of compatibility alone the

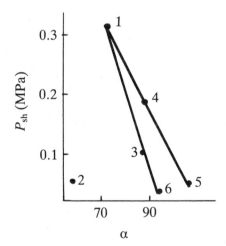

Fig. 3.51 – Shear strength of autohesive joints of elastomers (notation as in Fig. 3.50) versus complex criterion of their adhesive ability (adhesiveness) α.

specific features of adhesion as of a surface phenomenon are neglected. Evidently, despite the above mentioned close relationship between α and the molar volume V, their correlation is inadequate for the same reason. Hence follows the inadequacy of the advice to assess adhesive properties of polymers by the critical molar volumes of the pendant groups [799, p. 142] (Fig. 3.52). By way of example involving epoxy adhesives, it was demonstrated that even if the corresponding dependence is linear and the effect of the molecular volumes of amine curing agents on the strength of the joints is monotonic [1155], factors of thermodynamic origin still have to be taken into account [1156, 1157]. On the other hand, when relating α to the surface parameters of the contacting phases (and moreover, with various conventional characteristics such as critical surface tension [545] and contact angle [1156] it is only in an implicit way that the effect of the bulk properties of the phases on the properties of adhesive joints, deformative initially, are considered. Perhaps, it is only the complex criterion introduced that can provide for the physical sound differentiation of the contribution of inter- and intraphase characteristics to adhesive properties of polymers.

The latter suggestion was verified in the investigations involving polymers with surfaces modified by different functional groups, while the bulk properties remained the same. The results of studies with epoxy bonded adhesive joints of polytetrafluoroethene modified by grafting on to its surface polystyrene, poly(methyl methacrylate), poly(vinyl acetate) and polyvinyl alcohol [109, p. 170] are depicted in Fig. 3.53. As may be seen, the dependence between the shear strength of the joints and α values of the grafted polymers obeys an inverse linear relationship[‡] indicating that the bulk character-

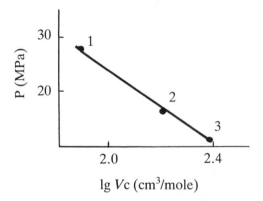

Fig. 3.52 – Strength of adhesive joints between polyorganosiloxanes and glass fibres (13 μm thick) versus critical molar volume of substituent at the silicon atom. 1, Methyl; 2, vinyl; 3, phenyl.

---

† It is not only the adhesive properties of polymers that can be evaluated by monitoring α, but others as well; for instance, hygroscopicity [1154].

‡ The fact that poly(methyl methacrylate) does not fit the dependence can be due to an error in calculating α by using a value for k_0 that is lower than the actual. $k_0 = 0.684$ reported in [471] (derived for $\rho = 1169 \text{ kg/m}^3$) was substituted into eq. (356). However, the actual value for ρ is higher: for a close-packed polymer $\rho = 1250 \text{ kg/m}^3$. Then, from eq. (199) $k_0 = 0.730$ and, according to eq. (356), $\alpha = 106.1$. The revised value is shown in Fig. 3.53 by the solid dot; as is seen, its fit to the dependence is much better.

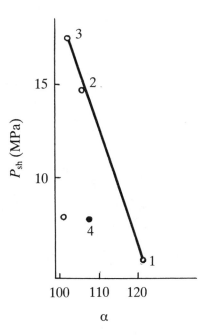

Fig. 3.53 – Shear strength of adhesive joints of graft-modified polytetrafluoroethene with epoxy-based composition versus complex criterion of the adhesive ability of the grafted polymers. Surface grafted polymers. 1, Polystyrene; 2, polyvinyl alcohol; 3, poly(vinyl acetate); 4, poly(methyl methacrylate).

istics of the grafted polymers are directly associated with the efficiency of interfacial interaction with the adhesive. Therefore, strictly speaking it is to the transition layers of the substrate rather than to the entire phase that the concept of bulk properties relates. The zone of interfacial interaction is confined to the thickness of these layers leaving the more distant layers beyond its range. Therefore, the above considerations are pertinent specifically to this region. Correspondingly, surface effects should be related to the nature and structure of the boundary layers. This approach makes it possible to avoid a number of contradictions and at the same time, it illustrates once again the significance of the effect of the substrate on the properties of the boundary and transition layers in adhesives and, in the final analysis, controlling the modes of interfacial interaction.

If the condition claiming the absence (or at least the minimal height) of the interfacial energy barrier is invalid, an inverse P versus α dependence would be substituted for the direct proportionality between these quantities. This might be the case in systems with polar substrates when the universal mechanism of interfacial interaction is affected by factors of molecular–kinetic origin and this is demonstrated by the regular decrease of the correlation coefficients in Table 2.12 observed for polar substrates. Fig. 3.54, depicting the peeling strength versus α plot for the adhesive joints of polychloroprene and butadiene–acrylonitrile elastomer SKN-40 with six low-polarity adhesives [704], provides direct evidence of this thesis. As is seen from Fig. 3.54, α and the strength of

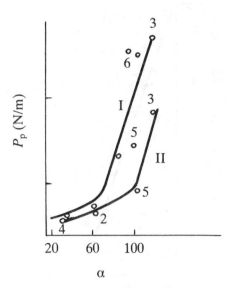

Fig. 3.54 – Peeling strength of the adhesive joints of polychloroprene (I) and butadiene–acrylonitrile SKN-40 (II) elastomers bonded at 423 K with polyethene (1), polypropene (2), polystyrene (3), polytetrafluoroethene (4), polyvinyl alcohol (5), and poly(methyl methacrylate) (6) versus complex criterion α of the adhesive ability of polymers 1–6.

systems[†] both follow a similar monotonically rising pattern, thus confirming the validity of these assumptions.

Our calculations demonstrate that there is actually no interrelationship between P_p and s, whereas the $P_p(A)$ function is extremal. Hence, if there is an activation barrier at the interface then interfacial interaction is due to the bulk effects in transition layers (α), but these, however, are related to the structural factors (A) rather than to changes in the mobility of macromolecular chains (s). Obviously, it is only up to a definite limit determined as the ratio between the polarities of the adhesive and the substrate that these effects are positive. In its turn, this limit can be expressed as the ratio between eqs (245) and (246), i.e. by the increments of the dispersion and polar components to the surface energy of adhesives, thus revealing the direct relationship of the kinetic features of interfacial interaction with the energy spectrum of the boundary and transition layers. Then, from eq. (251) it follows that the critical value of P_p is attained when the value for the energy of interfacial interaction coincides with the height of the activation barrier at the interface. Above and below this threshold (critical value) most of the known strength dependences are qualitatively different, either ascending or descending.[‡]

---

† In this case also the more accurate α value for PMMA (solid dot) provides a better fit to the strength dependence of the adhesive joints of poly(methyl methacrylate) and polychloroprene.

‡ Besides the purely applied aspect this conclusion reveals a deep physical analogy between the parameters defined by eqs (256) and (356) emphasizing the interrelationship between different aspects of adhesive interaction. In fact, maximum efficiency of the adhesive interaction can be deduced either in terms of σ^i, eq. (258) or α, eq. (354).

This conclusion is confirmed by the analysis of the variations in peeling strength of the adhesive joints of low-polarity (polyisobutene [702]) and polar (polyethylene−terephthalate [706]) substrates with adhesives of different polarity (ranging from poly-ethene to butadiene−acrylonitrile elastomer SKN-40). As is seen from Fig. 3.55, the P_p versus A and α plots are strictly linear; however, in agreement with the predictions, the sign of the slopes is different. Linearity of the $P(\alpha)$ dependences indicates that the predominant contributions to interfacial interaction are the bulk effects in transition layers, while linearity of the $P(A)$ graphs discloses that the distinct mechanism providing for these effects is determined by structural factors.

These concepts are supported by the analysis of the data on measurement of the peeling strength of adhesive joints involving polybutadiene SKD and three butadiene−acrylonitrile copolymers (SKN-18, SKN-26, and SKN-40) [1158] as adhesives and butadiene−styrene elastomer SKS-30ARK [1159] as a common substrate. In accordance with expectations the peeling strength versus A and α dependences (Fig. 3.56, a, b) are close to linear. A relatively large deviation is observed for SKN-18 (Fig. 3.56, a) and SKN-40 (Fig. 3.56, b) which is probably due to insufficient reliability of the reported values for refractive indexes used in the calculations. Note that these dependences (Fig. 3.56, a, b) hold true for all of the four adhesives examined, while the P_p versus solubility parameter or versus surface energy plots are linear for three butadiene−acrylonitrile adhesives only, i.e. for the set of closely related polymers. It is quite clear that the best fit is observed between P_p and the flexibility of macromolecular chains of elastomers

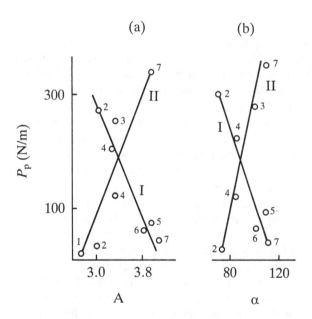

Fig. 3.55 − Peeling strength of the adhesive joints of polyisobutene (I), polytetrafluoro-ethene (II) bonded with polyethene (1), natural rubber (2), polyvinyl acetate (3), butadiene−styrene SKS-30 (4) and butadiene−acrylonitrile elastomers SKN-18 (5), SKN-26 (6), and SKN-40 (7) versus constant of Hildebrand's equation (a) and complex criterion of the adhesive ability of polymers 1−7 (b).

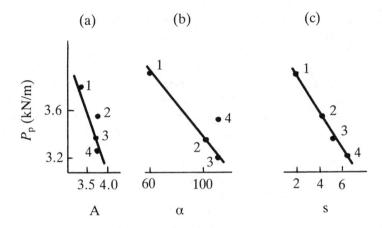

Fig. 3.56 – Peeling strength of the adhesive joints of butadiene–styrene elastomer SKMS-30ARK bonded with polybutadiene SKD (1) and butadiene–acrylonitrile elastomers SKN-18 (2), SKN-26 (3), and SKN-40 (4) versus constant of Hildebrand's equation (a), complex criterion of the adhesive ability (b), and equilibrium flexibility of the macromolecules of adhesives (c).

(Fig. 3.56, c). Hence, it is explicitly this factor that predominantly controls the efficiency of adhesive interaction. Similar results were obtained in relating to these same characteristics for poly(vinyl acetate), poly(methyl methacrylate), polyvinyl chloride, polystyrene, and polyethyleneterephthalate with tensile strength of the joints with polyvinyl alcohol formed at 293 and 393 K [1148].

Hence, the analysis undertaken confirms, on the whole, the validity of the concepts described concerning the characteristics of interfacial interaction between polymers during adhesive contact. The quantitative parameters A, s, a, and α describing the properties of polymers at interfaces in a physically unequivocal manner provide a foundation for the proposed theories.

At the same time it is necessary to define the range of applicability of this approach. From the energetic standpoint it is confined to materials predominantly involved in interfacial interaction due to dispersion forces. This condition implies that there are no valence bonds formed between the adhesive and substrate since even a small number of these results in a qualitatively different pattern of the processes in transition and boundary layers of polymers. The conclusions of this section relate to polymer–polymer systems, as a rigid substrate (metal, glass, etc.) would result in a different structure and therefore properties of the transition layers in the corresponding adhesives. In this case the macroscopic contact area between the elements of the system becomes the factor of primary importance. The modes of formation and its influence on the efficiency of adhesive interaction are discussed below. However, it is appropriate to precede the discussion of this subject by examining the specific features of rheological behaviour of high-molecular mass compounds during interfacial contact.

3.2 MACROSCOPIC CONTACT IN ADHESIVE JOINTS

3.2.1 Rheological aspects of adhesive joint formation

To complete the theory of adhesive joint formation, it is necessary to take into account the specific behaviour of polymers under the effect of an external force field. Due to the chain structure of high molecular mass compounds, most of the respective dependences are of non-linear character. Their analysis from the rheological viewpoint is, in our opinion, the foundation of a bridge between the theory (representing a formal approach) and practice (seeking to account for the reality of the observed effects) of adhesion processes. Moving across this bridge, in both directions, are ideas and facts, the most substantial of which assist in their mutual advance. In this section we will confine our attention to the problems associated with the formation of adhesive joints, since the rheological aspects of fracture are determined to a considerable degree by continuum mechanics, discussion of which is beyond the scope of this monograph.

Rheological aspects of the processes of adhesive joint formation are first of all related to the non-Newtonian behaviour of polymer flow. Hence the primary importance of viscosity η as a parameter which is essential both for the interpretation of non-linearity of intraphase properties of the materials and, as a result, for the specific features of the formation of interfacial contact. For instance, the time to achieve an equilibrium state in the surface layers of low-molecular mass substances (Newtonian fluids) is 1 ms [1160], reaching for polymers up to 15 min in the case of epoxy–oligomer wetting of a glass surface (at 293 K) and as high as 3 h in the case of epoxidized polybutadienes. Even in the instance of polymer melts this value is sufficiently high (some tens of minutes), e.g. 20 min for polyisobutene (at 393 K), 35 min for polyvinyl acetate (396 K), and 90 min for polybutyl methacrylate (432 K) [198].

The rheological characteristics of the variation of the contact angles of wetting of solid surfaces by polymers is determined by the role and influence of the factors of a molecular–kinetic nature, above all viscosity. The various approaches, as a rule, consider either η (eqs (171) and (174)) or θ (eq. (172)). Their combined application is associated with the necessity to adopt some simpliciations. For instance, assuming that general rules of wetting are governed by condition eq. (55), the relationship between the volume of a drop V_0 and the time to reach the equilibrium state of this sessile drop on a substrate can be expressed as follows [1161]:

$$\tau = 4(3/\pi)^{0.3} \, (\sigma_{sa} - \sigma_{sl} - \sigma_{la} \cos \theta)\eta \, \text{tg}^{1.3} \, V_o^{0.3}. \tag{359}$$

Such assumptions are justified when the development of side processes, for instance, of dissolution, is of low probability. In fact, the dependence eq. (359) satisfactorily describes the features of the wetting of polytetrafluoroethene and PMMA by poly-dimethylsiloxanes [1162], interfacial interaction being controlled by dispersion forces.

At the same time it should be borne in mind that spreading of the liquid on a solid surface may be developed by two mechanisms, diffusion and viscous. In the first case $\theta = $ const, while in the second $\theta \to \theta_\infty$ (condition eq. (169)) and, as was mentioned above, transfer of the molecules to the interfacial region should be taken into account, this process being hampered by viscous resistance of a liquid adhesive. Accordingly, the kinetic laws of the variation of the radius r of a spreading drop are different. Since in the case of the diffusion mechanism there is no need to account for η, r is determined by an inverse quadratic time dependence [1163]:

$$r = (r_o^2 + \Lambda\tau)^{0.5}, \tag{360}$$

where Λ is a constant, taking into account the influence of the character of spreading. The variation in the rate of r in the case of a viscous spreading mechanism is associated with the thickness of a liquid layer d and density ρ and, by analogy with expressions (172) and (173), with the values of surface energies:

$$\frac{4\pi\eta}{d} r \frac{dr}{d\tau} = g\rho^2 d^2 + 4(\sigma_{sl} - \sigma_{sa}) + 2\sigma_{la}/\cos\theta. \tag{361}$$

Evidently, three subsequent events may possibly take place, spreading, when from eq. (149) $dr/d\tau > 0$, a static state $(dr/d\tau = 0)$ and the contraction of a liquid to a drop $(dr/d\tau < 0)$. Note that the latter two conditions are not identical to the respective dependences describing the variation of spreading coefficient, since negative or zero values of χ only limit θ by certain boundary values.[†] Substituting all the aforesaid conditions into expression (361), one may obtain for each of the three cases, respectively:

$$g\rho^2 d^2 + 4(\sigma_{sl} - \sigma_{sa}) + 2\sigma_{la}/\cos\theta > 0, \tag{362}$$

$$g\rho^2 d^2 + 4(\sigma_{sl} - \sigma_{sa}) + 2\sigma_{la}/\cos\theta = 0, \tag{363}$$

$$g\rho^2 d^2 + 4(\sigma_{sl} - \sigma_{sa}) + 2\sigma_{la}/\cos\theta < 0. \tag{364}$$

The significance of eq. (362) is understandable. Equation (363) is useful mainly due to the possibility of calculating σ_{sl}. From eq. (364) it follows that contraction of a drop is taking place with decreasing θ values. This fact is in agreement with both the experimental data and theoretical ideas, the latter resulting, in terms of the scaling concept, in a simple dependence [1164]:

$$\sigma_{la} \cos\theta = \sigma_{la} + \chi. \tag{365}$$

Analysis of expression (361) disclosed that viscous spreading can proceed even at a certain constant value θ_∞ [1163]. However, it is hidden by the diffusion mechanism of liquid front propagation since

$$r = (r_o^2 + \Lambda_\eta\tau)^{0.5}. \tag{366}$$

In spite of the formal similarity of eqs (360) and (366) they are different as in the latter the influence of the liquid viscosity is taken into account through coefficient Λ_η. Hence, the meaning of Tanner's law, which was verified experimentally [1164], is clarified.

For polymers similar concepts may be expressed in the form of a function of distance d_c [1165]

$$\cos\theta_\tau/\cos\theta_\infty = f(\sigma_{la}\tau/d_c\eta), \tag{367}$$

over which the action of surface forces extend, i.e., according to the account in section 3.1, d_c can be identified with the thickness of the transition layer. According to Newman [1166], the integrated form of this expression is as follows

---

†   Generally, according to de Gennes [350], the increase of the spreading coefficient results in the decrease of the thickness of a layer of liquid wetting a solid polymer surface.

$$\cos \theta_\tau = \cos \theta_\infty \left[1 - b \exp \left(-c\tau\right)\right]. \tag{368}$$

where b and c are constants, the latter being the variable of the function eq. (367). Then, in terms of the Cherry–Holmes approach [360] and recalling eqs (169) and (172):

$$d(\cos \theta_\tau)/d\tau = \sigma_{la}\dot{\tau}/d_c \eta. \tag{369}$$

These results make it possible to distinguish between the influence of kinetic and viscous effects on variation of r. The only variable, which is somewhat difficult to account for, is d_c. It is clear that its magnitude is governed by the properties of both adhesive and substrate. For instance, for ethene–vinyl acetate copolymer (72 : 28) at 422 K $\sigma_{la} = 35$ mN/m and $\eta = 30$ kPa s. By contact angle measurements for this copolymer adhesive on an aluminium substrate constant c can be easily found. Then, $d_c = 360$ μm [360], being in rather good agreement with the value of $d_c = 200$–300 μm calculated from Fourier-transform IR-spectroscopy [1167]. Obviously, with a substrate of lower surface energy the value of d_c must be lower as well (it is this fact that explained the increase in spreading rate of oils on high-energy surfaces of polyhexamethylene adipamide as compared to that on low-energy polymethyleneoxide and its copolymer [1168]. In fact, according to their effect on the value of d_c for this adhesive, the substrates can be arranged in a series PTFE/aluminium/mica [1165].

The implications of the correctness of this approach are shared by a number of authors [1169, 1170] but Kaelble [1171], however, drew attention to the necessity of considering capillary rise forces, a factor which should be important in view of the well developed nature of the substrate surface. Nevertheless, corresponding analysis in terms of eqs (367)–(369) is hampered by a number of simplifications which have been introduced [1172]. For instance, it was suggested [1168] that the spreading coefficient of a liquid on a rough surface could be described by an empirical exponential function:

$$\chi = k_\chi \tau^x \tag{370}$$

where coefficient k_χ is positive and $0.3 < x < 0.6$. Therefore, attempts to seek other approaches to kinetic generalizations of the formation of adhesive joints appear to be fully justified.

One such approach is associated with the ideas of adhesive spreading on the substrate surface. The rate of this process at the geometric interface is hardly equal to zero and hence it is reasonable to think of it as having a certain rate gradient. Then, according to Fig. 3.57, the 'slipping' of adhesive can be characterized by parameter 'b' where its magnitude is commensurable with the size of the molecule of a liquid [1173]. For concentrated solutions,[†] however, as well as for polymer melts, the value of b must be significantly higher. De Gennes assumes that the stresses in the adhesive phase are preferentially concentrated near the interface rather than in the bulk and characterized by extensive chain entanglement [364]. Then

$$b = l_l \eta/\eta_o \tag{371}$$

---

† For those it is necessary to take into account the solvent-affected variation of the conformational states of macromolecules, in particular, one should consider the chain size and interaction parameters as a function of the distance from the substrate surface [1174].

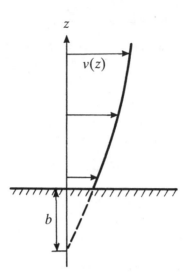

Fig. 3.57 – A schematic pattern describing the evaluation of the slip of an adhesive flowing
on a solid substrate surface.

where l_l is the size of a liquid molecule, η and η_o are the viscosity of systems with and without chain entanglement, respectively. To estimate the value of b, let us express η_o via the number of repeating units N related to a single entanglement N_I [368]

$$\eta_o = \eta N_I^2 / N^3 . \tag{372}$$

At N as high as approximately 10^4 then b is equal to about 1 mm. By thermoanemometric [366] and rheogoniometric [367] measurements (for polystyrene melt $b \approx 60\ \mu m$ [367]) this approximation was shown to be rather an overestimate. However, it is important that the need to account for the slipping effect was confirmed experimentally.

On the other hand, provided that the time-dependent character of the constant Λ in eq. (360) is taken into account, these dependences should approximate to a parabolic law. Then the motion kinetics of a point object can be written in the form of the ratio:[†]

$$x^2 = 2d\Delta\sigma\tau/\eta. \tag{373}$$

Examining the spreading of oleic acid on the surface of glycerol, it was demonstrated [1175] that the difference of surface energies in eq. (373) is simply the surface pressure π. Hence, the driving force of this process is a derivative of π with respect to time. Indeed, the spreading rate of alcohols over the surface of water is proportional to $d\pi/d\tau$, which is balanced by the friction force [1176]. For one-dimensional spreading of adhesive over a horizontal plane of width l

$$k_r \frac{l\rho\eta}{2M_o} x^2 \frac{dx}{d\tau} = \Delta\sigma + f_g(x, M_o/l) \tag{374}$$

---

† According to Adamson dependence eq. (373) is valid only at $d > 0.002$ m [111].

where M_O is the droplet mass, k_r is the roughness coefficient of the substrate surface and f_g is a function of the component due to gravity. However, for a drop spreading in the form of a rectangular parallelogram with decreasing thickness one has

$$\frac{M_O}{l} x^2 = f_g(x, M_O/l)/(2\sigma_{1a} + M_O g/2l). \tag{375}$$

When the spreading droplet has the shape of a spherical segment, the rate of its radius variation is determined by an exponential law [1177]:

$$r \approx V^{0.3}(\sigma_{1a}\tau/\eta)^{0.1}. \tag{376}$$

For instance, it was demonstrated for diamond–Vaseline oil (nuyole), copper–glycerol and polytetrafluoroethene–polydimethylsiloxane systems [186] that the exponent of the second term in eq. (376) is only slightly different from 0.1, being equal to 0.09, 0.10 and 0.12, respectively. A similar relationship is valid for systems in which selective wetting is achieved due to spreading over a solid substrate of a drop immersed in a liquid with viscosity other than of adhesive [1178].

The description of spreading kinetics by a parabolic law is valid for materials of different type [1164], including metals [359] and glasses [1179]. Polymeric adhesives, however, are usually spread forcibly under the effects of pressure and temperature. In this case wetting, in its thermodynamic meaning, does not play a decisive role, e.g. parallel plates partially immersed in a liquid are attracted to each other when moved colaterally if the distance between these plates exceeds half of the value of Laplace's capillary constant even when for one of the plates $\varphi < 90°$, while $\varphi > 90°$ for the other [249]. The direct consequence of the forced character of adhesive joint formation is the development of internal stresses within the joint. It is a rather complicated task to take into account the effect of this factor on the rheological features of the interaction between adhesive and substrate since it requires that the principles of mechanics be employed [1180], in particular, a 14-element model of a polymer by Voight has to be involved [1181]. Simplified solutions, as, for instance, those based on Gent's concept [1182] result in the calculated rheological characteristics of the rupture of highly-elastic adhesives being in good agreement with the experimental data (elasticity and loss moduli, and relaxation time spectrum) only in the low-strain range [1183].

At the same time attempts to take into account variations in the viscosity of the adhesive during its interaction with a substrate may be employed to provide a number of practically valid consequences. Note, as an example, the possibility of calculating the effective thickness of the adhesive layer d, relying on the variation mode of adhesive viscosity. Assuming the latter to be exponential

$$\eta = \eta' e^{\alpha\tau} \tag{377}$$

(where η' is the initial viscosity of the adhesive at the moment of application on to the substrate, α is a constant with the meaning of a proportionality coefficient of the exponential function) and [1184]:

$$d = l[3\alpha\eta'/2F(1 - e^{-\alpha\tau})]^{0.5} \tag{378}$$

where l is the length of the glue line and F the external load. For two-component adhesives it is necessary to consider the time τ_x beginning from application of the formulation up to the moment when an external load is applied to the system:

$$d = l(3\eta'^{\alpha\tau_x}/2F\tau)^{0.5}. \tag{379}$$

More general solutions [1185–1187] are based on the hydrodynamic characteristics of laminar flow of liquids in planar clearances or slits [1188]. These enable, in particular, account to be taken of the effect of the joint size (for instance, overlapping length l) on the variation of the adhesive layer thickness [1189]:

$$d_\tau = [d_o^{-2} + 2F(1 - e^{-\alpha\tau})/3\alpha\eta'l^2]^{-0.5}. \tag{380}$$

On the other hand, in some cases satisfactory agreement with experimental data may be secured by describing the behaviour of polymeric adhesives as of Newtonian liquids. In this case the following relationship is valid [1184]:

$$d = \left[\frac{2a + 1}{\tau(a + 1)}\right]^{a/a+1} \left[\frac{4bl^{a+1}}{F(a + 3)}\right]^{a/a+1} \tag{381}$$

where a and b are constants. This is confirmed by the data of Fig. 3.58. An increase in the number of components in the adhesive formulation (Fig. 3.59) results, however, in better agreement with eqs (377)–(381) [1184].

In spite of the undoubted usefulness of such approaches, it is obviously limited from the physico-chemical point of view. Analysis of rheological generalizations taking into account the influence of effects of hydrodynamic origin appears to lack these short-comings [84]. Let us elaborate on this problem using the model of a drop of viscous

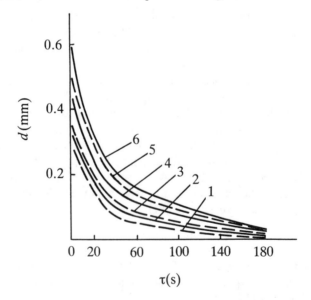

Fig. 3.58 – Kinetic curves describing the decline of the thickness of the glue line with $l = 20$ mm in adhesive joints produced under 0.012 MPa load. Epoxy adhesives of different viscosity were used, Epilox T 20-20 with η 0.38 Pa s (1, 2); Epilox A 20-00 with η 4.2 (3, 4) and 10 Pa s (5, 6). Solid lines, calculated according to eq. (381): dashed lines are experimental curves.

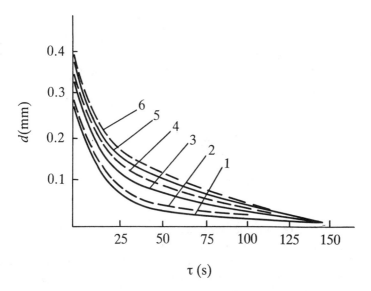

Fig. 3.59 — Kinetic curves describing the decline of the thickness of the glue line in adhesive joints produced under 0.01 MPa load. Epoxy adhesives Epilox A 20-00 (1, 2) and T 20-20 (3–6) filled to 20% content with powdered quartz (1, 2), silicon dioxide (3, 4), and aluminium (5, 6). Solid lines, calculated according to eq. (381); dashed lines are experimental curves.

liquid [1190] spreading over a smooth solid surface under the effect of a driving force f_σ owing to the gain in surface free energy at wetting, geometric parameters of this model are shown in Fig. 3.60.

Taking advantage of the canonic form of Young's eq. (55), f_σ at thermodynamic equilibrium can be written as follows:

$$f_\sigma = 2\pi r_1 \sigma_{la} \left(\frac{\sigma_{sa} - \sigma_{sl}}{\sigma_{la}} - \cos\theta \right) \tag{382}$$

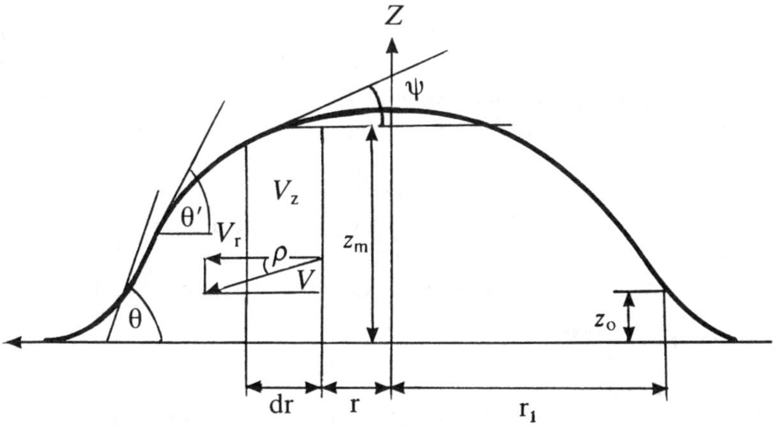

Fig. 3.60 — Schematic presentation of a droplet of viscous adhesive spreading on smooth surface of a solid substrate.

where r_1 is the radius of the droplet base and θ is the dynamic contact angle.

Placing the centre of the droplet base at the origin of the cylindrical system of coordinates, at a distance r trace out a ring dr thick and with the height $z_m = R(\cos \psi - \cos \theta)$. Then, from the definition of the viscous resistance force

$$f_r = \pi r^2 \eta \frac{dv}{dz} \tag{383}$$

(where dv/dz is the rate gradient of viscous flow) and for the liquid within the ring the viscous resistance force in the direction of R axis is as follows

$$-df_r(r) = \frac{2\pi\eta\, r\, dr}{\cos \varphi}\; \frac{d}{dz}\left[\frac{v_r(r,\,z)}{\cos \varphi}\right] \tag{384}$$

where $\varphi = z\psi/z_m$ and $v_r(r, z)$ is the component of the rate vector in the same direction. One can eliminate the term $v_r(r, z)$ from eq. (384) by solving it together with the equation for non-discontinuity. Then, for the case of $\theta \approx \theta' \approx \Psi$, when the curvature of the liquid surface at the droplet edge can be neglected, one may derive the dependence of viscous resistance force on the droplet volume V_0:

$$-df_r(r) = \frac{4\pi^2 R^6 \eta \sin^2\theta \cos^2\theta\, \theta^2\, dz_m}{3V_0 z_m^2}\; \frac{d\theta}{d\tau}. \tag{385}$$

Integrating eq. (385) over z_m from z_0 to $r_1\, tg\theta$ and neglecting $-1/r_1\, tg\theta$ as compared to z_m^{-1}, which is true for $10° < \theta < 90°$ when z_0 is small, at equilibrium of the viscous resistance force with the driving force by eq. (382) during flow, one may deduce the propagation rate of a droplet perimeter:

$$v_{r_1} = \frac{dr_1}{d\tau} = \frac{z_0\sigma_{la}}{2\eta r_0}\; \frac{\sin^4\theta\left(\dfrac{\sigma_{sa}-\sigma_{sl}}{\sigma_{la}}-\cos\theta\right)}{(r_1/r_0)^4\,\theta^2\,\cos^2\theta}\left[\left(\frac{r_1}{r_0}\right)^3\frac{2}{tg\,\theta}\right] \tag{386}$$

where r_0 is the base radius of a hemispherical droplet at $\theta = 90°$.

From the results of numerical integration of expression (386) at various $(\sigma_{sa} - \sigma_{sl})/\sigma_{la}$ ratios it follows that for low equilibrium contact angles over a wide range of dynamic contact angles linear kinetic dependences hold true on a logarithmic scale (Fig. 3.61). This fact accounts for a great deal of empirical dependences of the type:

$$\tau = x(r_1/r_0)^y. \tag{387}$$

Then the exponent in eq. (387) must be directly associated with the Young ratio of surface energies. The validity of this assumption is verified by the data of Fig. 3.62. Hence, in a general case spreading time can be expressed by a simple function:

$$\tau = A(r_1/r_0)^{B-C(\sigma_{sa}-\sigma_{sl})/\sigma_{la}} \tag{388}$$

where A, B and C are constants.

Considering the process of adhesive joint formation, one should bear in mind that it consists of three distinct stages, though quite often this is not taken into account. Strictly speaking, wetting and spreading of the drops of adhesive on the surface of a solid sub-

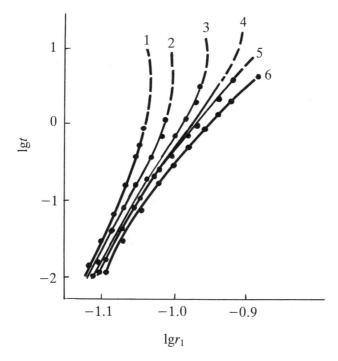

Fig. 3.61 – Spreading kinetics of a droplet of viscous adhesive at different values of the $(\sigma_{sa} - \sigma_{sl})/\sigma_{la}$ group: 1, 0.70; 2, 0.80; 3, 0.90; 4, 0.94; 5, 1.00; and 6 1.10.

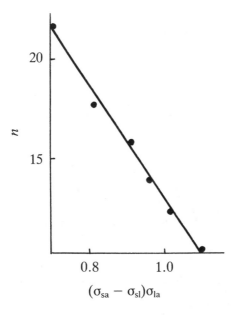

Fig. 3.62 – The power index of the spreading kinetics curve versus ratio between surface energies of the elements comprising a system.

strate are not the only phenomena responsible for the formation of interfacial contact. Because of the considerable difference in the temperatures of contacting phases and the tendency of a liquid to structuring it is necessary to consider the final stage of the joint formation process, namely coalescence of drops (which is especially important for melts and in cases when the minimal amount of adhesive is used). Development of this stage should be obviously governed by the influence of both the surface energy of adhesive and its viscosity.

Consider, in the context of a hydrodynamic approach, the model [1192] according to which the cross-section of a drop intersected by a plane normal to the substrate surface has an almost triangular form (Fig. 3.63). Some limitation may be applied by a rather reasonable approximation when the driving force of the coalescence process is approximated by Laplace's equation

$$f_+ = 0.5\ \sigma_{la}\ l(R^{-1} + R_1^{-1})$$ (389)

(where l is the length of the side surface of a drop, R and R_1 are radii of curvature), whereas the counteracting force by Newton's law:

$$f_- = -\ \alpha\eta\ \frac{dv}{d\beta}$$ (390)

where $dv/d\beta$ is the spreading rate gradient in the horizontal direction.

For the geometry of the model one has

$$1/R = 2(h - z)/a_0$$ (391)

and the varying drop height is given as follows

$$h = h_0 - 0.5z.$$ (392)

Then, substituting relationship (391) into eq. (389) and taking into account expression (392) one obtains:

$$f_+ = \sigma_{la}\ (h_0 - 1.5z)/a_0.$$ (393)

On the other hand, for a given time the values of f_- (eq. (390)) and f_+ (eq. (393)) should be equal and, thus:

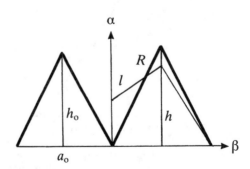

Fig. 3.63 – Schematic presentation of the droplets of viscous adhesive coalescing on a smooth solid surface.

$$- dv = \frac{\sigma_{la} \, (h_0 - 1.5z)}{a_0} \quad \frac{d\beta}{\alpha}.$$

(394)

Since

$$\alpha = z + \beta(h - z)/a_0$$

(395)

and

$$d\beta = -\frac{a_0 \, d\alpha}{h_0 - 1.5z},$$

(396)

expression (394) may be rewritten in the following form:

$$- dv = \frac{\sigma_{la}}{\eta} \frac{d\alpha}{\alpha}.$$

(397)

The boundary conditions are determined from the dependence (395) for $\beta = 0$ ($\alpha = z$) and $\beta = a_0$ ($\alpha = h = h_0 - z/2$; $v = 0$). Then, the left-hand side of eq. (397) should be integrated between the limits from v to 0, and the term on the right from z to $(h - 0.5z)$. As a result, one has [1193]:

$$v = \frac{\sigma_{la}}{\eta} \ln \frac{h_0 - 0.5z}{z}.$$

(398)

The results of experiments depicted in (Fig. 3.64) demonstrate that in the high temperature region (which is of most insterest in the practical use of adhesives) the kinetic dependence of drop coalescence is linear[†] for minute droplets only. At ambient temperature the linearity expands to the whole range of values for the neck height of a drop z and their initial heights h_0. One can see from eq. (398) that $z_{max} = 2/3 \, h_0$; however, in practice (Fig. 3.64) $z_{max} \leqslant 0.5 \, h_0$ and the best agreement with experimental data is observed at high values of z and small values of h_0.

The correctness of the final result is confirmed additionally by the fact that a similar proportionality to the ratio σ_{la}/η is characteristic of the rate variation of both coalescence and spreading of drops [1161]. Hence, the hydrodynamic approach is of general importance and it can be used as a foundation for the uniform interpretation of rheological principles of the formation of adhesive contact.

Relying on the fact that the kinetics of viscosity changes can be roughly approximated by an exponential function Torbin [1194], who elaborated on this concept, introduced the term spreadability of adhesive χ_0 representing the ratio between the time needed for the viscosity of a system to increase by e times and the initial viscosity. This parameter is expressed by the following equation,

---

† When minimal amounts of adhesives are used the need to consider the features of the coalescence of adhesive drops is demanded by technological reasons as well as by physico-chemical ones. These features should be taken into account in the case of poor wetting of the substrate, which might result in formation of the so-called 'starved' joints. On the other hand, when the minimally required amount of adhesive is exceeded then the strength of adhesive joint might be decreased since the latter parameter is inversely proportional to the thickness of the adhesive layer.

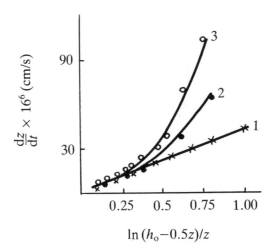

Fig. 3.64 – Kinetic pattern of the coalescence of the 1.3 mm high droplets of viscous adhesive at 1, 293 K; 2, 323 K; and 3, 423 K.

$$\chi_0 = d^2/k_f^2 A^2 F(1 - 0.01\gamma_0), \tag{399}$$

where γ_0 is the internal stress and k_f is the coefficient accounting for the shape of a substrate. Considering the spreading of adhesive between lenses, one may derive an expression for the pressure distribution over the interface and also the differential equations representing a generalization of the known features which describe the thickness variation of the layer of viscous liquid [1195]. For the case when the adhesive does not reach the lens edges

$$d = [3V^2/8\pi\chi_0 F(1 - e^{-\tau})]^{0.25}. \tag{400}$$

Then it is easy to find the minimal volume of adhesive which is required for the complete interfacial contact area to be formed between the plates of diameter D at a given load F:

$$V_{min} = (6\pi^3/64)^{0.5} \, (D/\chi_0^2 F^2 \cos 0.5\psi)^4 \tag{401}$$

where $\psi = \arcsin D/2R$. For $V < V_{min}$ structuring within the adhesive may start before complete interfacial contact has been secured, the time required to reach it (complete interfacial contact) being expressed by the simple formula:

$$\tau = -\ln[1 - (V/V_{min})^{-2}]. \tag{402}$$

For $V > V_{min}$ the following relationship is valid:

$$d/d_{min} = [2/(1 + \nu^{-2})]^{0.5} \tag{403}$$

where d_{min} is the thickness of the adhesive layer, corresponding to its minimal volume. Hence, for a given load the maximum thickness of the adhesive layer at infinitely large amount of adhesive is only 1.4 times greater than that of the minimum required to fill the gap between the substrates. At a considerable excess of adhesive, the ultimate layer thickness is close to d_{max}, which is practically independent of the amount. In the case of the adhesive leaking beyond the limits of substrate boundaries

$$d^2 - d_\tau^2 = \frac{\chi_0 F}{k_f D} e^{-\tau} \qquad (404)$$

where $k_f = 2\pi^2/\cos^2 0.5\psi$, the parameter ψ being taken from eq. (401). Then at $V \to \infty$ one gets

$$d_\infty^{max} = k_f D^2 (\chi_0 F)^{-0.5}. \qquad (405)$$

The final expression is, on the one hand, equivalent to eq. (399) solved with respect to d_{max} and, on the other hand, to Stefan's law, where factor k_f has the same meaning of the shape coefficient. However, instead of the varying thickness of a liquid layer its limiting value was used in eq. (405), while instead of the time variable τ the constant spreadability χ_0 was introduced. Besides, eq. (405) demonstrates the effect of geometric features of the area of substrate surface engaged in contact. For a rectangle with the length d and width l one can write $k_f = (5/12)^{0.5}[(l/d) - (l/d)^3]^{-0.5}$ and, in particular, for square plates $k_f = 0.456$. It follows that for the given load and interfacial contact area the thickness of the adhesive layer will be at a maximum when the substrates are of circumferential (round) shape. The thickness of a layer between rectangular surfaces is liable to decrease the more the rectangles are elongated. The well-known stress-distribution pattern over the area of an adhesive joint is indicative of the fact that the weakest spot in an adhesive joint is at its edges, and it can be explained by this proposition.

The prospects of a hydrodynamic approach in interpreting the patterns of adhesive interaction of liquids with solid substrates is additionally supported by the proportionality between the thickness of the adhesive layer and the diameter of the discs bonded being equal to the ratio between the load F and the interfacial contact area A. If the thickness of a layer were the parameter relating to equilibrium, it would have also been independent of the size of a substrate. For instance, when the linear dimensions of the samples being bonded are increased two-fold, the ratio F/A necessary to secure the same adhesive layer thickness should be increased by a factor of 4. Then, the widely used recommendations concerning the choice of the bonding pressure, stating a simple proportionality of the bonding pressure to the contact area, appears to be groundless. Another implication of the view-point developed [84] necessitates that the values of both V and V_{min} be considered. It is commonly assumed that the thickness of the adhesive layer can be decreased (thus increasing the strength of adhesive joint owing to the diminished contribution of the hazardous tangential stresses) by using a smaller amount of adhesive and by applying a higher load to ensure its spreading over the substrate surface. However, at the same time, according to eq. (401), the rise of F leads to a decrease in V_{min} thus causing a decrease in the thickness of the adhesive layer, whereas the adhesive is actually in excess. Apparently, the equilibrium thickness of the adhesive layer cannot be equal to zero; however, it is negligibly small as compared to the values of d observed in practice.

One should bear in mind that this approach is based on the analysis of the behaviour of Newtonian liquids. This is precisely the reason for the disagreement between experimental data on the thickness of an adhesive layer in real adhesive joints and theoretical calculations assuming the Newtonian character of a liquid flowing from the slit gap [1184, 1196]. In practice [1197] a liquid with viscosity of 90 Pa s fills all 100 nm-deep microvoids on the substrate surface within a 5 min time-span, whereas for unvulcanized rubber stock this effect will be attained only within a few hours. According to Zisman [545], a more viscous adhesive will not be able to fill such cracks of the solid surface at all. The shape of the defects is also of importance, and in an equation describing the advancement of a liquid within a crack the shape of the defects is accounted for by factor k_f [6]:

$$x^2 = k_f \, l \, \sigma_{sl} \cos \theta_A \, \tau / \eta \tag{406}$$

where θ_A is the maximum advancing angle and x is the distance. For the cracks of circumferential (circular) cross-section $k_f = 0.5$, for rectangular $k_f = 0.33$. However, for V-shaped groove-like microdefects of depth h, the equation is somewhat more complex,

$$h \ln (h - x) = l \, \sigma_{sl} \cos \theta_A / \sigma \eta h. \tag{407}$$

The pattern of kinetic dependence is even more complex when the form of the lens of a polymeric liquid filling the narrow capillary crevices on the substrate surface is taken into account [1198].

Since the formation of adhesive joints is inevitably associated with the deformation of micropeaks on the substrate surface, for non-Newtonian liquids it is also necessary to take account of the relaxation processes [1180]:

$$x^2 = k_f \, l \, \sigma_{sl} \cos \theta_A \, (B + \tau / \eta). \tag{408}$$

Here $B = A x_\infty [1 - \exp(\tau_\infty / \tau')]$, where x_∞ is the equilibrium advancement of the adhesive during an infinitely large time interval τ_∞ and τ' is the retardation (lag) time.

When considering the influence of rheological processes on the kinetics of adhesive joint formation it is necessary to take into account the displacement of air from micro-defects on the substrate surface. In a first-order approximation the rate of this process is estimated by the relationship [1199]:

$$v = 2g \, r_a^2 \, (\rho_a - \rho_1)/9\eta, \tag{409}$$

where indexes a and 1 refer to air and liquid, respectively. It is clear that the value of v is rather low. For instance, the rate of displacement of spherical bubbles of radius 10^{-4} m from a liquid viscosity of 1 Pa s is as low as 2×10^{-4} m/s. The air bubbles incorporated into an adhesive during its preparation for bonding, for instance during mixing of an epoxy oligomer with the curing agent, also displays a negative effect on the strength of adhesive joints [1200].

In addition to the factors discussed the formation kinetics of macroscopic contact between polymers is governed by the time-dependent variation of the viscosity of the adhesive. The general laws responsible for this variation are cited above.[†] These result

---

† These generalities are under continuous development owing to the progress in experimental technique; for instance, the method based on the relationship between the viscosity of an adhesive and the fluorescence intensity of the curing agents for epoxy adhesives [1204–1206] and which thus serve as fluorescent tags (probes) [1201–1203] appeared to be quite rewarding. In particular, it was established by this method that the viscosity of the ternary copolymer-adhesive comprising 2-ethylhexylacrylate, acrylic acid and vinyl acetate is at a maximum when the polymer layer is 30 μm thick [1207]. The abrupt fall in viscosity of a dilute polymer solution in a very thin layer of a thickness commensurate with the radius of gyration of a macromolecule can be presented as an example of a different kind [1208].

from both the interfacial interaction and the processes taking place within the adhesive phase due to structuring [1209] and causing, in particular, a decrease in the degree of polymer crystallinity. The latter effect is of particular importance for polyurethane and polychloroprene adhesives [1210] and results in the temperature dependence of the strength of adhesive joints obtained under bonding load F being described by an exponential function of the type:

$$P = k_f P_{\text{Coh}} \, T[(z + 1)F\tau/\eta T]^{y/(y+1)} \qquad (410)$$

where P_{Coh} is the inherent adhesive strength and the adjustment factor z varies from 1 to 2 [1211]. The effect of bonding temperature on P is similar; this is supported by the statistical treatment of a great number of relevant dependences for different plastics [1212] and elastomers [1213]. Porter [1214] asserted that these dependences are supported by the effects of energy dissipation in the course of the mutual motion of molecular dipoles under the effect of a shear force-field. He deduced the general relationships enabling calculation of the viscosity of polymeric liquids [1215] and of the elasticity modulus of glassy materials as a function of temperature [1216], and which are in good agreement with the experimental data.

On the other hand, the variation in the viscosity of an adhesive during its structuring can also be associated with the effect of the substrate surface. In general, the activation energy of the flow of a polymeric liquid over a solid surface is qualitatively different from that observed for conventional viscous flow of high molecular mass compounds. Effects of this kind are related [1217] to reorientation of the side chains and to the specific features of the motion of the fragments in the backbone, both resulting in a minimization of surface energy. For instance, the viscosity of a polyethene melt structured by dicumylperoxide does not depend on the thickness of the adhesive layer up to 400 μm, whereas the decreased (as compared to the bulk of the phase) value of the activation energy of structuring is caused by the catalytic effect exhibited by the surface of a steel substrate [1218]. Hence, it emphasizes the need to consider the specific features of viscous flow of polymers over solid surfaces whilst taking into account also the issues concerning both the flexibility of macromolecular chains and the free volume.

In terms of the classical hole model of liquids the mechanism of viscous flow over a solid surface is considered as an exchange, initiated by the shear force-field between the molecules and the 'holes'. Its initial step involves the overcoming of intermolecular forces, the corresponding work usually being identified with the activation energy of viscous flow. According to fundamental principles, the latter is related to viscosity via an exponential function:

$$\eta = k_\eta \exp (E_\eta^*/kT). \qquad (411)$$

The limiting values of E_η^* correspond to the fragments of macromolecules comprising 20–25 atoms in the backbone (but not to the infinitely long chains), these values being close to those of the segment. Hence, it is these fragments that should enable the viscous flow of polymeric adhesives by the mechanism of activated transport. This conclusion is commonly based on circumstantial evidence (for example, that cited in [1219]); however, direct proof can be obtained by using the number of repeat units in a segment s as a characteristic of segment flexibility [409]. For instance, comparison of this parameter within the values for the activation energy of viscous flow for various polymers [1220] revealed a practically linear dependence, as is illustrated in Fig. 3.65. The physical

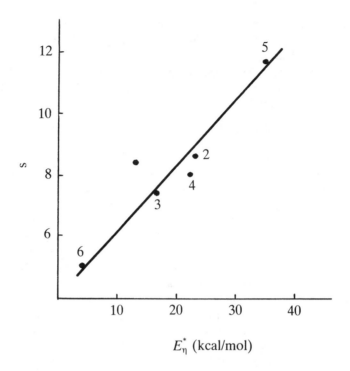

Fig. 3.65 – Activation energy of viscous flow versus number of repeating units in a segment of low-density polyethene (1), polypropene (2), polyisobutene (3), polystyrene (4), polyvinyl chloride (5) and polydimethylsiloxane (6).

meaning of this dependence is easily interpreted in terms of the concept of free volume an effective portion of which should be directly related to the activation energy of viscous flow. Such a relationship was indeed found for 11 polymers of different chemical structure (Fig. 3.66) and is described by a simple dependence [1221]:

$$lg\, E_{\eta}^{*} = 5.364 - 7.15\, V_f. \qquad (412)$$

This last deduction extends beyond the limits of a semiempirical interrelationship between the parameters of eq. (412). Rigorously, it traces the route bringing together the thermodynamic and molecular–kinetic approaches to adhesion postulated earlier as a major trend in the development of the basic theory of this phenomenon. Indeed, in eq. (412) E_{η}^{*} is of energetic origin, whereas V_f is a factor of structural type.

The theories discussed make it necessary to consider yet another factor arising from the viscoelastic behaviour of polymeric adhesives, namely the increase in the resistance to the action of mechanical forces in thin layers of a liquid phase during the final stages of spreading.

It is commonly assumed that a liquid droplet placed on the substrate surface initially acquires the shape of a spherical segment [1177] with the radius varying exponentially [1162, 1222]. In real systems the thickness of a liquid layer decreases in the course of spreading resulting in an increase in the forces retarding the process. According to Landau and Livshits [1223]:

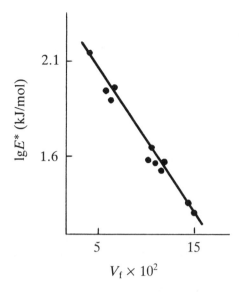

Fig. 3.66– Activation energy of viscous flow of adhesive melts versus effective fraction of free volume.

$$\frac{\partial}{\partial R} R \frac{\partial p}{\partial R} = 12R\eta v_z/h^3 \tag{413}$$

where v_z is the vertical rate-component and R is the distance from the symmetry axis. Assuming the rate gradient to be constant, integration of eq. (413) results in the relationship for the pressure drop in a droplet [1224]:

$$\Delta p = 0.375 \frac{\pi^2 \eta r^7 V_z}{v^2} \left[\frac{1}{r^2 - R^2} + \frac{3}{r^2} \ln (r^2 - R^2) \right]. \tag{414}$$

Then at $r = R$ we obtain an infinitely large value of Δp as evidence of the important role of this effect, testifying, on the other hand, to the inadequacy of the assumption of a spherical droplet. It is probably valid for low-molecular mass Newtonian fluids only. De Gennes [206] considers this deduction to be the main result of the work by Ogarev *et al.* [1192]. These investigators varied the viscosity of polydimethylsiloxane spreading on a mica surface by varying the temperature. De Gennes believes that deviation from the spherical shape is especially important for small droplets (according to eqs (413) and (414)) when the dome is surrounded by the projected base [206]. Then, at least in the final stages of spreading, the advancement of a liquid by the mechanism of viscous flow is somewhat retarded. These effects should be taken into account when choosing technological parameters for the formation of macroscopic contact in adhesive systems.

The dynamic character of the parameters which relate to the viscous flow of polymers substantially complicates the quantitative interpretation of the rheological aspects of the formation of adhesive joints and thus necessitates the differential type of the respective equations sometimes with unknown boundary conditions. Gul' [1225] suggested modelling the 'flowing in' of adhesives into microdefects of a substrate surface by polymer flow inside a capillary. He demonstrated that the permeation depth is proportional to the pressure at the slit entrance:

$$h \propto (p\tau/\eta)^{0.5}. \tag{415}$$

The validity of this approximation is supported by the similarity of eqs (415) and (360) provided that the viscosity of a liquid phase is taken into account in the latter. In fact, the peeling strength of the polyethene–cellophane system is related to parameter $(p\tau/\eta)^{0.5}$ [1225] via a linear relationship. These results once again emphasize the important role of microrheological processes [7] in the formation of polymer contact at macroscopic level.

These findings are evidence of the fact that the unambiguous rheological interpretation of the general laws of adhesive joint formation is a rather complex task [84]. For instance, even at $\theta < 90°$ wetting of the substrate surface by an adhesive with minimal viscosity may appear insufficient to provide a contact area that would be close to the theoretical. If an external load is applied to the equilibrium system then, in the majority of cases, the contact area will increase. This phenomenon is required by the complexity of the spreading mechanism, which remains quite complicated even in rather simple specific cases, such as for instance, the absence of evaporation of a liquid adhesive (or a solvent in multicomponent systems), of dissolution of a substrate, or of the formation of interfacial chemical bonds due to the effect of diffusion factors [1226], or of the morphology of substrate surface [1192] etc. Therefore the dependences of the area of molecular contact or of the adhesive joint strength on the duration of formation usually consistent of two parts, the initial one (high-rate) and the continuously ascending (almost linear) portion. The initial ascending section of the curve corresponds to the penetration of microhills of the substrate surface into the adhesive and the spreading of the adhesive over smooth areas and the later portion of the curve accounts for the filling of the microvoids by the adhesive. In the first stage the kinetics of the process is governed by the spreading rate of the adhesive entrapped between the solid surfaces of substrates, at the second one by the rate of microrheological gap-filling.

The absence of reliable information concerning the variation of the properties of adhesives and substrates in the course of forced formation of macroscopic contact together with the substantial mathematical difficulties, are the main reasons responsible for the fact that, in spite of certain progress in this field, rheology of adhesion is still one of the least developed branches of the science of polymer adhesion. Nevertheless, the phenomenology of the influence of relevant factors is, on the whole, clear, it being necessary to consider in any analysis both the theoretical principles of macroscopic contact formation and the applied problems of the technology of adhesive bonding.

3.2.2 Contact area in adhesive joints

A solid surface, polymeric in particular, is characterized by a developed relief, determined by the conditions of sample formation and processing. According to modern thought [25], the macroscopic non-uniformities (undulations) are of a size up to 10^3 μm and microdefects, are 2–3 orders smaller in height. The latter are of most practical significance. On the other hand, microscopic defects are much more difficult to control, as their role in the development of interfacial contact has to be examined when considering both spreading rheology and the deformation mechanics of microhills.

Microscopic defects on a real solid surface can be of various shapes. For instance, rod-like [1227], conical [1228], spherical [1229] and ellipsoidal [1230] microdefects have been distinguished in metals. This classification is quite conventional as, in reality, the profiles are rather complex and their quantitative treatment is complicated [1231]. Even within a small area microdefects can differ in their height by more than an order of magnitude [1232]. Thomas [1233] confirmed this conclusion by means of computerized mapping of the surface topography. Only the probabilistic approach provides for an adequate description of the distribution modes of microdefects [1231] involving, for instance, the theory of stochastic fields [1234].

The microrelief of polymeric surfaces is no less complex [27, 207]. The complexity is due to the fact that along with the effect of mechanical finishing of the surface the microrelief is also affected by the supermolecular structure of a polymer, specific features of polymer packing, degree of crystallinity and other factors, besides which, the substrate also produces its effect. It is very difficult to include each of the factors quantitatively. Today, it is premature to speak of the rigorous theory of surface topography of high-molecular mass compounds. Most of the ideas now viable in the field were generated in the physics of metals [25, 1231, 1234–1236] and which were then modified to match polymers mostly for the distinct cases of friction [26, 27, 29], the surface having been assumed to be isotropic [1237], while the shape of microdefects was somewhat arbitrary modelled by the system of rods [1238] or ellipsoids [1239]. Simulations of this kind confine the problem to that of mechanics and their examination is beyond the scope of this book; here we would only like to note that such an approach makes it possible to assess the effects of tensile loads normal to the plane, viscoelastic and elastic-flow deformations and microheterogeneities, as well as the stress distribution over the surface.

Nominal A_n (either termed theoretical, or geometric) and actual A_f (real, virtual) contact areas should be distinguished. A_f is considered as a multiplicity of sites over which the discrete contact between the interacting phases proceeds, hence A_f is always smaller than A_n. The ratio between A_f and A_n is governed by the separation h and the maximum height of surface defects $h_{\max}$ [1235]:

$$A_f/A_n = b(h/h_{\max})^c \tag{416}$$

where b and c are constants relating to surface microgeometry. Eq. (416) can be represented [1240] as the equality of two independent ratios involving the radius of the defects r_r:

$$h/r_r = h_{\max}(A_f/A_n)^{1/c}/r_r b^{1/c}. \tag{417}$$

The first one (i.e. on the left-hand side) serves as the basis for classification of the types of friction and wear, whereas the second is the most characteristic of surface roughness. Surface area of real contact is, in its turn, a complex function of the chemical type, primary and secondary structures of a polymer, and the interaction conditions. For instance, for spherical surfaces A_f is given as $\pi r_r N_r \Delta' f(\Delta'', d_r/\Delta')$ [1237], for ellipsoidal surfaces as $0.6\pi(k_r\Delta')^{0.5}N_r f(\Delta'', d_r/\Delta')$ [1239], where N_r is the number of defects, k_r is the roughness coefficient, d_r is the distance between the mean contours of the contacting surfaces, and parameters Δ' and Δ'' are characteristic of the mean-square deviation of the heights and the distribution function of the heights of the defects, respectively.

It is clear that direct application of such dependences when investigating the principles of adhesive joint formation is hardly plausible, though Rabinowicz [1241] reported an attempt to construct an energetic theory that defined the actual contact area via thermodynamic work of adhesion given by eq. (98):

$$A_f = \frac{F}{H}(1 - 2W_{\mathrm{Ad}}\, ctg\varphi/r_r H_0)^{-1} \qquad\qquad (418)$$

where H_0 is the hardness of the microdefects of the softer of the contacting phases and φ is the angle at the bottom of the trench between the hills. On the other hand, strong interfacial interaction results in non-uniformity of the properties within the thickness of the transition layers in polymers (see section 3.1.1). Naturally, this feature affects A_f, e.g. in metal–polymer adhesive systems microdefects within the oxide layer of a substrate and in its bulk produce a complicated effect on the value of A_f [1242]. Therefore, at the present state of the art it appears more reasonable to examine the influence imposed by major technological factors of the process of macroscopic contact formation (such as pressure, temperature and duration) on the actual contact area A_f. Since the reliability of the corresponding dependences is largely determined by the objectivity of the measuring procedure used to measure the actual contact area, it would be appropriate to elaborate first on the major experimental techniques involved.

Among a large variety of methods engaged in studies on the surface relief of adhesives and substrates the most unambiguous data is produced by optical methods. Strictly speaking, it is the area of optical contact that is measured using these techniques, rather than the actual contact area; however, there is practically a linear relationship [1243, 1244] between these two quantities.

To investigate polymers the method of Mekhau has been widely used. The method relies on photometric recording of the attenuated total internal reflection in the contact zones of a polished glass prism with the rough surface of the investigated sample. Total internal reflection is observed at a certain angle φ of the incident beam when the incident electromagnetic wave is reflected from the less optically dense medium (air) back into the first medium (glass). The equation for the propagation of an electromagnetic wave of the frequency ω is given as follows [1245]:

$$z = 2\pi k \lg[x_i\,(\sin^2\varphi/v_g^2\,\omega)/x_{i+1}] \qquad\qquad (419)$$

where v_g is the wave propagation velocity in the glass. Graphical presentation of eq. (419) in z/λ and x/λ ($\lambda = 2\pi \sin\varphi/v_g\omega$) coordinates depicted in Fig. 3.67 demonstrates that the line describing the propagation front does not overlap with the profile of microdefects, thus accounting for the somewhat limited sensitivity of the method. Indeed, the penetration depth d of the incident wave defined as the distance at which the amplitude of the wave is decreased by the factor e is given by the following relationship:

$$d = \lambda/2\pi\,[\sin^2\varphi - (n_a/n_g)^2]^{0.5} \qquad\qquad (420)$$

where n_a and n_g are the refractive indexes of the two media (air and glass, respectively). Then at $\varphi = 45°$ one has $d \approx 0.3\lambda$, which is around 200 nm for the wavelength $\lambda = 600$ nm.

In the majority of cases such resolution appears to be sufficient for evaluation purposes [22]; however, it should be borne in mind that for many polymers the range of

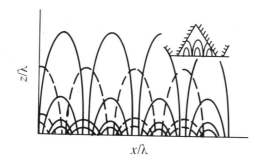

Fig. 3.67 – Schematic diagram describing propagation of electromagnetic waves in a clearance between rugous surfaces at total internal reflection.

the effect of intermolecular forces is an order of magnitude less — for instance, for elastomers it is only 5–10 nm [1246]. Therefore, certain attempts aiming to increase the resolution of Mekhau's method were undertaken. For instance, Sviridyonok *et al.* [1247] suggested using a polarized incident light beam, which enabled a deeper penetration of the beam into the microdefects of the surface and, consequently, provided for a decrease of 30% in the relative error of the method. More complicated is the case when attenuated total internal reflection from polymeric films displaying an absorption gradient is examined; however, for ethyene–propene–dicyclopentadiene ternary copolymer SKEPT-40 Stouchebryukov and coworkers [1248] demonstrated that it is sufficient to account for the dimensionless extinction coefficients in directions parallel and perpendicular to the polarization plane. A different approach is based on studies carried out in the IR range of the spectrum, provided that the substantial losses in dispersion spectrometers are decreased due to the sharp increase of the resolution ensured by Fourier-transform technique [1167]. This procedure makes it possible to evaluate d within the range of 1–10 nm.

When investigating the modes of macroscopic contact formation between polymers (namely of the actual contact that is generally, not accompanied by deformation of microdefects via their 'immersion' into the adhesive layer) is best provided by the measurement of the actual contact area along with the measurement of the strength of adhesive joint. The measuring device, based on a two-channel vectorelectrocardioscope (plotting the tensile strength along the horizontal axis and the actual contact area along the vertical one), is described in [1249]. The components of this equipment are shown in Fig. 3.68. The use of specifically designed tensile stress wire-detectors and of photo-elements provides for the simultaneous detection of P_t and A_f in the loading range, elongation rate, and temperature ranges 0.01–1.0 MPa, $(8–120) \times 10^{-4}$ m/s, and 293–473 K, respectively. Details of the use of the device are presented in Fig. 3.69 . The sample 13 is pressed to a prism 3 by the fixed load imposed by the lever 12. After applying the load it is then removed, the sample connected to the tensile stress detecting system and subjected to fracture by moving the carriage 6 with the mounted stress detectors 11.

By means of this method a number of technological factors was reliably evaluated with regard to their influence on the actual contact area and the strength of adhesive joints involving polymers.

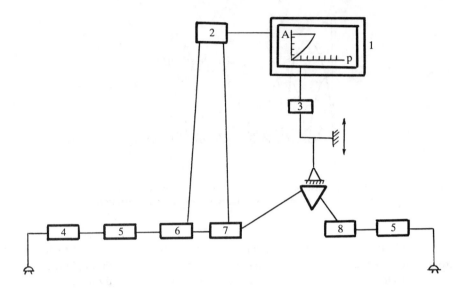

Fig. 3.68 – Diagram of the device for simultaneous measurement of the strength of an adhesive joint and of the actual contact area between the elements of the joint. 1, Vector-electrocardioscope; 2, cathode follower; 3, strain transducer; 4, voltage regulator; 5, transformer; 6, power supply unit; 7, photoelement; 8, light source.

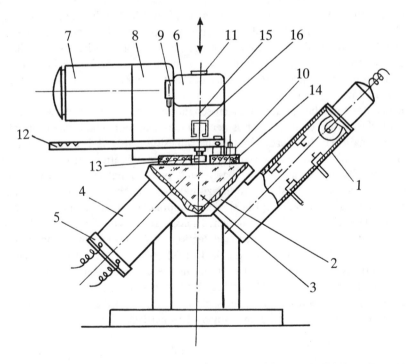

Fig. 3.69 – Construction of the device for simultaneous measurement of the strength of an adhesive joint and of the actual contact area between the elements of the joint.

One of the factors examined is the normal pressure p. It is clear that in the case of polymers its influence is related to Young's modulus E; the respective dependences are extremal, the highest corresponding to $p/E \approx 0.1$. The previously suggested unlimited growth of A_f with increasing p [1229, 1250] is, obviously, lacking any physical sense since the ratio between the actual (as defined above) and nominal contact areas cannot exceed unity. Hence, Thirion's formula

$$A_f/A_n = kp(1 + kp) \tag{421}$$

holds true only when p has a certain power y [1251] since only in this case will the actual contact area tend to reach the nominal one with the increase in p. Analogy with Herz equation

$$\lg (A_f/A_n) = \lg k + 2/3 \lg p \tag{422}$$

describing the exponential function clarifies the physical meaning of eq. (421). Voyutskii et al. [1251] demonstrated that y is indeed close to unity, as proposed by Thirion; however, it is not equal to it and, in any case, it depends on temperature and the Young's modulus of a polymer.

According to Bartenev [1252] the following relationship is valid

$$A_f = A_n (A_f^0/A_n + kp)/(1 + kp), \tag{432}$$

where A_f^0 is the actual contact area at $p \to 0$ and A_n is the nominal contact area. The ratio A_f/A_n defines the specific actual contact area $a_f = A_f/A_n$ in agreement with expression (416). This function must obey the condition $a_f < 1$. However, it does not involve parameters related to the physical state of a polymer. Having assumed the elastic character of deformations in the entire loading range (described by the elasticity modulus E) Bartenev and Lavrentiev [1253] performed the necessary manipulation

$$d(a_f) = k_r(1 - a_f) \, dp/E. \tag{424}$$

The term $(1 - a_f)$ accounts for the fact that, with an increase in p, the fraction of the actual contact area grows more steeply as the fraction of the surface area not involved in the contact decreases and the higher is Young's modulus at uniaxial compression. From eq. (424) it follows:

$$a_f = 1 - (1 - a_f) \exp (-k_r p/E). \tag{425}$$

In other words, with the increase of normal pressure the actual contact area increases only to a certain limit. The validity of eq. (425) and of the conclusions were verified under boundary conditions $a_f = a_f^0$ at $p = 0$ and $a_f = 1$ at $p \to \infty$ [1243, 1254].

The adhesive joint strength versus bonding pressure dependences obey similar patterns also [994, 1255]. However, in this case rheological aspects of the formation of macroscopic contact acquire significance. Indeed, usage of low-viscosity adhesives (e.g. of solutions or of monomeric adhesives) can result in the situation when microdefects on the substrate surface are filled with adhesive explicitly due to capillary forces, i.e. without the effect of an external load. This fact, i.e. that the joint strength is independent of applied pressure p, was observed in a number of studies, for instance, by Vakula and Voyutskii [1256], and it served as the keystone for the microrheological concept of adhesive joint formation [7, 1225]. Relying on Tobolsky's equation for polymer strength, Imoto et al.

[1257] suggested relating the number of contacts (crosslinks) with the bonding pressure by means of a relationship which is almost identical to eq. (425)

$$N_f(p) = k \exp (bp). \qquad (426)$$

Then

$$\ln (P_0/N_f^0) = \ln (k/P_0) + bp \qquad (427)$$

where P_0 and N_f^0 are the initial strength and the number of contacts at $p = p_0$, respectively and k and b are constants. It is clear that eqs (426) and (427) give an adequate description of only the initial stages of the process of adhesive joint formation, since the dense network of interfacial bonds developing during the process smooths off the differences between the relaxation properties of polymers. For instance, for filled poly(vinyl furfural) it was demonstrated [1258] that the filler affects the actual contact area only under loading. On the other hand, eqs (426) and (427) are valid for systems in which there are either no interfacial valence bonds or their effect has become constant due to the effect of the nature of the materials brought into contact. For instance, copper displays rather poor adhesive properties and, consequently, the strength versus bonding pressure dependence of its joints with reactive epoxy adhesives matches eq. (427) [1257]. Statistical calculations made by Lavrentiev et al. [1259] disclosed that the maximum number of molecular contacts in an elastomer–solid system is only slightly dependent on pressure.

Finally, one should bear in mind that the pattern of the actual contact area versus pressure dependence is most decisively affected by the physical state of the polymers involved. It was rather convincingly demonstrated by Kaelble [1260], who discovered the correlation between the derivatives $\mathrm{d}T/\mathrm{d}p$ and $\mathrm{d}T_g/\mathrm{d}p$, which is quite illustrative since T_g is, to a certain extent, characteristic of the efficiency of intraphase interaction. Hence, the formation temperature of adhesive joints is, like pressure, liable to control the forms of the actual contact area variation. In fact, the respective dependence is also an exponential one [1261, 1262]:

$$a_f = 1 - \exp[k_r p/E(T)], \qquad (428)$$

the physical state of a polymer being accounted for by Young's modulus. It is natural to assume that there should be an inflection point on the a_f versus temperature graph corresponding to the transition from the glassy to the rubber-like state (which is accompanied by a sharp change in Young's modulus). This conclusion was experimentally verified by Bartenev and Lavrentiev [1261]. For crystalline polymers a_f is increasing most dramatically at temperatures around the melting point.

Effects of this kind should be seen even more prominently around the glass-transition temperature T_g and in this case variation of the actual contact area is related to the number of contacting macromolecules N_c. Elkin [1263] demonstrated that at T_g, due to frozen elastic deformation, neither a_f nor N_c are affected by variation and even removal of the normal load. Calculations made by Midhailov and Nikolaev [1264] disclosed that for the natural rubber–steel system N_c remained constant in the temperature range below as well as above the glass–transition temperature. Yet this conclusion is even more note-worthy when examining the temperature dependence of the tensile strength of adhesive joints; as is seen in Fig. 3.70, the branches on both sides of the inflection point, corresponding to T_g, are actually parallel to the ordinate axis, thus providing evidence

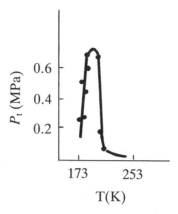

Fig. 3.70 – Tensile strength versus temperature for the adhesive joint between crosslinked natural rubber and steel produced at 293 K. Drawing rate, 16.6 μm/s.

that the actual contact area stays constant. Similar patterns were observed for the adhesive joints of elastomers [1265] and of polyvinyl acetate with glass [1266] where temperatures corresponding to maximum joint strength were found to correlate with T_g. On the basis of such regularities Levit *et al.* [1265] suggested a method to determine T_g from measurements of the peeling strength of adhesive joints. Huntsberger [1267] disclosed a smooth temperature dependence (temperature range 293–413 K) for the strength of adhesive joints of polyethene, polyvinyl acetate, and polybutyl methacrylate with steel at $T > T_g$ of the adhesive (adhesives were polystyrene and polymethyl methacrylate), while at $T \approx T_g$ the respective graph is extremal. Similar effects were also observed in the polymethyl methacrylate–glass system at decreasing temperature [1244].

At contact temperatures far from T_g the modes of a_f variation are governed predominantly by rheological factors, i.e. by the intensity of the thermally initiated spreading of adhesive on a substrate surface and its penetration into microdefects of the latter. Formation of strong adhesive bonds between polyethene and steel was observed at temperatures as low as 333–355 K [1268] and this was attributed to morphological changes in the adhesive in contact with the high-energy substrate surface. As long ago as in 1949 McLaren and Seiler [1269] reported extreme variations in the strength of polymer–cellophane joints in the range of bonding temperatures 263–343 K, a six-fold growth of the joint strength being observed during a temperature increase from 263–323 K owing, according to the authors [1269], to the growth in mobility of the macromolecular chains. It was found that the strength of polymer–metal systems was not affected when the bonding temperature $T > T_{\text{viscous flow}}$. This was observed, for instance, for polymethyl methacrylate [1270] and for polyvinyl chloride [1001]. The obvious inference is that the maximum available actual contact area is attained under these specified conditions.

Since formation of the actual contact area in polymers takes place at temperatures embracing a broad range, extreme strength versus bonding temperature dependences are usually observed in practice (as a rule no thought is given to transition temperatures in polymers), the observed pattern is of purely kinetic origin.[†] Indeed, at low fracture rates the respective maxima are degenerated [1275]. Bright [1276] demonstrated the kinetic equivalence of temperature and deformation rate in their effect on adhesive joint strength. To describe this dependence Gul [7] suggested the following relationship:

$$P = t_0 \nu \exp\left(U_a'/RT\right) \qquad\qquad (429)$$

where the preexponential term t_0 accounts for the nature of an adhesive and a substrate, while U_a' is the apparent activation energy of fracture. The reported ambiguous character of the observed temperature dependences [1037] is due to the fact that different counteracting processes are simultaneously induced in adhesives under the effect of temperature [22, 1255, 1277]. In some cases dependences of this kind were reported to display a number of maxima, as was observed, for instance, in epoxy bonded adhesive joints of crystalline and amorphous homo- and copolymers of trifluorochloroethene [1278]. The first maximum is attributed to the effect of thermodynamic factors, while the second is ascribed to the rheological features of the interaction between adhesive and substrate. Hvostik *et al.* [1213] examined a variety of strength versus temperature dependences of adhesive joints between elastomers (and rubbers) and metals. Fig. 3.71 depicts the generalized curve in coordinates $P/P_{\max}$ versus $(T - T_r)$, where $P_{\max}$ and T_r are reducing factors defined as the maximum peeling strength and the temperature at which a certain minimal strength of adhesives is still retained, respectively. The graph is

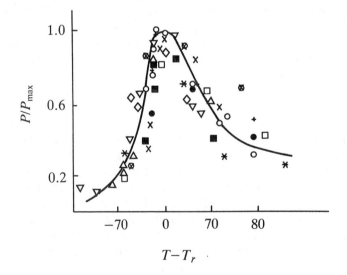

Fig. 3.71 – Generalized thermoadhesion dependence for a variety of elastomers: natural rubber (♦), SKI-3 (⊕), SKS-30ARK (□), BK-RT with mol. weights 5.0×10^5 (x) and 5.6×10^5 (+), BK-1675T (○), SKEPT-40 (◉); rubber stocks based on natural rubber (▾), SKI-3 (■), BK-1675T (∗), and BK-1675T with SKEPT-40 (85 : 15) (△).

[†] A detailed analysis of pertinent problems, which is beyond the scope of this monograph, was performed by Rayatskas [1271], who expanded the time–temperature [1272], rate–temperature [1273] and concentration–temperature [1274] superposition principles to the adhesive joints of polymers.

seen to confirm the validity of the concepts discussed describing the relationship between temperature and actual contact area and, correspondingly, the strength of adhesive joints involving polymers.

The kinetic character of the process of actual contact area formation is most vividly exhibited in adhesive joint strength versus time dependences. Taking into account the discussed features concerning the effect of pressure and temperature one may presume the exponential law to hold true in this case also. In fact, Lavrentiev [1279] reported the following relationship:

$$a_f \propto 1 - \exp(-\beta\tau). \tag{430}$$

However, parameter β is not constant, being a complex function of p and τ. The latter fact illustrates a closer connection of τ, rather than of p and T, with rheological factors. Hence, corresponding dependences must involve viscosity η and viscous flow strain ξ of polymers.

In the simplest case involving elastomers the equation suggested by Rehbinder [1280] is valid

$$d\xi = \frac{\xi_\infty}{\tau'} (1 - \xi/\xi_\infty) \, d\tau \tag{431}$$

where τ' is the after-effect period determined by viscosity and external load:

$$\tau' = \eta\xi/F. \tag{432}$$

Provided that there is proportionality between ξ and a_f (i.e. $\xi/\xi_\infty = a_f/a_f^\infty$) the kinetic dependence of the actual contact area acquires the following form

$$a_f + \ln(1 - a_f) = -F\tau/\eta \tag{433}$$

which is consistent with experimental data [1279]. It should be remembered that the limits of the initial eq. (431) restrict the applicability of expression (433) to elastomers only, whereas for plastics it is valid only at the initial stages of macroscopic contact formation. This is quite clear since an exponential law governs the variation of both the stress, provoking polymer flow [1281],

$$\gamma_\tau = \eta \exp(-\beta|F|) \, d\xi/d\tau \tag{434}$$

and the polymer viscosity

$$\eta = k \exp(U/RT). \tag{435}$$

Then

$$\xi = [\gamma_\tau \tau \exp(\beta|F|)]/[k \exp(U/RT)]. \tag{435}$$

Within the framework of a more general approach, based on statistical analysis of the actual contact area [1282], Bhushan [1283] expressed this area as a function of pressure and the duration of the process. Hence, at $p = 7-28$ kPa and $\tau = 10 \, \mu s$ the contact is elastic (plasticity index ψ of a polymer with a Young's modulus of 0.5 does not exceed 1.8) and the topography of a substrate surface is practically unaltered whereas the opposite effect is observed at $p = 1.38$ MPa and $\tau = 9$ days ($\psi > 2.6$).

It is, probably, unrealistic to distinguish between the effects of pressure, temperature and duration of contact on the formation modes of the actual contact area, nevertheless there were certain attempts reported. For instance, according to Bright [1276], at constant temperature the strength of adhesive joints is proportional to τ

$$P = P_\infty - \exp(-k/\tau). \tag{437}$$

Kanamaru [1001], having introduced the additional parameter τ_s, the time required for the segments to move from the free state to the bonded, obtained the following equation

$$P = P_0 + P_\infty [1 - \exp(-\tau/\tau_s)], \tag{438}$$

while Imoto [1257] suggested a linear relationship

$$P = P_0(1 + k\tau). \tag{439}$$

These considerations illustrate the inadequacy of such simplified approaches. These approaches are doubtlessly useful since they provide for a qualitative description of the influence of individual factors on the formation modes of the actual contact area; however, it should be borne in mind that in all of them the relaxation properties of polymers are neglected. In fact, this is the specific feature of the basic eq. (422). The situation is somehow improved by introducing the temperature dependence of Young's modulus, though such a procedure is a mere adjustment.

To get rid of this inherent deficiency it is worthwhile to make use of the hereditary theory of elasticity [1284]. Assume the surfaces covered by half-spheres of radii r_a and r_s as models of adhesive and substrate, respectively. Then the Herz equation will have the following solution [1285]

$$a_f(\tau) = \pi \left(\frac{\pi k_g F}{2R} \right)^{2/3} \tag{440}$$

where $R = (r_a + r_s)/2r_a r_s$, while the rigidity coefficient of a system is a function of Poisson's modulus v and shear modulus μ:

$$k_g = \frac{3}{8\pi} \left(\frac{1 - v_a}{\mu_a} + \frac{1 - v_s}{\mu_s} \right) \tag{441}$$

where the indexes a and s are used to refer to adhesive and substrate. Within the framework of the approach developed by Rabotnov [1285] the force-induced (F represents the imposed force) relaxation mode of the shear modulus obeys the following rule:

$$\frac{1}{\mu} = \frac{1}{\mu_0} \left(1 + \frac{3}{2(1 + v)} \right) \int_0^\tau k(t - \tau) F(\tau) \, d\tau. \tag{442}$$

The hereditary nucleus describing the relaxation behaviour of a polymer is determined as $k(t - \tau) = \sum_i \varphi_i \exp(t - \tau)/\tau_i$, where φ_i is the volume fraction of the phase characterized by the relaxation time τ_i, μ_0 is the shear modulus of a sample at $\tau = 0$ and $F(\tau)$ is the kinetic law describing the variation of F that demonstrates the universal character of eq. (442).

As a substrate is, as a rule, more rigid than an adhesive ($\mu_s > \mu_a$) the following relationship holds true

$$k_g \approx \frac{1 - v_a}{\mu_a} = \frac{1 - v_a}{\mu_0}\left(1 + \frac{3}{2(1 + v)}\int_0^\tau k(t - \tau)F(\tau)\,\mathrm{d}\tau\right). \tag{443}$$

Substituting eq. (443) into eq. (440) and performing the rearrangements one finally comes to a solution

$$a_f(\tau) = \frac{\pi}{8}\left[\frac{r_a r_s(1 - v_a)}{(r_a + r_s)\mu_0}\left(1 + \frac{3}{2(1 + v_a)}\int_0^\tau \sum_i e - \frac{t - \tau}{\tau_i}F(\tau)\,\mathrm{d}\tau\right)\right]^{2/3}. \tag{444}$$

The first thing to draw attention to is that the power indexes of eqs (422) and (444) are the same, thus serving as evidence of the validity of this approach. However, the final equation is, obviously, of much greater generality than eq. (422), since it enables clarification of the meaning of the various terms. Consequently, it appears feasible to establish in a physically sound manner the relationship between the kinetic modes of the interfacial contact area formation and the properties of contacting materials. For instance, relying on eq. (444) Zilberman succeeded in deriving the equation expressing the initial contact area a_f^0 via shear moduli and roughness characteristics of the samples:

$$a_f^0 = \frac{\pi}{8}\left(\frac{r_a r_s(1 + v_a)}{(r_a + r_s)\mu_0}\right)^{2/3}. \tag{445}$$

Hence, taking into account that within the molecular–kinetic frame of reference the efficiency of interaction between adhesive and substrate is an explicit function of the actual contact area, the concepts discussed here provide a non-contradictory basis for examining the effect of major technological and related factors on the formation modes of adhesive joints involving polymers.

3.2.3 Formation of polymer adhesive joints

The first consistent treatment of adhesion within the molecular–kinetic frame of reference, based on the classical approaches of Frenkel and Eiring, was undertaken by Hatfield and Rathmann [1286]. While examining the modes of the formation of adhesive joints involving polymers, they suggested exponential patterns to express the number of attraction molecules

$$N_a = \exp\left(-U_a^{(1)}/KT\right) \tag{446}$$

and repulsion molecules

$$N_r = \exp\left[(W_{\mathrm{Ad}} - U_a^{(1)})/KT\right] \tag{447}$$

via activation energy per molecule $U_a^{(1)}$ and the work of adhesion W_{Ad}. Then the number of unbonded molecules is as follows:

$$N_a' = N_r \exp\left(-W_{\mathrm{Ad}}/KT\right). \tag{448}$$

As the rate of disjoining the molecules is given by the corresponding derivative, the system formation kinetics is determined by the driving force F_{Ad} and the distance h between the initial position of the molecule and up to the region of interfacial interaction:

$$\frac{\mathrm{d}N_r}{\mathrm{d}\tau} = \frac{KT}{d} \left[\exp\left(W_{\mathrm{Ad}} - U_a^{(1)}\right)/KT\right] N_r \left[\exp\left(F_{\mathrm{Ad}}h/2N_rKT\right)\right.$$

$$\left. - \exp\left(-F_{\mathrm{Ad}}h/2N_rKT\right)\right]. \tag{449}$$

According to Hatfield and Rathmann [1286], the process proceeds at a constant rate equal to the starting value $(-\mathrm{d}N_r/\mathrm{d}\tau)_{\tau=0}$. The duration of the process is then given by the following relationship:

$$\tau = N_r/(\mathrm{d}N_r/\mathrm{d}\tau)_{\tau=0}. \tag{450}$$

From eq. (449) one finally comes to a hyperbolic sinusoidal pattern relating reciprocal duration of the formation process to the strength of the adhesive joint:

$$\tau^{-1} = \frac{KT}{d} \exp\left[(W_{\mathrm{Ad}} - U_a^{(1)})/KT\right] \left[\exp\left(F_{\mathrm{Ad}}h/2N_rKT\right)\right.$$

$$\left. - \exp\left(-F_{\mathrm{Ad}}h/2N_rKT\right)\right] = A\,\mathrm{sh}(BF_{\mathrm{Ad}}). \tag{451}$$

Brunt [1287] reported qualitative evidence of the validity of this result, Imoto [1199] demonstrated a quite satisfactory correlation between the calculated (eq. (451)) and experimental data for polyisobutene–polymethyl methacrylate (Fig. 3.72).

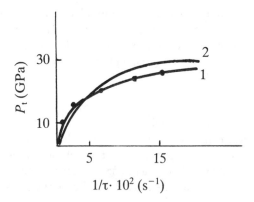

Fig. 3.72 – Tensile strength of adhesive joints between polyisobutene and polymethyl methacrylate versus contact time. 1, Experimental curve; 2, theoretical curve.

However, this approach did not gain acceptance, probably because of the difficulties in determining N_a and N_r, which, for polymers, seemed unlikely to be overcome. Nevertheless, taking into account the chain nature of high-molecular mass compounds and their ability to become involved in interfacial interaction on a segmental scale, it appears valid to substitute the parameters N_a and N_r for the numbers of interfacial bonds (within the framework of Bartenev's [1288] construction and later developed by Lavrentiev [1289]) or, more precisely [22], by the number of segments N_s.

The magnitude of N_s is determined by the number of segments in the boundary layer that are attached directly to the substrate surface N_b and by that of segments potentially capable of forming such bonds (but not yet bonded) N_f,

$$N_s = N_b + N_f. \tag{452}$$

A constant ratio of N_f/N_b corresponds to the equilibrium state, its distinct value being determined by the relaxation properties of a polymer and the efficiency of interfacial interaction with a substrate. In Eyring's approximation a single kinetic unit only is involved in the elementary process of a segment undergoing transition from the free state to the bonded one. This condition is valid, above all, for elastomers, for which the time gap between the formation and failure (breakdown) of the interfacial bond exceeds that between the two consecutive collisions of the segments due to thermal motion. Describing the probabilities of the formation and failure transitions by the frequencies v_f and v_b, respectively, one can easily deduce a simple kinetic relationship [1289]:

$$dN_f/d\tau = N_b v_b - N_f v_f. \tag{453}$$

For the equilibrium state at $\tau \to \infty$ the following expression holds true

$$N_\infty = (v_b/v_f)/(1 + v_b/v_f). \tag{454}$$

Solution of eq. (453) is as follows:

$$N_\infty - N_f = (N_\infty - N_0) \exp [-\tau(v_f + v_b)] \tag{455}$$

where the term in the square brackets is implied as having the meaning of a relaxation time τ_r. Clearly, the relationship (455) is valid if v_f and v_b are independent of N_s, i.e. when these frequencies are determined solely by the effect of external factors, e.g. processing conditions of adhesive joint formation.

In the simplest case parameters of e.g. (455) are independent of the contact duration; for polymers the contacting process at constant temperature proceeds spontaneously until equilibrium in the transition layers is achieved. Then the ultimate strength of a system is related to the relaxation time via the exponential relationship

$$P_\infty - P_\tau = (P_\infty - P_0) \exp (-\tau/\tau_r) \tag{456}$$

hinting that the time–temperature superposition appears to be viable. Indeed, at a fixed value of τ the value of P_∞ appears to be a function of temperature, while at fixed temperature a function of τ. Hence, the temperature dependences of adhesive joint strength are liable to be represented by curves characterized by a saturation plateau, as is exemplified in Fig. 3.73 for the polyisobutene–glass system. Then, the kinetic dependence of the generalized parameter $\ln[(P_\infty - P_\tau)/(P_\infty - P_0)]$ is described by a straight line, whose slope determines, according to eq. (456), the value of τ_r. The slope increases as the temperature is increased. For instance, the temperature-induced variation of the relaxation time of polyisobutene in its adhesive joint with glass is illustrated by the following pattern:

T, K	296	313	333	353	373	393	413
$\tau_r 10^{-3} s$	5.65	3.00	2.00	1.05	0.65	0.52	0.35

As is seen from Fig. 374(a), within the same temperature range only a three-fold rise of P_∞ is observed as compared to the 16-fold decrease of τ_r, these facts illustrating the

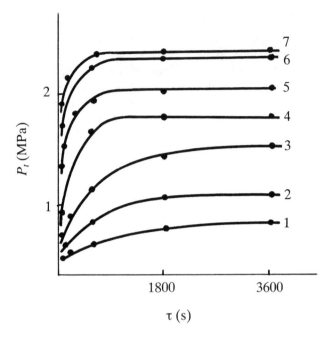

Fig. 3.73 – Strength of adhesive joints between polyisobutene and glass as a function of contact time. Bonding temperature: 1, 293 K; 2, 313 K; 3, 333 K; 4, 353 K; 5, 373 K; 6, 393 K; 7, 413 K.

greater effect of temperature on the relaxation properties of polyisobutene rather than on the adhesive.[†]

Hence, the generalized time–temperature relationship describing the strength of adhesive joints can be presented in exponential form

$$[P_\infty(T) - P_\tau]/[P_\infty(T) - P_0] = \exp\left[-\tau/\tau_r(T)\right] \tag{457}$$

as a function of the temperature dependences of parameter P and τ_r. Comparing this relationship with eq. (453), one can easily see that the frequencies of kinetic transitions of the segments are governed by the same temperature–dependent pattern,

$$\upsilon_f = \upsilon_0 \exp\left(-U_f^a/KT\right), \tag{458}$$

$$\upsilon_b = \upsilon_0 \exp\left(-U_b^a/KT\right) \tag{459}$$

where υ_0 is the frequency of segmental vibrations due to thermal motion only, and devoid of any substrate surface effect, while the activation energies U^a describe the

---

† When elaborating on the effect of bonding temperature on the strength of adhesive joints one should bear in mind that its manifestations are substantially dependent on the testing temperature. This is convincingly illustrated by the data of Fig. 3.74(b), depicting the respective graphs for adhesive joints between butadiene–acrylonitrile copolymer SKN-26 and glass [22]. The dependences presented can be explained using the molecular–kinetic features of the formation and failure of adhesive bonds and they are in good accord with the inferences ensuing from eq. (475). This remark is important when examining the strength of adhesive joints as a function of a temperature.

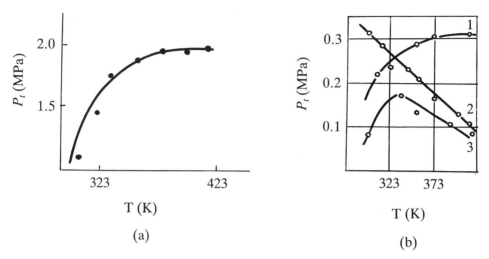

Fig. 3.74 – (a) Tensile strength of adhesive joints between polyisobutene and glass versus bonding temperature. (b) The strength of adhesive joints between butadiene–acrylonitrile elastomer SKN-26 and glass as a function of temperature. 1, Testing temperature const. 203 K, bonding temperature is scaled on abscissa; 2, bonding temperature const. 413 K, testing temperature is scale on abscissa; 3, bonding temperature = testing temperature.

height of the energy barrier at the interface. One can easily see that U_b^a should be commensurable with the activation energy of the diffusion process providing the transport of macromolecules or of their segments from the bulk of the phase into transition layers, while U_f^a must be comparable with the surface energy of a polymer,[†] i.e. with the energy utilized in interfacial interaction.

Combining eqs (455) and (459), one can express the relaxation time via these parameters:

$$\tau_r^{-1} = \upsilon_0 \exp\left(-U_f^a/KT\right) + \upsilon_0 \exp\left(-U_b^a/KT\right). \tag{460}$$

According to eq. (453), for the equilibrium state the following relationship holds true

$$N_\infty = \left[\exp\left(\Delta U_f^a/KT\right)\right] / \left[1 + \exp\left(\Delta U_b^a/KT\right)\right]. \tag{461}$$

Then, from eq. (460) it follows that

$$\tau_r^{-1} = \left[\upsilon_0 \exp\left(-U_b^a/KT\right)\right]/P_\infty. \tag{462}$$

As demonstrated above, N_∞ exhibits a much weaker dependence on temperature than on the term $\exp(U_b^a/KT)$. In this case the ratio υ_0/P_∞ can be assumed to be constant (τ_r') and the final expression is as follows:

$$\tau_r = \tau_r' \exp\left(U_b^a/KT\right). \tag{463}$$

[†] As a consequence, the contact area between viscoelastic bodies is related to the surface energies of the bodies [1290].

With the aid of this relationship one can evaluate U_b^a as the slope of the plot of the temperature dependence of relaxation time [1291]. For instance, for the polyisobutene–glass system, shown in Fig. 3.75, the activation energy was found to be 23.9 kJ/mol, which is significantly less than the 62.9 kJ/mol characteristic of relaxation processes within the bulk of adhesive. This observation provides direct proof of the validity of the ideas discussed in section 3.1.2 concerning the fact that it is the diffusion mechanism that is most probably involved in the formation of the macroscopic contact of polymers.

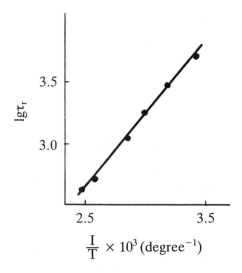

Fig. 3.75 – Relaxation time of the adhesive joint between polyisobutene and glass versus temperature.

In conformity with the data of the preceding section such features should be inherent to variations in the actual contact area in adhesive joints. According to the rearranged form of eq. (433), plotting the parameter $[a_f + \ln(1 - a_f)]$ versus time results in a straight line. Then the slope, i.e. η, characterizes the viscosity of the transition layers of an adhesive (Fig. 3.76). Taking into account that relaxation properties are governed predominantly by the nature and composition of a polymer, one might assume that η should be constant regardless of the effect of external factors, e.g. of the load. In fact, while the actual contact area of the polyisobutene–glass joint is a function of the load, as is illustrated in Fig. 3.77, η appears to be constant, according to the calculations [22], within a broad range of external loads (from 0.08 to 0.62 MPa) the value of η is around 2.4–2.8 kPa s. The situation is the same with cross-linked rubbers [1279].

We believe that these facts indicate that there are no structural rearrangements taking place in the transition layers of an adhesive during the formation of the actual contact area; the forms of the latter process are governed mainly by relaxation phenomena. Consequently, at elevated temperatures the maximum a_f value is achieved more rapidly.. For instance, when the contact temperature is $10°$ above the temperature of viscous flow of polyisobutene the maximum contact area is attained within 5 s [1292]. Naturally, the effect of temperature cannot avoid affecting η; however, it is only for adhesives with a loose network of intermolecular bonds that the role of this factor is substantial. For

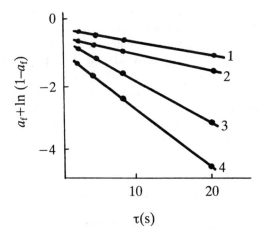

Fig. 3.76 — Anamorphosis of the kinetic dependence of the specific area of actual contact in adhesive joints of polyisobutene with glass, obtained at different bonding pressures. 1, 0.08; 2, 0.14; 3, 0.32; 4, 0.62 MPa.

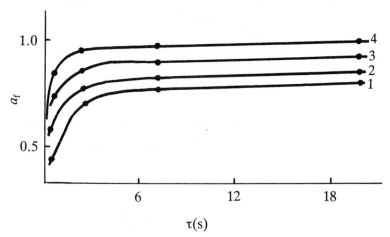

Fig. 3.77 — Specific area of actual contact in adhesive joints between polyisobutene and glass versus contact time. Bonding pressure: 1, 0.08 MPa; 2, 0.14 MPa; 3, 0.32 MPa; 4, 0.62 MPa.

instance, in the case of non-crosslinked elastomers raising the temperature from 295 to 413 K results in a decrease in η from 2.2 to 0.86 kPa s [22].

Hence, the nature of adhesives[†] and of interfacial interaction, as well as external factors (pressure, temperature, and duration of the process), bestow a different effect on the variation of the actual contact area on the one hand, and on the strength of adhesive joints on the other. Time-dependent aspects of both of these processes were simultaneously examined for the polyisobutene–glass joint as an example. As is seen in

---

†   In this case the role of this factor is determined by the indirect effect of the generalities pertaining to the thermodynamics of interfacial interaction (section 2.1.2) and the rheological aspects of the contact formation with polymers (section 3.2.1).

Fig. 3.78, P_{tensile} continues to increase even after the equilibrium values of a_f have been reached. This effect is observed over a broad range of pressures, demonstrating that the duration of the formation process produces a greater effect on the strength of a system than does the value of N_s. As is seen from Fig. 3.79, graphically such dependences are depicted by declining curves somewhat stretched along the time axis, or, alternatively, compressed along the pressure axis and it is noteworthy that explicit functions describing these dependences were deduced. Note (see Fig. 3.78) that the growth of adhesive joint strength is greatest during the first several seconds. It therefore appeared reasonable to examine this time range in greater detail.

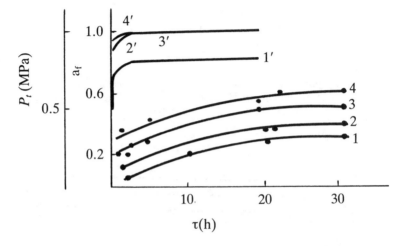

Fig. 3.78 – Tensile strength (1–4) and specific area of actual contact (1'–4') versus time of contact (at large contact times) for the joints between polyisobutene and glass. Bonding pressure: 0.08 MPa (1, 1'), 0.14 MPa (2, 2'), 0.32 MPa (3,3'), 0.62 MPa (4, 4').

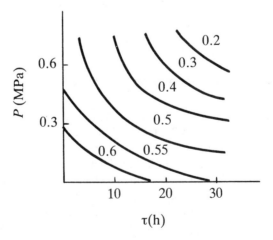

Fig. 3.79 – Response surface for the strength of adhesive joints between polyisobutene and glass at different contact times (figures at the curves denote tensile strength in MN/m²).

The data in Fig. 3.80 demonstrates that a_f reaches its equilibrium value within 20 s, while P_{tensile} remains constant. Consistant with expectations, application of pressure does not alter the pattern e.g. at 0.32 MPa, as well as at 0.62 MPa when macroscopic contact is complete (according to optical data), the kinetic dependence of the adhesive joint tensile strength tend to equilibrium and their limiting values at an equal rate. Hence, at small contact times the kinetic dependence of $[a_f + \ln(1 - a_f)]$ is linear (Fig. 3.76 and obeys eq. (433), whereas for a broad interval of T values the strength dependence shown in Fig. 3.81 reflects the expression

$$P = P_{\infty}[1 - \exp(-\tau/\tau_{\infty}^*)] \qquad (464)$$

which is similar to eq. (455), here τ_{∞}^* denotes the residence time of a macromolecular segment in a transient equilibrium state.

Hence, when examining the formation of the equilibrium actual contact area the aspects related to the effect of external load F appear to be of major interest. The general

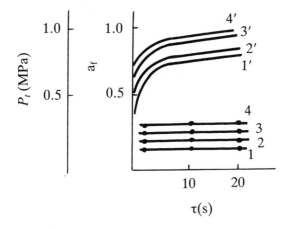

Fig. 3.80 – Tensile strength (1–4) and specific area of actual contact 1′–4′) versus time of contact (at small contact times) for the joints between polyisobutene and glass. Bonding pressure: 0.08 MPa (1, 1′); 0.14 MPa (2, 2′); 0.32 MPa (3, 3′), 0.62 MPa (4, 4′).

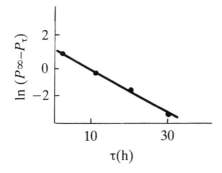

Fig. 3.81 – Anamorphosis of the kinetic dependence of the tensile strength of adhesive joints between polyisobutene and glass at large contact times (deduced from the data of Fig. 3.78).

laws of such influence were surveyed in section 3.2.2. However, it may be reasonable to give a detailed description with regard to the system under discussion, namely polyisobutene–glass, modelling the contact between a linear uncrosslinked polymer and a solid substrate.

For this system the kinetic dependence described by a curve reaching saturation, corresponding to $a_f \to A_n$ (Fig. 3.82, 1), is characteristic. At small loads such a trend is more sharply pronounced, evidence of the increasing deformation of microdefects (microhills) of the substrate surface with an increase in the load F. This is quite comprehensible as, since F is related to Young's modulus, the actual contact area can be derived, relying on the deformation law of the elastic hemispherical microdefect, as follows [1294]:

$$a_f = 1 - \exp\left[-\beta (F/E)^{2/3}\right]. \tag{465}$$

In Fig. 3.83 the experimental values of a_f are compared with those calculated by eq. (465). Satisfactory agreement between these data serves to conform the validity of the views regarding the role of deformation processes. Additional proof is provided by the fact that the value of the load F_∞ (0.22 MPa) calculated by the known [1243] relationship

$$F_\infty = 4\,E/\beta \tag{466}$$

and corresponding to the saturation portion of the kinetic curves (for sufficiently smooth surfaces $\beta = 1.3$), appears to be in good agreement with the data of Fig. 3.83, 1. Similar regularities were also found to hold true for the system polyethene–glass [1295].

Ideas of this kind support the types of adhesive joint strength versus bonding pressure dependences also. Fig. 3.84, 1, illustrating the effect of bonding pressure on the strength

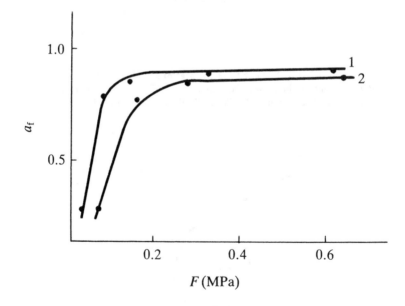

Fig. 3.82 – Specific area of actual contact between polyisobutene and glass versus applied bonding pressure. 1, Experimental curve; 2, curve calculated taking into account the reducing coefficient.

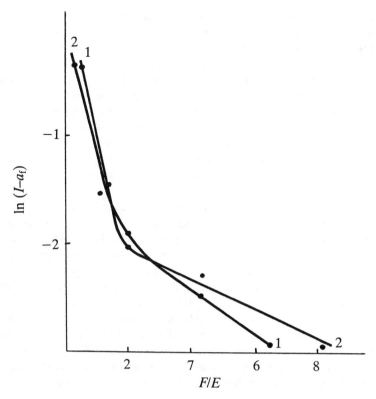

Fig. 3.83 – Specific area of actual contact between polyisobutene and glass versus ratio between applied bonding pressure and Young's modulus of the adhesive. 1, Experimental curve; 2, calculated curve.

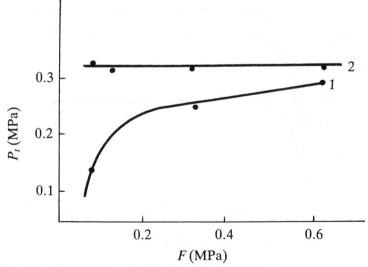

Fig. 3.84 – Tensile strength of the adhesive joints between polyisobutene and glass versus bonding pressure. 1, Curve calculated with respect to the nominal contact area; 2, to the actual contact area.

of adhesive joints, demonstrates that the latter characteristic, calculated as depending on the nominal contact area, undergoes an exponential rise with increase of F. However, the curves in Figs 3.82, (1) and 3.84, (1) are somewhat different. While a_f reaches saturation level at F_∞, P continues to increase even at higher loads (pressures). However, there is actually no contradiction in this observation. This is due to the fact that when measuring a_f by means of the Mechau procedure (see section 3.2.2) there is a certain error incurred owing to the penetration of the light beam into the second medium. For cross-linked rubbers the adjustment coefficient, defined as the ratio between the optical and the calculated contact areas, was found to be equal to 1.8 [1243]. The corrected a_f versus load dependence (Fig. 3.82, 2) coincides with the data in Fig. 3.84. Note that the fact that graph 2 in Fig. 3.84 is parallel to the pressure axis serves as direct proof of the validity of the concepts discussed.

Hence, attainment of the equilibrium real contact area serves only to create the necessary molecular–kinetic conditions for subsequent formation of the adhesive joint by establishing an equilibrium system of interfacial bonds [1296]. Though both of the processes proceed simultaneously, the second however, continues longer than the first. Since in the system under investigation the size of the microroughness elements of the substrate (glass) surface is an order of magnitude smaller than the size of supermolecular structures in the adhesive (polyisobutene), the polymer is to be deformed at the initial stage of the process. Owing to the relatively small Young's modulus of polyisobutene, even at loads for which $F/E \geqslant 2\, a_f \rightarrow 1$, the effect is ensured via the microrheological mechanism. This is demonstrated by the fact that (as is seen in Fig. 3.85), the maximum strength of the adhesive joint is reached at the same applied bonding load regardless of the bonding temperature (up to the temperature of the viscous flow). Insofar as the

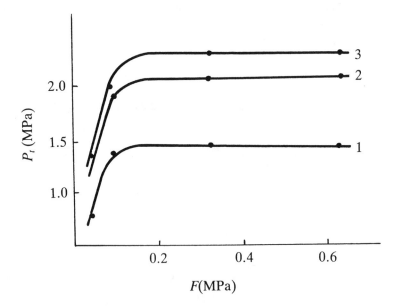

Fig. 3.85 – Tensile strength of the adhesive joints between polyisobutene and glass produced at different temperatures versus bonding pressure. Bonding time 30 min; bonding temperature: 1, 297 K; 2, 353 K; 3, 413 K.

conformational spectrum and flexibility of macromolecules are liable to be seriously affected only under more severe conditions, namely at $F/E > 10$ [1254] or at $F > 10$ MPa [1297], for a system to get to the equilibrium state, after the actual contact has been established, the segments will have to travel over the substrate surface to occupy energetically favourable positions.

Let us assess the driving force of this process. The activation energy can be determined with the help of the Williams–Landell–Ferry approach [1298], according to which the reduction coefficient describing the mobility of segments as a temperature-dependent function, or more rigorously [1271] defined as the ratio between the relaxation times at a given temperature and at a fixed temperature (as shown above, it is convenient to take the glass–transition temperature T_g as such) is given by the following relationship

$$\lg a_{(T)} = - [c_1^g (T - T_g)/(c_2^g - T - T_g)] \qquad (467)$$

where the constants c_1^g and c_2^g for polyisobutene are 16.56 and 104.4, respectively [1299]. Fig. 3.86 shows the graph corresponding to eq. (467). Since at $T > T_g$ relaxation processes in the polyisobutene–glass system are predominantly due to segmental and molecular mobility, the $\lg a_{(T)}$ versus inverse temperature dependence appears to be linear, indicating that there exists a certain mean (averaged) relaxation time that should fit eq. (463). Hence the activation energy of the processes in the bulk of polyisobutene, determined as the slope of the graph in Fig. 3.86, is 62.9 kJ/mol. This result is in good agreement with the reported values for activation energies of stress relaxation (81.7 kJ/mol) [1300], of creep (84.2 kJ/mol) [1301], and of viscous flow (56.6 kJ/mol) [1302] of polyisobutene determined by independent methods.

The validity of this approach is confirmed by the fact that the strength of adhesive joints between polyisobutene and glass [1249] follows a uniform generalized time–

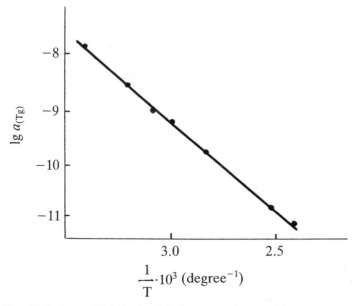

Fig. 3.86 – Reducing coefficient of polyisobutene at glass-transition temperature versus temperature.

temperature dependence pattern shown in Fig. 3.87. Hence, one can evaluate the
activation energies of the processes taking place in the transition layers of polymers.
Relying on the kinetic dependences describing the strength of adhesive joints between
various polymers and glass (Fig. 3.85 exemplifies the case with polyisobutene as the
polymer, whereas Fig. 3.88 that with butadiene–acrylonitrile elastomer SKN-40)
relaxation times were calculated by means of eq. (456) and, subsequently, activation
energies were derived from their temperature dependences depicted in Fig. 3.89. The
results are compiled in Table 3.7. As may be seen, the activation energy of the interaction
process within the surface layers, as well as at the interface, is greater the higher the
polarity of a polymer. This inference gains additional support in that adhesive joint
failure strength versus temperature dependences appear to be linear as illustrated in Fig.
3.90.

Since, according to the above calculations, activation energy of the intraphase interactions is, for polyisobutene, 2.3 times greater than U_f^q and 2.6 times greater than U_b^q, it
is natural to assume that as soon as the limiting real contact area is achieved the processes
of segmental transposition are developed, primarily within the transition and boundary
layers of the adhesive, so as to ensure a thermodynamically stable state; for instance, by
uncoiling the globular structures. It is clear that the processes are accompanied by an
increase in the number of interfacial bonds.

In fact, application of external pressure results in that the force per single interfacial
bond $F_b^{(1)}$ increases. The limiting value of this parameter corresponds to the maximum
strength of the adhesive joint [1303]. Since in general

$$P = N_b F_b^{(1)} \tag{468}$$

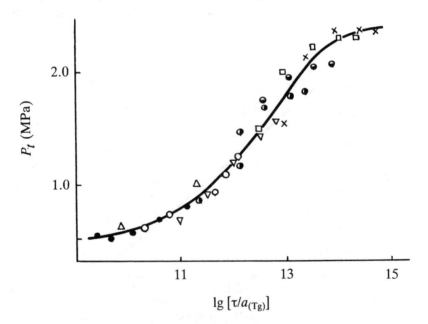

Fig. 3.87 – WLF curve of tensile strength of the adhesive joints between polyisobutene and
glass produced at 296 K (●), 313 K (○), 323 K (▲), 333 K ( ), 353 K (○), 373 K (○),
393 K (▫) and 413 K (×).

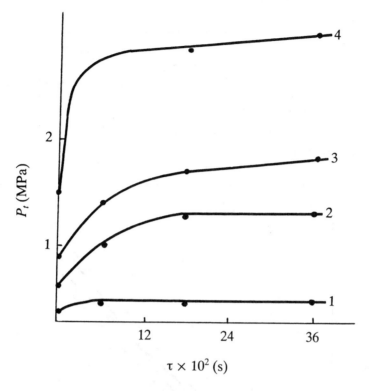

Fig. 3.88 – Tensile strength of the adhesive joints of butadiene–acrylonitrile SKN-40 with glass versus contact time at 1, 291 K; 2, 353 K; 3; 373 K; and 4, 413 K.

Table 3.7 – Major molecular characteristics describing adhesive interaction between polymeric adhesives and glass

Parameter	Adhesive			
	Polyisobutene	SKN-18	SKN-26	SKN-40
Activation energy of interfacial interaction U_f^a, kJ/mol	27.2	31.0	33.1	36.0
Activation energy of intraphase interaction within surface layers U_b^a, kJ/mol	23.9	27.2	28.9	31.0
Energy of a single interfacial bond $E_b^{(1)}$, kJ/mol	3.4	3.8	4.2	5.0
Relative number of interfacial bonds N_b/N_s	0.70	0.70	0.72	0.78
Number of interfacial bonds at 293 K N_b 10^{-16}, bonds/m^2	4.3	4.6	5.8	2.5

Table 3.7 (contd.)

Parameter	Adhesive			
	Polyisobutene	SKN-18	SKN-26	SKN-40
Number of interfacial bonds at 413 K N_b 10^{-16}, bonds/m^2	40.0	38.0	37.0	40.0
Force per single interfacial bond $F_b^{(1)}$, TN/bond	0.14	0.15	0.17	0.20

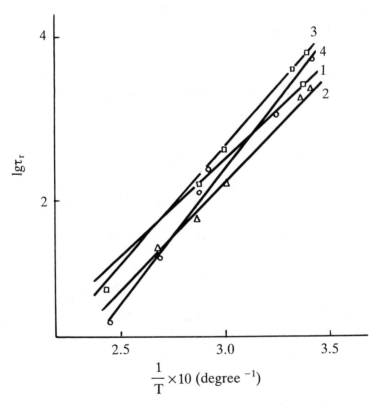

Fig. 3.89 – Relaxation time of butadiene rubber SKD (1), and butadiene–acrylonitrile elastomers SKN-18 (2), SKN-26 (3), SKN-40 (4) brought into contact with glass versus temperature.

(where N_b is the number of interfacial bonds per unit contact area) the kinetic dependence of adhesive joint strength can be written as follows [1304]

$$\frac{dP}{d\tau} = N_b \frac{dF_b^{(1)}}{d\tau} + F_b^{(1)} \frac{dN_b}{d\tau}. \tag{469}$$

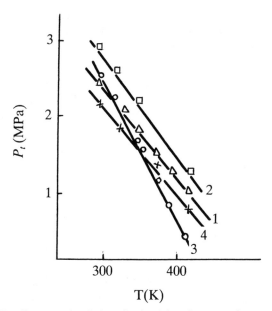

Fig. 3.90 – Tensile strength of the adhesive joints between elastomers and glass versus temperature (notation as in Fig. 3.89). Bonding temperature 413 K.

Obviously, failure of the joint equates the relationship to zero. On the other hand, the kinetics of the accumulation of the ruptured bonds is described by eq. (453). Substituting it into eq. (469), one comes to the following relationship

$$N_b \frac{dF_b^{(1)}}{d\tau} + F_b^{(1)}(N_b \nu_b - N_f \nu_f) = 0 \tag{470}$$

where $F_b^{(1)}$ is a constant characteristic of the process of adhesive joint fracture

$$\nu^* = (dF_b^{(1)}/d\tau)/F_b^{(1)}. \tag{471}$$

Since it is a well-known fact that applied pressure leads to the height of the energy barrier being decreased [1305], the frequency with which interfacial bonds are being ruptured, which is described by the exponential law

$$\nu_f = \nu_0 \exp\left[-(U_f^a - \upsilon P)/KT\right], \tag{472}$$

should be increased. However, eq. (459) fails to involve the external load as an inherent term of eq. (472). Hence, at each moment of time the fracturing load appears to be transferred only to those segments of the macromolecules of an adhesive that are attached to the substrate surface. Substituting eqs (458), (459) and (472) into eq. (470) and subsequently solving it with respect to P, one comes to the dependence

$$P = \left[U_f^a - U_b^a + KT \ln\left(\frac{N_f}{N_b} + \frac{\nu^*}{\nu_f}\right)\right] \Big/ \upsilon \tag{473}$$

which is depicted by a rectilinear graph.

In a first-order approximation v^* is evaluated by the rate of external load application similar to eq. (471). Then, for elastomers the following condition holds true

$$v^*/v_f \ll N_f/N_b \qquad (474)$$

where the v^*/v_f term in eq. (473) can be neglected. This is almost obvious, since, according to the definition eq. (471), $N_f/N_b = 0 \div 1$. The limiting values correspond evidently to the cases when there is either no interfacial interaction (i.e. it is equally probable for a segment to reside within the bulk of the phase or within the transition layer and from eq. (453) it follows that $N_f = N_b$), or there is a maximum density network of interfacial bonds (i.e. $N_f/N_b \to 0$). Hence, eq. (473) acquires the form

$$P = [U_f^a - U_b^a - KT \ln (N_b/N_f)]/v. \qquad (475)$$

The graph of this dependence, i.e. adhesive joint strength versus fracture temperature, is also a straight line [1304]. Fig. 3.91 gives an example of such a graph for the polyisobutene–glass system. Its slope is related to N_f/N_b, which, for a given system, is constant within the entire investigated temperature range. This is indicative of the equilibrium character of the fracture process of the adhesive joint produced under conditions leading to optimally close molecular contact. However, in general, the value of N_b/N_f is determined by the contacting conditions and the nature of the adhesive and substrate. Indeed, as is seen from Table 3.7, listing the respective values (these were calculated from the data of Fig. 3.90), for the high-polarity SKN-40 elastomer, as compared to polyisobutene, this ratio increases by 10%, but yet, even in this case $N_b/N_s < 1$.

Objectivity of this approach is provided by the deep physical implications of eq. (475). Comparing it with Frenkel's equation (316), one can easily perceive that the product of the Boltzman constant and the logarithm of the ratio between the numbers of the bonded and the free segments is simply surface entropy. Hence the grounds for seeking

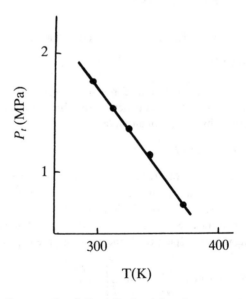

Fig. 3.91 – Tensile strength of the adhesive joints between polyisobutene and glass produced at 413 K versus temperature.

an analogy between expressions (473) and (30). Interpreting $(U_f^a - U_h^a)$ as the energy of a bond between the segment of an adhesive and a substrate, it seems reasonable, relying on this analogy, to identify the concept of adhesive joint strength as that of free surface energy, corresponding quantities being identical within the accuracy allowed by the constant v of eq. (475). This is a principle feature providing evidence of the validity of the above ideas concerning the interrelationships between adhesive properties and the surface energy of polymers. On the other hand, this feature serves to illustrate the soundness of the approach leading to eq. (475).

Within the framework of this approach it appears feasible to obtain a number of other important characteristics. If one assumes, according to Bartenev [1306], $v = 10^{-27}$ m^3, then the intersections of the graphs in Figs 3.90 and 3.91 with the ordinate axis can be identified with the values of the energy of a single interfacial bond. For butadiene–acrylonitrile elastomers this parameter is 10–50% larger than for polyethene (see Table 3.7). However, for all systems the values of $E_b^{(1)}$ are rather small thus supporting the adsorptive character of interfacial interaction between the adhesives examined and glass [1303]. Hence, the approach is supplementary to the ideas elaborated on in sections 2.2.2 and 3.1.2; however, it is not the total interaction energy, but only that fraction relating to a single interfacial bond that is evaluated.

From these data it is not difficult to deduce the rupture force per interfacial bond. Since the energy of an interfacial bond is that of van der Waals interaction, the range of which is within 0.4–0.5 nm, one can derive the value of $F_b^{(1)}$ from the relationship

$$U_f^a - U_b^a = kF_b^{(1)}. \tag{476}$$

As is seen from Table 3.7, this parameter of the interfacial bond is governed by the polarity of an adhesive and, in the final account, by the efficiency of its interaction with a substrate. Then, the number of interfacial bonds is given by the ratio

$$N_b = P_{max}/F_b^{(1)}. \tag{477}$$

For the model polyisobutene–glass joint its value is almost two orders of magnitude higher than that determined by Bartenev [1297, 1307] in frictional experiments. Such a discrepancy is due to the fact that N_b is constant during adhesive interaction, whereas in friction it continuously changes. With the rise of temperature segmental mobility within the transition layers increases, resulting in N_b increasing by an order of magnitude (see Table 3.7). Note that at 413 K the mobilities of given polymers reach a maximum, this fact providing for the maximum number of interfacial bonds being established, which (number) is actually independent of the nature of the adhesive. This fact may also serve as additional proof of the validity of the discussed molecular–kinetic concepts of adhesive joint formation.

The molecular characteristics of adhesive interaction obtained relate mainly to the processes taking place within the surface layers of polymers. At the same time, the physical states of the latter should not be subject to any variation with accompanying increase of the actual contact area. This is demonstrated by the data of Fig. 3.92 relating to the kinetics of the formation of the joint between butadiene–acrylonitrile elastomer (SKN-26) and glass examined by the dynamic method of Mandelstam–Khaikin [1303]. Molecular contact was assessed by the frequency shift $\Delta\omega$, which is, according to

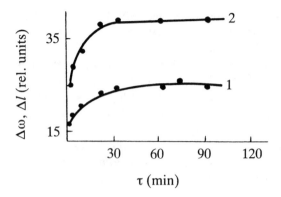

Fig. 3.92 – Kinetic curves of the frequency shift of the vibrations (1) and transmittance band broadening (2) of the resonance vibration curve in adhesive joints of butadiene–acrylonitrile elastomers SKN-26 with quartz obtained at 293 K under 0.02 MPa applied bonding pressure.

Lavrentiev [1308], characteristic of the elastic properties of surface layers; the number of interfacial bonds was evaluated by the broadening of the transmittance band Δl related to the decrement of the extinction of the excited vibrations [1309]. As is seen, at each point of the kinetic curves (i.e. at a moment τ) the mechanical loss tangent $\Delta\omega/\Delta l$ is constant and independent of τ, implying thus that there are no changes taking place in the surface layers of an adhesive with an increase in a_f. This inference is quite reliable, since the experimental procedure involved is so sensitive that it even makes it possible to observe such fine effects as to distinguish between the two types of interfacial (adhesion) bonds between polystyrene and polybutadiene phases in ternary block copolymers DST [1310]. The above effects are in good agreement with the studies of autohesive joint strength between these elastomers [1311].

It is of specific interest to compare the strength of adhesive joints with the molecular characteristics of interfacial interaction (Table 3.7) on the one hand, and with the properties of boundary layers (Table 3.2), on the other. Poly-ϵ-caproamide and glass were chosen as substrates representing two of the major types, while butadiene–acrylonitrile copolymers were used as adhesives (cf. the footnote on p. 000).

Fig. 3.93 shows the correlation curves applicable to the first group of adhesive joints. What is really noteworthy is that the peeling strength is linearly related to the packing coefficient (Fig. 3.93, 1) and to the density of the adhesive (Fig. 3.93, 2) in the bulk as well as in boundary layers. Hence, when the substrate bestows no orientating effect then the application of an external load in the course of adhesive joint formation smooths off the difference between the bulk of the phase and the interfacial zone, particularly those which are due to the variation of the mobility of macromolecular chains. This conclusion is confirmed by the fact that the linearity of the respective dependences is retained even at increased temperature (423 K), when the difference between the bulk and surface properties of adhesives should have been demonstrated most explicitly.

Clearly, as is seen from Fig. 3.44, macromolecular flexibility substantially affects the value of P_p. However, according to Fig. 3.49, variation of this parameter is due to the differences in polarity of elastomers resulting in the height of the activation barrier of the process of adhesive joint formation being affected and hence the activation energies of

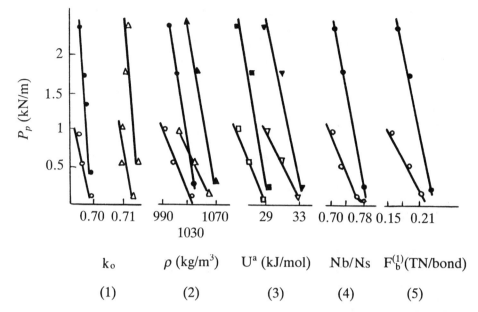

Fig. 3.93 — Peeling strength of the adhesive joints between butadiene–acrylonitrile elastomers SKN and poly-ε-caproamide versus molar packing coefficient (1) and density (2) of the adhesives in the bulk (○, ●) and in boundary layers (△, ▲); versus molecular character-istics of adhesive interation–activation energy of inter- (□, ■) and intraphase (▽, ▼) inter-action in surface layers (3), relative number of interfacial bonds (4), force per single inter-facial bond (5). Bonding temperature 293 K (open symbols), 423 K (solid symbols).

interfacial and intraphase interactions in surface layers are also affected. The latter parameter, according to the data of section 3.1.2, is related to the energetic character-istic of the diffusion process. In fact, it is linearly related to the peeling strength of the system (see Fig. 3.93, 3); however, the P_p versus U_b^a graphs follow a similar pattern. Hence, it implies that in systems comprised of a plastic and an elastomer then diffusion processes are not confined to within the transition layers, but are developed at the very geometric interface also and include the interpenetration of an adhesive and a substrate. The forms of the dependences actually remain the same when the temperature is increased to 423 K. Taking into account the linear character of the P_p versus s (see Fig. 3.44) graph, it appears quite natural to accept that adhesive interaction between materials proceeds via the mechanism of local segmental diffusion (since s was defined as the number of repeat units in a segment). Thus, the above observations appear to be the first quantitative verification of the suggested mechanism, contrary to the implicit considerations mainly associated with the visual monitoring of the process of adhesive joint formation.

The predominant development of diffusion processes should actually result in levelling off the effect of other factors of molecular–kinetic origin on the modes of adhesive joint formation. In particular, in this case one would hardly expect that the relationship between P_p and the number of interfacial bonds would be linear. Indeed, we have discovered that corresponding graphs are parabolic. It appears much more correct to correlate the strength of adhesive joints with the relative rather than the absolute, number of interfacial bonds since when the total number of bonds is evaluated the efficiencies of inter- and intraphase interactions are not distinguished. The statement is illustrated by the data of Fig. 3.93, 4 demonstrating that, while the graphs are almost rectilinear at 293 K, These become strictly linear at 423 K, when extensive diffusion processes are developed. Naturally, the P_p versus force per single interfacial bond dependences are linear at 423 K, as well as 293 K (Fig. 3.93, 5).

Correlations between these parameters and adhesive joint strength are quite different when the substrate is glass. This is quite a different case since in this group of systems diffusion hardly affects the consistencies of adhesive interaction. In fact, we have found out that for these systems the P_p versus k_0, ρ, N_b/N_s dependences are depicted by curves with maxima and minima. According to the calculations, the P_p versus s, A and U^a graphs follow similar patterns. We relate the levelling off of macromolecular mobility, attendant upon such effects, to the influence of factors of mocular–kinetic origin, which are especially essential in systems comprising a high-energy substrate.

Hence, analysis of the strength of adhesive joints as affected by various formation conditions confirms the validity of the theories presented concerning the molecular–kinetic nature of the mechanisms of the respective processes. The soundness and success of these theories are provided by the deep physical meaning of the quantitative parameters involved *per se* as well as of their interrelationships. A number of such parameters bearing a different meaning was surveyed in the final part of section 3.1 (in 3.1.2, to be precise). Here, it also appears sensible to finish the section by examining the characteristics of adhesion processes introduced. For the sake of accuracy corresponding correlations were examined for the same model systems involving butadiene–acrylonitrile elastomers and glass. Let us look at the following parameters, activation energy of intraphase interaction within the surface layers of copolymers and that of their interaction with the substrate (Fig. 3.94), packing coefficients in the bulk and within the boundary layers of adhesives (Fig. 3.95), and the relative number of interfacial bonds in the system (Fig. 3.96).

Characteristics applying to surface layers in adhesives should be governed by the orientating effect of a solid substrate surface causing the retardation of relaxation processes in elastomers. The latter circumstance provides for interfacial interaction being energetically more beneficial than the intraphase interaction, i.e. diffusion transfer of either macromolecules or their segments in surface layers. One can find direct proof of this deduction in Table 3.7; indeed, $U_b^a < U_f^a$. However, neither of the processes is independent, their interrelationship being verified by the strictly linear graphs of both U_b^a and U_f^a versus packing coefficients (Fig. 3.94, 1) and the density (Fig. 3.94, 2) of boundary layers of elastomers regardless of whether there is or is not a substrate.

The direct relationship between these parameters is due to the uniformity of the mechanism providing for the changes of the properties of transition and boundary layers in elastomers; this uniformity is stipulated by the fact that the characteristics considered are affected by variations in macromolecular mobility to an equal extent. In fact, for the copolymers under investigation s is linearly related to k_0 (Fig. 3.18, c) and ρ (Fig. 3.18, b). Hence, the absolute number of interfacial bonds appears to be a characteristic of the process of adhesive joint formation and which is independent of the physical state of an elastomer in the boundary and transition layers. However, the potential ability of an adhesive to interact with the substrate cannot avoid being dependent on these states. Therefore, contrary to the dependences on N_b, both functions $k_0(N_b/N_s)$ are strictly linear (Fig. 3.95, 1).

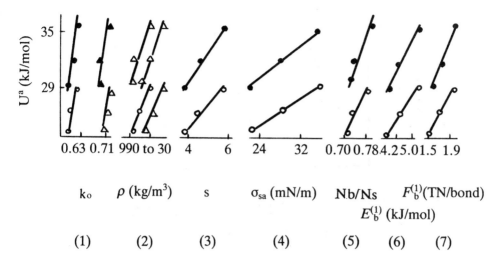

Fig. 3.94 — Activation energies of interphase (solid circles) and intraphase (open circles) interactions within the surface layers of butadiene—acrylonitrile elastomers SKN in contact with glass versus molar packing coefficient (1) and density (2) in the bulk (o, •) and in boundary layers (△, ▲); equilibrium flexibility of macromolecular chains (3), surface energy (4), calculated molecular characteristics of adhesive interaction—relative number of interfacial bonds (5), energy (6) and force per single bond (7).

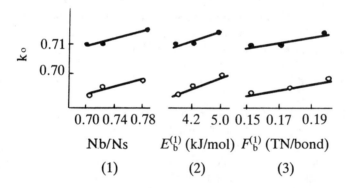

Fig. 3.95 — Molar packing coefficient of butadiene—acrylonitrile elastomers SKN in the bulk (open circles) and in boundary layers (solid circles) bonded with glass versus calculated molecular characteristics of adhesive interaction—relative number of interfacial bonds (1), energy (2) and force per single interfacial bond (3).

The consistencies described feature different aspects of the role of energetic and structural factors in a general complex of phenomena of adhesive interaction. On the one hand, these phenomena are governed by the mobility of macromolecular chains in adhesives, whereas, on the other hand, the measure of this mobility is linearly related to the activation energies of intraphase processes in surface layers and of interphase processes at the geometric interface (Fig. 3.94, 3), as well as with the relative number of interfacial bonds (Fig. 3.96, 1) and it is not by chance that both these parameters are mutually correlated (Fig. 3.94, 5). These considerations explain the linear interrelation-

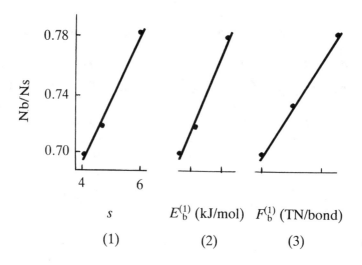

Fig. 3.96 – Relative number of interfacial bonds in the joints between butadiene–acrylonitrile elastomers SKN and glass versus equilibrium flexibility of macromolecular chains (1), energy (2), and force per single interfacial bond (3).

ships between the energy of a single interfacial bond and the packing coefficient (Fig. 3.95, 2), activation energy (Fig. 3.94, 6) and the relative number of interfacial bonds (Fig. 3.96, 2). Both the former and these considerations demand the validity of the thesis elaborated in section 2.2.2 concerning the direct relation between adhesive properties of polymers and their surface energy. In fact, strictly rectilinear dependences were found to exist between the latter parameter and activation energies of the intraphase interaction in surface layers and of the interphase interaction at the geometric interface; these are shown in Fig. 3.94, 4.

The resultant effect of the factors of different types is revealed in a complex parameter, force per single interfacial bond, describing the relationship between the parameters of developing processes with the ultimate (equilibrium) strength of an adhesive joint. One could anticipate $F_b^{(1)}$ to be an explicit function of the packing coefficient of the boundary layers in elastomers, of the height of the activation barriers of corresponding interactions, and of the relative number of interfacial bonds between an adhesive and a substrate. In fact, we observed such relations and the respective graphs are depicted in Fig. 3.94, 7, Fig. 3.95, 3 and Fig. 3.96, 3.

To illustrate the value of the developed molecular–kinetic concepts lets us examine the kinetics of adhesive joint formation assuming that the maximum full interfacial area A_{max} has been preliminarily attained. This situation is much more common than one might have initially contrived since it encompasses such significant cases as those of a substrate in contact with constant-tack and low-viscosity adhesives, and also the systems produced by chemical metallization.

For such systems the kinetic dependence of their strength appears to be a function of the number of the repeat units of a macromolecule $N(x, \tau)$ disposed at the moment τ at a distance x from the interface, as well as of the binding force between the repeat unit and the substrate $F_b^{(1)}(x)$ and of the length of the respective bond $l^{(1)}$ [1312]:

$$P(\tau) = \int_{l(1)}^{\infty} N(x, \tau) F_b^{(1)}(x)\, dx. \tag{478}$$

Taking into account that, according to Heavyside,

$$P(\tau) = \int_0^\tau H(t - \tau) A(\tau)\, d\tau \tag{479}$$

for the case under consideration, when $A(\tau) = A_{max}$, one has [1313]:

$$P(\tau) = \int_0^\tau A(\tau) \int_{l(1)}^\infty \frac{\partial N(t - \tau)}{\partial \tau} F_b^{(1)}(x)\, dx\, d\tau \tag{480}$$

Imagine the process of adhesive joint formation as proceeding via successive adsorption of repeat units by a non-polymeric substrate. Then, the time between the states described by $(N - 1)$ and N adsorbed units appears to be a function only of the states with $(N - 2)$ and $(N + 1)$ adsorbed units, and is independent of the states described by the number of units beyond this range. Consequently, the process can be described within the frame of reference of the theory of Markov chains with continuous time. For this case, as is known, Kolmogorov's equations are valid:

$$\left\{ \begin{aligned} \frac{dp_{ij}}{d\tau} &= \sum_k \lambda_{ik}\, p_{kj} \\ \frac{dp_{ji}}{d\tau} &= \sum_k \lambda_{ki}\, p_{jk} \end{aligned} \right. \tag{481}$$

where p_{ij} is the probability describing the transition of a macromolecule from the state with i adsorbed units to the state with j adsorbed units $(j = \{i - 1; i; i + 1\})$ and λ_{ij} is the expectation (mean value) relating to this transition. The boundary conditions having been taken into account, the solution of this system of equations is as follows:

$$\frac{N(\tau)}{N} = A^{-1} \left\{ \sum_{i=1}^{N/s} \sum_{j=i-1}^{i+1} [1 - c_{ij_1} \exp(-\lambda_{ij_1}\tau)] \right.$$

$$\left. + \frac{N}{A} \sum_{i=1}^s \left[1 - \left(\sum^s c_{ij_2} \exp(-\lambda_{j_2}\tau) \right) \left(\sum^{N/s} c_{ij_1} \exp(-\lambda_{j_1}\tau) \right) \right] \right\} \tag{482}$$

where N is the number of units in a macromolecule, c_{ij_1} is the eigenvector of the jth eigenvalue for the process of segmental adsorption; c_{ij_2} is the eigenvector of the jth eigenvalue for the process of intrasegmental mass transfer, $\lambda_{j_1} = \lambda_j^{-1}$ are the eigenvalues of the matrix $\|\lambda_{ij_1}\|$ for the process of segmental adsorption, $\lambda_{j_2} = \lambda_j^{-1}$ are the eigenvalues of the matrix $\|\lambda_{ij_2}\|$ for the process of intrasegmental mass transfer and s is the number of repeat units in the segment.

Since it is the calculation of the relative variations of the strength of adhesive joints that is of most importance, the problem is confined to a determination of the values of s and τ, and, consequently, to construction of the matrices $\|\lambda_{ij_1}\|$ and $\|\lambda_{ij_2}\|$.

An equilibrium segment value s determined by the known experimental [409] and theoretical [909] procedures is an overall parameter averaged over the given volume and temperature interval (see section 3.1.1), therefore it is not related to the magnitude of internal stresses within the sample. To eliminate this limitation in order to obtain the possibility of using parameter s in the kinetic dependences under consideration let us make use of scaling concepts. Within the frame of reference [368] a concept of a blob is introduced instead of a segment. A blob is defined as a fragment of macromolecule which is independent, in its motion, of other fragments, while the real macromolecular chain is described as a freely jointed chain of blobs. Internal stresses within the transition layer (and even the more within the boundary) of a polymer affect the interblob relaxation time as well as the blob dimensions. When a macromolecule is stretched by the force F, the variation of the end-to-end distance L of the chain at $(Fl/KT) \gg 1$ is given through

$$<|\bar{L}|> = Nl \left(\frac{Fl}{KT}\right)^{2/3}. \tag{483}$$

The number of blobs in an extended macromolecule is as follows

$$g_f = \left(\frac{KT}{Fl}\right)^{5/3} \tag{484}$$

while the length of each blob is defined as

$$l_g = l \, g_f^{3/5}. \tag{485}$$

The unperturbed dimensions L of the blob g_l can be derived as follows. From the Tailor's series one has

$$L^2 = l^2 \, N\gamma. \tag{486}$$

In accord with the scaling principle one may write:

$$L^2 = l^2 \, N^\nu. \tag{487}$$

Solving eqs (486) and (487) together one has:

$$\nu = \frac{lg\gamma + 2 \, lgN}{lg^2 N}. \tag{488}$$

For the three-dimensional case [368]

$$g_l = \nu/6. \tag{489}$$

Combining eqs (484), (488), and (489) one comes to

$$g = \begin{cases} g_l = 10 \, \dfrac{lg^2 N}{2(lg\gamma + 2 \, lgN)}, & g_l < g_f \\[3mm] g_f = \left(\dfrac{KT}{Fl}\right)^{2/3}, & g_l > g_f \end{cases} \tag{490}$$

$$L = l_g^2 N/g. \tag{491}$$

Considering the blob model, de Gennes suggests making use of the spline-function of the distribution of the end-to-end distance of the chain [368]. Consistent with this approach region I in Fig. 3.97 corresponds to the motion of the units within the blobs; the respective distribution function is of non-Gaussian character which accounts for the real features of the molecules

$$P_N^I(\bar{L}) = (\bar{L}/l_g)^g \, P^I(l_g). \tag{492}$$

Region II corresponds to the motion of the blobs approximated by a Gaussian distribution function:

$$P_N^{II}(\bar{L}) = \exp\left(-3\bar{L}^2/2L^2\right) P^{II}(l_g) \tag{493}$$

where $P^i(l_g)$ are the respective reducing functions. On the basis of this spline-function Fridman *et al.* [1312] carried out preliminary calculations of the probabilities and of the times (which are the reciprocal quantities of probabilities) of intra- and interblob relaxation (τ_1, τ_2).

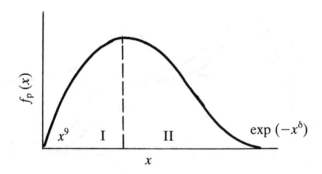

Fig. 3.97 – Distribution function of the end-to-end distance of a polymeric chain.

The effect of internal stresses on the kinetics of adhesive joint formation was accounted for by the function

$$\tau_1 = \tau_0 \exp\left[(U_\alpha - V_g \gamma_i)/KT\right] \tag{494}$$

which is similar to a well-known equation of Aleksandrov–Gurevich. Here V_g is the volume of a blob $(V_g \approx l_g^3)$, γ_i is the magnitude of internal stresses and U_α is the activation energy of the α-process. Within the framework of this approach the temperature dependence of the relaxation time is usually given by

$$\ln(\tau/\tau_g) = \frac{U_\alpha(T - T_g)}{RTT_g} - \frac{b(T - T_g)/f_g}{f_g/\Delta c + T - T_g} \tag{495}$$

where T_g is the glass transition temperature of a polymer, τ_g is the relaxation time at $T = T_g$, f_g is the free volume fraction at $T = T_g$ and Δc is the variation of the thermal expansion coefficient at $T = T_g$.

Substituting eqs (490), (494) and (495) into eq. (482) makes it possible to take account of the effects of internal stresses and temperature within the transition layer on the formation rate of adhesive contact. The kinetics of adhesive joint formation under different conditions was computer-simulated for systems comprised of butadiene–styrene copolymers of various molecular weight and metal.

Fig. 3.98 depicts the graph of the dependence of $\lg \lambda_1$ on reciprocal temperature. At $T = T_g$ a sharp (10–100-fold) rise of λ_1 is observed. The maximum is reached at $T = T_g + 60°$; however, subsequently λ_1 substantially decreases. This fact is in agreement with the general theories discussed in section 3.2.2 pertinent to the attainment of maximum contact area. However, in the case under consideration the interfacial area remains constant and, consequently, the pattern of the temperature dependence of the time of interblob relaxation is determined by the effect of macromolecular mobility. Obviously, at $T = T_g + 60°$ the processes within the phase (intraphase processes) become more substantial than the interphase processes and hence, at higher temperatures λ_1 declines. Note that at different T_g the graph is correspondingly shifted along the abscissa, this fact serving to prove the adequacy of the mathematical description presented of the process of adhesive interaction.

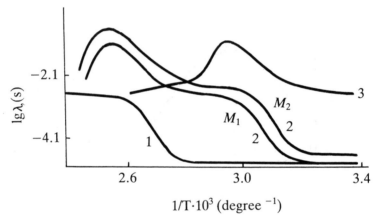

Fig. 3.98 – Interblob relaxation time for butadiene–styrene copolymers with molecular mass $M_1 = 8.9$ and $M_2 = 44.5$ at glass–transition temperature: 1, 273 K; 2, 333 K; and 3, 383 K versus temperature.

For a system in an elastic state there is no need to consider the interaction of units within the blob. However, at $T < T_g$ neglecting this kind of interaction results in serious inaccuracies. Taking into account the mutliplicity of relaxational transitions in butadiene–styrene elastomers [1314], the number of units in the blob must be an explicit function of temperature. This inference is confirmed by the data of Fig. 3.99.

At $T_g \ll T < T_{vf}$ (T_{vf} is the viscous flow temperature) the blob dimensions are rather small and segmental motion is actually unhindered. In this case (provided that interfacial contact has been formed) the formation kinetics of adhesive contact follows the description provided by the molecular–kinetic concepts discussed. However, as the temperature approaches closer to the glass-transition temperature the cooperative character of segmental motion prevails and the suggested mechanism of the process involving intra- and interblob adsorption becomes predominant.

To verify the validity of the approach developed we have plotted the experimental dependences of reduced strength ($P^* = P/P_{max}$) of adhesive joints between butadiene–

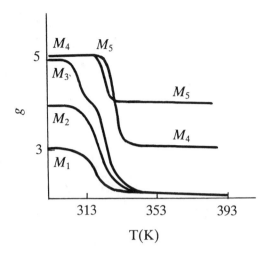

Fig. 3.99 – Number of repeating units in a blob at $M_1 = 8.9$; $M_2 = 17.8$; $M_3 = 26.7$; $M_4 = 35.6$ and $M_5 = 44.5$ versus temperature.

acrylonitrile–styrene copolymers and chemically precipitated copper versus the annealing temperature of the joint, as well as that of the dynamic loss modulus G'' versus temperature and these are shown in Fig. 3.100. As can be seen, the modes of the dependences in Fig. 3.100, 1 and Fig. 3.100, 3 are actually the same. Since G'' is supposed to be characteristic of the energy dissipation of mechanical vibrations due to internal friction and may serve as a measure of the intensity of molecular motion, the formation kinetics of the adhesive contact in the system under examination is fully governed by the mobility of macromolecules. From the data on the parameters of the phase structure of the copolymers Fridman et al. [1315] deduced the dependences of λ_1, λ_2 and g on the annealing temperature of the systems. When combined with eq. (482), these make it possible to derive the formation kinetics of the interfacial area. The results of this undertaking are presented in Fig. 3.100. As is seen, the discrepancy between the calculated and experimental data is within 10%, which is quite acceptable [1313, 1316].

Hence, both the results described and the reasoning demonstrate the validity of the approach developed and the prospects of using it for a correct description of the formation kinetics of the contact area between adhesives and substrates. We believe that this approach may serve as a foundation for the generalized algorithm [1317, 1318] providing for the computation of the strength of adhesive joints from the molecular characteristics of polymers [1319].

Summing up, we may claim that consistent application of the molecular–kinetic concepts in the examination of the phenomenon of adhesion creates a uniform and non-contradictory basis, stemming from which the chemical nature and structure of polymers, as well as the conditions of interfacial interaction with various substrates could be accounted for in their effect on the formation and behaviour of adhesive joints.

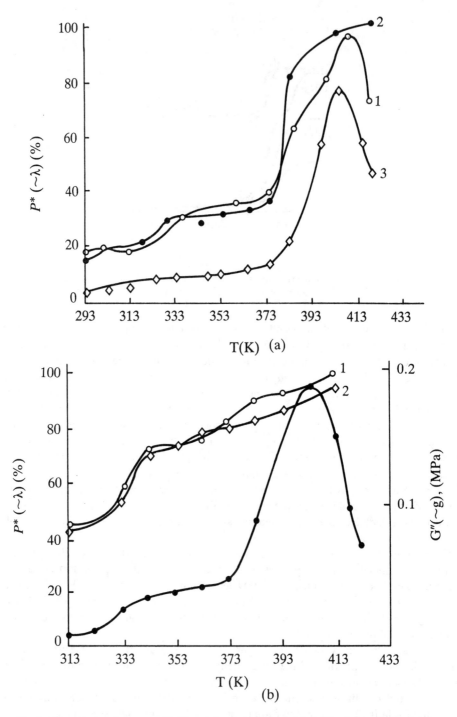

Fig. 3.100 – Reduced strength (1, 2) and dynamic loss modulus (3) of the adhesive joints of acrylonitrile–butadiene–styrene copolymers Lustran PG-299 (a) and ABS-2020 (b) with electroplated copper versus annealing temperature. 1, Experimental curve; (2), calculated curve.

4

Adhesive properties of polymers

The thermodynamic and molecular–kinetic concepts concerning the nature of adhesion discussed in preceding sections create the foundations for the development of a general approach to the control of the processes of formation and of the properties of adhesive joints. In essence, this problem can be formulated as a physico-chemical problem of distinct theoretical implication. However, compared to the problems envisaged above, this one is of much greater consequence with regard to the practical aspects related to the formation and usage of adhesive joints and, moreover, its verification is predominantly associated with technological studies. In fact, we believe that this final section should provide a link between the fundamental and the applied aspects of polymer adhesion, acting as an interpreter between these two. We are aware of the fact that the huge quantity of the data accumulated (which, in addition, are often contradictory) relating to only one of the aspects of adhesion, namely adhesive bonding, prevents one from reaching an unambiguous interpretation of the observed facts; at the present state of the art even the description of the technological regularities of bonding in terms of the physical chemistry of adhesion encounters serious obstacles. Nevertheless, theoretical analysis would be incomplete if we did not make an attempt to apply the results discussed above to the practical aspect concerning the control of the adhesive properties of polymers and primarily their enhancement.

4.1 PHYSICAL PRINCIPLES AND THE CONTROL OF ADHESIVE PROPERTIES

Achievement of the equilibrium strength of an adhesive joint appears to be the final result of adhesive interaction. Therefore, the problem of evaluating the efficiency of adhesive interaction is confined to testing the strength properties of an adhesive joint. Obviously, from the physico-chemical point of view, the most important part of the data is obscured by effects dependant upon factors of mechanical origin. Indeed, the effect of sample loading on the deformation patterns of both an adhesive and a substrate makes the assessment of the adhesive properties of polymers (enabling the efficient binding between the contacting elements) a qualitatively more difficult task. Complex criteria describing the tendency of a polymer to form the strongest possible adhesive joints or, alternatively, the potential for efficient binding can be termed adhesive ability or adhesiveness.[†]

The basic parameters comprising such criteria are varied. The first group of pertinent parameters involves those that govern the efficiency of interfacial attractive interaction between a substrate and an adhesive in energetic terms (microscopic approximation), while the second refers to the phenomenon in force terms (macroscopic approximation). Hence, it is clear that the concept of adhesive ability is of much broader implication than that of the adhesive joint strength and its quantification is basically difficult. However, not everything can be expressed numerically (and, in fact it is not always necessary to do so) and this is particularly true when speaking of a complex multiparameter criterion.[‡] It is because of the multifaceted nature of the problem that this approach is reasonable in the analysis of the control of the adhesive properties of polymers.

For this purpose, and treating adhesion as a surface phenomenon, it is necessary to examine the possibilities of affecting the characteristics of the contacting surfaces, polymeric ones in particular. According to molecular physics, the basis for such examination is provided by the zone theories describing the regularities of electron transitions from the bulk phase to the surface. Within the frame of reference of these theories the concept of surface states (electronic states localized at the surface) is common in solid-state physics, and, as indicated by Deryagin *et al.* [109, p. 95], is of the utmost significance in adhesion.

In this regard it is appropriate to recall that electrons can have only rigorously defined (permitted) energies, the energy ranges being grouped together to form permitted zones (energy bands). These encompass the whole of the phase only in the case of metallic conductors. Even for semiconductors there is a forbidden zone (band) separating the permitted ones, its width corresponding to the minimal energy required to raise an electron to a higher level, i.e. from the valency bond to the conduction band. In this case the first of the zones is fully occupied. However, for dielectric materials, which constitute the majority of polymers, a wider forbidden zone, compared to semiconductors, is intrinsic (not less than $0.32-0.48 \; 10^{-18}$ J) its mean statistical occupation being determined by the Fermi level. Fig. 4.1 illustrates the described energy band diagrams, shaded areas corresponding to the situation with all levels occupied by electrons.

To get over the forbidden zone (energy barrier) the electrons must be energized from an external source. The energizing forces can be the electric potential (and the resultant current of electrons is the electric conductance), or light (photoconductance) or heat (thermoelectric conductance). In the absence of an external energy source semi-conductors behave as dielectrics. The transition of electrons from the valence band via the forbidden zone (electron conductance) is accompanied by setting free the respective

---

† Here we will distinguish these terms in that the first refers to the substrates (which are solid-state bodies, polymers in particular), while the second is applied to adhesives (which are usually those polymers capable of flowing). In unspecified cases or for general reference we will use the term adhesive ability or adhesive properties.

‡ Many notions and concepts only benefit when relevant definitions are not restricted to obeying certain limiting conditions. For instance, the concept of 'reactivity' does not cease to be valid because it is impossible to describe it entirely quantitatively. The situation is clearly similar with adhesive ability.

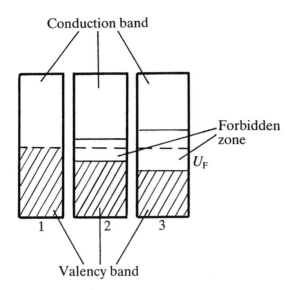

Fig. 4.1 — Energy band diagrams for metals (1), semiconductors (2) and dielectric materials (3). U_F = Fermi level.

energy levels in the valence band (appearing as vacancies), bearing in mind the physical meaning of electrons of opposite charge (the holes). This type of conductance is observed in polymers with a system of conjugated bonds, for instance, in polyenes, polyimides, polyvinyl carbazoles, polybenzoxazoles, and other polyheteroarylenes [1321], in homo-polynucleotides and homopolypeptides, characterized by a wide valency band [1322], the corresponding quantity is at a maximum in polyacetylene $(0.61 \ 10^{-18} J)$ with the minimal width of the forbidden zone $(0.12 \ 10^{-18} J)$ [1323].

In polyvinyl polymers, as the degree of saturation of the main chains increases, the forbidden zone widens, for example, to $1.6 \ 10^{-18} J$ in the case of polytetrafluoroethene depending on the energizing procedure in the visible or UV-range [1324]. According to the calculations of Duke [1325], in such polymers, as distinct from polyconjugated high molecular mass compounds, transfer between the sites of the localized states (described by Kiess and Rehwald [1326] as jumps) provides for the mechanism of electron conductance.† Such an approach, which is valid in the first instance for amorphous polymers, leads to the possibility of regulating their physical properties (including adhesive) by means of modification procedures, for instance, the incorporation of antimony or phosphorous fluorides into polyacetylene results in parallel metallic zones oriented as normal to the macromolecular axes [1328].

Having recalled these initial concepts, it should be emphasized that they are also valid for the analysis of adhesion phenomena. Three major significant features should be noted.

Firstly, the energy band diagrams of the type shown in Fig. 4.1 (2 and 3) can be applied to describe the surface of any solid body. This statement is confirmed by the data of Fig. 4.2 depicting the surface of anthracene (a molecular crystal) in vacuum [1329];

---

† The simultaneous effects of both high pressure and shear deformation also leads to forced injection of electrons from the metal into the dielectric [1327], the resultant metal–polymer adhesive joints acquiring a highly conducting state.

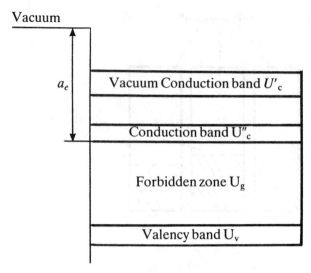

Fig. 4.2 – Energy band diagram for the surface of anthracene in vacuum at $U_v \approx U_c =$ 0.016–0.08 10^{-18} J and $U_g = 0.64$–1.28 10^{-18} J; a_e is the electron affinity.

one can see all of the three types of energy bands, the valency band, the forbidden zone and two conduction bands. The energy band diagrams of semicrystalline (partially amorphous) [1330] and wholly amorphous [1331] polymers shown in Figs 4.3 and 4.4 are very similar, but with only a single conduction band.

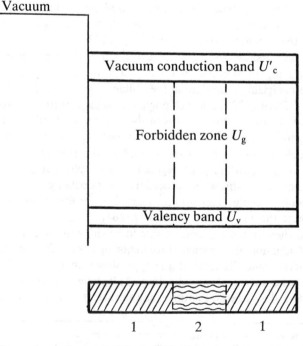

Fig. 4.3 – Energy band diagram for the surface of a polymer in vacuum. 1, Crystalline regions; 2, amorphous regions.

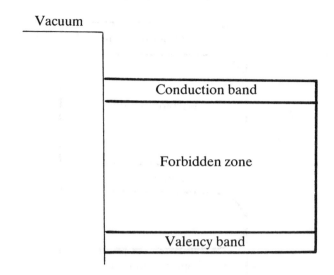

Vacuum

Conduction band

Forbidden zone

Valency band

Fig. 4.4 – Energy band diagram for the surface of amorphous polymer in vacuum.

Secondly, these concepts provide a reliable foundation to describe the characteristics relating surface energy and the work of an escaping electron discussed in section 2.2.1. In accounting for the above data this latter parameter is interpreted as the energy needed to raise an electron from the Fermi level to the apex of any potential barrier feasible for the surface. Then the surface energy is proportional to the work of an escaping electron, which is, in its turn, a function of the potential barrier (mainly determined by the nature and concentration of the sorbed compounds, i.e. by the electrical double layer) and the height of the Fermi level (determined by the structural characteristics of the phase within the boundary layers).

Thirdly, the contact between the functional group of an adhesive and the substrate surface [109, p. 120] is identified with a defect at the substrate surface (i.e. the contact can be treated as the defect). Such an analogy creates grounds to apply the generalities of intraphase electronic transitions to the study of adhesive joints. Indeed, Fabish et al. [1332] established that the energy band diagram of the interface between a polymer and a metal obeys a classical pattern, the sign of the inflection of the valency bands and of the conduction bands being determined by the ratio between the work of an escaping electron from each phase, of a polymer W_e^P and of a metal W_e^M (Fig. 4.5). Theoretical analysis of these features [1333] lies within the framework of conventional zone theories of solids.

The concepts described account for the electret properties of polymers in adhesive joints. Indeed, in polymer–metal systems produced in the absence of an applied external electric field an electric charge was registered, which was attributed to the formation of transient organometallic compounds [1334], and their diffusion into the bulk of a polymer was detected [1335]. These results were verified in studies with a broad range of polymers (polyalkenes, polyethyleneterephthalate, polycarbonate, penton) and metals (copper, aluminium, lead, gold, titanium). More definite data were reported by Mizutani et al. [1336] for polyethyleneterephthalate in contact with copper and aluminium; electronic transitions were energized by means of a photoinjection procedure. For these two systems the width of the forbidden zones was determined as 0.46 and 0.45 10^{-18} J,

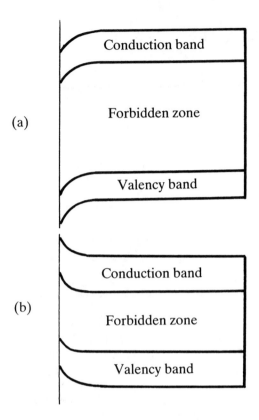

Fig. 4.5 – Energy band diagram describing the metal–polymer interface. (a) $W_e^M < W_e^P$;
(b), $W_e^M > W_e^P$.

respectively; the difference between these values is significantly lower than that between the respective functions derived from the data on external photoemission into vacuum (0.73 and 0.55 10^{-18} J, respectively), providing evidence for the existence of surface states in polyethyleneterephthalate.

These data also make it possible to assess the effect of the environment on the width of the forbidden zone, this being supposed to simulate the effect of substrate. In an inert atmosphere of helium the width of forbidden zone increases up to 0.47 10^{-18} J, while under oxygen (a potential electron 'trap') it reaches the value of 0.49 10^{-18} J [1336]. Consequently, adhesive interaction must result in the number of surface states, surface potentials etc. varying. These conclusions were confirmed in studies with semiconductors coated with gelatin [1337]. In this respect the results describing the variation of the surface charge density (μC/m^2) on germanium at the interface with various polymers [109, pp. 139, 143] are interesting:

polystyrene	−6.0	chlorinated polyvinyl chloride	38.0
polyvinyl alcohol	−1.0	ethyl cellulose	46.0
polyvinyl acetate	1.5	methyl cellulose	60.0
nitrocellulose	14.0	benzyl cellulose	134.0
acetyl cellulose	35.0		

These data provide direct evidence that the chemical nature of an adhesive affects the characteristics of the substrate surface $(Q_{Ge}^s = -25.0\,\mu C/m^2)$; this effect can be accounted for within the concepts discussed in sections 2 and 3, as well as within the framework of the zone theories of solids.

Hence, surface states do appear to play a major role in the processes of adhesive interaction. According to the modern view, two types of electronic states localized on an ideal surface can be distinguished.

The first type was introduced by Tamm [1320] who examined the question of the energy spectrum of electrons for a unidimensional model of a finite crystal, the variation of a periodic potential in the unit cell being taken into account. The major conclusion of this theory establishes that the energy spectra of the infinite and the finite crystals are substantially different. This difference is seen clearly in that the permitted energy levels fall within the forbidden zone of an infinite crystal when there is a discontinuity in the crystal lattice. The introduced states are surface states since the respective wave functions decay exponentially when moving away from the crystal boundary (face) into vacuum and when moving into the crystal the decay is oscillatory [1338].

For a unidimensional finite sequence of atoms in the absence of variation in the potential at the boundary then the energy spectrum is governed by the Schokley's state [1339]. Within the frame of reference of such a model the discrete energy levels are split into bands and, moreover, even when the latter is forbidden, at certain lattice parameters the permitted levels are obtained. In this case also the corresponding wave functions decay with increasing distance from the boundary, i.e. from the surface of the phase.

Hence, Schokley's states, arising because of band superposition, are characteristic when the variation in potential is strictly periodic (including the last unit cell), whereas Tamm's states result from the deformation of the potential in the unit cells. The difference between these states is illustrated in Fig. 4.6, which demonstrates that, strictly

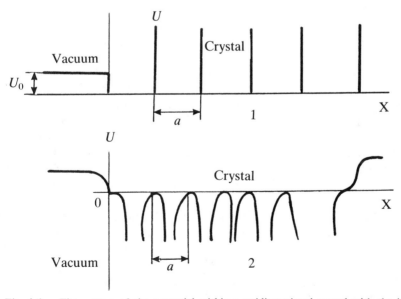

Fig. 4.6 — The pattern of the potential within a unidimensional crystal with the lattice constant a in vacuum, 1, Tamm's approach; 2, Schokley's approach.

speaking, localized energy levels on the surface of a solid body result from discontinuity of the regular structure arising at the interface. Schokley's states are characteristic primarily for substances with covalent bonds and in which substantial overlapping of the valency orbitals of the constituent atoms is inherent. The related exchange energy provides for the configurational stability of the latter. Tamm's states arise when the affinities for electrons of the atoms at the surface and in the bulk do not coincide.

The energy levels of the surface states are affected, to quite a reasonable extent, by the nature of interfacial bonds. This is easily illustrated by way of a simple example involving surface complexes resulting from adsorption. Their energy level is different from that of the free uncoupled bond at the surface of an adsorbent as well as of the free molecule of an adsorbate. This observation reminds one that the energetics of interfacial interaction and the identification of the nature of the forces responsible for the interaction are problems of major significance. On the other hand, as shown above, the nature of the adhesive affects the surface change of substrates; these can be either metals, or semiconductors, or dielectrics.

However, such an obvious classification of solids is not very productive when discussing the physical chemistry of surface phenomena. According to Morrison [1340], it is much more promising to consider the difference in electronegativity of the cations and anions, rather than the mechanism of electron transition. Then, one group comprises ionic objects, namely compounds like alkali metal halides and those ionic-covalent compounds in which covalent bonds dominate. Their electric properties vary in a broad range encompassing those (properties) of the insulators as well as those of typical semiconductors with a moderate forbidden zone width. The second group comprises covalent and metallic materials, for instance intrinsic semiconductors, extrinsic semiconductors involving the elements of Groups III and IV of the Periodic table, and metals. In this case interfacial interaction is associated neither with the Lewis (involving an electron pair) nor electrostatic (polar) mechanisms, but is due to free (uncoupled) bonds, i.e. unpaired electrons localized on the orbitals so oriented with respect to the surface that in an infinite crystal they would have been the binding orbitals.

This classification is beneficial in that the distinction between the different types of interfacial interaction appears to be its natural feature. As a result the energy band patterns are substantially simplified, hence only the two cases shown in Fig. 4.7 have to be considered. As a first-order approximation the energy of ionic surface states coincides with the energy of electron entrapment (in the case of a cation) and electron escape (in the case of an anion). Neglecting the effect of surface heterogeneity, one may assume all surface ions to be identical, i.e. characterized by a total overlapping of the wave functions. According to Pauli principle the surface states widen and finally superimpose to form isolated bands (Fig. 4.7, 1). For covalent solids the existence of uncoupled free bonds should result in the surface states involving two bands — occupied and vacant (Fig. 4.7, 2).

These concepts provide reliable grounds to target control of the adhesive ability of solids, especially of polymers, by modifying the surface in such a way so as to affect its energy. Theoretically, surface energy can be raised either by increasing the potential barrier, or by reducing the Fermi level. However, this route is hardly acceptable for polymers. Hence, the approach of Schokley appears preferable, since under normal conditions Tamm's levels are difficult to change.

Application of Schokley's concepts to polymers is made easier since the overlapping

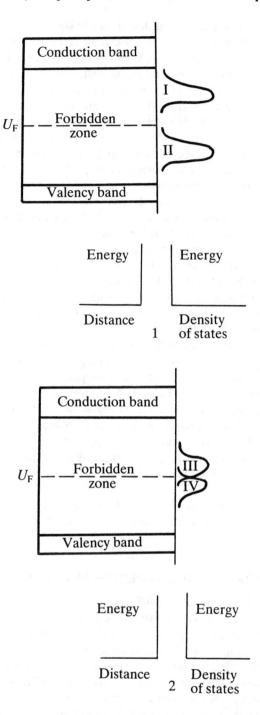

Fig. 4. 7 – Energy band diagrams describing the surface states of a solid of ionic (1) and covalent (2) type. I and II are the bands of surface levels that are close to the conductance and valency bands, respectively; III and IV are the bands of surface levels corresponding to the antibonding and bonding orbitals, respectively.

of energy levels in non-overlapping (isolated) bands should result in a polyradical character of the surface. Indeed, an isolated carbon atom has four valency electrons localized on a single s- and three p-orbitals. When considering a carbon atom bonded with other atoms one usually restricts the analysis to that of tetrahedral sp^3-hybridization. Consequently, taking the spin of the atom into account, there are eight states of the carbon atom, four of which are engaged in interatomic interaction, while the other four are of substantially higher energy. However, for the carbon atom to interact with a solid surface three orbitals would suffice. The fourth orbital remains vacant, providing for the free valences [1339], hence Schokley's states may be regarded as responsible for the free radical characteristics displayed by the surface [1341]. Moreover, within the framework of these concepts the role of over-stressed and uncoupled bonds in interfacial interaction of solids becomes more comprehensible. In particular, one should bear in mind the possibility of the recombination of Schokley's free-radical states, due to overlapping and coupling of the uncoupled bonds [1342] and induced by disturbance of the surface atomic layers of high-energy electronic configuration.

The above approach establishes a sound physical basis on which to interpret the concepts of zone theories in terms of the physical chemistry of surfaces and vice versa. It is clear that the problem of increasing the surface energy of polymers, associated with the enhancement of their adhesive properties, is reduced to that of increasing the Schokley levels. The possible route to achieve this involves rehybridization of the orbitals of macromolecules within the boundary and transition layers. For this purpose it was suggested that either the free valences proper [344, 1343, 1344], or the functional groups or atoms with unbonded (unshared) electronic pairs (containing N-, S-, P-, O- etc.) and/or π-electrons [695, 1343, 1344] should be incorporated within the layers. Within the frame of reference of modern ideas concerning surface states and their relationship with surface energy, and consequently, with the adhesive ability of polymers, unpaired electrons can be considered as a source of the polyradical character of the surface. Unshared electrons (unbonded pairs of electrons) may be regarded as a source of the additional interfacial 'phantom' bonds of length 1.6 nm and with dipole moment 1.62 D [1345].

Let us examine how these ideas [346] match the experimental data. For this purpose it appears reasonable to distinguish them as applied to adhesives and substrates individually.

4.2 IMPROVEMENT OF ADHESIVE PROPERTIES OF POLYMERS

4.2.1 Chemical nature and structure of adhesives

The number of studies related to the synthesis of polymers with enhanced adhesive properties and to the related development of adhesive formulations is so enormous that it is almost impossible to survey (we believe it sufficient to refer to the vast bibliography compiled in [12, 14, 15]). However, to the present time, attempts to establish unequivocal correlations between the adhesive properties of polymers and their chemical nature cannot be accepted as very realistic. Firstly, this is due to the multiplicity and diversity of physical methods involved in adhesive joint testing and moreover, interpretation of the data obtained is often ambiguous. Secondly, one cannot judge on the authenticity and reliability of the results and inferences reported since statistical treatment of the experimental data has been neglected in the majority of publications.

However, in line with the subject of this treatise we have ventured to make use of the data that assist the physico-chemical interpretation, avoiding, as far as possible, the empirical and technologically oriented studies. For many of the items examined there were no thermodynamic and molecular–kinetic parameters of the type discussed in sections 2 and 3 reported; hence, some of the examined dependences are only of a qualitative character.

At least two of the trends in the studies related to this subject are associated with the name of H. Mark. As far back as 1942 he suggested that the stickiness and tackiness of rubbers (i,e. in modern terminology, their adhesion) are in some way connected with the mobility and diffusive ability of macromolecules [5]. These ideas were later developed and experimentally verified in a series of studies by Voyutskii and co-workers [115, 22]. Here it was convincingly demonstrated that to form strong adhesive joints the polymers have to display high macromolecular mobility ensuring their diffusion either across the interface or along it [22].

Almost 20 years later Mark extended [1347] the initial concepts [5] by taking into account interfacial interaction between an adhesive and a substrate. Having assessed, from these grounds, a large number of polymers, he distinguished polyacrylates as the most promising adhesives. Recently Mark reconsidered the problem of choosing an effective adhesive and suggested that the cohesive properties of polymers are related to their ability to form donor-acceptor bonds [1348, 1349]. As a result the prediction regarding poly-acrylates as promising adhesives was confirmed. Moreover, an 'ideal' adhesive composition combining the benefits of common monomeric and anaerobic adhesives was proposed which comprises acrylonitrile with isooctylacrylate (5 : 1) and methylolacrylamide with polymethyl methacrylate (5 : 1) taken in equal amounts and ethyleneglycol dimethacry-late (10% of the amount of acrylonitrile). Polymethyl methacrylate ensures the necessary rheological characteristics, while the nitrile and hydroxy functions provide for the efficient interfacial interaction. However, this approach has not gained practical usage because of the inadequate strength of the resultant polymers.

Deryagin and co-workers [109] developed another approach relating the chemical nature of polymers and their adhesive properties. Having assumed that there is an electrical double layer at the adhesive–substrate interface he suggested that after the rupture of the joint the fracture surface becomes charged due to separation of the charges comprising the donor–acceptor pairs at the interface. Glass was chosen as the standard substrate and the hydroxy functions on its surface were substituted by others by means of chemical modification. Subsequently, a polymer, chosen so as to ensure maximum probability of the formation of interfacial donor–acceptor bonds, was layered onto a modified surface, the equilibrium system obtained was fractured, the charges and their sign being registered for both the adhesive and the substrate. As a result the functional groups were arranged in the following series: $NH_2 > OH > OCOR > C_6H_5 > Cl > COOH > CN > C = C$. It was suggested that for an adhesive with universal properties it is better that its molecules incorporate a set of groups from the extremities of the listed series.

The benefit of this concept involves the possibility of a quantitative interpretation. However, one should bear in mind that this approach relies on a semiphenomenological assessment of the role of interfacial donor–acceptor bonds. In spite of the fact that the molar cohesion and dipole moments of the functional groups were involved in the evaluation, the listed series of functional groups reflect only the sign of the surface, and fails to account for the fundamental characteristics. In other words, the ideas discussed are valid only when surveying the final result of adhesive interaction, whereas their interpolation to the initial stages of the respective processes can hardly be admitted as methodologically correct. This statement can be illustrated by way of the example comparing the values and the signs of surface charge with the strength of adhesive joints between fluorocarbon polymer films 100 μm thick [1350]. The respective data are compiled in Table 4.1 (the difference between '+' and '−' is only 5 N/cm). As is seen, the results presented display a random pattern revealing no connection with the chemical nature of polymers. At the same time, as a first-order approximation, this approach gives a true picture of the relative variation of the adhesive properties of polymers and in this regard it is appropriate to emphasize the coincidence of the listed donor−acceptor series with that arranged according to the magnitude of parameter α discussed in section 3.1.2.

Table 4.1 − Comparison of the electrical and adhesive strength properties of the joints between fluorocarbon films[a]

| Fluoro-carbon polymer | Q_e^s (pC) | | | | | | | | | | |
| | −1000 | −500 | −220 | −26 | +20 | +45 | +54 | +62 | +90 | +130 | +140 |
	F-42	F-2	F-2M	F-1	F-10	F-3M	F-40	F-50	F-4Mb	F-400	F-100
F-42	+	+	+	+	−	+	−	−	−	−	−
F-2	+	+	+	−°	−	−	−	−	−	−	−
F-2M	+	+	+	−	−	−	−	−	−	−	−
F-1	+	−	−	+	−	−	−	−	−	−	−
F-10	−	−	−	−	+	−	−	+	+	−	+
F-3M	+	−	−	−	−	+	−	−	−	−	−
F-40	−	−	−	−	−	−	+	−	+	+	−
F-50	−	−	−	−	+	−	+	+	+	−	+
F-4MB	−	−	−	−	+	−	−	+	+	−	+
F-400	−	−	−	−	−	−	+	−	−	+	−
F-100	−	−	−	−	+	−	−	+	+	−	+

a Fracture time 100 s.

However, Deryagin's approach, like that of Mark, is not able to take a direct account of the effect of the cohesive strength of cured adhesives and substrates on the strength of the adhesive−substrate system as a whole. The difference between the concepts of adhesive ability and adhesive joint strength, which is a concept incorporating both 'interphase' and 'intraphase' implications, has already been stressed above; however, this fact does not imply that, within the physico-chemical aspect, the latter should necessarily be neglected. Hence the significance of the almost obvious recommendations to improve adhesive ability (adhesiveness) by reinforcement of the polymers proper. Clearly, these recommendations are of considerable technological significance. However, it is also clear that these rely on procedures with an apparent physico-chemical background such as, according to Lipatov [1351], filling of composites, modification with surfactants and formation of interpenetrating networks between polymeric components. Only the strengthening of the transition layers results in a regular increase of the adhesive joint strength [933, 1352].

Summing up these data and those of the preceding sections, one can deduce two possible approaches to controlling the chemical nature of polymers and providing enhanced adhesive ability. The first involves the increase of the surface energy of polymers and for this purpose the atoms and functional groups listed in section 4.1 have to be incorporated into their structure. However, one should bear in mind that at least two components are engaged in adhesive interaction and the higher the surface energy of each of the components involved the better. However, according to the second approach ensuing from the principle of interfacial energy minimization, to provide improved adhesive ability the energy barrier at the interface should be minimal.

Both of these conditions has its own significance though they become even more effective when combined. Apparently, the separate individual approaches, in the absence of each other, would inevitably be limited. Indeed, one can easily imagine a case when the surface energy of an adhesive has to be lowered instead of increased. This might be the case when the substrate is a low-energy polymer, for example, poly(methyl methacrylate) containing the atoms of fluorine in the pendant groups ($\sigma_{sa} = 10.1 - 14.1$ mN/m [1353] as compared to 3.6 mN/m of PMMA) or the copolymers of methyl methacrylate with 0.8–1.0% perfluoroakyl acrylates (which display minimal σ_{sa} values [1354]). However, decrease of the surface energy of the adhesive inevitably leads to decreased cohesive characteristics of the bonded species and, finally, to decreased adhesive joint strength. This feature should be taken into account in bonding practice.

Using these basic principles the following general formula for the repeat unit should describe adhesively-active polymers [695, 1343, 1344, 1346]:

$$A-[-R^= -B-R'-X-R'-B-R^= -]_n -A \qquad \text{(model A)}$$

where $R^=$ is the unsaturated group as a source of π-electrons (for instance, allyl group or any of its higher homologues comprised of not more than five carbon atoms), R' is either a methene or an ethene group ('spacer'), A is the terminal group providing for the formation of interfacial bonds of maximum energy, B is the function ensuring enhanced macromolecular mobility (for instance, the ether function) and X is the function containing the atoms of nitrogen, phosphorus, halogens etc. in the most labile and active form.

In this model of a macromolecule (model A) function X involves atoms with unbonded electron pairs permitting rehybridization of the orbitals according to Schokley and implying a higher surface energy of the polymer and, finally, its improved adhesiveness (adhesive ability). Note that it not just heteroatoms that comprise the X-function (as could have been assumed) but explicitly atoms with unbonded electron pairs that are not involved in bonds with other atoms of the molecule and it is for this reason that the amino-group-containing polymers display a consistently higher adhesiveness than those containing nitro groups. The functions B, positioned sufficiently far from each other and separated from X-functions by a spacer R', provide for the adequate mobility of the main chains, i.e. for the ability of a polymer to wet the substrate surface effectively and, consequently, to establish the maximum interfacial contact. The double bonds in $R^=$, in addition to the main feature of being a source of π-electrons, stimulate the same process by ensuring certain macromolecular flexibility. Besides this, the double bonds of $R^=$ are most liable to become involved in destructive processes in macromolecules, induced by external action, and resulting in the formation of radicals thus supporting the effect of X-functions. A similar effect is also realized by the presence in a macromolecule of any weak, i.e. low-energy, bond, e.g. on that is thermally unstable. Finally, the terminal A-groups provide an increase in the reactivity of a macromolecule which is seen as an increase in the probability of forming a network of interfacial bonds.

Considering the problems related to the synthesis of adhesives, a model adhesive containing X-functions in the side chains, the said functions being attached to the main chain via alkene spacer R'' ($R'' = 2$–$3\ R'$) to ensure their enhanced mobility, appears to be more worthwhile [344, 1343, 1344, 1346]:

$$A-\left[\begin{array}{c}-R^{=}-B-R'-B-R^{=}-\\ |\\ R''\\ |\\ X\end{array}\right]_n -A \qquad \text{(model B)}$$

Those polymers whose structure is identical to models A or B or is a close approximation can be distinguished as a group of specialist adhesive polymers. The need for such products is beyond any doubt and it is from these that the adhesives of the 'third generation' should derive [1346]. At the present state of the art, general-purpose polymers (for instance, those for engineering) are used for adhesives and adhesiveness (adhesive ability) is not on the list of basic characteristics. To ensure the requirements imposed by modern adhesives the general-purpose polymers are used in combination with a large number of special additives and the large number of components in the system drastically reduces any prediction of its properties. Perhaps, it is more justifiable, technically [14, 986], as well as economically [1355], and for each individual case, to use materials which display a narrower but more targeted spectrum of properties. This is particularly so with adhesives.

The design and synthesis of such polymers displaying specific adhesive properties presents a sophisticated task, since the macromolecules of such polymers would involve a large number of separate fragments incorporating various functional groups. It is therefore appropriate to present a survey of some general routes to the synthesis of such polymers (which are, by the way, essential in problems other than adhesives).

Hierarchical analysis of the structural organization of polymers, and this involves molecular (monomeric), segmental, macromolecular, supermolecular, and composite levels [1356], reveals that the greatest potential in designing the chemical structure of a macromolecule is achieved at the first of these levels[†] (contrary to the traditional approach which involves manipulations at the chain-growth stage). Indeed, at this level, monomers of almost any structural and chemical sophistication can be produced. Then, to control the chemical structure of complex polymers it is clear that the fragments of various structure should be incorporated within the monomer during synthesis, i.e. at the molecular level rather than at the chain-growth stage (which is already the macromolecular level) [1357]. Such a multifragment monomer (comer) is obviously a more complex structure than a conventional monomer of vinyl or diene type, each of its fragments corresponding to a single comonomer so that the final content of these

---

[†] This hierarchy as well constitutes the basis for the uniform algorithm for calculation of the strength of adhesive joints of polymers [1317, 1319].

fragments in the macromolecule appears to be predetermined. Let us illustrate this statement by way of the following examples.

The traditional route for the production of ethene–vinyl chloride copolymer involves copolymerization of the two comonomers. Since the reactivity ratios of the comonomers are quite different (by an order of magnitude in radical copolymerization [1358] and by two orders in ionic–coordination copolymerization [1359]) the composition of the final product is far from equimolar with respect to the comonomer content. The copolymer is enriched with the more reactive one, the relative content of the fragments in the copolymer being governed by statistical rules. This drawback is easily avoided by making use of the two-fragment comer instead of two comonomers, each fragment of the comer corresponding to a chain unit of definite chemical structure. In accord with this approach to produce poly(ethyene-co-vinyl chloride) = 1 : 1 one has to subject polychloroprene to hydrogenation [1360]:

$$n\ H_2C=CH_2 + n\ HC\underset{\underset{Cl}{|}}{}=CH_2$$

$$\left[H_2C-CH=\underset{\underset{Cl}{|}}{C}-CH_2 \right]_{2n} \xrightarrow{\ 2n\ H_2\ } \left[H_2C-CH_2-\underset{\underset{Cl}{|}}{CH}-CH_2 \right]_{2n} \qquad (496)$$

The product obtained is characterized by the predetermined explicitly defined composition and displays, as compared to the conventional copolymer, better physicochemical properties.

Copolymers other than statistical can be produced likewise. For instance, copolymers of ethene and propene are produced by hydrogenation of polyisoprene, while the ethene–ethene–acrylate copolymers are obtained by hydrogenation of the copolymers of butadiene with methyl methacrylate and methyl acrylate [1361]. This approach is most useful for the synthesis of polymers which, in principle, cannot be obtained by traditional copolymerization. For instance, the copolymers of isobutene with dichloropropene cannot be obtained other than by homopolymerization of 5-methyl-1,3-dichlorohexadiene-2,4 [1362, 1363]; similarly, the copolymers of vinyl monomers with substituted butadienes can be produced by homopolymerization of hexatrienes-1,3,5 via 1,6-addition polymerization and subsequent hydrogenation of the products [84, 1343]:

$$n\ \underset{A\ \ B\ \ C\ \ D\ \ E\ \ F}{HC=C-C=C-C=CH} \longrightarrow \left[\underset{A\ \ B\ \ C\ \ D\ \ E\ \ F}{HC-C=C-C=C-CH} \right]_n \xrightarrow{\ n\ H_2\ }$$

$$\left[\underset{A\ \ B\ \ C\ \ D\ \ \ E\ \ \ F}{HC-C=C-CH-CH-CH} \right]_n \cdot \qquad (497)$$

The synthesis of complex multifragment polymers by means of polycondensation procedures appears to be just as promising. For instance, on heating the tetramethylol derivatives of phenol [1364] phenol aldehyde products are formed, which are, however, of higher molecular weight and of a more regular structure than the commercial products. In this case the groups liable to become involved in the condensation reaction are incorporated within the structure of a single monomer (comer) as terminal functions, the other fragments remaining the same:

$$n \, \text{A–X–B} \longrightarrow \left[\text{X} \right]_n + n \, \text{AB}. \tag{498}$$

Even if X is a heterocyclic fragment, a polyheteroarene, otherwise synthesized by cyclization of polyfunctional compounds and thus preventing polynucleic heterocycles from being introduced in macromolecules, can be produced by homocondensation. For instance, polyquinoxaline was reported to have been synthesized from the corresponding comer [1365]

$$n \; \text{(quinazoline–CH}_2\text{Cl)} \longrightarrow \left[\text{(quinazoline–CH}_2\text{)} \right]_n + n \, \text{HCl} \tag{499}$$

Hence, designing the chemical structure of a polymer at the level of a multifragment monomer substantially extends the possibilities for the production of multifunctional high molecular compounds [1357], and thus creating the basis for the synthesis of complex adhesives.

The most thoroughly investigated applied example of this approach is suggested by the homopolymerization of substituted hexatrienes-1,3,5, which are the structural analogues of butadiene–(vinyl monomer) copolymers: actually, these cannot be obtained by traditional polymerization procedures. In fact, the ethene–chloroprene copolymer is chemically and structurally similar to poly–2–chlorohexatriene-1,3,5; propene–iodoprene copolymer to poly–2–methyl–4–iodohexatriene-1,3,5 etc. In a study of 24 monomers of this class comprising alkyl [1366, 1367], aryl [1368–1370], halide [1366, 1371], and heteryl [1372] substituents it was disclosed that peroxide (persulfate) initiated polymerization of hexatrienes proceeds via 1,6-addition mechanism illustrated by eq. (497) and the resultant product is an unbranched chain containing π-conjugated blocks [84]. Such results are consistent with the known features relating to the polymerization of unsubstituted hexatriene in the presence of transient metal ethers [1373], as well as with the studies of the structure of poly–2,3,4,5-tetramethylhexatriene [1374, 1375].

All synthesized polyhexatrienes exhibit high adhesive properties, which is attributed to the fact that their chemical structure conforms the requirements of model A [1376]. In accord with the predictions nitrogen-containing heteryl-substituted polyhexatrienes display the best adhesiveness. In fact, while the shear strength of the rubber–steel adhesive joints bonded with polyhexatrienes as adhesives is, on the average, around 10 MPa, it is as high as 21.4 MPa in the case of poly–1–phenyl–6(2–quinolinyl)hexatriene. Even poly-1,5-diphenyl-6(2-pyridyl)hexatriene (12.7 MPa), which is the least effective of the nitrogen-containing heteryl-substituted polyhexatriene adhesives, is higher than the average values.

The tensile strength of the adhesive joints is actually the same for all polyhexatrienes; this is probably due to cohesive fracture of the transition layers of the adhesives. It is therefore, incorrect to assess the adhesiveness of the latter by the tensile strength properties of the joints. It is much more appropriate to make use of the characteristics of physico-chemical origin, i.e. surface energy, equilibrium flexibility of macromolecular chains, and the complex criterion α, rather than of the deformative ones. These parameters were calculated with the aid of eqs (217), (221), (324) and (356) based upon experimental data on the refractive indexes of substituted hexatrienes-1,3,4 n_m and of the respective polymers n_p, refractions,[†] van der Waals' volumes, cohesion energies of the repeating units and of segments calculated from the respective increments r_i [490, p. 200], ΔV_i [471], ΔE_i^* [477], and ΔE_i^s [910]. Solubility parameters of polymers δ were calculated from eq. (202) and the constant of Hildebrand's eq. (195) was reported [395] to equal 3.8 for all compounds except for alkyl substituted compounds where $A = 3.3$.

The respective data (experimental and calculated) are compiled in Table 4.2. One can see that, according to the variation of parameter α, the polyhexatrienes are arranged in a series which coincides with the experimentally observed arrangement of polyhexatrienes with regard to their adhesiveness [1379]. Moreover, the correlation coefficient between α and s is 0.788, thus providing additional support for the relevance of this approach.

Note also the good consistency between the values for σ calculated from the refracto-metric data for polymers by using eq. (217) and those obtained for monomers by using eq. (221). Having taken this into account, it is reasonable to assess the adhesive properties of polyhexatrienes-1,3,5 by their surface energies, provided that the substrates are highly polar polymers and that the fracture is examined in peeling tests. To prove the validity of this assumption we have calculated, by making use of eq. (305), the values for P_p^* and, subsequently, relying on the data for the constant a listed in Table 2.11, calculated the peeling strengths of the adhesive joints of 24 polyhexatrienes with polychloroprene (Nairit — trade mark) and a butadiene–acrylonitrile copolymer SKN-40. In Table 4.2 these are compared with the experimental data [1379]. As can be seen, the calculated and experimental characteristics of the adhesive properties of polyhexatrienes are in good accord with each other, providing evidence of the reliability and usefulness of the basic theoretical approach.

Additional enhancement of the adhesiveness of polyhexatrienes can be obtained by halogenation. For instance, it was found [1380] that the rubber–steel adhesive joints (butadiene–acrylonitrile SKN-40, butadiene–styrene SKS-30, and polyisoprene SKI-30 rubbers were used in the study) bonded with chlorinated, brominated and iodinated halogen-substituted polyhexatrienes (listed in Table 4.2) undergo cohesive fracture involving the rubber element of the joint.

High adhesive properties of chlorinated halogen-substituted polyhexatrienes can be substantially improved by modifying the formulations with radical-producing compounds. For instance, introduction of even minute amounts of dicumylperoxide with bromine (see Table 4.3) results in a reasonable rise in the shear strength of the respective adhesive joints under static as well as under dynamic loading modes [1381] (Table 4.4).

[†] Apart from the role discussed in section 2.2.1, this parameter acquires special significance in view of the concepts of section 4.1 relating the chemical nature and adhesive properties of polymers; it follows from the fact that refraction was shown to be in direct proportion to the degree of the unsharedness of the electron pairs [1377]. Hence, it is not by chance that the molecular refraction exaltation and the number of conjugated bonds in macromolecules of the type considered obey a linear relationship [1378].

Table 4.2 — Physico-chemical characteristics of substituted polyhexatrienes-1,3,5

Substituent	M_1	n_D^{20} monomer	n_D^{20} polymer	$\sum_i m_i a_i r_i$ (cm³)	$\sum_i \Delta V_i$ $\left(\dfrac{cm^3}{mol}\right)$	$\sum_i \Delta E_i^*$ $\left(\dfrac{cal}{mol}\right)$
1. 2-methyl–	94.16	1.5051	1.5990	31.670	67.27	3685.9
2. 2,5-dimethyl-	108.19	1.5121	1.5082	36.278	77.36	4332.0
3. 2-chloro-	114.57	1.5417	1.7255	32.071	65.85	4392.4
4. 2-bromo-	159.02	1.5338	1.7297	34.923	67.92	5198.1
5. 2-methyl-4-chloro-	128.60	1.5350	1.7533	36.679	76.15	5038.5
6. 2-methyl-4-bromo-	173.05	1.5220	1.7384	39.531	77.68	5844.2
7. 2-methyl-4-iodo-	220.05	1.5907	1.7409	44.534	91.72	6961.2
8. 1-phenyl-	156.23	1.5980	1.6493	51.453	102.43	7247.8
9. 1-phenyl-6(p-nitrophenyl)-	277.33	1.5943	1.7735	82.116	159.58	12898.3
10. 1-phenyl-6-o-tolyl-	246.35	1.5190	1.6730	80.514	160.49	12101.9
11. 1-phenyl-6-m-tolyl-	246.35	1.5894	1.6492	80.794	160.49	12101.9
12. 1-phenyl-6-p-tolyl-	246.35	1.5894	1.6321	80.824	160.49	12101.9
13. 1-phenyl-6(p-methoxyphenyl)-	262.35	1.5786	1.6855	82.594	164.16	12244.5
14. 1-phenyl-6(2,4-dimethyl phenyl)-	260.38	1.5644	1.6277	85.984	173.07	12748.0
15. 1-phenyl-6(3,5-dimethyl phenyl)-	260.38	1.5629	1.6239	86.264	173.07	12748.0
16. 1,6-diphenyl-	232.33	1.5971	1.6493	75.844	147.90	11455.8
17. 1,6-diphenyl-2-methyl-	246.35	1.5779	1.6474	81.488	158.20	12101.9
18. 1,6-diphenyl-3-methyl-	246.35	1.5779	1.6467	81.488	158.20	12101.9
19. 1,6-diphenyl-2,3-dimethyl-	260.38	1.5607	1.6427	87.132	168.50	12748.0
20. 1-phenyl-6(2-pyridyl)-	232.30	1.6123	1.6537	66.817	147.66	12924.7
21. 1-phenyl-6(2-quinolinyl)-	283.38	1.7171	1.6547	86.577	180.66	15983.6
22. 1-phenyl-4,5-dimethyl-6(2-pyridyl)-	260.37	1.5971	1.6564	76.115	168.25	14216.9
23. 1-phenyl-4,5-dimethyl-6(2-quinolynyl)	311.43	1.6040	1.6440	95.875	201.26	17275.8
24. 1,5-diphenyl-6(2-pyridyl)-	308.4	1.6149	1.6548	91.307	193.31	17132.7

$\sum\limits_i \Delta E_i^s$ $\left(\dfrac{cal}{mol}\right)$	δ $\left(\dfrac{cal^{0.5}}{mol^{1.5}}\right)$	σ (mN/m)		α	s	P_p^*	P_p^{cal}(kN/m)		P_p^{exp}(kN/m)		Ref.
		From eq. (217)	From eq. (221)				To Nairit	To SKN 40	To Nairit	To SKN 40	
20411.1	7.40	32.1	33.0	116.0	5.5	−1.5927	2.45	0.34	1.42	0.29	[1366]
28805.4	7.47	30.9	35.5	138.7	6.6	−1.4698	1.34	0.31	1.30	0.27	[1367]
40901.2	8.17	31.8	34.4	142.3	9.3	−1.5618	1.43	0.33	1.39	0 29	[1366]
	8.78	47.4	50.3	118.2		−3.3361	3.05	0.71	2.96	0.65	[1366]
49295.5	8.13	33.6	35.2	115.0	9.8	−1.7491	1.60	0.37	1.49	0.24	[1371]
	8.67	45.4	49.7	137.0		−3.0896	2.82	0 65	2.76	0.37	[1371]
	8.71	56.0	58.1	136.2		−4.4600	4.07	0.94	3.88	0.81	[1371]
103923.4	8.41	35.8	36.9	204.6	14.3	−1.9843	1.81	0.42	1.74	0.35	[1368]
151350.1	8.99	53.2	54.7	249.6	11.7	−4.9827	3.73	0.86	3.62	0.79	[1369]
160883.1	8.68	45.9	47.1	267.8	13.3	−3.1507	2.88	0.67	2.72	0.61	[1369]
160883.1	8.68	44.7	46.4	273.5	13.3	−3.0047	2.74	0.63	2.68	0.55	[1369]
160883.1	8.68	43.9	46.4	277.5	13.3	−2.9084	2.65	0.62	2.57	0 54	[1369]
153096.3	8.64	46.1	47.9	270.2	12.5	−3.1753	2.90	0.67	2.84	0.58	[1369]
152567.7	8.58	44.1	46.3	291.4	12.0	−2.9324	2.68	0.62	2.59	0.51	[1370]
152567.7	8.58	43.8	46.2	293.0	12.0	−2.8964	2.64	0.61	2.52	0.48	[1369]
169198.5	8.80	44.2	46.8	261.0	14.8	−2.9444	2.69	0.62	2.58	0.50	[1370]
177592.8	8.75	44.3	46.2	274.8	14.7	−2.9564	2.64	0.63	2.54	0.57	[1370]
177592.8	8.75	44.3	46.2	274.8	14.7	−2.9564	2.64	0.63	2.54	0.57	[1370]
185987.1	8.70	44.3	46.8	289.0	14.6	−2.9654	2.71	0.63	2.64	0.54	[1372]
110699.2	9.36	56.8	59.7	234.2	8.6	−4.5698	4.17	0.97	3.97	0.88	[1372]
162537.5	9.41	58.7	61.5	277.6	10.2	−4.8342	4.41	1.02	4.28	0.92	[1372]
127487.8	9.19	57.1	59.2	256.4	9.0	−4.6112	4.21	0.98	4.07	0.83	[1372]
179326.1	9.26	58.4	61.3	301.3	10.4	−4.7921	4.37	1.01	4.18	0.90	[1372]
214608.8	9.41	60.9	64.7	287.6	12.5	−5.1466	4.70	1.09	4.58	0.98	[1372]

Table 4.3 — Adhesive formulations based on chlorinated halogen-substituted polyhexatrienes (composition in parts by mass):

Halgenated polyhexatriene-1,3,5	Formulation									
	1	2	3	4	5	6	7	8	9	10
Poly-2-chlorohexatriene	10	10								
Poly-2-bromohexatriene			10	10						
Poly-2-methyl-4-chlorohexatriene	1				10	10				
Poly-2-methyl-4-bromohexatriene							10	10		
Poly-2-methyl-4-iodohexatriene									10	10
Dicumylperoxide		1		1		1		1		1
Bromine		1		1		1		1		1
Xylene-toluene (1 : 1)	40	40	40	40	40	40	40	40	40	40

Table 4.4 — Strength properties of rubber–steel adhesive joints bonded with chlorinated halogen-substituted polyhexatrienes

Characteristic of joint strength	Formulation[a]									
	1	2	3	4	5	6	7	8	9	10
Shear strength MPa										
at 293 K	7.7	9.4	7.2	9.9	7.8	9.5	7.7	9.6	7.9	9.9
at 373 K	5.8	7.6	5.7	7.4	5.1	7.9	5.5	7.7	5.5	7.3
at 423 K	3.4	5.2	3.1	4.9	2.8	5.1	3.2	5.2	3.2	5.7
Shear strength after 35×10^3 strain cycles (frequency 5 Hz, deformation 50%) MPa	7.4	10.9	8.1	11.4	7.5	10.1	9.1	11.8	10.2	12.7

a   Numbered as in Table 4.3.

The data presented indicate that enhanced adhesive ability is a feature inherent not only to the polyhexatrienes proper but also to the respective monomers. In fact, they are characterized by high refractive indexes (Table 4.2) and, for this reason, should be described by high surface energy values also, hence 1,6-diphenylhexatriene is much more ordered at the surface of polyvinylchloride than within the volume of the phase [1382]. Consequently, bonding of rubber to rubber or to metallic substrates with 2-methyl-4-halogenohexatriene-1,3,5 ensures cohesive fracture of the systems obtained either at room or at curing temperatures [1383]. The strength of the adhesive joints produced in this way is higher than that of the substrates, which is, apparently, due to the diffusion of monomers into the transition layers of the latter and subsequent polymerization. Introduction of peroxide initiators (0.5%) reduces the bonding time by half [84]. Hexatrienes-1,3,5 can be used as effective solvents for polyhexatrienes as well as, for instance, for polychloroprene, the respective solutions being highly efficient in bonding rubbers of any type [1383].

Relatively high adhesive properties are demonstrated by poly(5-methyl-1,3-dichloro-hexadiene-2,4) already mentioned above. When additionally chlorinated, it provides rubber–steel adhesive joints of higher strength than the most efficient of the industrial chlorine-containing elastomers, chlorinated polychloroprene, making it possible to bond even the styrene rubbers to steel (Table 4.5) [1363, 1363].

Table 4.5 – Tensile strength of rubber–steel adhesive joints bonded with chlorine-containing adhesives, MPa

	Adhesive	
Elastomer	Chlorinated poly(5-methyl-1,3-dichlorohexadiene-2,4)	Chlorinated polychloroprene
Polyisoprene SKI-3	4.8–5.0	1.5
Butadiene-acrylonitrile SKN-26	5.0–7.0	3.5
Butadiene–styrene SKS-30-ARKM-15	4.5–6.2	does not bond

Besides the above, there is yet another class of adhesive polymers corresponding to model A, comprising polyurethanes with reactive end groups, for instance, isocyanate or epoxy. Such adhesives were produced by reacting excessive amounts of 2,4-tolylene-diisocyanate with polytetramethyleneglycol (SKU-PFL) or with polyethene- (SEF-3A), polyoxypropene- (SEF-PG), or polyoxybutene- (PEF-3A) glycol adipates additionally treated with glycidol [1384]. The curing agents were m-phenylenediamine (PhDA) or a eutectic mixture of aromatic diamines (EMAD) in the case of isocyanate-containing adhesive SKU-PFL, and polyethenepolyamine (PEPA) in the case of polyurethanedie-poxies. Listed in Table 4.6 are the peeling strengths of adhesive joints of filled rubbers and synthetic leathers with polyvinyl chloride and polyurethane coatings. Taking into account that the peeling strength P_p ensured by the best of the commercial adhesives does not exceed 1 kN/m after 30 min exposure and 2.7 kN/m after 1 h (provided that the substrates were pretreated mechanically) the data of Table 4.6, pertaining to the untreated substrates, demonstrate high efficiency of the adhesives under discussion.

The application as adhesives of polymers similar in structure to model B is just as effective. These comprise the polymers with amide and urethane groups in side chains and which are known to enhance the adhesiveness of the respective polymers. However, when incorporated within the backbone they hinder the mobility of adhesively active fragments. Therefore, carbochain amide- and urethane-containing polymers are more effective than the 'classical' heterochain polyamides and polyurethanes. The monomers for these polymers comprise two double bonds with different degrees of substitution and, correspondingly, different reactivity. In the course of polymerization of the molecules of the general formula $R^=-X-R_1^=$ (where X is the amide [1385] or urethane [1386] group) a more reactive ethene bond forms the backbone, while the less reactive resides as

Table 4.6 — Peeling strengths of adhesive joints involving polyurethanes with isocyanate and epoxy end-groups, kN/m

Polyurethane (PU)	Curing agent (CA)	PU : CA	Substrate coating[a]	
			PVC	PU
SKU-PEL	PhDA	1.0 : 1.0	2.2/2.6	3.1/3.3
SKU-PFL	EMAD	1.0 : 0.9	2.0/2.9	3.1/3.7
SKU-PFL	EMAD	1.0 : 0.5	2.5/2.8	1.8/3.2
SEF-3A	PEPA[b]	—	1.5/2.8	1.9/3.7
SEF-PG	PEPA[b]	—	1.4/2.5	1.5/3.4
PEF-3A	PEPA[b]	—	1.1/2.7	1.4/3.6

a Numerator refers to the strength after 30 min exposure; denominator, to that after 1 h.

b The content is determined by the concentration of epoxy end-groups in the adhesive.

the pendant group. When the reactivities of $R^=$ and $R_1^=$ are very similar the monomers are most liable to undergo cyclopolymerization. Of the synthesized carbochain polymers listed in Table 4.7 such is the structure of the first three polymers, while the others have the free (less reactive) double bonds incorporated in the side chains.

Because of these differences the polymers obtained are distinctly different in their adhesive properties. As is seen from the data in Table 4.7, the cyclic-chain polymers provide less strong bonding than the linear ones, which is apparently due to the fact that there are no amide groups and double bonds in the side chains, i.e. they do not match model B. It is, however, quite clear that the adhesiveness of even the cyclic polymers is higher than that of poly-N-methylacrylamide (which is the only commercial polymer of a close chemical structure, this fact providing additional evidence of the validity of the initial concepts relating the adhesive properties of high molecular mass compounds with their chemical structure.

Table 4.7 — Shear strength of adhesive joints of steel 3 bonded with carbochain amidogroup-containing polymers, MPa

Polymer	Testing temperature (K)		
	298	353	423
Poly-N-isopropenyl acrylamide	10.5	9.5	7.0–8.0
Poly-N-vinyl methacrylamide	12.5	9.5	7.5–9.0
Poly-N-(α-propenyl)acrylamide	13.0	11.0	10.0
Poly-N-(α-phenyl vinyl)acrylamide	15.0	12.8	11.2
Poly-N-vinyl cinnamamide	17.5	13.5	12.5
Poly-N-methylacrylamide	9.0	—	—

As a theoretical support to this experimental relationship in the carbochain amido-group-containing polymers we have calculated the physico-chemical characteristics of adhesive ability by means of the procedure described above. Typical non-polar (polyiso-butene P-118) and polar (polychloroprene, butadiene—acrylonitrile elastomer SKN-40) were chosen as substrates. The pertinent data are compiled in Table 4.8. Correspondence of the listed poly-N-alkenylacrylamides to model B is confirmed by the relatively high refractive indexes and surface energies as compared to known polyamides.

However, most unequivocally the validity of this approach is supported by the agreement between the calculated and experimental values for the peeling strength of adhesive joints, as is illustrated in Fig. 4.8. On the one hand, the absolute values are rather large to allocate the polymers listed to a particular class of adhesives. On the other hand, the impressive agreement between the calculated and experimental data demonstrates the reliability of these theoretical concepts of adhesion. Quite illustrative in this respect are the close correlations observed between different physico-chemical characteristics of carbochain amide-containing polymers. As is seen from the data in Table 4.9 the paired correlation coefficients are rather large despite the different origin of parameters σ, α and s [1388].

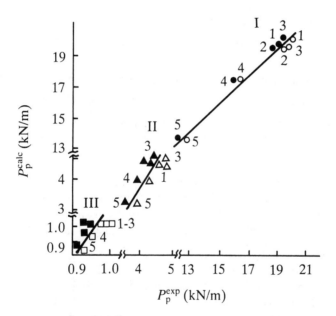

Fig. 4.8 – Correlation between calculated (open symbols) and experimental (solid symbols) peeling strengths of the adhesive joints of polyisobutene P118 (I), polychloroprene NT (II) and butadiene—acrylonitrile SKN-40 (III) elastomers bonded with poly-N-isopropenyl acrylamide (1), poly-N-vinyl methacrylamide (2), poly-N-(α-propenyl)acrylamide (3), poly-N-(α-phenylvinyl)acrylamide (4) and poly-N-vinylcinnamamide (5).

Investigations with urethane-containing polymers lead to similar inferences. As with the amide-containing polymers, these can be either linear, with urethane groups and double bonds in the side chains, or cyclic-chained if the reactivities of $R^=$ and $R_1^=$ in the dialkenylurethanes subjected to polymerization are close. It is also clear that in the

Table 4.8 − Physico-chemical characteristics of carbochain polymers for use as adhesives

Polymer	M_1	n_D^{20}		$\sum_i m_i a_i r_i$	$\sum_i \Delta V_i$
		Mono-mer	Poly-mer	(cm^3)	$\left(\dfrac{cm^3}{mol}\right)$
Carbochain amido-group-containing polymers					
Poly-*N*-isopropenyl-acrylamide	111.14	1.6065	1.7640	28.368	62.36
Poly-*N*-vinylmethacrylamide	111.14	1.6065	1.7548	28.368	62.42
Poly-*N*-(α-propenyl)-acrylamide	111.14	1.6063	1.7764	28.371	62.36
Poly-*N*-(phenyl vinyl)-acrylamide	173.21	1.5979	1.7225	51.380	100.00
Poly-*N*-vinylcinnamamide	173.21	1.5930	1.7417	51.577	102.65
Carbochain urethane-containing polymers					
Polyisopropenyl (γ-chlorocrotyl)-urethane	161.59	1.5415	1.8560	38.369	81.90
Polyvinyl(cinnamyl)-urethane	203.24	1.5732	1.7165	57.867	118.73
Polyisopropenyl(allyl)-urethane	141.17	1.5752	1.7784	34.659	78.44
Polyvinyl(allyl)-urethane	127.14	1.5875	1.8218	30.050	68.14

first case the adhesive properties of the polymers should be higher than in the latter case. In fact, this assumption is supported by the measurements of the shear strength of adhesive joints of steel 3 bonded with urethane-containing polymers [1386], of which the first two are cyclic-chained, while the other two are linear [1387], MPa:[a]

polyisopropenyl(allyl)urethane	8.5−9.2
Polyvinyl(allyl)urethane	10.0−11.5
polyisopropenyl (γ-chlorocrotyl)urethane	18.5−19.0
polyvinyl(cinnamyl)urethane	15.0−16.0

a testing temperature, 298 K.

The polymers listed correspond to the requirements of model B by virtue of the urethane groups in addition to the double bonds and chlorine atoms (polyisopropenyl (γ-chloro-crotyl)urethane) and phenyl substituents (polyvinyl(cinnamyl)urethane) being all in the side chains.

The differences in composition of carbochain urethane-containing polymers make it possible to distinguish the physico-chemical characteristics governing the adhesive ability of the polymers. As in the case of amide-containing polymers these characteristics were derived according to the procedure described above. The respective data are compiled in

$\sum_i \Delta E_i^*$ $\left(\dfrac{cal}{mol}\right)$	$\sum_i \Delta E_i^s$ $\left(\dfrac{cal}{mol}\right)$	δ $\left(\dfrac{cal^{0.5}}{mol^{1.5}}\right)$	σ, mN/m		α	s	P_p^*
			From eq. (217)	From eq. (221)			
6704.1	77098.2	10.37	56.0	63.0	118 0	11 5	−4.9612
6704.1	77098.2	10.36	60.2	63.0	117.1	11.5	−5.0465
6704.1	77098.2	10.37	59.3	62.9	118.5	11.5	−4.9187
9943.0	138947.0	9.97	55.7	59.4	182.4	14.0	−4.4190
9943.0	138947.0	9.84	48.3	58.0	181.6	14.0	−3.4489
6253.3	76532.3	8.74	45.8	49.6	144.6	12.2	−3.1385
19731.5	139554.5	9.51	54.7	58.5	201.8	13.0	−4.2835
7492.8	77705.7	9.77	58.1	61.5	136.3	10.4	−4.7502
6846.7	69113.1	8.91	64.1	60.9	124.5	10.1	−5.6132

Table 4.8. Of particular note are the higher values for refractive indexes and surface energy of the synthesized polymers as compared to those of common polyurethanes. However, the absolute agreement between the calculated and experimental values for the strength of adhesive joints of polyisobutene, polychloroprene, and butadiene—acrylonitrile elastomer [1388], illustrated in Fig. 4.9, provides the most unambiguous proof of the validity of the theoretical concepts.

On the other hand, the objective character of the approach is supported by the high correlation coefficients between s and σ (−0.719), as well as s and α (0.875). For polyhexatrienes-1,3,5 and for the carbochain amido-containing polymers (Table 4.9) the signs of the coefficients were the same. As in the latter case, the equilibrium flexibility of macromolecular chains was found to decrease when cyclic fragments were introduced, the respective parameter being related to the strength of the adhesive joints via a rather close inverse relationship, for polyisobutene, polychloroprene, and butadiene—acrylonitrile elastomer the corresponding correlation coefficients are almost identical being −0.732, −0.733 and 0.721, respectively.

Phenol aldehyde oligomers of the vinyl acetylene series comprise a class of polymers matching model B. These constitute one of the most widely used groups of adhesives. With regard to the above concepts one may suggest that the introduction of the mobile

Table 4.9 — Paired correlation coefficients for different physico-chemical characteristics of carbochain amidogroup-containing adhesives

Characteristics				Peeling strength of the joint with:		
	σ	α	s	Polyiso-butene	Polychlo-roprene	SKN-40
α	−0.845	–	0.987	−0.851	−0.854	−0.860
s	−0.848	0.987	–	−0.855	−0.857	−0.863

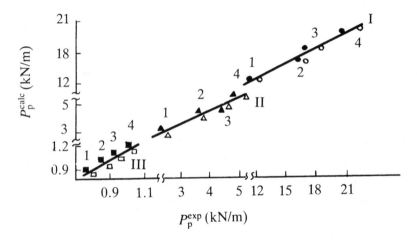

Fig. 4.9 — Correlation between calculated (open symbols) and experimental (solid symbols) peeling strengths of the adhesive joints of polyisobutene P118 (I), polychloroprene NT (II) and butadiene−acrylonitrile SKN-40 (III) elastomers bonded with polyisopropenyl (γ-chlorocrotyl)urethane (1), polyvinyl(cinnamyl)- (2), polyisopropenyl(allyl)- (3) and polyvinyl(allyl)urethanes (4).

unsaturated substituents into the side chains would result in a substantial rise in oligomer adhesiveness.

Raevsky *et al.* [1391] proposed an improved one-step procedure for the synthesis of *p*-dimethylvinylethynylphenolaldehyde oligomers, which, contrary to the known procedures [1389, 1390], leads to products in which alkene 'bridges' (spacers) between phenolic nuclei do not impose any steric hindrance to the mobility of vinylethynyl substituents [1393], as confirmed by IR-spectroscopic data [1392]. Consequently, these synthesized oligomers display better adhesive properties than conventional resins (excessive amounts of unreacted phenol are reduced by treating the condensation products with epichlorohydrin [1393]), as is illustrated by the data in Table 4.10. Note the effect of a substituent in the internucleic spacer on the adhesiveness of oligomers. The smaller the substitutent, the higher should be the chain mobility and, consequently, the strength of the adhesive joints. Indeed, adhesive joint strength is at a maximum in the case of formaldehyde oligomers, whereas it is minimal for those of butyric aldehyde

Table 4.10). On the other hand, in agreement with the basic theoretical concepts, introduction of double bonds, π-electron donors, produces a positive effect on the adhesiveness of the oligomers, as is exemplified in Table 4.10 for the condensation product of phenol and acrolein.

Table 4.10 — Shear strength of the adhesive joints of metals bonded with phenolaldehyde oligomers of the vinylacetylene series (numerator) or with their epoxy-derivatives (denominator), MPa

Aldehyde	Substrate	
	Steel 3	Aluminium alloy D-16
Formic	35–48/42–54	7–9/9–11
Acetic	30–35/36–42	5–7/6–9
Benzoic	20–25/24–30	3–5/5–6
Butyric	20–25/24–29	3–5/3–6
Acrolein	25–28/28–32	4–6/7–10

Phenol aldehyde oligomers in general, and those under discussion in particular, are rather stiff-chained, this feature being quite unfavourable in respect of the durability of the adhesive joints. To eliminate this drawback the oligomers discussed were blended with polychloroprene [1394] or butadiene–acrylonitrile elastomer [1395]. In the first case the temperature range at which rubber–metal adhesive joints were reliably performing was extended to 373 K (for the widely used analogue based on *p*-tert.butylphenol-formaldehyde oligomer the respective threshold does not exceed 333 K) [1396] and in the second case corresponding formulations (Table 4.11) appear to be efficient, as demonstrated in Table 4.12, in adhesive joints under a dynamic loading pattern. These features are of substantial practical importance [1396, 1397].

Table 4.11 — Butadiene–acrylonitrile-modified adhesive formulations based on epoxydimethyl vinylethynylphenol-formaldehyde oligomer

Formulation	Modifier content (mass %)		
	SKN-26-1	SKN-26	SKN-40
11	15	–	–
12	2	–	–
13	20	–	–
14	–	15	–
15	–	–	15

Table 4.12 – Strength properties of the adhesive joints of steel 3 bonded with modified formulations based on epoxydimethyl vinylethynylphenolformaldehyde oligomer

Characteristic of the joint	Formulation[a]				
	11	12	13	14	15
Shear strength, MPa					
at 298 K	41.5	52.0	34.8	41.0	47.5
at 473 K	13.5	11.8	6.0	13.5	11.5
Non-uniform loaded pull strength, MPa					
at 298K	5.8	6.4	9.0	5.5	7.2
at 473 K	1.4	1.9	3.7	1.4	2.4
Torsional shear strength, MPa					
10^3 cycles	38.5	37.8	38.6	38.1	39.2
10^4 cycles	37.3	37.1	37.5	37.0	38.1
10^5 cycles	35.5	34.6	34.9	35.4	36.0
10^6 cycles	34.5	33.2	34.8	34.2	34.5
Non-uniform loaded torsional strength, MPa					
10^3 cycles	40.5	39.7	41.8	40.0	39.2
10^4 cycles	39.0	37.5	40.0	38.4	38.7
10^5 cycles	39.0	38.1	40.3	38.7	38.1
10^6 cycles	37.0	35.9	37.7	37.3	34.9

a   Numbering as in Table 4.11.

Finally, ending this survey of polymers whose chemical nature fits the requirements discussed in section 4.1, it is necessary to note that to a very large extent these are met by compounds with free valences. Stable organic radicals should obviously be on the list of such compounds. Bruck [1398] developed procedures for the synthesis of such radical-containing polymers involving polymerization of the reaction products between methacrylic acid and 2,2,6,6-tetramethyl piperidine and their subsequent oxidation to N-oxides [1398]. In full agreement with the concepts developed, the polymers synthesized displayed better adhesive properties than the known non-radical analogues.

Thus, selecting the adhesive suggested by the models for adhesive polymers makes it possible to ensure a significant rise in the strength of the corresponding adhesive joints. These facts seem to offer additional confirmation of the validity of this approach to the control of adhesive ability. However, one should bear in mind that this deduction relates in the main to studies with model adhesives. In the case of commercial adhesives, which, in fact, are quite complex formulations, theoretical interpretation of the experimental results is hindered because it is impossible to distinguish the effects due to individual components of adhesive formulations on the performance of adhesive systems. Nevertheless, the satisfactory qualitative (and sometimes even quantitative) consistency between the calculated and experimental results persuades one that the concepts developed are quite applicable to the analysis of the properties and performance of adhesive joints abundantly reported in the literature.

One of the direct and explicit pieces of evidence supporting the validity of the models discussed for polymers to be used as adhesives involves the enhancement of the adhesive ability of a polymer resulting from the 'attachment' of terminal carboxy functions (A-groups in models A and B). Indeed, Hofrichter and McLaren as far back as in 1948 [1002] demonstrated that the corresponding dependence is described by a rather sharp exponential function:

$$P = x \, [\text{COOH}]^y \tag{500}$$

where y is a parameter determined by the bonding temperature and the nature of a substrate [1399]. Such effects were observed for a variety of joints comprising, for instance, cellophane, polyethene [1400], polyurethanes [1401] and polycarbonate glycols [1402]. The fact that it is the concentration of carboxy functions in polymers that determines the variation of their adhesive properties was directly confirmed by monitoring the emission processes accompanying the fracture of adhesive joints between glass and poly(butyl methacrylate-co-methacrylic acid) 90 : 5 [1403]. Varying the content of COOH-groups by associating them with different quaternary ammonium salts, Andreev et al. [1404] succeeded in showing the monotonic character of the respective concentration dependences (Fig. 4.10).

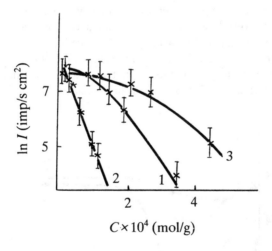

Fig. 4.10 – Emission of high-energy electrons accompanying the fracture of adhesive joints between glass and poly(butyl methacrylate-co-methacrylic acid) 95 : 5 versus concentration of $[(\text{CH}_3)_3\text{NC}_{12}\text{H}_{25}]^+\text{Cl}^-$ (1), $[(\text{CH}_3)_2\text{C}_{12}\text{H}_{25}\text{NCH}_2\text{CH}_2(\text{OH})]^+\text{Cl}^-$ (2), $[(\text{CH}_3)_3\text{NC}_{18}\text{H}_{37}]^+\text{CH}_3\text{COO}^-$ (3) added to associate with carboxy functions.

The optimal content of carboxy functions in polymers is around 3–10%. The upper limit is characteristic for polyalkenes [1405] and polyacrylates [1406], which Aubrey and Ginosatis [1406] associated with the maximum density of interfacial hydrogen bonds producing a 1.5-fold rise of the adhesive joint strength. Within this range C_{COOH} is linearly related to the dielectric permittivity of the samples, recalling eqs (207) and (208)

[493] , and the data of Fig. 2.31 [496]. The effect of COOH-groups is already discernible at the 3% concentration; however, when the polarity of the major component is higher this figure has to be significantly lowered. For instance, only 0.9% of acrylic acid is required to achieve maximum adhesive ability with chlorinated polypropene [1407]. Hence, one may infer that the carboxylation of non-polar or low-polarity polymers results predominantly in an intensification of the intraphase processes in adhesives (structurization) [1400], while with high-polarity polymers it is the interfacial processes of adhesive interaction that are intensified, as can be observed by examining the effects in polyethene [1400] and polyvinylbutyral [1408].

Let us elaborate on the last effect, which, in fact, is of greatest interest. In full agreement with the theoretical concepts, the concentration of carboxy groups in isotactic polypropene modified with poly(ethene-*co*-acrylic acid) 92 : 8 was shown to be linearly related to the logarithm of the surface energy of the system [1405]. In more complex materials the dependences are not that unequivocal and can be described by curves with inflections. For instance, as is illustrated in Fig. 4.11, this was the case observed in the studies of the effect of polyurethane ionomer content in a polyurethane network based on oligooxy-tetramethylene glycol and trimethylolpropanetolylene diisocyanate on the surface energy of the formulation. Lipatov *et al.* [1401] associates the observed pattern with the effect of internal stresses in the system. However, up to 7% content the dependence is found to be exponential, evidence of the prevalent role of the interfacial factors rather than of the intraphase ones. In fact, this was the pattern observed for the adhesive joint strength of carboxy-containing polymers versus carboxy function content (for instance, in [1405, 1406]) and moreover, such a pattern is in accord with eq. (500).

On the other hand, the data in Fig. 4.11 not only make it possible to reveal the purely thermodynamic reasons for the enhanced adhesiveness of COOH-containing polymers, they also provide for the analysis of the distinct mechanism of interfacial interaction involving, obviously, formation of hydrogen, or, in the limiting case, of ionic (salt) bonds.

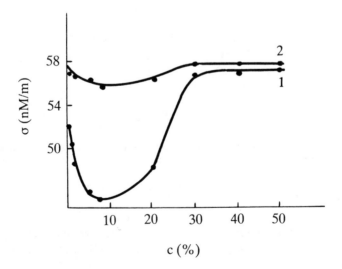

Fig. 4.11 – Surface energy of a urethane adhesive case in air (1) and on the glass surface (2) versus concentration of ionomer in adhesive.

Apparently, it is not always the case that the latter are formed. However, substitution of hydrogen atoms in COOH-groups during interaction between an adhesive and a substrate appears to be, perhaps, the most general and favourable mechanism for the control of interfacial interaction ensuring a regular rise of adhesive joint strength. Clearly, this conclusion is true primarily for metal–polymer joints, in which the formation of interfacial ionic (salt) bonds is most probable. This is supported by the data in Table 4.13, illustrating the effects of the cation in the ionomer and of the metallic substrate on the strength of the respective adhesive joints [1401].

Thus, carboxylation of polymers appears to be an important pathway leading to enhanced adhesive ability. However, in real COOH-containing systems under real service conditions one can anticipate the simultaneous effect of most of, if not all, of the factors listed. Hence, eq. (500) cannot be expected to hold true in practice and, consequently, to be suitable for forecasting the strength of adhesive joints.

Table 4.13 – The effects of the cation in polyurethane ionomer and of the metallic substrate on quasiequilibrium strength of adhesive joints (N/m)

Cation	Substrate		
	Steel	Aluminium alloy D-16	Brass
Li	52	40	55
Na	46	32	41
K	47	27	40

At the same time, the carboxy function is one but not the only representative of the series of oxygen-containing functional groups. Hence, oxidation of the samples should, obviously, result in a similar effect on the chemical nature of polymers and their adhesive ability. This approach is very common with substrates (see section 4.2.2), but there is no reason to ignore it in the case of adhesives. For instance, Roumyantsev et al. [1409] demonstrated that the peeling strength of the adhesive joints between steel and ethene–vinyl acetate copolymers is an explicit function of the degree of oxidation of the latter. The effect of oxygen-containing groups on adhesive properties is vividly illustrated by studies involving epoxy adhesives. These comprise hydroxy, ether (ester) and oxirane groups. In accord with stoichiometric considerations, their concentrations should be close to each other. However, according to X-ray photoelectron spectroscopic data, this inference is valid only for the bulk of the adhesive. In the transition layer, which is up to 10 nm thick, the content of oxirane groups is essentially greater than that of the hydroxy [1410] or of the ether groups, the ratio between the contents of these being 3 : 1 : 1 [1411]. This is the major reason which accounts for the enhanced adhesiveness of this type of adhesive. On the other hand, curing epoxy oligomers involves the oxirane groups, resulting in the total concentration of the hydroxy and ether groups being raised from 20 to 33% [1411, 1412] and, as a result, cured epoxides display poorer adhesive properties than the uncured ones.

When comparing adhesive ability within the series of methacrylate [1413] or of cyanocrylate [691] copolymers very clear and unambiguous results are obtained. The following illustrates the peeling strength of the adhesive joints of steel 3 with the co-polymers of methyl methacrylate and $H_2C=C(CH_3)-COO-CH_2R$ (10% mol. content) as a function of the nature of the radical R, N/m:

CH_3	170	CH_2CN	170
C_6H_5	350	$CH_2C_6H_4$	290
CF_3	90	$CH(OH)-CH_3$	260
CH_2Cl	220	$CH_2-O-C_2H_5$	360
CH_2Br	260	$CH_2N(CH_3)_2$	375[a]
CH_2OH	270	$CH_2NHC(CH_3)_3$	350[a]

a Comonomer content, 5% (mol).

It follows from these data that enhancement of the adhesive ability of the copolymers is associated with the increase of the number of carbon atoms in the radical containing unbonded electron pairs, as is the case, for instance, with halogen-, hydroxy-, phenyl- or nitrogen-containing methacrylates (even when their concentration is only 5% mol.), as well as with those with the ether group in the radical which increases the flexibility of polymeric chains. This latter effect is supported by the measurements of the shear strength of the adhesive joints of aluminium alloy D-16 bonded with cyanoacrylates of the general formula $H_2C=C(CN)-CO-OR$, MPa:

CH_3	18.8	C_4H_9	11.2
C_2H_5	16.7	C_5H_{11}	8.2
C_3H_7	13.4	$CH_2CH_2-O-C_4H_9$	9.0

Indeed, as the alkyl radical increases in length from the methyl to the *n*-amyl the shear strength P_{sh} decreases. The hexyl radical would lead to a still lower value of P_{sh}. However, when the oxygen atom is incorporated in the radical the resultant increase in the internal rotation in the chains leads to a reversal of the trend.

Such dependences are easily explained within the frame of reference of the theoretical concepts of adhesion discussed here. As applied to α-cyanoacrylates the respective analysis is given in section 2.2.2. We have made use of the same approach to interpret the variation of adhesive properties of the higher poly methacrylates described in [1414].[†]

To carry out the task, for polymeric octyl, decyl and dodecyl methacrylates, we have calculated the values of the surface energy, the equilibrium flexibility of macromolecular chains, and of the complex adhesion criterion by means of eqs (221), (324), and (356), respectively. The reported [1414] peeling and shear strengths were plotted versus the calculated quantities. The respective graphs are shown in Fig. 4.12. The smooth character of the graphs serves to confirm the validity of the basic concepts relating molecular

† Since the molecular weights of the examined polymethyl methacrylates are different, the values of P_{sh} [1414] were reduced by means of eq. (344) to the molecular weight of $M = 6000 - 8000$.

characteristics and the adhesive ability of polymers. Note that the shear strength of aluminium—aluminium joints shows a linear relationship for each of the three parameters examined. Such an effect is probably due to the stable composition and stable properties of the boundary layers of the substrate. For polyethene, which, as is known, exhibits a tendency to oxidation, it is only the P_{sh} versus the flexibility (parameter s) dependence that is linear (contrary to σ and α, flexibility is not susceptible to variations in composition of the polymer surface). The opposite and entirely consistent picture is observed for the adhesive joints of polyethene, in this case the P_{sh} versus σ and α graphs are linear, while it is the P_{sh} versus s plot that is non-linear. All of the described features, including the apparent discrepancies of the adhesive joint strength in peeling and in shear tests, are consistent with the concepts discussed in Sections 2 and 3.

On the basis of these principles the variation in the adhesive ability towards copper of polystyrene modified with various functional groups can be interpreted. While the initial non-modified polymer provides for a peeling strength of 0.07 kN/m, when the surface was modified with carboxy functions a peeling strength as high as 0.36 kN/m was obtained. Increasing the concentration of atoms with unshared electron pairs, i.e. when diazo-, hydroxydiazo- and amino-functional groups are introduced, then values of 0.84, 1.00 and 1.07 kN/m, respectively are found [1415]. It is also interesting to note that the surface energy of sulfonated polystyrene appears to be a direct function of the content of sulfogroups [1416], another source of atoms with unbonded electron pairs.

It is clear that other atoms similar to those discussed above have an analogous positive effect on the adhesive ability of polymers. Hence, it is not by chance that nitrogen- and halogen-containing adhesives are on the list of those most commonly used.

An increase in the nitrogen content results, in complete agreement with the ideas developed, in an increase of the surface energy of a polymer. For polyamides this effect amounts to 31.25% [1417]. Correlation of the adhesiveness of synthetic polyamino acids to metals with the chemical nature of polyamino acids revealed that the highest adhesiveness was demonstrated by poly-L-lysine (2.1 MPa), while the lowest by poly-glutamic acid and polycysteine [1418]. In view of the ideas discussed this result appears to be predictable, polylysine is the only polyamino acid of those examined that has two amino groups. It is, moreover, significant that of the 20 essential amino acids lysine is characterized by the highest isoelectric point (pI = 10.76) of all, except for arginine which contains four nitrogen atoms. This effect of nitrogen can be seen by the grafting of diisopropylazodicarboxylate on to polybutadiene resulting in σ_{sl}^{d} increasing to a level at which interfacial hydrogen bonds become feasible [1419]. Hamed and Shieh [1420] reported a similar effect of hydroazoether substituents which produced an increase in the autohesive joint strength of polybutadiene.

In such cases not only should the effect resulting from the rise of surface energy be taken into account but also that of the purely chemical reactivity of the samples modified with nitrogen-containing compounds. For instance, treatment of ethene-vinyl acetate copolymers with tolylenediisocyanate [1421] leads to a rise of polymer adhesiveness due both to incorporation of the amide groups and of the highly reactive isocyanate groups. The approach described is also valid for the curing components of adhesives, e.g. cyano-ethylation of diethylenetriamine results in enhanced efficiency of epoxy formulations including this component in bonding metals and glass [1422].

One of the most common monomers used for the synthesis of polymeric adhesives is acrylonitrile. Its grafting on to various polyalkenes, which are, as is known, the polymers

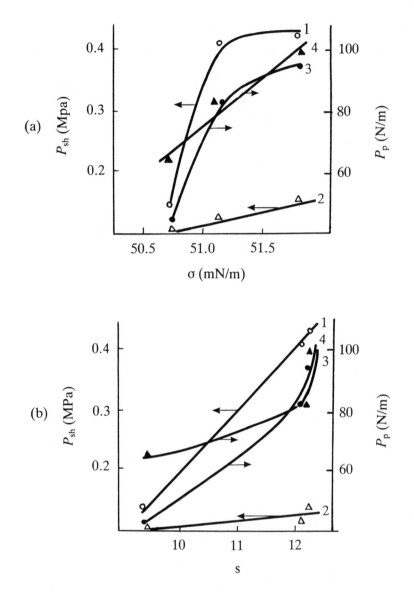

Fig. 4.12 – Shear strength of adhesive joints of polyethene (1), aluminium alloy D-16 (2), and fabric (3); peeling strength of adhesive joints between polyethene and fabric (4) bonded with polyoctyl- (POMA), polydecyl- (PDMA), and polydodecylmethacrylates (PDDMA) versus calculated values of surface energy (a), equilibrium flexibility of macromolecular chains (b) and complex criterion of adhesiveness (c) of the adhesives.

with low surface energy, produces a substantial rise in their adhesive ability. This effect was reported for low- and high-density polyethenes and for the copolymers of ethene with α-butene and hexene-1 [1423]. When acrylonitrile was copolymerized with propene [1424] the respective dependence was found to be linear, as is illustrated in Fig. 4.13. Fig. 4.14 demonstrates the strength of the adhesive joints of copper as a function of the

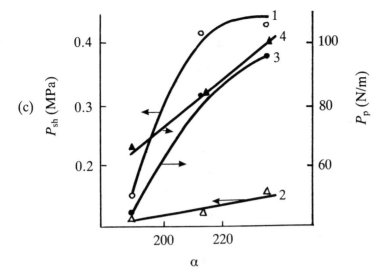

(c)

Fig. 4.12 (contd.)

chemical nature of polyalkenes and of the length of the grafted polyacrylonitrile chains (to be precise, of their molecular weight M_p). Note the inverse dependence between the size of the grafted chains and the adhesive ability of the resultant copolymers; for $M_p \leqslant 3000$ $P_p = 93$ N/m, while for $M_p > 10000$ $P_p = 65$ N/m, the content of grafted polyacrylonitrile being 35–40% and 20–27%, respectively. The effect is in full agreement with the concepts relating the adhesive properties of polymers to their molecular mass and discussed in sections 2.2.1 and 3.1.2. In the case considered the distinct mechanism of the effect of M_p on P_p is associated with the rise of the probability for intermacro-molecular interaction of nitrile groups with increasing length of the grafted chains and, consequently, with the ensuing decrease of the surface energy of polyalkenes.

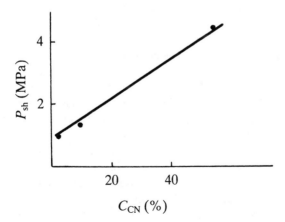

Fig. 4.13 – Shear strength of the adhesive joints of steel bonded with poly(ethyene-*co*-propene) versus content of acrylonitrile grafted to macromolecular chains of the adhesive.

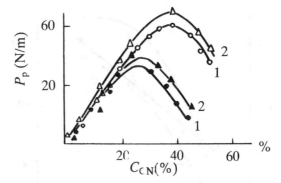

Fig. 4.14 – Peeling strength of the adhesive joints between copper and low-density (1) and high-density (2) polyethenes with grafted acrylonitrile versus concentration of the grafted comonomer (acrylonitrile) in adhesives. Molecular mass of the grafted chain. (○), 2900; (△), 3000; (●), 10800; (▲), 12300.

However, the most studied nitrile-containing adhesives are the butadiene–acrylonitrile copolymers. Their adhesive properties were thoroughly discussed in terms of the developed theoretical concepts in section 3. Here it will suffice to note that, besides the presence of nitrile functions, adhesiveness of such materials is largely determined by the double bonds as efficient donors of π-electrons. It is largely this feature that determines the effeciency of polydienes used as adhesives (along with the increased flexibility of macromolecular chains incorporating double bonds). Indeed, the work on peeling of polybutadiene from poly(ethene-*co*-propene) substrate showed the adhesive ability to be proportional to the double bond content of the adhesive [1425]. In this context and quite illustrative, are the results concerning the variation of the refractive index of poly-butadiene in the course of its intramolecular cyclization with titanium tetrachloride [1426]. Since, in keeping with the general features of the cyclization of diene elastomers [1427], this process is accompanied by an exponential decrease of the double bond content (Fig. 4.15), it would be reasonable to anticipate that the adhesive properties would also vary. If one adops surface energy to assess adhesive ability (and this is, as we have already demonstrated, quite correct), then, in conformity with section 2.2.1, a linear relationship should be observed between the double bond content and the refractive indices of cyclo-rubbers. Fig. 4.16 depicts the respective experimental curve, illustrating the validity of the above assumption.

Incorporation within the structure of a diene elastomer of functional groups comprising atoms with unbonded electron pairs appears to be the best prospect for enhancing their adhesiveness. In fact, in full agreement with this oligodienes containing nitrile, hydroxy, carboxy [1428], carbamide [1429], and, especially, allyl [1430], urethane [1431] and isocyanate [1432] functional groups are the ones which exhibit maximum adhesiveness. Macromolecules incorporating urethane or isocyanate groups within the backbone relate to model A of an adhesive polymer. Those (elastomers) containing urethane groups in side chains relate to model B, urethane functions providing for a substantial rise in the strength of the corresponding adhesive joints [1433]. The results of investigations of the donor properties of high molecular weight compounds confirm the inference that it is the atoms with unbonded electon pairs that provide the enhancement of adhesive ability. Shmourack [1434] suggested that, by using maleic anhydride as a typical electron acceptor, the decrease of electric resistivity in systems of

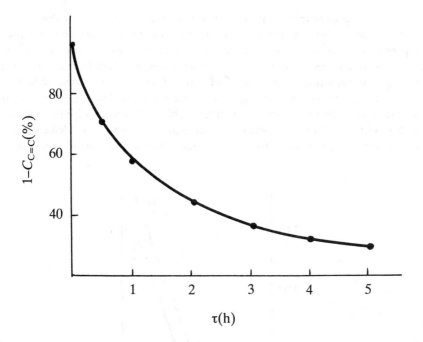

Fig. 4.15 – Variation of double bonds content in *cis*-polybutadiene-1,4 SKD in the course of cyclization caused by $TiCl_4$.

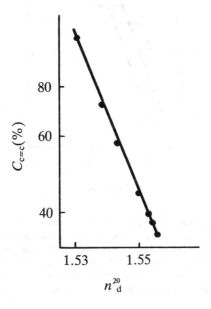

Fig. 4.16 – Refractive index versus double bonds content in cyclized *cis*-polybutadiene-1,4 SKD.

oxygen- or nitrogen-containing polydiene elastomers be used to assess the donor properties of these polymers. The substrate used with all of the elastomers examined was polyisoprene. The measured adhesive characteristics involved the peeling strength of vulcanizates and the covulcanization degree[†] (this was evaluated with the aid of radio-active tracers by measuring the amount of adhesive per unit substrate area remaining after the joint had been disbonded). In Fig. 4.17 these are plotted versus electric resistivity and, as can be seen, the graphs are strictly linear, which is in accord with that anticipated. Hence, the existence within the polymer structure of atoms with unbonded electron pairs may be considered a prerequisite for these polymers to display enhanced adhesive ability.

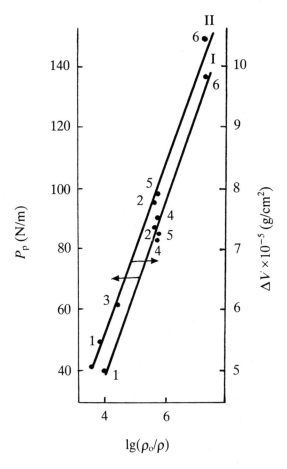

Fig. 4.17 – Peeling strength of adhesive joints (I) and covulcanization degree (II) of carboxylated SKD-1 (1) and epoxidized (SKDE) (2) butadiene elastomers, of butadiene–acrylonitrile (SKN-5A) (3), butadiene–diethylaminoethylmethacrylate (SKD-15E) (4), butadiene–(2-methyl-5-vinylpyridine) (DMVP-10C) (5), and butadiene–styrene–(2-vinyl-pyridine) (DSVP-15) (6) elastomers with polyisoprene SKI-3 versus decrease of elecric resistivity on interacting adhesives (1–6) with maleic anhydride.

---

† This characteristic causes one to recall the analogy between adhesive ability and chemical reactivity treated in section 4.1. This analogy is directly applicable to such systems of interacting objects.

Three general consequences follow from this proposition, first proposed in 1971 by Raevsky and Pritykin [695], though it is only now that its validity has been proved. These are firstly, a physically sound basis is created making it possible to apply the knowledge of charge transfer complexes, which are adequately developed in physical organic chemistry, to the problems of adhesive interaction; secondly, the ideas of Deryagin and co-workers [109, p. 66] concerning the donor series and those of Berlin and Basin [10, p. 42] concerning molecular complexing acquire the essential importance with regard to adhesion. Finally, the third and most important consequence and the subject of this monograph especially of the current section, is that the physicochemical requirements are revealed that enable enhancement of adhesive interaction by regulating the chemical nature of the contacting phases so as to directly affect their donor and acceptor properties.

Besides the nitrogen-containing groups, atoms of halogens, according to the ideas discussed in section 4.1, should also contribute much to adhesive ability. In fact, polychloroprene and the products of its halogenation, as well as chlorinated natural rubber (polyisoprene) are amongst the most common adhesives. On the other hand, the search for efficient modifiers for adhesives depends on certain principles which guide the choice of comonomers or on special additives with regard to their chemical nature, as discussed in section 4.1. In a number of cases complete identity of the chemical nature of a modifier with the one suggested by the theory was observed. For instance, adhesiveness of polychloroprene was improved by copolymerizing it with halogen-substituted hexatrienes, for example, with 2-methyl-4-chlorohexatriene-1,3,5 [1435]. Polymers of 1,1,2,3-tetrachloro-, 1.1.2.4.4-pentachloro- [1436], and, especially, of 1,1,2-trichlorobutadienes [1437] were shown to exhibit enhanced adhesive ability to rubbers, metals, and glasses. The effect of unbonded electron pairs in governing the donor–acceptor mechanism of interfacial interaction was verified in a number of studies, in which it was demonstrated, in particular, that the increase of electron donor properties of elastomers provides for their better adhesive ability [1438]. The validity of this approach can also be illustrated by way of the example describing the use of adhesives involving chlorinated polychloroprene performing as an electron acceptor and butadiene-(2-methyl-5-vinylpyridine) copolymer as the donor. In conformity with the basic concepts, introduction of the latter component provides an increase of the joint strength between steel and butadiene–acrylonitrile rubbers (1.7, 2.3, and 3.6-fold increase for SKN-40, SKN-26, and SKN-18, respectively), or polyisoprene SKI-3 (a 2.4-fold increase) [1439]. As is seen from Fig. 4.18, for the non-polar system comprising polyisoprene the graph of the joint strength versus the donor component content in the adhesive shows a linear pattern up to a certain threshold value, determined, together with other factors, by the nature of the solvent [1440].

Incorporation of atoms with unbonded electron pairs provides for the enhancement of adhesive ability of even such non-reactive polymers as polyethene. For instance, substitution of a single hydrogen atom in a methylene repeating unit by a polar group results in that σ_c rises from 31 mJ/m^2 to 39 mJ/m^2, when the polar group is the hydroxy function (polyvinyl alcohol), and 44 mJ/m^2 for the nitrile function (polyacrylonitrile) [1417]. Similar effects can be achieved by introducing methylester groups within the surface layers of polyethene [1441], or by modifying it (PE) with acrylated epoxy oligomer [1442].

This last result is of major significance since it serves to show that the basic theoretical

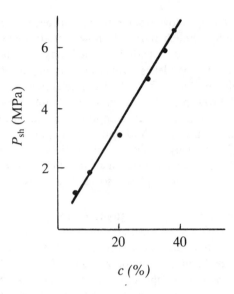

Fig. 4.18 – Shear strength of the adhesive joints between vulcanized rubber based on poly-isoprene SKI-3 and steel 3 versus content of butadiene-(2-methyl-5-vinylpyridine) copoly-mer SKMVP-15 in chlorinated polychloroprene adhesive.

considerations concerning the chemical nature of advanced adhesives are just as valid for the additives in adhesive formulations as for the major components. In fact, when even 2% of polyethene oxidized by γ-irradiation in air to the 0.013% content of carbonyl groups is added to a commercial non-modified product (PE) a 5-fold rise in the adhesive joint strength is observed with even such a metal as copper (difficult to bond to) [1443]. Taking into account the practical importance of such approaches, let us elaborate in more detail on the problem of selecting the major types of low-molecular mass additives liable to ensure enhanced adhesive ability of polymeric adhesive formulations.

Let us first examine the general features governing the physico-chemical relationships of the interactions between low- and high-molecular mass compounds. Such relationships are usually reported in technological experiments as semi-empirical dependences; however, these can be interpreted in terms of thermodynamics [1444]. For this purpose it is necessary to consider changes of the surface energy of the polymeric components of adhesive compositions. In fact, in accordance with eq. (151) the polar compounds should tend to migrate into the transition (in the limiting case into the boundary) layers of polymers. This was verified experimentally in studies of the migration of low-molecular weight modifiers into polyethene, polytetrafluoroethene and poly(ethyene-co-acrylic acid) (92 : 8) by X-ray photoelectron spectroscopy and ATR IR-spectroscopy [1445]. The time-variation of the surface energy of the modified phase was found to be dependent on the volumetric concentration of the modifier C_v and of the diffusion coefficient D, the ultimate expression being as follows [1446] :

$$\sigma_c = \sigma_{sa} - 2\,RTC_v\,(D\tau/\pi)^{0.5}.\qquad\qquad(501)$$

Note that the same expression describes the wettability of polyethene [1447] and of vinylidenechloride—acrylonitrile copolymer [1448] modified with aliphatic amides.

Calvert and Billingham [1032] suggested that one should also consider the equilibrium solubility S_* and the modifier elution rate

$$v_* = v_0 C_S / S_*, \tag{502}$$

where C_S is the surface concentration of an additive. Making use of the formalism of diffusion, with certain simplifications, the time during which 90%(mass) loss of the modifier ΔM from the polymer takes place, is as follows

$$\tau_{0.9} = 0.2 \, h^2/D \tag{503}$$

where h is the thickness of a polymer sample. Fig. 4.19 illustrates this relationship as observed in polyethene films in the course of migration of acrylate additives [1449].

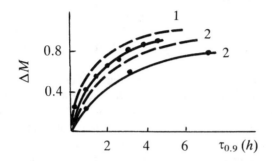

Fig. 4.19 – Migration kinetics of perfluoropropyl- (1) and perfluorobutylacrylates (2) from the 0.1 mm thick film of low-density polyethene. Dashed curve calculated according to eq. (503).

At the same time, in addition to the general case, two extreme situations should be considered as feasible. The first concerns the migration of additives in vacuum [1450], while the second relates to the more real performance conditions of adhesive joints under which not only the cohesion characteristics of polymers are important, but also the effect of external pressure. The situation is that, when the sample has been stored under pressure, e.g. at 0.4–0.5 MPa, and is then freed to the conditions of normal pressure, a modifier starts to migrate to the surface of the sample, whereas in 10 days the direction of the process is reversed [1451].

When examining the low-molecular weight components of adhesives, in the first instance curing agents and modifiers should be distinguished. Such a division relates to the type of interaction with the polymeric matrix, rather than the purpose for which the component is introduced (either curing an adhesive or modification of its properties, respectively). The molecules of curing agents are incorporated within the three-dimensional network structures during curing of the composition via covalent bonds. They should be chosen therefore, so as to be chemically reactive with regard to the polymeric component and to match the requirements discussed above for compounds designed for adhesives. Modifiers are distinguished in that they form with the components formulated as adhesives, polymeric in particular, a broad spectrum of bonds, which are not necessarily of the strongest type. Surface-active compounds (surfactants), various general-purpose additives such as plasticizers, promoters, stabilizers, antipyrenes etc., as well as fillers are the most common types of modifiers in use. Within the framework of the present treatise there is no need to examine the general-purpose additives since their effect on these principles; however, the procedure is the same as that of selecting a formation. On the contrary, choosing a filler is largely associated with considering its effect on these regularities; however, the procedure is the same as that of selecting a substrate. Therefore, we will discuss the related problems in the following section. Here it is reasonable to confine discussion to the effects of the structuring and the surface-active components of adhesive formulations.

As stressed above, besides matching the complex of adhesives design requirements, structuring agents should exhibit high reactivity. Isocyanates are the first to satisfy these requirements. However, the reported [84] isocyanate curing agents for adhesives lack atoms with unbonded electron pairs in their 'skeleton'. The only exception are the triiso-cyanate derivatives of biuret of the general formula HN(RNCO)–CO–N(RNCO)–CO–HN(RNCO), where R is the alkyl, aryl or alkaryl group. The atoms of nitrogen conjugated with carbonyl functions ensure an increase in adhesiveness, while the introduction of substituents R with a relatively large number of carbon atoms provides for the mobility of reactive groups and for the general flexibility of the final three-dimensional networks.

These compounds were synthesized by reacting 1,6-hexamethylene- or tolylene-2,4-diisocyanates with water [1452]. In the procedure suggested by Pritykin *et al.* [1452] the small amounts of unreacted diisocyanate and of the precipitated carbamide (0.5–1.0% and 4.5–5.0%, respectively) were not eliminated from the reaction mixture. The first performs as a modifier decreasing the interfacial surface energy and, conse-quently, stimulating the diffusion of the excessive amounts of the curing agent into the polymeric substrates being bonded, and the second behaves as an active filler. Hence, these impurities should demonstrate a positive effect on the adhesive properties of the synthesized systems denoted B-1 and B-2, respectively. These were developed to replace the most common triisocyanate, i.e. triphenylmethane triisocyanate (leukonat), which does not contain atoms of nitrogen with unshared electron pairs. The effect of the latter feature was shown in comparative studies on the efficiency of using each of the three triisocyanates for rubber–metal adhesive bonding in the course of the vulcanization [1453]. When using B-2 and, especially, B-1 to bond vulcanized rubbers a certain toughening of rubbers was observed (the effect is due to the diffusion of triisocyanates as well as of diisocyanates into the transition layers of substrates and the subsequent structuring [986] providing for the cohesive fracture of rubber–rubber and rubber–metal adhesive joints [1454].

High adhesiveness of B-1 and B-2 formulations is confirmed by the efficiency in structuring adhesives based on various polymers, for example, polychloroprene-based formulations (Table 4.14), particularly, those incorporating an alkylphenol–formalde-hyde modifier (Fig. 4.20), (chlorinated polychloroprene)-based, (natural rubber)-based formulations (Table 4.15) [1455], and compositions based on butyl rubber [1456]. However, the effect is most convincingly demonstrated in the case of polyurethane-based adhesives [1457] for high-efficiency bonding of synthetic leathers [1458, 1459] and fabrics [1460].

There is another trend in controlling adhesive ability permitting a physico-chemical interpretation and this involves the use of surfactants. Their introduction into adhesive formulations can either decrease [1461], or increase [1462] the strength of adhesive joints. One of the most common viewpoints concerning the nature of the effect produced

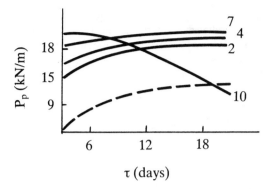

Fig. 4.20 — Peeling strength of the adhesive joints of steel 3 with butadiene–acrylonitrile (SKN-40) copolymer-based vulcanized rubber bonded with polychloroprene–phenol-formaldehyde adhesive modified with B-1 versus bonding time. Dashed line refers to the unmodified adhesive; figures denote concentration of modifier (%).

Table 4.14 — Pull strength of the adhesive joints produced with polychloroprene adhesives subjected to vulcanization, MPa

Substrate	Triisocyanate curing agent for the adhesive		
	B-1	B-2	Leukonat
Rubber base[a]			
Natural rubber	7.2	6.8	7.0
Polychloroprene elastomer			
(Nairit A)	8.5	8.1	7.9
(Nairit NT)	8.7	7.7	8.2
Butadiene–styrene elastomer			
SKS-30	7.5	7.3	5.9
Butadiene–acrylonitrile			
elastomer SKN-40	9.2	8.6	9.1
Metals[b]			
Steel HN-40	7.6	6.5	7.2
Brass	7.6	6.8	7.4
Aluminium alloy D-16	6.4	5.2	4.7

a   Autohesive joint.

b   Second substrate, vulcanized natural rubber.

Table 4.15 − Pull strength of the (steel 3)-rubber adhesive joints bonded with adhesives based on chlorinated elastomers subjected to vulcanization, MPa

Substrate base	Triisocyanate curing agent for the adhesive[a]		
	B-1	B-2	Leukonat
Natural rubber	6.8/6.4	6.9/6.4	7.5/7.2
Polychloroprene elastomer (Nairit NT)	8.0/6.8	8.2/8.2	6.3/6.7
Butadiene−styrene elastomer SKS-30	7.0/6.8	6.2/6.2	5.2/5.2
Butadiene−acrylonitrile elastomer SKN-40	9.5/8.9	9.2/9.0	8.6/7.4

a   Numerators refers to adhesives based on chlorinated natural rubber; denominator to those based on chlorinated polychloroprene.

by surfactants suggests that the molecules of the latter block the active sites at the substrate surface, resulting in decreased efficiency of the adhesive interaction. However, this does not exhaust all the possible effects of surfactants and it also neglects the effect of factors of a more general origin.

The analysis performed by Pritykin and Dranovsky [1463] demonstrated that adhesiveness can be improved either by introducing substantial amounts of surfactant into the bulk of the polymer phase, or by incorporating small amounts within the transition layers. In the first case, according to Gibbs' rule, surfactant migrates to the interface (recalling that the surface energy of a surfactant is always higher than that of even highly polar polymers [1464]) resulting, ultimately, in the formation of boundary layers of low cohesive strength — this feature is known to affect adhesive joint strength in an unfavourable way. To verify the validity of this suggestion we have studied the effects of the sodium alkylbenzenesulfonate modification of a polyvinyl chloride film. We have found that when the surfactant concentration exceeded a certain critical value, corresponding to total blocking of chlorine atoms at the substrate surface (as identified by IR-spectroscopy), the peeling strength of the systems fell even if a urethane adhesive, displaying donor properties with regard to the atoms of halogen, was used. The effect was reversed when the surfactant concentration was less than the critical value. Moreover, a 18−25% rise of adhesive joint strength due to the increase of the interfacial contact area was registered with the aid of the device described in Fig. 3.69. These results were confirmed by the data of Markov *et al.* [1465] who demonstrated that the highest adhesiveness of rubbers based on butadiene−styrene elastomer SKMS-30ARK was attained at a surfactant concentration which provides for maximum contact area between adhesive and substrate.

However, the most general approach must, obviously, rely on affecting the surface energies of distinct adhesive components by varying the chemical nature of surfactants. For this purpose, in accord with the data of section 4.1, atoms with unshared electron pairs should be incorporated into their structure. The assumption was verified by comparing two surfactants i.e. alkyl–benzenesulfonates derived from unsubstituted and chlorinated n-paraffins, which were used to modify polychloroprene adhesive formulations [1466]. As is seen from the data in Table 4.16, the use of a potentially more adhesive surfactant containing halogen atoms produces a 15–65% increase of the adhesive joint strength depending on the nature of substrate [1467].

Table 4.16 – Peeling strength, kN/m, of the adhesive joints produced with polychloroprene adhesives modified with sodium alkylbenzenesulfonates derived from unsubstituted (numerator) and chlorinated (denominator) n-paraffins

Substrate	Modifier content (%)		
	0.5	1.0	3.0
High-density polyethene,[a] after			
5 days	6.7/11.7	6.7/10.0	6.9/9.0
1 month	7.1/14.5	7.4/12.5	7.4/12.2
2 months	6.2/13.0	6.1/13.5	6.3/10.8
4 months	5.0/16.0	5.0/13.0	5.1/13.7
Glass-reinforced polyamide, after			
5 days	5.0/5.0	6.0/6.4	7.1/7.7
1 month	5.4/6.1	5.0/5.7	4.8/5.4
2 months	8.7/10.1	5.5/6.0	5.6/7.6
4 months	10.1/11.1	8.8/8.9	9.3/10.0
Glass-reinforced polycaproamide, after			
5 days	5.7/6.4	6.4/7.0	6.4/5.5
1 month	12.7/14.2	12.8/14.3	12.8/11.5
2 months	10.0/12.4	10.0/11.4	10.0/14.5
4 months	11.7/15.1	10.8/16.7	10.3/15.8

[a]　Pretreated with hot flame.

Further increase of the surface energy of a surfactant can lead to the modifier becoming reactive towards the polymeric component of the adhesive, whereas the critical micellar concentration within its (adhesive) phase is decreased, thus preventing the formation of weak boundary layers. The effect is achieved by introducing within the structure of surfactants either more atoms with unshared electron pairs, or double bonds as a source of π-electrons. The reactive surfactant (RSAC) of the first type is exemplified by oligoamidoamine [1468] and of the second by polymerizable surfactants [1469]. When oligoamidoamine reacts with the basic component of adhesive formulation macromolecular chains incorporating surface-active fragments are formed. As such macromolecules are accumulated in the system, the solubility of RSAC increases, finally resulting in destruction of their micelles at a rate proportional to the concentration of RSAC [1468, 1470].

As a result, surfactants and reactive surfactants should demonstrate different effects on the surface energy of modified polymers. This observation was confirmed for epoxy adhesives cured with polyethenepolyamine [1468]. As is seen from Fig. 4.21, the patterns of the dependences depicting the surface energy of modified adhesives versus concentration of either common surfactant modifier (oxyethylated ether of *p*-alkylphenol OP-10), or RSAC (oligoamidoamine) are essentially different with respect to the position of the maxima. For the first modifier the maximum corresponds to 6–7% content, whereas for the second the maximum is shifted to lower concentrations, viz. 1.5–2.0%, the curve being almost parallel to the abscissa at higher concentrations. It is quite important that while the introduction of the surfactant produces a 50% increase of σ_{la}, the use of RSAC results in a two-fold rise of the respective quantity (Fig. 4.21, a). In full conformity with the basic theoretical concepts, effects of this kind should also affect the adhesiveness of the formulations. In fact, the effect on the strength of adhesive joints with steel of increasing the concentration of surfactant follows a complex pattern, since in the case of RSAC the respective dependence is monotonic [1468, 1470], while the ultimate effect is twice as great (Fig. 4.21, b). Hence, reactive surfactants are amongst the most effective modifiers for polymeric adhesives [674, 1468, 1471].

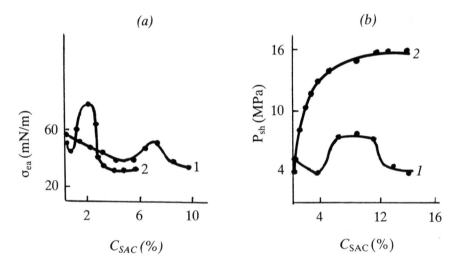

Fig. 4.21 – Surface energy of the epoxy-based (oligomer ED-20) adhesive cured with polyethenepolyamine (a); shear strength of the adhesive joints with steel 3 (b) versus concentration of modifier–*p*-alkylphenol (1) or oligoamidoamine L-19 (2).

There is one more consequence, important when making a choice of the common components in adhesive formulations, and which follows from the data presented above. Since surfactants cause changes in the conformational state of modified polymers, solvents should produce the same effect which would inevitably affect the adhesive characteristics of polymer samples. Investigations performed with the poly(methyl methacrylate)–glass system illustrate the validity of this proposition [1472]. By reverse phase gas-chromatography the films of adhesive PMMA layered on to the substrate (glass) from solutions in solvents with increasing Flory–Higgins parameter χ_1 (reflecting the worsening of the solvent quality) were shown to display a denser packing than those layered from solutions in good solvents. Fig. 4.22 depicts the respective dependences. The effect is due to the fact that in bad solvents macromolecules acquire more compact conformations and, therefore, a smaller number of contacts with polar groups on the sub-strate surface is feasible. Hence, the density of the phase of an adhesive on a solid surface will be determined by the ratio between the energies of intermacromolecular and inter-facial interactions; the more the latter quantity exceeds the former, the higher the corresponding packing coefficient. This trait appears to be quite important in view of the concepts concerning the surface properties of adhesives (section 2.2.1) and the interfacial contact area (section 3.2.2).

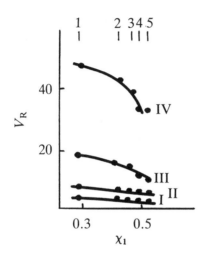

Fig. 4.22 – Sorption of chloroform (I), heptane (II0, toluene (III) and *m*-xylene (IV) by poly(methyl methacrylate) films at 317.5 K as a function of thermodynamic quality of the solvent from which the film was coated. 1, Dichloroethane; 2, toluene; 3, chloroform; 4, ethyl acetate; 5, butyl acetate.

Hence, the data presented demonstrate that the above approach to the improvement of adhesive ability of polymers can be of use in the selection of the basic (polymeric) component, as well as of the additional components of adhesive formulations. However, since, in the final analysis, such an assessment relies on the comparison of the strength properties of adhesive joints produced with the adhesives under consideration, one cannot avoid asking the question as to how the characteristics of adhesive joints are related to those of adhesives proper.

When undertaking such a comparison the strength properties of adhesives were described by ultimate tensile strength, elasticity modulus, and relative elongation ($P_{fracture}$, $E_{fracture}$ and ϵ_{rel}, respectively), while those of adhesive joints — by shear strength (P_{sh}) and peeling strength (P_p). Table 4.17 compiles the respective paired correlation coefficients, three major types of adhesives (plastomeric, and elastomeric, structurized with reacto- and thermoplasts) and three major types of substrates of different susceptibility to deformation (soft, semi-rigid and rigid) being involved in the study. As is seen, the coefficients are rather high, in some cases even getting close to 1.000 [1473] (this deduction is valid, on the one hand, for cohesively fractured systems; for those undergoing adhesive fracture, the deformative–relaxational properties of

Table 4.17 — Paired correlation coefficients[a] of the basic strength characteristics of adhesives with the shear (numerator) and peeling (denominator) strengths of adhesive joints

Adhesive and its strength characteristics		Substrate		
Composition	Parameter	Polyvinyl-chloride	Cellulose	Cellophane
Polychloroprene elastomer + p-tert.butyl-phenolformaldehyde oligomer	P_{fr}	-0.650 ± 0.055 / -0.547 ± 0.067	-0.668 ± 0.053 / -0.601 ± 0.061	-0.035 ± 0.095 / -0.319 ± 0.086
	E_{fr}	-0.883 ± 0.021 / -0.853 ± 0.026	-0.933 ± 0.012 / -0.884 ± 0.021	-0.485 ± 0.073 / -0.714 ± 0.047
	ϵ_{rel}	0.753 ± 0.041 / 0.860 ± 0.025	0.936 ± 0.012 / 0.936 ± 0.012	0.880 ± 0.022 / 0.918 ± 0.015
Butadiene—acrylo-nitrile elastomer SKN-40 + polyvinylchloride	P_{fr}	0.881 ± 0.027 / 0.822 ± 0.039	0.843 ± 0.034 / 0.858 ± 0.031	0.841 ± 0.035 / 0.988 ± 0.003
	E_{fr}	0.841 ± 0.035 / 0.788 ± 0.045	0.804 ± 0.042 / 0.820 ± 0.039	0.791 ± 0.045 / 0.983 ± 0.004
	ϵ_{rel}	-0.878 ± 0.023 / -0.891 ± 0.027	-0.963 ± 0.009 / -0.919 ± 0.019	-0.898 ± 0.023 / -0.721 ± 0.057
Polyvinylchloride + dibutylphthalate (plasticizer)	P_{fr}	0.981 ± 0.006 / 0.970 ± 0.010	-0.999 ± 0.000 / -0.893 ± 0.033	-0.843 ± 0.047 / -0.958 ± 0.013
	E_{fr}	0.890 ± 0.034 / 0.917 ± 0.026	-0.950 ± 0.016 / -0.758 ± 0.070	-0.956 ± 0.014 / -0.936 ± 0.020
	ϵ_{rel}	-0.940 ± 0.019 / -0.882 ± 0.031	0.881 ± 0.036 / 0.840 ± 0.048	0.596 ± 0.105 / 0.829 ± 0.051

a　Mean square error $\Delta r = N^{-0.5}(1 - r^2)$; N — number of measurements.

adhesives should be taken into account [1317]). Hence, the contradictions that might arise (when considering the concepts of physical chemistry and mechanics as applied to adhesive joints), are not ones that cannot be overcome. Moreover, the former are to be regarded as providing reliable grounds for the control of the strength properties of the joints governed by the generalities of adhesion [1346, 1474].

Summing up, the various aspects of the theory of interfacial interaction of condensed phases, here described, offers a non-contradictory and fruitful prospective basis for the analysis of relationships between the chemical nature and adhesive ability of polymers, thus creating a basis for the controlled improvement of the latter.

4.2.2 Chemical nature of the substrate surface

The concepts described in section 4.1 provide consistent grounds making it possible to elaborate also on the problem of improving adhesive ability of substrates. This aspect of the problem, as it relates to that of increasing adhesive joint strength, has only recently begun to receive adequate attention comparable to the concern shown in activities associated with developing new adhesives. The progress achieved in the field is beyond doubt; however, contemporary reviews (see, for instance, [1475–1478; 1479, pp. 91–121]) predominantly treat the technological aspects of the problem. At the same time, one must bear in mind the view stressed by Vol'kenshtein [1480]: 'the foundation on which rests surface science implies that the properties and performance of the surface are governed by both what is on its one side and what is on the other'. In other words, from the physico-chemical point of view the efficiency of adhesive interaction is associated with the effect of the chemical nature and of the state of the substrate surface to no lesser an extent than with the respective characteristics of the adhesives.

Analysis of the variety of experimental data makes it possible to reduce the many procedures reported for the pretreatment of substrates for adhesive bonding to two types, namely activation of the substrate or its modification [1481]. The first kind of procedure is aimed at changing the morphology and energy state of the substrate surface without affecting its chemical nature to any significant extent. On the contrary, modification involves incorporation of various functional groups into boundary and transition layers of the substrates. Table 4.18 gives a general classification of these methods. Besides the known methods, the table also includes those that have not yet been of concern in bonding, application of coatings, sealants and related technologies [1482]. Detailed analysis of each of the listed methods is beyond the scope of this book, hence here we will elaborate on those of only major implication, paying special attention to its physico-chemical aspect, rather than to the technological.

Among the reduction methods the procedure involving the treatment of substrates with alkali metals is the most common, despite the difficulties in carrying it out even when the reducing agent is used in an organic solvent (for example, in tetrahydrofuran [1483–1486]), or in liquid ammonia [1487]. Apparently, such active procedures cannot be avoided when pretreating polymers of low polarity, like fluorocarbon polymers. The treatment is accompanied by defluorination, resulting in formation of double bonds. These are liable to produce at least two effects. First, reacting with the oxygen in air they give rise to the formation of oxygen-containing functional groups [1484, 1487]. In this case the procedure may be considered as one producing an oxidizing effect on the substrate (note that treating the substrate with a mixture of nitric and chloric acids removes the oxidized layer thus decreasing the adhesive ability of fluoroplastics to the initial level [901, section 9.6.2]). Second, sorption of water by unsaturated compounds leads to the formation of 'bridges' between isolated π-conjugated blocks [1486], resulting in the formation of charge-transfer complexes. Hence, the systems become electroconductive, the semi-logarithmic plot of electroconductivity versus humidity (in the range from 0 to 70%) showing a linear relationship [1486].

Table 4.18 – Classification of major substrate-pretreatment procedures prior to adhesive bonding

Activation

| Mechanical | Abrasive materials |
| | Abrasive materials in transferring medium |

Solvational	Inert solvent
	Active solvent
	Intermediate solvent

| Vacuuming | Without elimination of migrating products |
| | With elimination of migrating products |

| Radiation | Electromagnetic radiation |
| | Bombardment with ionic beam |

Activation involving the complex effect of different factors

Modification

Reduction	Inorganic reducing agents	Salts
		Complexes
	Organic reducing agents	Individual compounds
		Complexes
	Alkali metals	Solutions
		Melts
		Complexes
	Electro chemical	In the presence of reducing agent
		In the absence of reducing agent

Oxidation	Inorganic oxidizing agents	Peroxides
		Acids and salts
		Complexes
	Organic oxidizing agents	Oxygen-containing
		Oxygen-free
	Electric discharge	In oxygen-free medium
		In the medium with oxygen
		In plasma

Table 4.18 (contd.)

Adsorption	Physical adsorption	Functionally substituted adsorbates
		Polyradical adsorbates
	Chemosorption	Addition
		Substitution
Grafting	Addition	Radical processes
		Ionic processes
	Substitution	One-stage processes
		Multi-stage processes
Structuring	Addition	Radical processes
		Ionic processes
	Substitution	One-stage processes
		Multi-stage processes
	Irradiation	Electromagnetic radiation
		Bombardment with ionic beam
Migration from the phase	Low-molecular products	One-stage processes
		Multi-stage processes
	High-molecular products	Individual modifiers
		Modifying mixtures

Modification involving the complex effect of different factors

Another effect produced by alkali metal-treatment of polymers involves substantial development of their surface resulting from destructive processes. And, finally, by the overall effect of the oxidation-reduction processes and destruction, free-radical states were shown to be generated [1483].

In view of the ideas discussed in section 4.1, it is clear that all of these effects favour the enhancement of the adhesive ability of polymers. However, the detrimental performance conditions of the alkali metal-treatment procedures demand the obvious need to use other metals and, more generally, other reducing methods. For example, Baumhardt-Neto et al. [1488] suggested a procedure involving the surface treatment of polytetrafluoroethene with a 10% solution of iron pentacarbonyl in butanol and subsequent conversion of the metal to oxides by reaction with acidic or alkaline permanganate systems. Iron and manganese oxides are primarily incorporated within the amorphous regions of the substrate, substantially altering the contact angles on wetting. An electrochemical method is used for pre-bonding treatment of polytetrafluoroethene. This involves cathodic reduction of the surface layers of the polymer brought into contact with a platinum element in an aprotic solvent (dimethylformamide) containing inert electrolyte (tetrabutylammoniumtetrafluoroborate) [1489] and was shown to be comparable in efficiency with the procedure involving sodium-naphthalene complexes. The modified phase, up to 40 μm thick, formed during the treatment demonstrates adequate adhesive ability, thus enabling the use of common adhesives.

Oxidation with mineral acids and their salts can be observed as a classical method of affecting the chemical nature of polymer surfaces. Comparison of the efficiency of grinding, degreasing in a vapour bath and acidic etching procedures for substrate pre-treatment revealed the superiority of the latter [1490]. The same conclusion was shown to be true for mechanical treatments with emery-paper and sandblasting, treatments with flame, plasma or dichromate solution in sulfuric acid at 339–344 K [1491]. Further-more, the maximum strength of epoxy-based composites reinforced with carbon fibres was achieved when, prior to interaction with epoxy adhesives, the carbon fibres were treated in succession with potassium nitrate and sulfuric acid [1492]. Subjecting poly-ethene to the direct effect of sulfuric acid within 3 h resulted in the introduction of one sulfo-group per 60 repeat units of the macromolecules, while maximum content of the groups was attained, at room temperature, in 20 h [1493] and as a result the critical surface tension of the substrate is substantially increased. However, such treatment encourages the development of destructive processes rather than producing modification *per se*. However, under more mild conditions, such as treatment of polyalkenes with a mixture of SO_2 in air, produces a similar improvement in adhesive ability [1494].

In practice the effect of acidic treatment is enhanced by chromium compounds added to sulfuric acid. For instance, the combined action of the compounds of Cr(VI) and $K_2Mn_2O_7$ (as another oxidizer) was shown to produce a 2.5–3.0-fold increase in the strength of adhesive joints between polyethene and metals [1495] and the use of potassium or sodium dichromate [1496], or, indeed, of chromium Cr(VII) oxide [1497] was shown to increase the adhesive ability of even polytetrafluoroethene. Exposure of polyethene to the effect of saturated sodium dichromate solution in sulfuric acid was found to increase the thermodynamic work of adhesion by six times [1498], whereas in 10 min at 343–368 K the surface energy increased from 32 mN/m to 72 mN/m [1499]. The adhesive strength of electroplated butadiene–acrylonitrile–styrene copolymers was demonstrated to be proportional to the square root of the rate of oxidation etching with saturated dichromate solution in sulfuric acid [1500].

To investigate the mechanism by which such treatments affect adhesive ability Briggs [1501] examined the accompanying changes in polyalkenes by X-ray photoelectron spectroscopy. In Fig. 4.23 we have compared the data of Briggs [1501] concerning the concentrations of the atoms of carbon, oxygen and sulfur, as well as of the sulfo-substituted carbon atoms, with the strength of adhesive joints bonded with epoxy adhesives as dependent on the exposure time of the substrate to the effect of the oxidizing medium. As can be seen, the P_{sh} and O/S ratio might be assumed to be most closely related since both $P_{sh}(\tau)$ and $O/S(\tau)$ have very similar, but inverse, dependences. However, this is not the case with sulfo-groups, for which there is no such dependence. Hence, we believe that the effects of dichromate solution in sulfuric acid are due to the atoms of oxygen other than those comprising HSO_3-groups. Indeed, as is seen in Fig. 4.24, the ratio of the total number of carbon atoms to that of oxygen not associated with sulfur is linearly related to the strength of adhesive joints. The validity of this conclusion was supported by investigations of the kinetics of the reaction between polyethene and

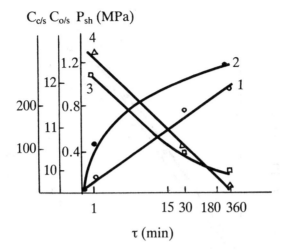

Fig. 4.23 – Shear strength of the adhesive joints of low-density polyethene (1) and poly-propene (2); the ratios between the concentrations of the atoms of carbon and sulfur (3), and carbon and oxygen (4) versus exposure time of the substrate to $H_2SO_4 - K_2Cr_2O_7 - H_2O$ (150 : 7 : 12).

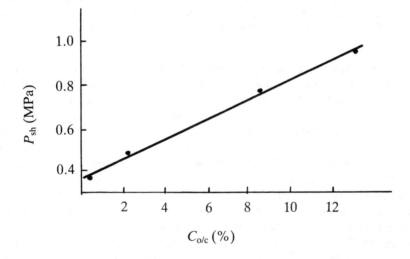

Fig. 4.24 – Shear strength of the adhesive joints of low-density polyethene treated with dichromate solution in sulfuric acid versus ratio between the number of oxygen atoms other than those of sulfo-groups and the total number of carbon atoms. Adhesive, epoxy-based formulation.

an aqueous sulfur-free solution of CrO_3. Both the strength of the adhesive joints and the intensity of the signal induced by the atoms of oxygen [1502] were studied. As is seen in Fig. 4.25, both curves are almost identically affected by the treatment.

Two consequences that are significant follow when choosing the types of acidic treatment for polymers. Firstly, since the process involves oxidation reactions, the

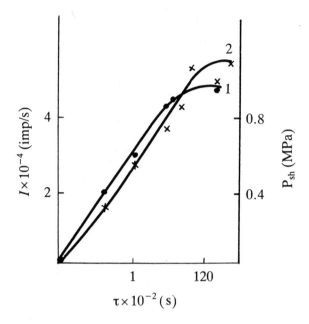

Fig. 4.25 – Shear strength of the adhesive joints of low-density polyethene bonded with epoxy-based formulation (1); intensity of the signal from oxygen atoms O 1 s (2) versus time of exposure of the substrate to an aqueous solution of CrO_3 at 298 K.

presence of inhibiting impurities should affect the course of acidic treatment. Water is known to be the most common source of inhibiting compounds. For instance, using CrO_3 solution in triple distilled water instead of monodistillate [1502] produces an increase in adhesive joint strength and, moreover, ensures a more rapid treatment leading to attainment of the maximum stable level, as is illustrated in Fig. 4.26. Secondly, the oxidizing efficiency appears to be a function of the degree of substitution of the carbon atoms in the macromolecular backbone. Hence, polyethene and polypropene should differ in their response to acidic treatment. Indeed, under the same etching conditions the oxidized transition layer in polypropene is substantially thinner than that in polyethene. In the first case the formation of this layer proceeds more rapidly, whereas in the second its thickness gradually increases as a function of the exposure time [901, section 9.6.2]. In fact, Piiroja [1503] demonstrated that for polyethenes of different density the depth of the zone involved in oxidation by sodium dichromate solution in sulfuric acid was linearly related to the peeling strength of autohesive joints, except for the regions encompassing boundary layers up to 40–50 μm thick. However, it is the boundary layers that are most likely to be affected by the oxidizers. Therefore, treatment for 2 min. suffices to etch through the boundary layers, whereas it takes 6–8 min. to oxidize the polymer 70–80 μm deeper, the rate of this process increasing with the increase of the double bond content in the polymer, the overall depth of the oxidized zone remaining the same.

The data presented demonstrate that oxidation with mineral acids generates a complex succession of transformations resulting in the formation of hydroxy, carbonyl (aldehyde, or ketone) and carboxy functions [1504, 1505], as well as hydroperoxide [1506], within

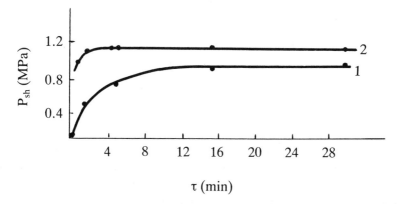

Fig. 4.26 – Shear strength of the adhesive joints of low-density polyethene versus time of exposure of the substrate to the effect of CrO_3 solutions in mono- (1) and tridistilled (2) water.

the surface layers of polymeric substrates. For instance, treatment of polyalkenes with chromous anhydride (as the most active component of the aqueous modifying mixture[†]) proceeds via the addition of chromic acid to the tertiary carbon atoms and subsequent hydrolysis of intermediate compounds to the products containing keto functions [1504]:

$$
\cdots-CH_2-\overset{\overset{R}{|}}{CH}-CH_2-\cdots \xrightarrow{HCrO_3} \cdots-CH_2-\overset{\overset{R}{|}}{\underset{\underset{Cr(IV)-(OH)_3}{|}}{\underset{O}{|}}}{C}-CH_2-\cdots \xrightarrow{H_2O} \cdots-CH_2-\overset{\overset{R}{|}}{\underset{\underset{OH}{|}}{C}}-CH_2-\cdots \longrightarrow
$$

$$
\longrightarrow \cdots-\overset{\overset{H}{|}}{\underset{\underset{O}{||}}{C}} \; ; \; \cdots-CH_2-\overset{\overset{R}{|}}{\underset{\underset{O}{||}}{C}} \; ; \; \cdots-CH_2-\overset{\overset{O}{||}}{\underset{\underset{CH_2}{||}}{C}} -\cdots \tag{504}
$$

The mechanism of this process, as examined by X-ray photoelectron spectroscopy [901, section 9.6.2], was found to proceed via stages involving the variation of the valency (degree of oxidation) of the initial atoms of chromium

$$R_3CH + H_2CrO_4 \longrightarrow R_3C-OH + Cr(IV) \tag{505}$$

$$R_2CH_2 + H_2CrO_4 \longrightarrow R_2CH-OH + Cr(IV) \tag{506}$$

$$Cr(VI) + Cr(IV) \longrightarrow 2\,Cr(V) \tag{507}$$

$$R_2CH-OH + Cr(VI)/Cr(V) \longrightarrow R_2C{=}0 + Cr(IV)/Cr(III) \tag{508}$$

---

† Sokolov *et al.* [1507] describes a peculiar modification of the procedure involving oxidation of the surface of polyethene during its contact with metals on to which 2–4 chromium trioxide monolayers were applied by a molecular coating technique. When glass was pretreated by this method its interaction with epoxy oligomers was shown to proceed via formation of π-complexes, as disclosed by means of electron spectroscopy of diffuse reflection [1508].

and complex formation resulting in the product of the structure

$$\ldots -CH_2-C(CH_3)-CH_2-\ldots$$
$$|$$
$$O$$
$$|$$
$$(HO)_2-Cr(III)\,(H_2O)_n$$

Briggs [901, section 9.6.2] calculated that a maximum of 3 atoms of chromium were available per 1000 carbon atoms. In low-density polyethene the number of branching points is an order of magnitude higher (2–3 per 100); however, they are non-uniformly distributed between the amorphous phase and the less accessible crystalline phase. Hence, the complexes comprising Cr(III) are predominantly concentrated within the boundary and transition layers of polyethene[†] and their hydrolysis should result in the formation of hydroxy groups and recovery of the chromic acid:

$$\overset{|}{\underset{|}{-C}}-O-Cr(III) \xrightarrow{H_3O^+} \overset{|}{\underset{|}{-C}}-OH + HO-Cr(III). \qquad (509)$$

As a result of the consecutively-parallel reactions described by eqs (505)–(509) the dependency of adhesive joint strength of polymers pretreated with sodium dichromate solution in sulfuric acid on the duration of the pretreatment (i.e. exposure time to the effect of the oxidizing solution) give rise to a complicated pattern exemplified in Fig. 4.27 for the polyethene–poly(vinyl chloride) system [1477]. Hence, mathematical

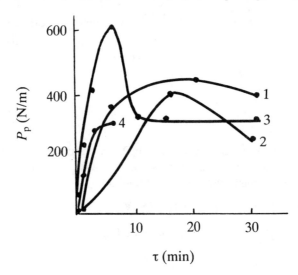

Fig. 4.27 – Peeling strength of the adhesive joints of low-density polyethene with poly(vinyl chloride) versus time of exposure of the substrate to the effect of dichromate solution in sulfuric acid at 1, 293 K, 2, 313 K, 3, 333 K, and 4, 353 K.

---

† When complexes of other metals are involved, insoluble (non-hydrolyzable) under specified conditions are formed and an inverse effect is observed. Golander and Sultan [1509] reported the relevant facts in a study treating the surface of polyethene successively with acidic (H_2SO_4) solution of $KMnO_4$ and solutions of $CaCl_2$, $BaCl_2$ or $ZnCl_2$.

simulation of the pretreatment process, even in the case of simple systems $CrO_3.H_2SO_4.H_2O$, is confined to adjustment of the regression equations [1510] and lacks deep physical meaning. However, from the purely applied aspect the value of such equations is beyond doubt. In particular, the relative contribution of individual character-istics of acidic solutions and of the technological parameters of the treatment procedure to the ultimate adhesive joint strength can be evaluated with their aid. For instance, Piiroja [1503] demonstrated that the peeling strength ($[P_p] = N/m$) of the adhesive joints of low- and high-density polyethenes is given as the function of the density ρ of the polymer, concentrations of the active components of the sodium dichromate solution in sulfuric acid ($C_{H_2SO_4} = 60-85\%$ and $C_{CrO_3} = 2-5\%$), temperature ($T = 60-90°C$) and duration of the treatment ($\tau < 10$ min)

$$P_p = 0.2326 \, \rho^{-10.55} \, C_{H_2SO_4}^{0.2419} \, C_{CrO_3}^{0.3485} \, T^{1.03} \, \tau^{0.7251} + 68. \tag{510}$$

One may infer from this equation that the properties of the polymeric substrate and the duration of the treatment are the features of primary influence on the ultimate efficiency of the procedure, whereas the effects associated with concentrations of active components are less pronounced and are actually identical for both components. Note that eq. (510) assumes a linear temperature dependence for the strength of adhesive joints and, together with the data discussed above (for example, those in Fig. 4.27) this fact stresses the limited value of the empirical approach.

When discussing the mechanism of substrate oxidation one must remember that not only the main chains, but also the side chains of macromolecules can be subjected to attack by acids. This aspect is important for polymers that are more complex than the presently discussed polyalkenes. For instance, treatment of butadiene−acrylonitrile elastomers with dilute acid solutions was found to proceed via hydrolysis of nitrile functions to carboxy groups [1511]. As a result the strength of their adhesive joints produced with epoxy-based formulations was observed to increase, as well as the durability of the joints in saline solutions (at temperatures up to 323 K).

At the same time, the procedures and systems described are not the only means of introducing atoms of oxygen into the structure of polymeric substrates. At elevated temperatures air appears to be the most common oxidizer. For instance, the adhesive ability of extruded polyethene was found to increase in proportion to the concentration of oxygen atoms in surface layers [1512]. Using specific reactions of the hydroxy functions with trifluoroacetic anhydride and of carbonyl functions with pentafluoro-phenylhydrazine Delamar, Zeggane and Dubois [1512] detected that it was the hydroxy functions at 578 K, and hydroxy and carbonyl functions at 698 K, that determined the oxygen content in surface layers. At the same time, no carboxy functions were detected (pentafluorobenzylbromide was used as the specific reagent), and hence direct oxidation on heating in air appears to be less efficient than the treatment described with acid−salt systems. In the absence of oxygen (for example, when polymers are heated under inert protective liquids) heating results in the opposite effect. For instance, thermal treatment of high-density polyethene and polycarbonate at 370 K in oxygen-free paraffin oil produces, respectively, an 18 and 3.6% drop of the surface energy, predominantly, owing to the decrease in σ_s^p [1513]. On the other hand, the adhesive ability of polymers is increased when oxidation is promoted by introducing the necessary co-catalysts. Permanganate oxidizers, free halogens, hypochlorites, metaperiodates and trichlorosym-triazinetrion [1514] are used as such promoters substantially increasing bonding efficiency even of such difficult meterials to bond as fibre glass reinforced epoxy-based plastics and copper.

A variety of compounds can be used as oxidizers. For instance, ozone, at a concentration in air of 0.01–0.1%, gives a 1.6–1.7 times increase in the adhesive ability of the rather inert carbon fibres towards the matrix-forming binders of corresponding composites [1515]. Treating the reinforcing fibres with oxidizers appears to be a trend of general applicability; however, oxidation with acids is a more technologically sound procedure than that involving ozone [799, 1516, 1517]. Indeed, treatment of carbon fibres ($\sigma_c = 40$ mN/m) with nitric acid for 0.5, 3 and 18 h results in an increase in σ_c and, consequently, so does the strength of the epoxy-plastics reinforced with these fibres (50–60% fibre content), the respective increase of the joint strength being 34.8%, 56.5% and 142.2% [1518]. Similar regularities were found when oxidation of boron fibres was investigated [1518, 1519] and Fig. 4.28 shows the observed relationship between the adhesive joint strength and critical surface tension of the fibres.†

However, hot-flame treatment of the substrate surface is still the simplest and most widely used procedure associated with oxidation of the substrate in air. Depending on the ratio between propane (prop) and oxygen or air (ox) the flame can be neutral (which is a rare case), or it can have either a reductive ($V_{ox}/V_{prop} = 17$–18), or, and this is the most common case, an oxidative effect [1522]. The general mechanism of the oxidation of polyalkenes involves the following stages [1523, 1514]:

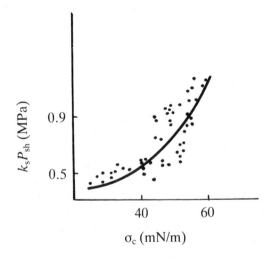

Fig. 4.28 – $k_s P_{sh}$ versus critical surface tension of the substrate (boron fibres) treated with HNO$_3$, where P_{sh} is the shear strength of the adhesive joints between boron fibres and epoxy-polyisocyanate matrix, k_s is the coefficient describing the completeness of the interfacial contact.

---

† Specific surface area of the boron fibres is not affected by treatment with nitric acid [1520], contrary to the case with the carbon fibres for which an 18-hour treatment results in the 2-fold rise of the roughness coefficient [1518]. However, for the carbon fibres the ratio between the number of generated carboxy functions and specific surface area is also proportional to the duration of the oxidative treatment [1521].

$$\dots -CH_2-CH_2-\dots \; + R^{\cdot} \longrightarrow \dots -CH-CH^2-\dots \; + RH$$

$$\downarrow O_2$$

$$\dots -CH-CH_2-\dots$$
$$O-O^{\cdot}$$

$$\dots -CH=CH-\dots \; + \; ^{\cdot}HO_2; \quad \dots -C-CH_2-\dots \; + \; ^{\cdot}OH; \quad \dots -CH + HOCH-\dots\,;$$
$$O \qquad\qquad\qquad\qquad O$$

$$\dots -CH_2-\dots \; + \; \dots -O-OH; \; \dots -CH-CH_2-\dots \longrightarrow$$
$$C \qquad\qquad\qquad OOH$$

$$\longrightarrow \dots -CH-CH-\dots \; + \; ^{\cdot}OH \longrightarrow \dots -CH-CH_2-\dots$$
$$O^{\cdot} \qquad\qquad\qquad\qquad OH$$

(511)

Piiroja and Dankovics [1525] demonstrated that the hot-flame treatment of polyethene produces an increase in the degree of branching of its macromolecules and, what is even more important, raises the concentrations of the double bonds, of hydroxy, carbonyl, and carboxy functions according to the routes described in scheme (511). Consequently, the ratio between the concentrations of the atoms of oxygen and carbon within the surface layers of substrate grows by 67.6 times within the time-span of 1.2 s and by 124 times in 4.8 s [1526]. As a result, the surface energy of the substrate undergoes a regular rise and so does (as is illustrated in Fig. 4.29) the adhesive ability towards poly(vinyl chloride) [1525]; the strength of autohesive joints of polyethene increasing from 0.55 to 6.6—7.2 MN/m² [1526].

Most importantly the efficiency of hot-flame treatment is affected by the properties of the polymeric substrate treated [1503] and the temperature of the process, the increase of temperature from 323 to 403 K was shown to increase the peeling strength of adhesive

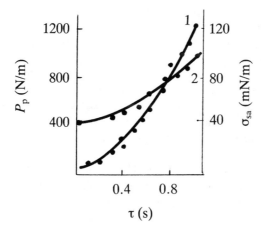

Fig. 4.29 — Peeling strength of the adhesive joints of polyethene with chlorinated poly(vinyl chloride) (1) and surface energy of the substrate (2) versus duration of the hot flame treatment.

joints from 0.1 to 1.5 kN/m [1527]. The effect of the modes of hot-flame treatment is less pronounced. However, two of the important parameters, namely concentration of oxygen in air C_{O_2} and duration of the treatment τ, are directly related to the oxidation efficiency and, therefore must be controlled. In fact, the peeling strength of the adhesive joints between polyethene and chlorinated poly(vinyl chloride) can be described by the following regression equation [1527]

$$P_p = 1541 \, C_{O_2} + 1045 \, \tau + 3.93 \, G_g + 32.2 \, r - 2373 \qquad (512)$$

where $C_{O_2} = 0.6-1.1\%$, $\tau = 0.3-1.1$ s, gas consumption G_g is in l/h and r is the distance between the funnel and the substrate surface varying in the range 10–30 mm.

Irradiation with UV light presents a special case of the oxidation procedure since it does not affect the magnitude of σ^d but raises σ^p [1529]. From the technological standpoint this procedure is more convenient than the direct action of strong oxidizers and is most acceptable for polyalkenes [1530], though it was also used to graft allyl monomers on to poly(ethyleneterephthalate) [1531]. In the latter case carboxy and phenolic functions were detected at large exposures [1532], the ratio between the concentrations of the atoms of carbon and oxygen increasing in the course of the treatment from the stoichiometric 1.502 and 1.515. The ultimate efficiency of the procedure, as judged by the magnitude of this ratio, is comparable to that of the oxidation in an electric discharge. Somewhat smaller exposures are required to modify polyalkenes; however, they still have to be adequately large, e.g. after 80 min. exposure of polyethene to UV-radiation the content of hydroxy functions is still outside the detection range of IR-spectroscopy, the adhesive ability being unaffected by the treatment [1533].

To shorten the time required to affect the substrate UV oxidation is stimulated by introducing photosensitizers, e.g. benzoin ethers, thioxanthones, ketals, acetophenones [1534], diphenylsulfide [1535], anthraquinone [1533] and its 1-methyl-3-nitro-derivative [1536], and most commonly, benzophenone [1534, 1535, 1537–1541]. The mechanism of their effect relies on the classical concepts of radical processes [1534]. It is essential that at large exposure times the efficiency of photosensitization is actually the same regardless of the photosensitizer used and photosensitization produces little effect on the strength of the adhesive joints of polyethene [1536]. However, at low exposure times the proper choice of photosensitizer is very significant.

When benzophenone was introduced into a polyurethane composition coating ethene-propene-diene copolymers a 60 s radiation with UV-light resulted in a substantial rise of mutual adhesive ability [1535]. However, a more reliable and common way to introduce a photosensitizer is to apply it from solution directly on to the substrate surface. In this way, adhesive ability of various substrates, e.g. elastomers [1537, 1541] and plastics [1538, 1540], was improved. In the case of elastomers the accumulation rate of oxygen-containing functional groups was raised by 3–4 times [1541], in the case of plastics a 10 s UV-irradiation of polypropene produced an increase in the strength of its adhesive joints from 0.1 to 5.3 N/mm [1538] and a 30 s irradiation of poly(methyl methacrylate) resulted in the 4.1-fold rise of the adhesive joint strength and in a 22% increase of σ_c [1540]. Similar effects were observed when surface layers of polyethene were modified with another common photosensitizer anthraquinone [1536]. Fig. 4.30 illustrates the effects of photosensitizers on the adhesive joint strength of polyethene, as related to the exposure time of the substrate to UV-radiation. The effect of photosensitizers can be promoted. For instance, benzophenone can be promoted with

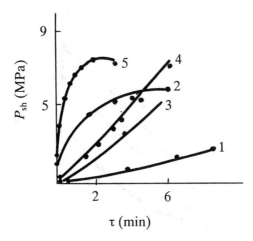

Fig. 4.30 – Shear strength of the adhesive joints between high-density polyethene and glass reinforced plastic bonded with epoxy formulation versussduration of the substrate photo-oxidation (1). Substrate modified with sensitizer (0.035 ± 0.002 mol/kg) — tetrachloro-benzoquinone (2), anthraquinone (3), 1-methyl-3-nitroanthraquinone (4); a 90 ± 10 μm thick film of low-density polyethene modified with anthraquinone (0.0145 mol/kg) and bonded with high-density polyethene.

alkylamines containing hydrogen atoms in the α-position with respect to nitrogen. Indeed, Gaske [1542] observed a synergistic action of benzophenone and methyl-diethanolamine or dialkylaminoalkylbenzoates and the following scheme was proposed:

$$\langle\bigcirc\rangle\text{-}\overset{\text{C}}{\underset{\overset{\|}{\text{O}}}{}}\text{-}\langle\bigcirc\rangle + R_2NCHR \xrightarrow{h\mu} \langle\bigcirc\rangle\text{-}\overset{\cdot}{\underset{\overset{\|}{\text{O}}}{}}\text{-}\langle\bigcirc\rangle + R_2N\overset{\cdot}{C}H_2R$$

$$\downarrow O_2$$

(513)

$$\begin{array}{c} R_2NCH_2R \\ -R_2NCHR \longleftarrow \overset{\cdot}{R_2NCHR} \quad R_2NCHR \\ \underset{\overset{\|}{\text{HO-O}}}{} \quad -R_2N\overset{\cdot}{C}HR \quad \underset{\overset{\|}{\text{O-O}}}{} \end{array}$$

As formation of one alkylamine radical involves the binding of no less than 12 molecules of oxygen [1543] the reaction described by eq. (513) proceeds with high oxygen consumption. This is why mixtures of benzophenone with amines are less sensitive to the presence of oxygen than other photosensitizers, an obviously advantageous feature.

Kiyushkin et $al.$ [1533] carried out a detailed investigation of the UV-photosensiz-ation of polymeric substrates. The system involved in the study comprised polyethene as the substrate and anthraquinone (0.13%) as photosensitizer, the other substrate was fibre glass reinforced epoxy-based plastic and an epoxy-based formulation was used as adhesive. Fig. 4.31 shows the graphs depicting the dependences of the strength of adhesive joints, concentration of carbonyl functions, and concentration of hydroxy functions on the exposure time of the substrate to UV-irradiation. As is seen, all three graphs are very similar, except for the initial portion of the curves. Note the linear pattern

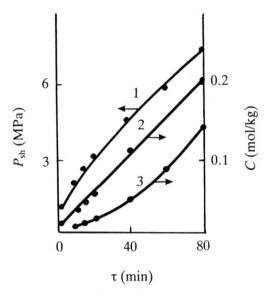

Fig. 4.31 − Shear strength of the adhesive joints of low-density polyethene bonded with epoxy formulation UP-5-213 (1), concentrations of the hydroxy (2) and carbonyl (3) functions versus time of exposure of the substrate containing anthraquinone (0.13%) to UV-light.

for the last part of the dependences (Fig. 4.31, 2) which persuades one to conclude that enhancement of the adhesive ability of the substrate is due predominantly to the generation of HO-functions. Within the framework of the modern view photo-oxidation is described as a chain reaction with degenerative photobranching on hydroperoxides and carbonyls and with second order chain termination. In fact, anamorphosis of the data in Fig. 4.31, presented in the corresponding coordinates of Fig. 4.32, follows a strictly linear pattern. Hence, the shear strength of adhesive joints can be deduced to be a function of the concentrations of oxygen-containing functional groups:

$$P_{sh} = 11.0 \, (C_{>C=O} + C_{OH})^{0.5} + 0.35. \tag{514}$$

These data provide a direct confirmation for the view which regards the atoms with unbonded electron pairs as having a favourable effect on the adhesive ability of polymeric substrates.

It is not only through the direct effect of oxidizers that oxygen-containing groups can be generated at the surface of polymers. Their formation can be also induced by oxygen entrapped within the 'cavities' and 'craters' between the non-wetted surface of the substrate and the adhesive, or by oxygen adsorbed at [1545], or dissolved in [1546] the substrate phase. Oxygen at the interface initiates contact thermo-oxidative processes (CTOP) which are seen when adhesives are brought to interact with glasses [1547]; however, they are of special significance with metals as substrates [622, 656, 1548]. With simplifying assumptions the time-span of such a process can be related to the thickness of the phase and temperature via an exponential dependence [1549]:

$$\tau_c = \tau_o h^2 \, \exp \, (E_c/RT) \tag{515}$$

where $\tau_o = 0.112 \pm 0.035$ s/cm^2.

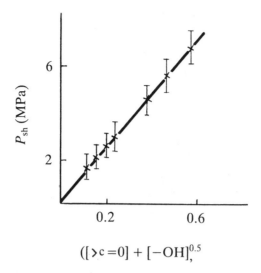

Fig. 4.32 – Shear strength of the adhesive joints of low-density polyethene containing anthraquinone (0.13%) bonded with an epoxy-based formulation versus total concentration of the hydroxy and carbonyl functions generated in the substrate by UV-light.

In some cases contact thermooxidation can lead to the generation of carboxy functions which are subsequently transformed into salts which are soluble in the polymer. This mechanism was assumed to provide for diffusion of the ions of the metallic substrate into the boundary and transition layers of the polymeric adhesive [1550, 1551]. Indeed, the mode of the distribution of iron cations into the phase of polyethene under various conditions, depicted in Fig. 4.33, can be associated with the surface rather than with the volumetric processes [1552, 1553]. Chemisorption of oxygen by the surface of the metallic substrate appears to be the initial step in the interaction between metals and polyamides. This is why polyamides form strong adhesive joints with steel, aluminium, titanium, zinc and even copper, but not with the highly inert gold [1554].

There are two routes along which contact thermo-oxidative processes (CTOP) can develop. This first involves destruction of macromolecules [1555], resulting in the decreased viscosity of adhesives (primarily of those in melts) and an intensive development of interfacial contact area; it also gives rise to an accumulation of low-molecular weight products within the bulk of the adhesive, as well as in the zones adjacent to the substrate surface [1551]. In the second route, chain destruction occurs resulting, in the final analysis, in their crosslinking [1556]. In fact, thermooxidation of polyethene at 423–503 K leads to a 5–15-fold reduction of the molecular mass in transition layers 200–660 μm thick in the initial stage of the process, whereas in the final stages crosslinked structural regions penetrating 30–160 μm deep into the transition layer are formed [1557].

Because of the simultaneous occurrence of both processes the contact time between adhesive and metal as well as the contact temperature can affect the strength of adhesive joints quite ambiguously [656, 1548]. Kalnin' [1558] suggested that the parameter $X_m = \bar{M}_\tau (m_\# + 1)/\bar{M}_0$, where $\bar{M}_\tau/\bar{M}_0$ is the ratio between the intermediate molecular mass and the initial one (molecular mass of the soluble fraction), $m_\#$ is the mass fraction

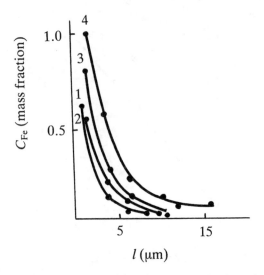

Fig. 4.33 – Distribution of iron in an oxidized layer of low-density polyethene as a function
of the distance from the surface of steel 08 kp at different contact temperatures and contact
times. Contact temperature: 1, 448 K; 2–4, 473 K; Contact time: 2, 15 min; 3, 30 min;
1, 4, 90 min.

of the crosslinked (insoluble) fraction, could be used as a measure of the contributions
of the destructive and the structuring processes. To assess the joint strength, instead of
the experimentally determined quantity (peeling strength P_p), he used the fracture work
per unit volume of the transition layer of the adhesive $W_v = \gamma_p/2E = P_p/h$, (where γ_p is
the ultimate strength in the form of boundary peeling stresses, E the elasticity modulus
and h the phase thickness). Fig. 4.34 shows the plot of the relative work of fracture of
the adhesive joints of steel with polyethene or ethene–vinyl acetate copolymer versus
X_m a composite parameter characteristic of the contact thermooxidative processes. As is
seen, the relationship is represented by the S-shaped curve revealing a smooth, though
complex, character of the dependence. Note the linearity of the plot near $X_m = 1.0 \pm$
0.05, suggesting a basis for predicting the influence of thermooxidative processes on
adhesive joint strength.

   Investigation of the features of the contact thermooxidative processes carried out by
Kapishnikov [1559] for the adhesive system comprising polyethene and steel resulted in
the major conclusion that the peeling strength of adhesive joints versus contact time and
the kinetic plots of the thermo oxidative characteristics of the adhesive are identical
[1560]. The only difference relates to the absolute values of the corresponding constants.
Indeed, both temperature dependent processes are described by the common exponential
equation [1552]:

$$Y(T) = Y_o \exp (E_c^*/RT), \tag{516}$$

where E_c^* is the effective activation energy of CTOP and Y denotes any adhesive–
sensitive parameter. For instance, Y could be the varying rates of the strength of adhesive
joint v_p' of the concentration of oxygen v_O' or of carbonyl functions $v_{>C=O}'$, or of the
mass of the oxidized layer v_m'. For the adhesive joint between polyethene and steel the
constants of eq. (516) relating to the processes described are as follows:

$v_{p'}$	$v_{O'}$	$v_{>C=O'}{}^{a}$	$v_{m'}$
N/m s^{-1}	mmol/g s^{-1}	s^{-1}	(mass parts) s^{-1}

	$v_{p'}$	$v_{O'}$	$v_{>C=O'}{}^{a}$	$v_{m'}$
Y_o	2.1×10^5	1.5×10^4	5.0×10^3	6.1×10
E_c^* kJ/mol	35.6	60.7	56.5	55.6

a Determined as the ratio of optical densities of the IR absorption bands at 1740–1710/ 4350 cm^{-1}.

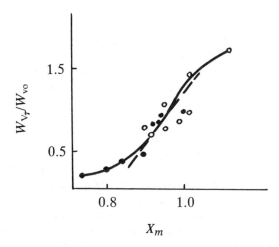

Fig. 4.34 – Relative work of fracture of the adhesive joints of polyethene (open circles) or ethene-(vinyl acetate) copolymer with steel 08 kp versus complex characteristics of contact thermooxidative processes.

The values listed for E_c^* show that it is energetically more favourable to form interfacial bonds between adhesive and substrate than to accumulate oxygen-containing groups or to become involved in destructive processes. Hence, the approach which provides enhancement of the adhesive joint strength by means of thermooxidation of the elements of the joint appears to be well founded.

Hence, by taking into account the features of thermal treatment of substrates in contact with an adhesive it appears possible to affect directly the adhesive ability of polymers. From an examination of their mechanisms these processes may be controlled either by introducing into the adhesive a specific filler which can sorb the low molecular weight degradation products [1561, 1562] or directly by introducing reducing agents [1563] or thermostabilizers [1564, 1565]. It should be borne in mind, however, that, when in excess, thermostabilizers can inhibit CTOP [1566], the opposite result being obtained. In this case the bonding and service conditions of adhesive joints have to be considered along with the effects due to the chemical nature of the adhesive, substrate and stabilizer [1567].

Oxidation processes in polymeric substrates can also be initiated by the mechanical loading of adhesive joints. According to the modern view [1568–1570] applied stresses stimulate the degradation of macromolecules, initiate oxidation and hinder the attainment of equilibrium conformations, the latter retarding those chemical reactions that require structural rearrangements in the chains. Such an effect, which may be considered as mechano-physical, as distinct from mechano-chemical, results in the packing density and orientation of macromolecules being altered [1571], and these are the precise parameters which were shown in preceding sections to govern the adhesive ability of polymers. It is not only the initial stressing of polymers but also the internal stresses which give rise to the mechano-physical effect [1572]. Relaxation of internal stresses has been shown to affect the oxidation kinetics of individual polyalkenes [1573] and elastomers [1574], including their application in composites. For instance, during the oxidation of polypropene the fraction of deformed bonds within transition layers diminishes [1576]. The higher the overstress, the more rapid is the scission of carbon–hydrogen bonds. The influence of these factors is substantially increased as the loading of adhesive joints is increased. At the same time, it should be borne in mind that the susceptibility of polymers to stresses is different in the initial stages of the oxidation process [1571]. Therefore, in the general case, the effect of mechanical loading on the adhesive ability of polymeric substrates is uncertain, calling for specific consideration of each individual case.

It is a quite complicated task to control conversion of oxidation processes. Hence, more technological procedures of substrate oxidation involving treatment with various types of electric discharge, e.g. corona discharge, have recently attracted special interest [1577]. To avoid the development of secondary oxidation processes the corona discharge is generated under high vacuum. Treatment with high-frequency electric discharge also involves bombardment of the substrate surface with charged particles, deactivation of the electronically excited states in gases, as well as, initiation of ultraviolet irradiation. The efficiency of this procedure depends essentially on electrophysical parameters, namely electric current I [1578, 1579] and voltage U [1580]. As is seen from Fig. 4.35, the latter parameter produces a direct effect upon the strength of adhesive joints. Erykalova and Vladychina [1477] suggested a complex criterion defined as $\zeta = IU/S_S(\mathrm{d}\theta/\mathrm{d}\tau)$, where S_S is the area of the substrate surface subjected to treatment. One should recall that there is no direct relationship between contact angles and adhesive joint strength. Indeed, the contact angle on polyethene or poly(vinyl chloride) treated with corona discharge shows a smooth variation as a function of the treatment duration; whereas the corresponding dependence of the peeling strength of adhesive joints displays maxima, as is illustrated in Fig. 4.36. Hence, mathematical models of these processes are quite complex since they have to accommodate the effect of a large number of factors [1581].

Despite the tendency to use corona discharge treatment under the highest vacuum one cannot totally eliminate the development of the oxidation processes. Therefore, oxygen-containing functional groups should inevitably be formed in the course of the treatment of polymeric substrates with corona discharge. This was confirmed in the X-ray photo-electron spectroscopic studies of the surface of carbon fibre reinforced plastics with the polyetheretherketone matrix [1582]. It is also supported by the increase of the strength of welded autohesive joints of polyethene observed regardless of the activity of the gas medium — be it air, oxygen, nitrogen, argon or helium [1583]. Stradal and Goring [1583], who discovered the latter effect, associated it with the substrate acquiring the properties of an electret. However, in recent investigations with X-ray photoelectron spectroscopy [1501] oxidative processes were detected even under inert atmosphere. As is demonstrated by the data in Fig. 4.37, there is a smooth relationship between the peeling strength of autohesive joints and the degree of oxidation of polyethene in air and

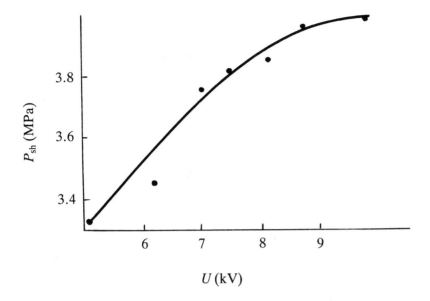

Fig. 4.35 – Shear strength of the adhesive joints between glass reinforced plastics and the LDPE–HDPE welded film versus voltage of the corona discharge with which the film was treated.

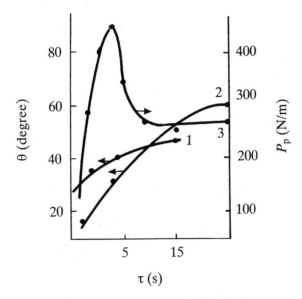

Fig. 4.36 – Contact angles on low-density polyethene (1) and poly(vinyl chloride) (2); peeling strength of their adhesive joints (3) versus duration of the substrate treatment with corona discharge.

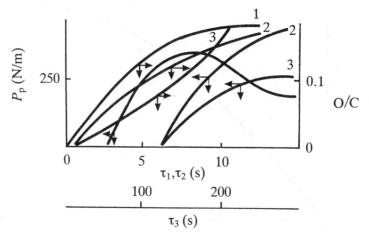

Fig. 4.37 — Peeling strength of autohesive welded joints of low-density polyethene (left ordinate axis) and degree of oxidation of low-density polyethene (right ordinate axis) versus duration of the substrate pretreatment with corona discharge in air τ_1 (1), under nitrogen τ_2 (2) and argon τ_3 (3).

nitrogen media measured as the ratio between the intensities of the peaks due to oxygen and carbon atoms, e.g. at equal oxidation degrees then equal peeling strengths are realized. The distinction between the effect of inert argon and that of oxygen or nitrogen is that it takes 20 times longer to achieve the corresponding result (oxidation degree and peeling strength).

On the other hand, when the surface of polyethene (pretreated with corona discharge) was subsequently exposed to bromine solution, or alkaline ethanol solution, or phenyl-hydrazine it was only in the latter case that the strength of autohesive joints remained unaltered (as compared to that of the joints with only the corona discharge treatment of the substrate) [1584]. In this latter case interfacial interaction was found to be due to hydrogen bonding. A similar effect was observed in adhesive joints between polyethene and poly(ethleneterephthalate) where formation of hydrogen bonds between phenolic hydroxy functions of PET and carbonyl functions in PE was found to result in the rise of σ_{sa}^h from 23 to 50 mN/m and the consequent increase of the joint strength by almost an order of magnitude [1585]. Generation of oxygen-containing functions in polyethene giving rise to interfacial hydrogen bonding was found to be caused even by neutral activated components of the gas medium in corona discharge [1586], this observation being in accord with the data of Fig. 4.37. It is important to note the inverse relationship between, on the one hand, the concentrations of carbonyl groups and unsaturated fragments, and, on the other [1587], the discharge energy. When low-energy corona discharge is used variation of the wetting hysteresis of polyethene with water is a linear function of σ_{sa}^p, whereas at high energy it appears to be dependent on the heterogeneity of the substrate surface [1588]. Hence the explanation for the proportional relationship between the polar component of the surface energy of polyethene pretreated with corona discharge and thermodynamical work of adhesion [1589] becomes clear.

Quite obviously, among the polar groups resulting from substrate oxidation there should also be peroxide groups, the last of the series of oxygen-containing groups to be considered. These were indeed detected and the relationship between their concentration and the strength of adhesive joints of polyethene was reported [1578].

By means of X-ray photoelectron spectroscopy and with the aid of specific reagents Briggs and Kendall [1590] identified the types of functional groups that were formed in low-density polyethene under corona discharge and their numbers per 10^4 repeat units were also reported [1590]:

—OH (as in alcohols)	150
>C=O	110
—COOH	110
—OH (enolic form)	60
>C=C<	53
—OOH	47

As is seen, contrary to the effect of direct oxidizers, carboxy and, particularly, hydroperoxide functional groups are not the only ones to result from the corona discharge treatment and, moreover, hydroxy and carbonyl functions, are in apparent excess. It is to the effect of highly active functional groups that the reported short 'life-time' of enhanced adhesive ability of the corona discharge-treated polyethene [1503, 1581] can be ascribed.

Briggs [901, section 9.6.2] suggests the following general scheme for the processes taking place during the discharge treatment:

$$-CH^2-CH_2-CH_2- \xrightarrow[-H]{I^+,\ h\mu,\ e^-} -CH_2-CH-CH_2- \xrightarrow{O_2} \underset{\overset{|}{O-O^\cdot}}{-CH_2-CH-CH_2-} \xrightarrow{H}$$

$$\xrightarrow{H} \underset{\overset{|}{O-CH}}{-CH_2-CH-CH_2-} + \underset{\overset{\parallel}{O}}{-CH_2-CH-CH_2-} + \underset{\overset{|}{OH}}{-CH_2-CH-CH_2-} \qquad (517)$$

Additionally, in polyethene subjected to corona discharge treatment the presence of free radicals was detected by both ESR and spectrophotometry with diphenylpicrylhydrazyl label for several seconds as well as one hour after the treatment [1578]. Similar data were obtained with homo- [1522] and copolymers [1591] of propene. Corona discharge treatment of polypropene, polyacrylonitrile, polyamide, viscose [1592] and carbon [1593] fibres was shown to substantially enhance their wettability by epoxy adhesive, as well as increasing the strength, impact viscosity and water durability of the corresponding composites.

When polyethyleneterephthalate was subjected to corona discharge treatment phenolic hydroxy functions were formed [1594] according to the following scheme:

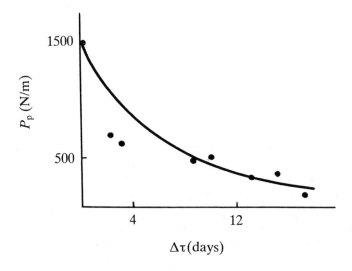

As a result radicals and oxygen-containing groups are formed; these are subsequently reorientated from the direction normal to the substrate surface to acquire a planar orientation with respect to the surface [1595]. According to Owens [1596], hydrogen bond formation is the dominating mechanism of the interfacial interaction of the pre-treated substrates with adhesives and this assertion is verified by the experimental data of Kagan *et al.* [1585]. As a consequence of these transformations, the 'life-time' of enhanced adhesive ability of PET should be shorter than that of polyalkenes: indeed, according to the data of Fig. 4.38, a three-fold drop of the adhesive joint strength is observed in adhesive joints formed 2 days after the pretreatment [1595].

Fig. 4.38 – Peeling strength of autohesive welded joints of poly(ethyleneterephthalate) versus time lapse between corona discharge treatment and welding.

To avoid this drawback the duration of the corona discharge treatment has to be extended. However, it is associated with 'amorphization' of the boundary and transition layers in PET [1597], their oxidation involving local microdefects. Nevertheless, no additional development of the substrate surface relief is associated with this procedure (this is doubtlessly its advantage as compared to acid-treatment procedure which is accompanied by formation of cracks), the enhancement of adhesive ability being due to morphological factors. Apparently, this effect is of general implication. In fact, by focusing the electric discharge with the grooved support placed on the back side of a polyethene film Luc [1598] succeeded in substantially increasing the adhesive joint strength. The fact that adhesive ability of polymers varies whilst the concentration of active functional groups remains constant does not contradict the theoretical ideas of adhesion since not only has the chemical nature of macromolecules to be considered, but also their mobility, a factor of molecular–kinetic origin. For instance, no substantial changes in polymer concentration were detected in the course of the activation of polyethene film with the barrier-slipping discharge (this technique reduces the probability of spark discharge through the substrate because of the relatively greater ionization). Intensities of the absorbance bands due to carbonyl functions and double bonds with respect to the standard band ($1300 \, \text{cm}^{-1}$) are unaffected by the 30-minute treatment [1599]. However, for methyl groups the intensity increases by 9.7% [1600]. At the same time, IR-spectroscopic data indicate that the mobility of macromolecules in transition layers increases [1600]. Having come to the same conclusion, Stradal and Goring [1601] associated it with changes in polymer density and, correspondingly, with the nature of its packing.

It is a technically complicated task to perform electric discharge-treatment under high vacuum. Hence, in practice polymeric substrates are preferably treated with glow discharge in a gaseous atmosphere, i.e. by the plasma method [1478, 1579, 1602–1604].

Two factors are important regarding the problem of increasing adhesive ability of polymers; these are the plasma-induced changes in the chemical nature and in the morphology of the substrate surface. Let us elaborate on each of these separately.

The first of these factors is associated with the composition of the gaseous medium. For instance, Yasuda [1605] demonstrated that for a large group of polymers treated with argon or nitrogen plasma both oxygen and nitrogen containing groups were formed at the substrate surface only when treated in nitrogen. While the formation of nitrogen-containing groups appears to be obvious (Fig. 4.39 depicts the kinetic curve of the respective process in polyethene [1606]), formation of oxygen-containing groups is somewhat unexpected, as polyalkenes and polytetrafluoroethene were thoroughly deoxygenated prior to plasma treatment. However, it appears impossible to completely eliminate all traces of oxygen dissolved in a polymer [1607], or present as an impurity in the gaseous medium. These results are quite reliable. There might be another explanation when oxygen-containing polymers are subjected to plasma treatment. In this case oxygen-containing groups are generated in the course of plasma-induced destruction of the polymer and are located mainly within the boundary layers of the substrate. Indeed, oxygen-containing groups were detected [1609] predominantly on the surface of polyetheretherketone [1608] subjected to the effect of argon plasma. Consequently, the adhesive ability of substrates rises sharply. For example, treatment of polyether–sulfone with argon plasma produces a maximum increase in the shear strength of its adhesive joints bonded with an epoxy formulation — namely 24.1-fold, the effect of nitrogen

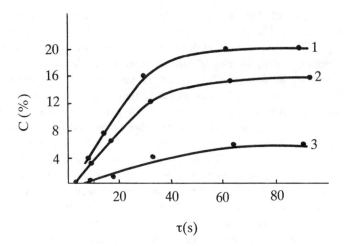

Fig. 4.39 – Total nitrogen content (1), concentrations of single (2) and double (3) carbon–nitrogen bonds in polyethene treated with nitrogen plasma as a function of the treatment time.

plasma being very close namely 21.7-fold [1610]. When the substrate contains both the atoms of oxygen and nitrogen, it is treatment with ammonium plasma that produces the best effect. In the case of poly-p-phenyleneterephthalamide this procedure produces a 2.1-fold rise in the strength of its adhesive joints with Kevlar-49 fibre [1611], whereas for joints with polymethyleneoxide the increase is by an order of magnitude [1612]. For polymethylene chains under ammonium plasma the related effects are ascribed to plasma-induced generation of $\dot{N}H$ and $\dot{N}H_2$ radicals [1613] (see also Fig. 4.39), reminding one of the positive role of atoms with unshared electron pairs discussed in section 4.1.

It is clear that the effects reviewed should be enhanced when oxygen is introduced into the plasma forming gas. Indeed, a 10 s exposure of polypropene to the effect of nitrogen plasma containing 25% of oxygen reduces the contact angle from 68° to 50° [1614]; this is equivalent to a 3–4-fold rise in surface energy at the start of the treatment and to a 2-fold rise at the equilibrium state [1615]. In the case of polyethene σ_s was shown to increase from 33 to 46 mN/m [1616]; for polytetrafluoroethene σ_s increased from 16–20 to 72 mN/m [1617], contributed to by the growth of the polar component prevailing over the simultaneous decrease of the dispersion component [1609, 1618]. Similar behaviour is characteristic of the oxygen-containing polymers, e.g. polyethylene-terephthalate or polyhexamethylene adipamide [1619]. For PET treated with oxygen-containing plasma the $\sigma_{sa}^p/\sigma_{sa}$ ratio, reflecting the increase in the polar component, was shown to rise from 0.138 to 0.544 [1620].

When only oxygen is used as plasma forming gas a broad spectrum of corresponding functional groups arises, ranging from hydroxy to carboxy ones [1621, 1622]. Fig. 4.40 depicts the kinetics of the corresponding processes [1606]. However, this treatment is ineffective in that, on the one hand, it is accompanied by extensive degradation, whereas on the other hand, the introduced oxygen-containing functions are easily eliminated by heating the treated substrates [1623]. Therefore, it appears simpler to use water vapour instead of oxygen and this procedure was shown to produce an increase in the surface

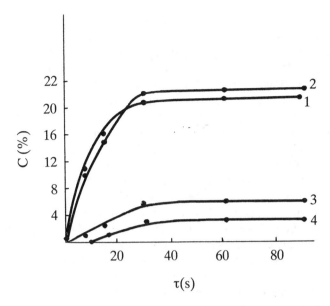

Fig. 4.40 – Total oxygen content (1), concentrations of ether (2), ketone (3) and carboxy (4) groups in polyethene under oxygen plasma as a function of the treatment time.

energy of polyethene up to 62 mN/m [1622]. The optimal ratio of the oxygen index to the inert one in 10 nm thick transition layers of polyalkenes [1624] and polyesters [1625] subjected to plasma treatment was determined to be 1.5–3.5.

Similar considerations should be valid when other gas media are involved. For instance, treatment of a silicone elastomer with hydrogen plasma containing trace amounts of oxygen resulted in the maximum variation of the contact angle [1607]. The following contributions may be involved in the overall processes, UV-irradiation by plasma (leading to the generation of free radicals and double bonds), bombardment of the surface with metastable particles emitted from the discharge zone, bombardment of the substrate with hydrogen atoms, and accumulation of excessive electric charge at the surface. By successively excluding the effect of each of these factors it was found that bombardment of the substrate surface with hydrogen atoms contributed the most to the variation of the contact angle. Oxygen-containing groups were even detected on the surface of polyethene treated with helium plasma [1626]; however, their concentration was much lower than in the case of dichromate in sulfuric acid treatment. The modified boundary and transition layers can be easily removed by mechanical treatment, or by the effects of solvent or even heating [1623]. Small amounts of carbonyl groups were detected when phenolformaldehyde oligomer and poly(methyl methacrylate) were subjected to the effect of tetrafluoromethane plasma; in this case F_3C-, F_2C-, and $FC-$ groups were the products of the modifying procedure [1627]. Similar results were obtained when polyethene was treated with fluorine plasma [1628]. For a variety of substrates the plasma treatment described was shown to result in an increase of the contact angles. The substrates reported were polytetrafluoroethene, polyvinylfluoride [1629], polyvinylidenefluoride [1629, 1630], low-density polyethene, poly(ethene-*co*-propene) [1630], polystyrene [1631], polymethyleneoxide [1629] and polyesters [1632]. For example, in this latter case a 10–20 min treatment resulted in θ increasing from 70° to 100–110°; for poly-tetrafluoroethene the final θ value was 108° [1632], i.e. actually equalling that of the polyester. Hence the surface energy of polyesters, which are rather polar compounds, decreases to 20 mN/m [1633], i.e. to the value characteristics of fluoroplastics. Naturally, the composition of the plasma-forming gas is quite important. The effect of CF_4 was shown to be somewhat intermediate between that of SF_6 and C_2F_6 [1631]. The presence of oxygen assumes a quite specific role. It is clear that oxygen-containing groups would be formed if the treatment involving oxygen succeeds that with tetrafluromethane; in this case σ_s of polyesters was found to be as high as 58 mN/m, whereas in the case of the reverse sequence treatment σ_s is only 20 mN/m, as already mentioned above [1633].

On the other hand, the simultaneous action of both fluoro- and oxygen-containing gases produces an increase in the surface energy [1634], which is, however, ensured by the additional rise of the concentration of oxidized substrate fragments. For instance, by varying the composition of the gaseous mixture of C_3F_8 and C_2H_4O Haque and Ratner [1635] managed to increase the surface energy of various substrates from the values in single figures to 45 mN/m. Dorn *et al.* [1634] demonstrated that maximum strength of the adhesive joints between polyethene and steel bonded with an epoxy formulation was obtained with equal consumption of tetrafluoromethane and oxygen. However, the real benefit of adding oxygen is not confined to the observed increase of the joint strength (one should bear in mind that, regardless of the gas composition, fluoro-containing substrates are less active in their response to plasma treatment than polyalkenes [1636]); in fact, the rate of plasma treatment was shown to increase also, for aromatic polymers the rise being 3–4-fold [1637].

Hence, as oxygen-containing groups are formed in polymers in the course of their treatment with plasma of any composition, it appears most sensible to plasma-treat the substrates in air. For polyethene the concentration of oxygen-containing groups and free radicals resulting from such treatment was shown to vary within 10^{17}–$10^{19}\,g^{-1}$ [1638] depending on the gas pressure, the strength of adhesive joints produced with epoxy formulation being increased from 1–2 to 14 MPa [1639]. For polyacetals, polybutene-terephthalate, polyphenyleneoxide, polycarbonate, polyarylate and polyesterimide a 4-fold rise in the adhesive joint strength was observed when an epoxy formulation was used as adhesive and a 12-fold in the case of polyurethane adhesive [1640]. If one of the joint elements is a metal, the flux density of revaporized electrons, considered in section 2.2.1, has an effect also. As is demonstrated in Fig. 4.41, the flux density of revaporized electrons is related to the adhesive ability of polyethene, although not so obviously and unambiguously as the substrate composition, in particular the concentration of carbonyl functions [1641]. In practice such consistencies were observed not only for various polymers [1642–1646], but even for silicon carbide fibres [1647] and these are most clearly seen in the initial stages of substrate treatment procedures [24, pp. 94–95]. On the one hand, this is due to the effect of oxygen-containing groups; for polyethene the concentration of atoms of oxygen in the boundary layers was shown to increase from 0.2 to 15.8 mol% [1648]; for polyethyleneterephthalate both the ratio between the concentrations of the atoms of oxygen and carbon as well as the surface energy in the range $\sigma_{sa} = 37$–55 mN/m were shown to obey similar kinetic patterns [1649]. On the other hand, degradation processes are inevitably initiated by plasma in the transition layers of substrates (to protect them the use of metallic net screens has been recommended [1628, 1650]).

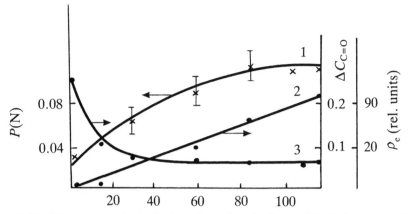

Fig. 4.41 – The strength of adhesive joints between polyethene and lead (1); concentration of carbonyl groups (2); flux density of revaporized electrons in the polymer (3) versus duration of the treatment of polyethene with glow discharge.

Indeed, the rate of weight loss of polypropene in the course of the air plasma treatment was reported as 13×10^{-7} kg/m s, of polymethyleneoxide 28×10^{-7} kg/m s [1651]; for natural rubber, butadiene–acrylonitrile, ethene–propene, siloxane and fluorocarbon elastomers this was shown to vary within the range from 6 to 33×10^{-7} kg/m s [1652]. This is due to the elimination of the weak boundary layers from the substrate surface which was observed in many cases (even when poly-p-xylylene was treated with the argon plasma [1653]). At the same time the development of competitive structuring processes should be expected. For instance, plasma treatment of polytetrafluoroethene can result in the scission of the atoms of fluorine and in crosslinking [1654], providing for the regular rise of substrate wettability [1655]. In general, according to Westerdahl, Hall and Levi [1656] structuring prevails in hydrocarbon and fluoro-containing polymers, whereas in polyamides plasma-induced destruction of the boundary and transition layers dominates. Such factors cannot avoid affecting the surface morphology of polymeric substrates. To elaborate on this aspect of plasma action it is necessary to emphasize that the structural changes in polymers extend to depths commensurable in dimensions with the thickness of the boundary layers [1621, 1622, 1624, 1625, 1629, 1638, 1646, 1657, 1658]. Amorphous regions are subjected to the most vigorous assault, as was demonstrated by Friedrich and co-workers [1659, 1660] for individual polymers and composites [1661]. Because of the 'deamorphization' processes protracted ordered regions are formed in the bulk polymers [1662], as well as in films [1663] and even in gel-like polymers [1664]. As a result of plasma treatment the substrate surface becomes more developed. This feature can result in the selective sorption of the separate components of an adhesive, resulting in an unfavourable effect on the strength of composites. Indeed, the degree of curing of the epoxy matrix reinforced with plasma-treated polypropene was shown to decrease due to sorption of the curing agent by the filler [1665].

The variety of factors acting simultaneously produces a situation where the efficiency of the plasma treatment in controlling the adhesive ability of substrates appears to be largely dependent on the technological procedures (Muscia, Carr and Kember [1666] claim that these are even more significant in the effect on adhesive joint strength than the nature of the plasma forming gas). This is why practical recommendations concerning the most thoroughly studied cases, e.g. treatment of polyethene, are quite contradictory [1667]. Perhaps, sufficiently unambiguous deductions can be drawn from the corresponding regression equations (taking into account the limitations of the physical implications discussed above in such an approach). For example, the peeling strength of adhesive joints involving fluorocarbon polymer film (10 μm thick) was shown to be governed predominantly by the voltage U, whereas the treatment duration τ and pressure p produced an equal but substantially less profound effect [1668]:

$$P_p = 284.1 + 73.5U + 25.73\tau + 19.43p. \tag{520}$$

The determined optimum values for the technological parameters are as follows $U = 1650$ V, $\tau = 80$ s, $p = 53.2$ Pa then, from eq. (520) $P_p^{max} = 4.2$ N/cm.

Hence, the action of plasma on polymers produces [1613, 1669–1671] an increase in their surface energy due to the formation within the boundary and transition layers of oxygen-containing groups both in the main chains of polymers [1652], and in the side chains [1605, 1607, 1626, 1627] and of π-conjugated systems [1672]. Also of importance is that radical states are generated in the course of the plasma treatment [1607, 1638, 1653, 1657, 1669, 1673] and these accumulate predominantly in the boundary layers (they were detected in the boundary layers of even powdered polymers [1674]). According to ESR data the concentration of radicals in a polymer varies as a function of the gas medium composition in the following order $CF_4 = CO > H_2 = Ar > CH_4 > N_2 > O_2$ [1675], the more active the gas medium the higher the probability of recombination. Along with polymer modification sealing of microdefects in the boundary layers [1676] and 'deamorphization' of these zones in polymers [1659, 1661, 1662, 1665] also take place resulting in the formation of rather ordered regular regions [1676]. Also one cannot exclude the possibility of the substrate surface being charged in the course of plasma treatment [1669, 1677], the strength of adhesive joints of butyl rubber being shown to be related to the magnitude of the charge potential [1678].

According to the ideas of section 4.1, all of the factors discussed should produce a favourable effect on the adhesive ability of polymers, as was indeed observed in practice. In fact, it is the plasma method that ensures the minimum ratio between concentrations of the atoms of carbon and oxygen in the boundary layers of substrates as detected by X-ray photoelectron spectroscopy. For polyetheretherketone (stoichiometric ratio $C_C/C_O = 6.3$) the ratio obtained is 1.9, as compared to the 2.7 characteristic of etching with chromates [1609].

Another substrate pretreatment procedure prior to adhesive bonding involves modification of polymers by grafting various functional groups on to their surfaces. Along with the chemical nature of polymeric surfaces this method makes it possible to regulate their charge (for example, by forming coatings comprising polyelectrolyte complexes, particularly, those involving surfactants [1679]).

Relying on the established physico-chemical concepts, those compounds whose molecules comprise atoms with unshared electron pairs appear to be most suitable as modifiers of this type. These are, firstly, nitrogen-containing compounds, e.g. poly-acrylamide [1680] and its dimethyl derivative [1681]. Grafting of polyacrylamide to polymethyleneoxide was shown to increase the shear strength of its adhesive joints by an order of magnitude [1680]. Aniline, its mono- and dimethyl derivatives, benzonitrile, benzamide and pyridine, were reported to be useful for this purpose. In these cases also, the ultimate effect is of the same order even for polytetrafluoroethene [1682]. Hence, the general route to increase adhesive ability of various polymers involves their surface modification with compounds of this type.

Oxygen-containing compounds comprise the second class of modifiers with unshared electron pairs. Of these acrylic acid is most commonly used. For polyethene modified by the grafting of acrylic acid the surface energy acquires stable values at a grafting ratio exceeding 15% [1683]. Consequently, a 50 μm thick layer of polyacrylic acid grafted on to polyethene ensures an 11-fold rise in the peeling strength of the joint between modified polyethene and aluminium[†] [1684]. The corresponding dependences of the peeling strength versus grafting ratio display maxima [1686]. This feature is due to the effects of a chemical nature as well as of substrate surface topography. In the latter instance a uniform distribution of the grafted functional groups was shown to bestow a positive effect on the adhesive ability of a polymer [721]. The results of modifying poly(ethylene terephthalate) by grafting acrylic acid [1681] were similar. The beneficial effect conferred by 1-butyne-3,4-diol is due to the same reasons. However, in this case it is not only the presence of oxygen atoms in the modifier which determines the increase of adhesive ability but groups with π-electrons also contribute to the overall effect. As a result of such modification the surface energy of polytetrafluoroethene was shown to increase from 18 to s72 mN/n [1687]. Obviously, epoxidation of the substrates should produce an even greater effect. Indeed, this was the case observed with the polyamide 'Kevlar' treated with the diglycidyl ether of butanediol-1,4 [1688]. One of the oxirane groups of the modifier reacted with the macromolecules of the polymeric fibre, while the other, remaining unreacted, served to enhance the reactivity and, consequently, the adhesive ability of the substrate.

In principle, such a two-step procedure has more potential than the traditional one-step method as it makes it possible to introduce into the surface layers of substrates a substantially greater variety of functional groups. In fact, there are no impediments to the introduction of any sufficiently reactive group into the surface layers at the first stage followed by subsequent transformation by various chemical reactions at the second. A well-known procedure involving primary halogenation of the surface of a polymer and its subsequent amination provides an example to illustrate the route. This approach has been used to attach oxygen-containing groups to polytrifluorochloroethene. The procedure involved the treatment of polytrifluorochloroethene with lithium substituted reagents resulting in the substitution by protected groups, which, in the second stage were converted to hydroxy, aldehyde, and carboxy functions [1689].

Silane compounds, a class of common reagent for the modification of glass fibres, are rarely used for polymers. This is associated with the low reactivity of most polymers towards these modifiers, a situation which is common for many products potentially capable of bestowing high adhesive ability on polymers. Therefore, the choice of the procedure to initiate the modification becomes a problem of special significance [1690]. Plasma treatment was shown to be quite effective [1680, 1682, 1686, 1687] as it ensures a relatively high concentration of the active sites e.g. 10^{-11}–10^{-8} mol/cm^2 [1691]. Ultraviolet radiation [1681, 1692] and especially atomic radiation [1683], are also

---

[†] Formation of a monomolecular coating of polyacrylic acid on aluminium provides for the 4-fold rise of the strength of adhesive joints with polyurethane; however, a 4–5 nm thick layer is assumed to be the optimum [1685].

rather widely used. The effect of the latter is relatively easy to predict [1693] and since it is accompanied by the accumulation of radicals, is quite favourable for improving the adhesive ability of polymers. Thermally initiated modification is less beneficial [1684]. Using vinyl acetate and methyl acrylate as monomers the peeling strength of the adhesive joints of the modified polythene were an order of magnitude higher when radiation grafting in the vapour phase was employed as compared to those obtained by immersing the pre-irradiated substrate in a solution of the monomer [1694].

Within the processes under discussion halogenation of polymeric substrates attracts substantial interest. The enhancement of the adhesive ability of polymers resulting from such treatment was not initially reported. A comparison of the efficiency of oxidation, bromination, chlorination and chlorophosphorylation of polyethene demonstrated that the last two procedures produce the maximum effect on the magnitude of the contact angles [1695]; however, increased concentration of bromine in rubbers was also shown to result in increased surface energy values [1696]. Hence, the adhesive ability of rubbers is substantially improved by even small amounts of halogens. The treatment of a variety of substrates† with 0.05% $NaClO_3$ solution [1698], 0.1% chlorine solution [1699], 2–4% solutions of dichloroamine [1700], N-chlorosulfamide, dichlorophenylacetic acid, tert.butylhypochlorite [1701], and hexachloromelamine [1702] were shown to result in substantial growth of their adhesive joint strengths. For instance, by treating vulcanized rubbers with hexachloromelamine solution in ethyloctane the peeling strength of their adhesive joints bonded with polyurethane adhesive was increased from 1.2 to 10.3 kN/m.

A similar effect was observed when polyethene was subjected to ultraviolet radiation [1703] in an atmosphere containing bromine [1704], or either tri- [1705] or tetra-chloroethene [1705, 1706] as was also the case with corona discharge treatment of polyethene substrate containing not less than 0.5% of chlorinated polyethene with the chlorine content up to 45% [1707, 1708]. Prior to plasma treatment it was found much more beneficial to expose the polymeric substrates to the effect of the vapours of di,- tri-, or pentachloroethanes, or tetrachloroethene [1709], rather than to that of the hydro-carbon solvents. The treatment with bromine was shown to enhance the adhesive ability of polyethene [1710], butadiene–acrylonitrile [1711] and bromochloroprene [1712] elastomers, as well as of polyamide fibres [1713]. In general, fluorination affects predominantly the frictional properties of polymers; however, modification of polytetra-fluoroethene with 0.5% of various perfluoroethenes was shown to improve the strength of its autohesive joints and of the adhesive joints with ceramics [1714, 1715]. Increasing the length of the fluoro-containing substituent enhances the effect, as was observed on treating polyvinyl alcohol films with chloroanhydrides of perfluorinated valeric, enanttic and pelargonic acids [1716]. Note, that such effects are observed if the halogens are attached by substitution rather than by addition mechanisms, as in the latter case the concentration of the double bonds is, as a rule, decreased, thus reducing the adhesive properties of polymers.

---

† One should bear in mind that such treatment of adhesives results in the opposite effect attributed to the substantial energy barrier arising at the interface. For instance, surface bromination of butadiene–acrylonitrile copolymer-based vulcanized rubbers produces a 2–10-fold decrease of the peeling strength of their autohesive joints as compared to the joints of non-modified elastomers [1697].

Differing from oxidation, halogenation does not actually affect the bulk of a polymer [1695] and this appears to be the essential advantage of this method. Hence, a large number of procedures has been proposed involving substrate modification with the atoms of fluorine, chlorine and bromine (iodination was shown to decrease the adhesive ability of polymers because of their oxidation which results in an accumulation of low-molecular weight degradation). However, as a technological procedure halogenation is unpopular in that effective protection against the hazardous effect of halogens and halogen-containing compounds is essential.

To overcome the technological drawbacks it was suggested that the substrates should be treated with systems evolving bromine [1481, 1717] and this was shown to ensure a substantial acceleration of the modification of the boundary layers. By using the optimum formulation involving bromine salts Orlov and co-workers [1481, 1718] obtained very strong adhesive joints even with the low-polarity rubbers (see Table 4.19). The data of Table 4.20 illustrate the effect of bromination on rubber-to-polymer bonding [1719]. As is shown in Fig. 4.42, similar considerations are valid regarding the adhesive ability of wood. Hence, the effects reported are quite general.

Additional to the routes involving chemical reactions, the surface of polymeric substrates can also be modified by sorption of the corresponding compounds. Monocarboxylic acids of the aliphatic series [1720], e.g. oleic, stearic [1721] and p-chlorophenyl

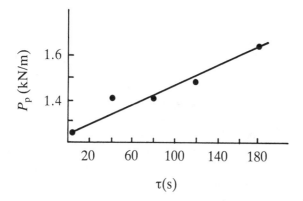

Fig. 4.42 – Peeling strength of the adhesive joints of poly(vinyl chloride) with birch wood bonded with a polybutadiene–acrylonitrile formulation modified with chlorinated poly(vinyl chloride) versus bromination time of the substrate.

Table 4.19 – The strength of adhesive joints between steel 3 and brominated vulcanized rubbers

Vulcanized rubber based on:	$P_{p'}$ (kN/m)	P_{pull} (kN/m)	$P_{tensile}$ (MPa)
Natural rubber	8.2	10.4	5.8
Polyisoprene SKI-3	7.4	6.2	5.4
Polyisoprene (SKI-3) and polybutadiene (SKD)	9.3	9.9	6.4

Table 4.20 – Peeling strength, kN/m, of adhesive joints between vulcanized rubbers modified with bromine and various polymeric materials[a]

Substrate coating	Polychloroprene adhesive		Polyurethane adhesive	
	Bonding time		Bonding time	
	(3 min)	(24 h)	(3 min)	(24 h)
Poly(vinyl chloride)	$\dfrac{0.7-1.8}{1.4-2.5}$	$\dfrac{1.9-3.2}{5.4-7.2}$	$\dfrac{1.1-1.7}{1.8-2.3}$	$\dfrac{3.2-4.4}{7.4-8.0}$
Polyurethane	$\dfrac{0.8-1.6}{1.9-2.5}$	$\dfrac{1.8-2.9}{5.5-7.2}$	$\dfrac{1.5-2.0}{1.8-2.2}$	$\dfrac{3.1-4.4}{7.6-7.9}$

a   Numerator refers to unmodified rubber and denominator to the modified rubber.

substituted fatty acids [1722] (which display the maximum effect due to the atoms of chlorine) are classic examples[†] of these. Another modifier of the same type is dimethylol-p-cresol which Kitamura, Hate and Ohira [1723], applying it from the gaseous phase to the substrates, succeeded in increasing their adhesive ability to polyamide.

According to the ideas discussed above, increased efficiency should be the feature inherent to the modifiers which possess nitrogen atoms along with those of oxygen. Indeed, the treatment of poly-γ-benzyl-L-glutamate with 2-aminoethanol resulted in a rise of σ_{sa}^p of the substrate from 17.2 to 44.3 nN/m, σ_{sa}^d having been simultaneously decreased from 31.0 to 26.5 mN/m [1724]. Naturally, when a modifier contains both types of atoms the modified materials have an enhanced adhesive ability. For instance, when imidosilane (produced by reacting benzophenonetetracarboxylic acid dianhydride with aminophenyltrimethoxysilane) was used as a substrate modifier instead of the closely related γ-aminopropyltriethoxysilane the strength of adhesive joints after4-h exposure to 673 K in air or nitrogen decreased by 59 and 1%, respectively, as compared to 72 and 25% [1725].

Evidently, this approach is effective due to the increase of surface energy, but not to the increased reactivity of substrates. To avoid the appearance of this latter feature the atoms of nitrogen in modifiers should be protected sterically; for example, with bulky substituents. Hence sterically hindered aminocompounds were suggested as effective modifiers for polymeric substrates [1726]. For instance, using the dimer of 3,3-dimethylene-2,2,6,6-tetramethyl-4-oxopiperidine Pritykin et al. [1727] increased the shear strength of the adhesive joints of poly(vinyl chloride) bonded with phenol–epoxy adhesive by 30% and with polyurethane adhesive by 140%. An even greater effect might

---

† Anti-surfactant properties exhibited by these modifiers may contribute additionally to the increase of the adhesive ability.

be expected from the sterically hindered amino-compounds containing unsaturated bonds as a source of π-electrons. Indeed, modification of poly(vinyl chloride) with oligovinyl-2,2,6,6-tetramethyl-4-ethynyl-Δ^3-piperidine [1728] resulted in the increase of the adhesive joint strength by an additional 5–15% as compared to the preceding case [1727]. Comparative data [1726] compiled in Table 4.21 confirm these suggestions. These were shown to be valid for other sterically hindered nitrogen-containing modifiers as well, for example, for quaternary phenylester ammonium salts [1729]:

$$[R\text{--}COOCH_2CH_2N(CH_3)_2CH_2 \text{---} \bigcirc]Cl$$

Additional enhancement of the effect is obtained by the increased concentration of nitrogen atoms, especially, when in combination with other atoms with unshared electron pairs. Consequently, application of 6-diallylamino-1,3,5-triazine-2,3,-dithiol on to the copper substrate resulted in a 2-fold rise of the strength of its adhesive joint with polyethene [1730].

Organosilicon modifiers have become widely used and they are indispensible in glass reinforced plastics production. The ideas discussed in section 4.1 reveal the reasons for the observation of Ahagon and Gent [1731] that vinylsilane, possessing a double bond, ensures a 35-fold stronger adhesive bonding as compared to the saturated ethylsilane. In a

Table 4.21 – Modification of substrates with sterically hindered amino compounds

Modifier	Substrates				
	PE[a]	PVC[b]	Metals[c]	PE-[a] steel	PVC-[c] metals
Nil	0.12	0.57–0.84	0.43–0.49 / 0.37–0.39	0.08	0.47–0.59 / 0.32–0.49
	0.19	0.68–0.97	0.51–0.53 / 0.43–0.47	0.14	0.59–0.64 / 0.49–0.57
2,2,6,6-Tetramethyl-4-oxo-piperidine	0.32	0.94–1.38 / 7.9–14.3	0.65–0.79 / 0.61–0.74	0.25	0.64–0.70 / 0.59
Dimer of 3,3-dimethylene-2,2,6,6-tetramethyl-4-oxopiperidine	0.33	0.98–1.39 / 8.2–15.1	0.63–0.82 / 0.60–0.71	0.29	0.66–0.85 / 0.51–0.63
Oligovinyl-2,2,6,6-tetra-methyl-4-ethynyl-Δ^3-piperidine	0.35	0.95–1.42 / 9.9–14.7	0.62–0.85 / 0.58–0.79	0.91	0.69–0.80 / 0.54–0.67

a   Shear strength, MPa.
b   Numerator, shear strength, MPa; denominator, peeling strength, kN/m.
c   Shear strength, MPa; numerator, steel 3; denominator, aluminium alloy D16AT.

variety of organosilicon compounds used for surface modification of polyethene the unsaturated ones were shown to be the most effective. These were vinyltriethoxysilane, γ-methacryloxypropyltrimethoxysilane, vinyl-tris(γ-methoxy)-silane and vinyl-tris(γ-methoxydiethoxy)-silane [1732]. Their effect was related to an intensification of interfacial interaction due to the increase of substrate surface energy. Indeed, using a ^{14}C radioactive tracer Dreyfuss et al. [1733] demonstrated that in the system involving polybutadiene and modified glass the number of interfacial bonds and the peeling strength of the system were interrelated. It appears reasonable to anticipate that incorporation of halogen atoms into the structure of a modifier would increase the ultimate effect. In fact, the work on peeling of carboxylated polybutadiene from modified glass fibre was 13.2 times higher and the durability of the system in pentane was 900 times greater when p-bromomethylphenyltrichlorosilane was used as a modifier instead of p-tolyltrichlorosilane [1734] (the latter observation concerning the durability of the system is in accord with the data of section 2.2.2). Similar results disclosing the effect of substrate modification on the surface energy and thermodynamic work of adhesion were obtained for polychloroprene reinforced with Vollastonite® [1735]. The data presented illustrate the approach relating the chemical nature of surface modifiers with the adhesive ability of the modified substrates.

When considering the methods intended to enhance the adhesive ability of substrates, along with the routes aimed at affecting the chemical nature of the surface, one should not ignore the role of structural factors. Of these changing the extent of microrelief is a priority. The increase of this parameter is accompanied by an increase in the interfacial contact area and an intensification of the microrheological flow of an adhesive into substrate defects. In real systems changes due to this factor are comparable with the effect of separate types of interfacial bonds [1736]. For a variety of polymers the surface topography is directly related to the strength of their adhesive joints, the shape of the microdefects (microhills and microtrenches) [1737] together with the mode of their distribution (specifically its uniformity) [1738] were shown to be the features of primary significance. Variation of surface topography provides an important potential source of a more efficient substrate pre-bonding treatment. For instance, the shear strength of the adhesive joints between polyethene and aluminium was increased when the arrangement of the extended structures in the boundary layer of the polymer coincided with that of the pores in the oxide layer of the metal [1739]. Consequently, subjecting the latter to the effect of various treatments provides the means to control the strength of its adhesive joints [1740], e.g. the roughness of a brass substrate was shown to be directly related to the critical fracture energy of its adhesive bonds with rubbers [1741]. It is the height of the microhills on the surfaces of various substrates, polished with emery paper of different grades, that governs the strength of their adhesive joints with poly(methyl methacrylate) or poly(vinyl chloride) [1742]. The degree of development of the substrate surface can also affect the strength of composites. For instance, the ultimate shear strength of the epoxyplast reinforced with carbon fibre (56–60% filler content) was found to be closely related to the surface roughness of the reinforcing fibre [1518]. The corresponding correlation coefficient calculated from the experimental data of [1518] is as high as 0.989. Fig. 4.28 provides indirect proof of this conclusion.

In principle, besides mechanical treatment, the surface roughness of a polymer can be affected by subjecting the substrates to the effect of active liquid media. For instance, microrelief of the surface of aramide fibres can be developed by treatment with bromine, which results in a 20% rise in the strength of the corresponding adhesive joints (composites) [1743]. For polyethyleneterephthalate films maximum roughness corresponds to the 20–30 μm thick etched layer. In this case, as well as for poly(vinyl chloride) and butadiene–acrylonitrile–styrene copolymer [1744], the magnitude of the effect is independent of temperature, or of the chemical nature of the medium, be it active solvent [1745], or an inert medium [1746]. However, it appears to be determined by the structural features of the boundary layers. For the substrate which is reactive with regard to the liquid not only the effect of macroscopic non-uniformity (distribution of microdefects) can be produced, but also that of microscopic factors (such as arrangement of macromolecular chains). For instance, depending upon the degree of orientation of polycarbonate films [1747] their roughness can be varied to a depth of 0.5–1.0 μm [1748] (for biological objects adhesive ability was also shown to be related to their elongation [95, 98]). Furthermore, as shown in Table 4.22, alkaline treatment of poly-carbonates is a function of temperature and of the treatment duration. Having processed these data, Chotorlishvili [1749] proposed a rigorous function $h(T, \tau)$, where h is the depth of the microdefects (roughness) of the substrate surface. For the ranges of T and τ specified in Table 4.22 the discrepancy between the calculated and experimental values of h does not exceed 18%.

Taking into account the role of structural factors, one might expect that physical excitations would produce a greater effect on the substrates than do liquid media. For instance, by means of pulsed laser irradiation [1750] the substrate microrelief can be given geometrical features which correspond to those of the incident beam [1750, 1751]. Despite its potential such approaches are quite specific and have not yet become widely

Table 4.22 — The roughness of polycarbonate films h (μm), as dependent on the temperature and duration of the treatment with the 52% solution of potassium hydroxide

τ (min)	T (K)						
	308	318	323	328	333	343	353
0	0.14	0.14	0.14	0.14	0.14	0.14	0.14
2	0.14	0.14	0.15	0.15	0.16	0.17	0.22
4	0.14	0.14	0.15	0.16	0.18	0.26	0.54
6	0.14	0.14	0.15	0.17	0.19	0.43	1.00
8	0.14	0.15	0.16	0.20	0.22	0.60	–
10	0.14	0.16	0.18	0.24	0.26	0.82	–
12	0.15	0.20	0.21	0.27	0.31	–	–
14	0.16	0.22	0.23	0.32	0.40	–	–
16	0.17	0.24	0.26	0.40	0.56	–	–
18	0.18	0.28	0.30	0.48	0.80	–	–
20	0.19	0.31	0.35	0.62	–	–	–
30	0.24	0.47	0.74	–	–	–	–
40	0.34	0.66	–	–	–	–	–
50	0.45	–	–	–	–	–	–

used. In present day practice mechanical methods of affecting the surface roughness of polymeric substrates are still the most common [1476, 1752] and these were shown to be highly effective in increasing the adhesive ability even of metals [1753].

The problem of regulating roughness is associated with the need to consider the characteristics of the abrasive instrument as well as the effect on the nature of the substrate. For instance, Zorll [1754] considers that the maximum strength of the adhesive joints of metals results from their treatment with 'the coarse abrasive particles of irregular shape'; spherical particles produced a somewhat worse result, whereas elongated abrasive particles were the least effective of all. Wartusch and Saure [1755] suggested air-washing of the substrate covered with an abrasive as the mechanical treatment procedure for polymers. In this case the practical variables of the process are quite important, e.g. the pressure of the air stream, the distance from the surface being treated and the incidence of the air stream with regard to this surface; however, the size of the abrasive grains appears to be decisive [1756]. When the size of the latter (which was corundum) was around 120 μm the depth of the defects in the transition layers of polymeric materials was 1–7 μm, the enhanced adhesive ability of the electroplated copper substrates treated in this way being retained for 10 months [1755]. Gilbert (1757] advises the use of corundum particles with an average diameter of 169 μm, the measured depth of the microdefects on the substrate surface being 3.47 μm. To eliminate the effects due to the chemical nature of an abrasive Goloubkov and Savel'eva [1758] suggested the use of the frozen polymer itself.

One should bear in mind that, along with the development of microrelief, the positive effect on the adhesive ability of substrates of the types of mechanical treatment discussed is also due to atoms with unbonded valences (and thus exhibiting increased activity), being generated at the faces and on the defects in crystalline polymers or in the separate zones of amorphous polymers. Moreover, in the course of abrasive treatment the boundary layers of polymers can become charged, the intensity of the electric field produced can be as high as $10^8–10^9$ V/m [24, p. 74]. And, finally, one should take into account that weak boundary layers are inevitably removed from the surfaces in the course of such mechanical treatment, this feature producing an undoubtedly positive effect on the adhesive joint strength.

A peculiar example of the effect of mechanical treatment of substrates on the adhesive bond strength was observed in adhesive joints between metals bonded with an elastomeric adhesive [1759]. In such a system the rates of interfacial interaction are different on each side (boundary) of the interface, these being balanced so that the decrease of one (effected by polishing the corresponding surface of the metallic substrate) results in an increase of the other. In this way the strength of rubber–metal adhesive joints was substantially increased. Apparently, such effects are common to a large number of similar examples found in practice. These are due to the non-simultaneous diffusion of the individual components from the adhesive phase (for example, of phenolaldehyde oligomers [1760] and isocyanates [1761]), which is especially dramatically observed in adhesive joints of elastomers with metals [1760, 1762].

Because of the effects described which are inherent in the mechanical treatment of polymeric substrates the chemical nature of their boundary layers is changed. This feature was noted [1763] at the same time as the mechanical concept of adhesion was stated. However, firm principles concerning the effect of this treatment on the adhesive ability could not be derived until the mechanically-induced transformations in substrates were investigated. In polyethene these transformations were shown to result in the generation of oxygen-containing groups [1572]. Table 4.23 compiles the quantitative data obtained by IR-spectroscopy [1764]. Note that mechanodestruction of polyethene in air as an inert atmosphere is accompanied by the formation of those products that demonstrate, according to section 4.1, enhanced adhesive ability. Taking into account that the formation of oxygen-containing and of unsaturated compounds proceeds predominantly via a free-radical mechanism, it appears proper to claim that the mechanical treatment of polymeric substrates should be accompanied by generation of free-radical states.

Table 4.23 — The products of mechanodestruction of polyethene

Composition	Concentration, $C \times 10^{-12}$, m^{-3}		
	Prior to destruction	Destruction in air	Destruction in helium
$RR'C=CH_2$	1.6	3.2	2.9
$RCH=CH_2$	3.0	6.0	4.9
$RCH=CHR'$	0.7	6.4	2.0
RCH_3	17.0	32.0	24.0
$RCOOH$	0.4	0.5	1.0
$RCH=O$	0.6	2.0	5.0
$RCOOR'$	0.5	1.0	2.1

This conclusion is in full accord with the modern principles of mechanochemistry. For instance, generation of macroradicals during the wear of polymeric substrates in vacuum was deduced from the intensity of electron emission which decreased on the introduction of a radical acceptor (hydroquinone) [1765]. With the aid of a similar but a more effective acceptor, 1,2-diphenylpicrylhydrazyl, radicals were detected on the surfaces of polyesters and polytetrafluoroethene subjected either to mechanical treatment or to the effect of glow discharge [1766]. In ESR studies macroradicals with lifetimes from 10^{-6} to 10^{-3} s in volumetric concentrations of up to 10^{15} mol^{-1} were registered [1767]. Free radicals were also detected by IR-spectroscopy in the friction zones of various polymers with metals [1768]. If their concentration is additionally decreased by modifying the surface with N-isopropyl-N'-phenylphenylenediamine-1,4 the friction coefficient of the modified polyamide or polycarbonate decreases by 5 and 9 times, respectively [1769].

Such effects are associated with variation of the Schockley levels, hence formation of radical-containing products should ensure the enhancement of the adhesive ability of polymers. Indeed, no less than a two-fold increase in the concentration of paramagentic centres on the surface of carbon fibre was shown to provide for the increase of the strength of epoxyphenol composites [1770]. The fact that this effect is related to the free-radical states generated is confirmed by the correlation between the pull strength of the fibre from out of the matrix and the ratio of the concentration of paramagnetic sites to the specific surface of the substrate. A more vivid demonstration of the validity of this conclusion is supplied by the data of Kobets [1518] relating the strength (described by the same parameter as in Fig. 4.28) of the systems under discussion to the concentration of paramagnetic sites in the boundary layers of substrates, Fig. 4.43 depicts the corresponding pattern. Note that at all treatment temperatures covered in the experiments, the straight lines describing these dependences converge to a single point, which is obviously indicative of the uniform mechanism of interfacial interaction. The relationship between the strength of adhesive joints and the efficiency of the generation of free radicals during the mechanical pretreatment was shown to be valid for a large variety of substrates, from silicates [1771] to wood [1772]. Hence, to ensure greater joint strength it appears sensible to choose the pretreatment and bonding conditions under which recombination processes within the boundary and transition layers of substrates are either inhibited or completely terminated [1773–1775]. The main impediment to this is associated with the short lifetime of radicals thus making it difficult to retain the properties of the treated substrate until it is brought to interact with an adhesive.

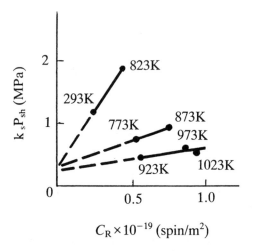

Fig. 4.43 – $k_s P_{sh}$ versus concentration of paramagnetic sites on the surface of substrates treated with air at different temperatures; where P_{sh} is the shear strength of the adhesive joints between carbon fibres and epoxy–anilinephenolformaldehyde matrix, k_s is the coefficient describing the completeness of the interfacial contact.

To eliminate this drawback Lerchental *et al.* [1773] and Genel', Vakula and Fokin [1776] independently suggested applying the mechanochemical action on the substrate covered with the layer of adhesive, which in this case performs as an effective reagent for atoms with unbonded free valences generated in the boundary layers. Sufficiently high concentrations of radicals can thus be achieved, e.g. for sand-papered polyethene and polypropene these were 14×10^{17} and 0.99×10^{16} spin/m^2 [1773]. Modifications of this method have been reported. Mechanical treatment of the substrate was performed in the presence of vinyl monomer [1773], the variety of the grafted functional groups being thus substantially increased. Instead of treating the substrate covered with adhesive it is even more advantageous to subject the adhesive joint as a whole to mechanical action provided that the adhesive contains an abrasive [1777] and hence, reducing the number of steps in the bonding procedure. The latter treatment can be further intensified by ultrasonic waves [1778]. At the same time introduction of the additional abrasive component into the adhesive formulation is, of course, the undesirable feature. It appears reasonable to use one of the basic components of the composition provided that the latter is in the solid state. For this purpose solid carbon dioxide was introduced into the adhesive, the system then being subjected to mechanical treatment [1779]. The shear strength of the adhesive joints involving steel and an epoxy-based adhesive formulation was thus increased up to 3.9 MPa.

Hence, using the methods which utilize these ideas makes it possible to enhance the adhesive ability of polymers [1780, 1781]. The data of Genel' and Vakula [1782] compiled in Table 4.24 serve to prove the validity of this conclusion. The adhesive joints listed involve a number of both non-polar as well as polar polymeric substrates bonded with an epoxy-based adhesive formulation (epoxy oligomer ED-20 cured with oligoamide L-20), polychloroprene-phenolformaldehyde adhesive (88-N), polyurethane

Table 4.24 — Tensile strength, MPa, of the adhesive joints of polymers activated by mechanical treatment

Adhesive joint		Mechanical method	Mechanochemical[a] method
Substrate	Adhesive		
Low-density polyethene (LDPE) with:			
LDPE	ED-20/L-20	0.41	2.5/4.2
LDPE	PU	0.42	2.3/3.4
LDPE	88-N	0.48	2.7/3.3
Polypropene	ED-20/L-20	0.43	2.4/4.2
Polyvinyl chloride	ED-20/L-20	0.42	2.2/3.4
steel	ED-20/L-20	0.43	2.8/4.8
Aluminium alloy	ED-20/L-20	0.41	3.4/5.0
copper	ED-20/L-20	0.39	2.4/3.9
High-density polyethene (HDPE) with:			
HDPE	ED-20/L-20	1.20	4.3/5.3
LDPE	ED-20/L-20	0.43	2.7/4.3
Polypropene	ED-20/L-20	0.61	5.2/6.7
Polyvinyl chloride	ED-20/L-20	1.90	2.4/3.2
Polytetrafluoroethene (F-4) with:			
F-4	ED-20/L-20	0.34	2.6/4.0
F-4	LFE-26x	0.39	0.9/1.4
steel	ED-20/L-20	0.37	3.1/3.9
copper	LFE-26x	0.44	1.1/1.7

a Numerator refers to mechanochemical treatment of the substrate coated with unfilled adhesive, denominator refers to mechanochemical treatment of the substrated coated with adhesive filled to the 50% (vol) content with silicon carbide of particle size of 160 μm.

adhesive (polyurethane PU derived from polyester K-24 cured with 2,4-tolylenediiso-cyanate) and a fluoropolymer-epoxy (LFE-26x) formulation.

The fact that the effect observed is associated with the formation of free radicals at the surface was confirmed in studies with ESR- and ATR–IR-spectroscopy. Radicals were detected at temperatures up to the polymer melting points, indicating that these (radicals) were protected by the layer of adhesive [1783]. The high durability of the adhesive joints during long-time storage [1782] or under the effect of active media [1784], illustrated by the data of Table 4.25, provides, according to section 2.2.2, indirect evidence of the free radical mechanism of interfacial interaction ensuring the formation of covalent interfacial bonds. As far as durability can be described by exponential dependences, (according to the concepts of section 3.1.2), the structural features of polymers responsive to the change of their adhesive properties should be accounted for by means of parameter λ of eq. (331). Then, the strength of mechano-chemically-produced adhesive joints must be related to the molecular mass, degree of orientation and crystallinity of the substrate. This kind of relationship was observed in adhesive joints between low- and high-density polyethenes [1785]. Hence the shear strength of adhesive joints involving polyethene appears to have higher values when the substrate is pretreated mechanochemically rather than by the conventional mechanical procedure [1786]. Likewise, an increase in the degree of crystallinity of substrates was shown to result in an increase of the adhesive joint strength [1781]. Finally, dependent upon the treatment temperature, the efficiency of the mechanochemical procedure was found to follow a pattern exhibiting multiple peaks [1779, 1787], these satisfactorily corresponding to the relaxation transitions in the substrates and, what is even more important, to the maximum yields of paramagnetic sites. Maximum joint strengths correspond to the low-temperature transition region associated with freezing of the segmental mobility [1787].

At the same time, one should remember that the paramagentic sites observed in ESR-investigations are secondary products resulting from the interaction (taking place during mechanochemical treatment) of the primary macroradicals (whose own life-times are rather short ca. $10^{-3}–10^{-6}$ s) with an adhesive or with the substrate boundary layer. This conclusion is supported by the fact that the efficiency of the mechanochemical procedure is explicitly related to such atomic parameters as ionization energies, electronegativity, and affinity for electrons [1788] of the adhesive and the substrate.

Table 4.25 – Shear strength, MPa, of the adhesive joints of polytetrafluoroethene bonded with a fluoro-containing epoxy formulation

Substrate pretreatment procedure	Initial joint strength	Shear strength after exposure of the joint to the effect of:[a]		
		Air	Water	1% nitric acid
Glow discharge	5.7	4.6	0.8	0.3
Mechanical	0.4	0.2	0	0
Mechanochemical	4.1	4.4	4.0	3.8

a Exposure conditions, 1000 hr at 373 K.

Thus, the mechanochemical method of enhancing the adhesive ability of polymers seems to match the requirements ensuing from the physico-chemical concepts here being developed. Hence the firmly based efficiency of this procedure, which has become useful in the production of strong and durable adhesive joints of even such hard-to-bond polymers as polyethene and polyterafluoroethene [1782, 1789]. However, there is a more direct way to increase the 'radical density' of the substrate surface and to achieve the direct radical mechanism of interfacial interaction, instead of the interchange mechanism. According to section 4.1, this might be the direct incorporation within the boundary and transition layers of compounds possessing free valences. Stable organic radicals were suggested for this purpose [1790]. Table 4.26 lists typical representatives layered on to the substrate surface from organic solutions.

Table 4.27 compiles the data on the strength of adhesive joints of various materials modified with stable radicals [1790, 1791]. As is seen, for each pair of substrates the values for the joint strength are quite close regardless of the chemical nature of these radicals. We believe this feature to be indicative of the uniform mechanism of interfacial interaction and, consequently, a certain limiting adhesive ability of substrates appears to be attained owing to the fact that the substrates utilize identical adhesive ability-stimulating structural fragments, namely the free valences.

This effect is also seen in other types of adhesive joints, e.g. paint and varnish coatings and composites. As is illustrated in Table 4.28, modification of the substrates with stable radicals ensures the attachment strength of polydivinylacetylene (PDVAc), silicone (Sil), chorosulfopolyethene (CHSPE) and nitrocellulose (NC) coatings (as tested by the common scratch procedure). For composites reinforced with glass fibre modified by various stable radicals (and for the sake of comparison by the most common organo-silicon coupling agents) the measured strengths in static flexure were as follows (MN/m^2):

γ-Aminopropyltriethoxysilane	2590
Vinyltrimethoxysilane	3990
Methylmethacryloxytriethoxysilane	4780
Glycyloxypropyltrimethoxysilane	5300
Aliphatic radical III	5740
Biradical XII	6000
Unsubstituted radical IV	6440
Oxosubstituted radical VI	6590
Chlorosubstituted radical VII	6700
Triradical XIII	6850
Organosilicon radical XIV	7140

These data demonstrate [1790] that stable radicals are substantially more effective (in increasing the adhesive joint strength) than the known modifiers, particularly those with unsaturated fragments. Note the increasing adhesive ability of the substrates modified with biradicals and, moreover, with triradicals, emphasizing the effect of the concentration of unbonded electrons in a modifier molecule.

Even under rather severe conditions involving the effect of strong acids, which are required in polymer electroplating [1792], the introduced stable radicals do not recombine at the substrate surface. Consequently, adhesive ability of substrates in the course of electrochemical coating by nickel is increased [1793]; indeed, when the concentration of the oxyradical V was increased from 0 to 0.1 g/l the bond strength of

Table 4.26 − Stable radical modifiers of substrate surfaces

Modifier No.	Stable radicals	Application conditions	
		Solvent	C (%)
1	2	3	4
	Hydrocarbon		
I	Perchlorotriphenylmethyl	CCl$_4$	5
	Phenoxyl		
II	Bis(diphenylmethane)-benzene-2,4-diisopropyl-phenoxyl	Acetone	5
	Iminooxyl		
III	Di-*tert*.butyliminooxyl	−	−
IV	2,2,6,6-Tetramethylpiperidine-1-oxyl	Acetone	5
V	2,2,6,6-Tetramethyl-4-hydroxypiperidine-1-oxyl	Acetone−hexane (1 : 1)	3
VI	2,2,6,6-Tetramethyl-4-oxopiperidine-1-oxyl	Hexane	10
VII	2,2,6,6-Tetramethyl-4-chloropiperidine-1-oxyl	Hexane	5
VIII	2,2,6,6-Tetramethyl-4-bromopiperidine-1-oxyl	Hexane	5
IX	2,2,6,6-Tetramethyl-4-nitrilepiperidine-1-oxyl	Benzene	3
X	2,2,6,6-Tetramethyl-4-benzoyloxypiperidine-1-oxyl	Benzene	3
XI	Bis(2,2,6,6-tetramethyl-1-oxyl-4-piperidyl)-succinate	Acetone	5
XII	Bis(2,2,6,6-tetramethyl-1-oxyl-4-piperidyl)-terephthalate	Dimethyl−sulfoxide	3
XIII	Tris(2,2,6,6-tetramethyl-1-oxyl-4-piperidyl)-trimesinate	Chloroform	5
XIV	Tris(2,2,6,6-tetramethyl-1-oxyl-4-piperidyloxy)-phenylsilane	Benzene	10
XV	3,3-dimethylene-2,2,6,6-tetramethyl-4-oxopiperidine-1-oxyl dimer	Benzene	20
XVI	Oligovinyl-2,2,6,6-tetramethyl-Δ^3-piperidyl-1-oxyl	Hexane	3
	Hydrazyl		
XVII	1,1-Diphenyl-2-picrylhydrazyl	Benzene	5
XVIII	1,1-Diphenyl-1,3,4-triacetoxy-phenylhydrazyl	Acetone	3
	Verdazyl		
XIX	Triphenylverdazyl	Acetone	5

Table 4.27 — Shear strength, MPa, of the adhesive joints of materials modified with stable radicals

	Modifiers													
	I	II	III	IV	V	VI	VII	VIII	IX	X	XI	XII	XIII	XIV
Steel 3														
Steel 3	20.4	33.8	33.9	30.4	33.7	35.9	29.7	28.9	37.3	34.5	34.1	33.8	32.4	35.1
Caprolon	10.0	12.1	12.9	14.2	16.8	8.4	9.3	12.8	13.1	12.7	13.0	11.2	13.4	14.5
Styrene rubber[a]	CFT	CFB	CFB	CFB	CFB	CFB	CFB	CFB	CFB	CFB	CFB	CFB	CFB	CFB
Nitrile rubber[a]	CFT	CFB	CFB	CFB	CFB	CFB	CFB	CFB	CFB	CFB	CFB	CFB	CFB	CFB
Aluminium alloy D16AT														
D16AT	12.6	21.2	20.6	22.8	27.6	19.6	20.0	20.7	21.4	21.2	21.1	20.9	22.0	22.8
Caprolon	5.7	11.4	12.0	12.9	14.8	9.4	10.0	10.1	12.1	11.7	11.9	10.7	12.8	13.2
Nitrile rubber[a]	CFT	CFB	CFB	CFB	CFB	CFB	CFB	CFB	CFB	CFB	CFB	CFB	CFB	CFB
Polyvinylchloride[b]														
Polyvinylchloride[b]	5.7	10.2	10.2	—	—	—	—	—	3.4	10.0	10.0	10.9	11.2	12.4
Caprolon	3.0	11.8	12.7	14.3	16.5	9.7	10.2	12.1	12.7	12.4	12.5	10.7	12.9	13.7
Polyamide[c]	14.5	19.0	17.9	20.7	22.9	18.2	19.0	19.1	19.4	18.7	16.9	16.4	19.8	20.4
Rubbers[a,d]	CFT	CFB	CFB	CFB	CFB	CFB	CFB	CFB	CFB	CFB	CFB	CFB	CFB	CFB

a CFT, cohesive fracture involving transition layers of the substrate; CFB, cohesive fracture through the bulk of the phase.
b Peeling strength of autohesive joints of polyvinyl chloride (kN/m).
c Autohesive joints of glass-reinforced polyamide CP-68.
d Autohesive joints of rubbers based on polyisoprene (SKI-3), butadiene–styrene (SKS-30), polychloroprene (Nairit NT) and butadiene–acrylonitrile (SKN-40) elastomers.

Table 4.28 — Relative number (%) of the samples of steel 3 (numerator) and aluminium alloy D16AT (denominator) where coatings detached.

Modifier[a]	Coating			
	PDVAc	Sil	ChSPE	NC
No.	34/19	37/44	27/19	47/67
VI	24/27	26/37	17/11	38/52
V	27/33	30/37	24/10	44/49
III	16/17	17/25	8/4	21/21
IV	33/36	35/40	24/16	43/58
VII	27/32	28/35	16/10	37/37
VIII	32/30	33/33	20/9	39/40
XII	25/31	28/33	19/14	34/42
XIII	19/24	21/29	11/7	27/29

a   Numbered as in Table 4.26.

the coating increased on the average by 17% and, when further increased to 0.4 g/l, a further 6% was added to the final bond strength. The effect was similar when the modifier was incorporated directly within the substrate boundary layers [1794] rather than introduced into the electrolyte solution.

The efficiency of modifying the substrates with stable radicals is seen most markedly in the adhesive joints of polytetrafluoroethene or polyethene [1795]. Peeling strength of the joints involving these polymers pretreated with the compounds V, VI, IX–XI, XIV and XVI (as listed in Table 4.26) was shown to increase by 175% for LDPE and by 160% for polytetrafluoroethene F-4, as compared to the 40% rise ensured by acidic etching or modification with one of the most active modifiers, i.e. chromium-containing 'Volan' [1726]; the corresponding values for the peeling strength are as follows (kN/m):

Modifier	LDPE	F-4
Nil	1.2	1.7
Dichromate solution in sulfuric acid	1.5	1.9
'Volan'	1.7	1.8
Hydroxysubstituted monoradical V	2.2	3.9
Oxosubstituted monoradical VI	2.5	4.3
Nitrile-substituted monoradical IX	2.7	4.4
Phenyl-substituted monoradical X	2.8	3.7
Biradical XI	5.3	5.7
Silicon-containing triradical XIV	5.4	5.1
Oligomeric monoradical XVI	2.5	2.0

It can be seen that the maximum increase of the adhesive joint strength is achieved when using biradical XI; radical XIV provides for the structuring of the transition layers of polyethene rather than for an increase in their reactivity. The effect is minimal for radical V; nevertheless, even in this case it amounts to an increase of 80–85% in the peeling strength. Similar improvements were shown in the bonding of polytetrafluoroethene [1726] and of poly(vinyl chloride) [1728] premodified with stable radicals.

Investigations into the effect of the nature of the substituent in the 4-th position in the iminooxyl radicals of the piperidine series (representing the most thoroughly examined class of such modifers) support the validity of the concepts concerning the role of atoms with unbonded electron pairs or of halogens in adhesive interaction (see Tables 4.27 and 4.28). Radicals containing atoms of chlorine (as VII) or of bromine (as VIII) ensure a higher adhesive joint strength than the oxygen-substituted iminooxyl radicals. Comparison of the efficiency of radicals V, VI and IX demonstrates that the adhesive ability of the correspondingly modified substrates is primarily governed by the 'reactivity' of the free valence in these radicals. For these modifiers the pattern of the adhesive joint strength variation is the inverse of that characteristic of hydroxy-, oxo- and nitrile-substituted non-radical modifiers. Taking into account that bi (XI and XII) and triradicals (XIII and XIV) lead to a larger increase of the adhesive ability as compared to mono-radicals (and the density of the free valences in these is obviously lower), one can infer that the strengthening effect is due to unbonded electrons. This conclusion is also confirmed by the data on the refractive indexes of the boundary layers in substrates modified with stable radicals, as shown in section 2.2.1 and these are directly related to the surface energy of polymers.

Today stable radicals are the most effective of the surface modifiers intended to enhance the adhesive ability of polymers. This feature is due to the fact that they fulfil the requirements which follow from the theoretical ideas of section 4.1. In ESR-spectroscopic studies it was demonstrated that the free valences of stable radicals were unaffected by their sorption by the substrates. Hence, the observed strengthening of the adhesive joints is associated with the addition or substitution mechanism of interfacial interaction being exchanged for a free-radical one. The recombining character of the reactions involving free radicals accounts for the latter mechanism being independent of the chemical nature of the modified substrates and ensures a universal efficiency of the corresponding modifiers.

When considering the enhancement of the adhesive ability of substrates one cannot avoid discussing the problem of fillers. This problem was partly treated earlier when analysing some aspects of the subject of this section. However, considering its significance the problem calls for a separate discussion. Indeed, considering the reinforcing elements of composites as fillers within the broad meaning of the term and taking into account that any filled polymeric composite can be treated as the sum of elementary polymer-filler adhesion systems, it appears justifiable to extend to these the same physico-chemical treatments relating to the adhesive ability of substrates.

Using these ideas Voyutskii et al. [1796] suggested an adhesion theory for elastomer reinforcement [1797]. We believe that its development on the basis of the concepts of section 4.1 creates reliable grounds rather than being an empirical approach, for constructive selection of the ways to enhance the adhesive ability of fillers. In fact, concerning the consistencies of interfacial interaction in composites and polymer solutions [1804] empirical approaches reveal the predominant role of the surface energy of a filler (e.g. bulk microelements [1798, 1799], finely dispersed particles [1800–1802], pigments [1803] etc.). However, they do not indicate any general way of specifically controlling this parameter. To a certain extent this is explained by the difficulties in measuring the bond strength in the matrix-filler systems [799, 1805] making it difficult to analyse the experimental dependences in terms of adhesion. At the same time such analysis becomes feasible if one surveys the adhesive ability of substrates as a feature associated with atoms with unbonded electron pairs or free valences. Let us consider the most typical results.

Adhesive ability of such a widely used filler as carbon-black substantially increases when phenolic hydroxy-functional and, especially, carboxy-functional groups are introduced into its boundary layers [1806]. A much greater effect is demonstrated by other atoms from the list discussed in section 4.1, for example nitrogen. In a comparative study of the efficiency of various pretreatment procedures for fillers, carbon-fibre substrate was used which was subjected to direct oxidation or treatment with ethylene glycol or monoethanolamine. It was found that the force required to pull the modified carbon filament out of a phenolformaldehyde matrix was, as compared to the unmodified substrate, 1.61, 1.53 and 2.23 times higher, respectively [1807]. Simultaneous introduction of the atoms of oxygen and nitrogen by combining oxidation of the substrate and its reaction with monoethanolamine resulted in an increase of this parameter by a factor of 2.42. The validity of the initial concept is supported by the data on halogenation. Indeed, Messick, Progar and Wightman [1808] disclosed a strictly linear dependence (which was inversely linear in full agreement with the thermodynamic considerations of section 2) between the concentration of fluorine atoms in the boundary layers of graphite fibres and their critical surface tension. While modification of synthetic ceolyte with polyethyleneoxide or poly(ethene-*co*-vinyl acetate) solutions was shown to result in an increase of the shear strength of the adhesive joints of steel bonded with a ceolyte-filled furane-epoxy formulation by 42.3 and 46.0%, respectively, ceolyte modification with chlorosulfopolyethene produced a 55.9% increase [1809].

The isocyanate function is known as one of the most reactive functional groups, hence the observed substantial reinforcement of composites filled with carbon black [1810] or titanium dioxide [1811] and pretreated with isocyanates. In the case of glass reinforced plastics the efficiency of adhesive interaction between the epoxy matrix and glass fibre coated with coupling agent was shown to be directly related to the number of uncompensated valence bonds at the substrate surface [1812]. However, quite naturally, the maximum effect is observed when free radical states are involved in the interaction. This was the case with vulcanized rubber powders treated with *p*-nitrosodiphenylamine as the source of radicals [1813].

Taking into account the data presented it appears correct to expect that the adhesive ability of various fillers would be enhanced by treating them with certain nitrogen—oxygen-containing compounds, as well as with triisocyanate derivatives of biuret, or stable organic radicals.

To illustrate the first case finely dispersed silicon dioxides (aerosils) were used as fillers where the surface layers contained 0.5—1.4 mmol/g of aminoethoxy groups (AEA), 0.05—0.3 mmol/g of anilinomethyldiethoxy groups (AM), 0.1—0.6 mmol/g of amino-silazane groups (ASA), and 0.6—1.0 mmol/g of diethylene glycolic groups (DEG). The effective reinforcement of these fillers was found to be comparable to the effects produced by the common curing promoters, *p*-oxyphenyl-β-naphthylamine (POPhNA) [1814], *N,N*-dimethyl-*p*-phenylenediamine (DMPhDA) [1815], and copper salicylalimine (CSI) [1816]. Introduction of these promoters into compositions filled with non-modified aerosil was shown to strengthen the adhesive joints [1817] due to acceleration of the hot curing of adhesives [1818–1820]. Compositions filled with the modified fillers display enhanced mechanical properties [1821], whereas their use as adhesives ensures a significant rise of the shear strength of the adhesive joints of metals [1822]. As is seen from the data of Table 4.29, for the three groups of adhesives containing different curing promoters, modification of the filler with only the hydroxy functions (DEG) produces an increase in strength of 60, 20 and 92.3%, respectively, whereas with fillers modified with compounds comprising both the atoms of nitrogen and oxygen (AEA, AM and ASA) the increase (on the average for the three promoters) is by 2.9, 1.9 and 2.3 times, respectively.

Table 4.29 – Shear strength, MPa, of the adhesive joints of steel 3 bonded at 423 K with promoted adhesives filled with modified aerosils

Functional groups within the surface layers of filler	Promoter of adhesive		
	POPhNA	DMPhDA	CSI
Nil	4.5	7.5	5.2
ASA	11.5	12.5	15.5
AM	15.6	15.4	18.0
AEA	10.0	14.3	16.5
DEG	7.2	9.0	10.0

Similar effects were achieved when the fillers were modified with isocyanate functions introduced with the help of N',N'',N'''-tris/ω-isocyanatoalkyl(alkaryl)/-biurets. The benefits of using these as curing agents for adhesives were illustrated in section 4.2.1, i.e. increased mobility of isocyanate functional groups, ensured by their remote position at the ends of alkyl groups, e.g. hexamethylene chains, was shown to be of special significance. It seemed reasonable to anticipate that this feature would be valuable with regard to the reinforcement of filled polymeric compositions. To verify this proposal carbon black (DG-100) or aerosil (A-175) premodified with triisocyanate biurets B-1 and B-2 were introduced into a standard system based on butadiene–acrylonitrile rubber SKN-26. The modifiers B-1 and B-2 are quite different with respect to the mobility of isocyanate functions and in the second one it is substantially hindered due to steric factors. Rubbers of the same composition filled with non-modified fillers or with those treated with 2,4-tolylenediisocyanate (whose structure offers a much poorer fit to the adhesives design requirements than those of B-1 and B-2) [1810] were used for comparison. The use of the fillers suggested produces good ultimate strength in the filled rubbers and a definite increase in the relative elongation, the modulus at 300% elongation being somewhat decreased [1823]. However, the values of these parameters are still within the normal range, whereas, according to the data of Table 4.30, the shear strength of rubber–steel adhesive joints increases because of the considered use of the correspondingly modified fillers [1824]. The effect is seen to a much greater extent when the fillers are modified with compounds exhibiting free valences [1825], the observed increase of the joint strength being associated with the fact that the free-radical mechanism of interfacial interaction is more effective than that involving substitution reactions (Table 4.30) [1826].

Table 4.30 — Shear strength, MPa, of the adhesive joints between steel and vulcanized rubbers reinforced with modified fillers

Modifier	Modified filler	
	Carbon black	Aerosil
Nil	0	0
2,4-tolylenediisocyanate	3.0	4.0
B-1	5.2–5.5	5.7/6.2
B-2	3.6–3.8	6.9/7.2
Hydroxysubstituted monoradical V	5.0	8.5
Oxosubstituted monoradical VI	6.0	8.5
Benzoyloxysubstituted monoradical X	5.7	6.4

Thus, by varying the surface composition of a filler, so as to meet the requirements discussed the expected regular rise of the adhesive ability of the substrate (filler) is achieved in this case also, hence the increased efficiency of the fillers in reinforcing the composites. The validity of this approach, as applied to reinforced elastomer compositions, was demonstrated by way of the example involving model systems comprising elastomers as adhesives and, as substrates, fabrics coated with carbon black or aerosil modified with triisocyanate biuret derivatives or stable radicals [1827]. As is seen from Fig. 4.44, the plots of the peeling strength of the corresponding model adhesive joints versus relative elongation at tear of the filled rubbers are linear. Hence, it is clear that the adhesive properties of the modified fillers and the resulting reinforcement effect in elastomers have the same mechanism (the slopes of the lines in Fig. 4.44 reflect the relative effectiveness of the individual modifiers and these are in accord with the data of Table 4.30). The results discussed confirm the validity of the approaches taken to determine ways of increasing the adhesive ability of substrates and to reveal the possibilities of applying these concepts in bonding practice.

The generality of the ideas discussed makes it possible to apply them to the analysis of the performance and behaviour of a variety of materials interacting to form adhesive joints in the broadest meaning of the term. An impressive example supporting this conclusion can be seen when observing the modes of carcinogenesis. In section 1.1 we demonstrated that in principle this process should obey the general features of adhesion, adhesive ability of the tumour cells being not less than an order of magnitude lower than of the normal cells. In fact, using an improved technique for the measurement of the force for cell separation (which in terms of physical chemistry corresponds to their adhesive ability) Modyanova [1828, 1829] discovered that this characteristic was regularly lower for spontaneous and induced tumour tissue *in vivo* as well as *in vitro*. For instance, for mouse hepatocytes in ontogenesis from 2.5 to 16 months the separation force per one cell reduced from 0.075 ± 0.001 to 0.040 ± 0.001 mg, whereas in *o*-amino-

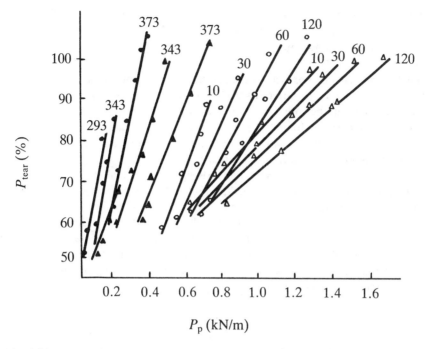

Fig. 4.44 — Relative elongation at tear of vulcanized rubbers based on butadiene–acrylo-nitrile elastomer SKN-40 filled with carbon black modified with N',N'',N'''-*tris*(6-isocyanate-hexyl)biuret (○, ●), filled with silicon dioxide modified with 2,2,,6,6-tetramethyl-4-oxo-piperidine-1-oxyl (△, ▲) versus peeling strength of the corresponding model adhesive joints at different bonding times (open symbols) and bonding temperatures (solid symbols); the values of τ (min) and T (K) are shown in the figure.

azotoluene-induced hepatocarcinogenesis during the same period of time it reduced from 0.078 ± 0.002 to 0.016 ± 0.001 mg [1830], i.e. the decrease is more substantial than in the control experiment. Besides, adhesive ability of the parenchimic liver cells and of the lung alveolar epithelium cells was found to be related to the genetic tendency to tumour growth [1831].

These facts were perceived within the concept of a system of mechanical integration of tissue, however, we believe that they can be much more correctly accounted for within the framework of the theoretical concepts of adhesion. Consequently, modification of the external membranes of the tumour cells thus increasing their adhesive ability from the physico-chemical standpoint (neglecting the effect of genetic factors) seems to be a prospective way to retard tumour growth [54] or at least hamper proliferation [1832].

DNA specific preparations of contactant type (macromolecular protein-containing fractions of calcium-free saline tissue extracts) [1833, 1834] were suggested as agents to increase the strength of intercellular contacts [1835, 1836]. Injection of contactants in an early post-natal period to mice deficient in cell adhesive ability and, therefore, dis-playing tendency to spontaneous blastogenesis in the corresponding internal organ was shown to result in the increase of the cells' separation force and to reduce the occurrence of spontaneous tumours [1837]. Injection of contactant (isolated from lung tissue) either simultaneously with a carcinogenic agent (urethane) or a month prior to it reduced the occurrence of lung adenomas, antiblastogenic activity was displayed by only those doses of contactant that increased the adhesive ability of cells [1832]. By such treatment the occurrence of lung and liver tumours in mice was reduced by 50 and 32%, respectively [1838].

A much more general approach to the solution of the problem discussed may involve modification of only the cell surface (wall) but not of the cell as a whole; this may appear especially beneficial if the modifiers are radicals [54]. In fact, in tumour cells concentration of free radicals was found to be substantially lower than in normal cells [1839]. The validity of this assumption was supported in comparative studies of the anti-tumour activity of the known preparations 'Urethane' and 'tioTEF' with their analogues displaying free valences manifested as a stable ESR-signal, these analogues were shown to be more effective than the non-radical agents [1840]. Similar effects were observed in a number of other biological subjects [1841] which apparently rely on the physically non-contradictory principles of controlling the adhesive ability of condensed phases.

Summing up, the trends in regulating the adhesive properties of polymers developed on the basis of generalized concepts of adhesion can be assumed to be a reliable basis on which many practically important problems concerning intensifying the processes of interfacial interaction can be solved.

Conclusion

Apparently, any conclusion has to involve two elements, i.e. a summing up and a description of the prospects in the development of the field of science under discussion, at least in the way the authors conceive it.

In the introduction we declared our intention to interpret the phenomenon of adhesion, and the rationale of doing this, relying on the two traditional physical approaches, thermodynamic and molecular–kinetic. We have striven along this course throughout the monograph, and have ultimately concluded that neither the thermo-dynamic nor the molecular–kinetic approach are, as was assumed, closed systems with a language and formalism of their own and with their own range of concepts and ideas. Alas, they form a cluster of interrelated and interwoven concepts making it possible to account for the phenomenon of adhesion (at least for polymers) from a common position.

This main conclusion of the monograph is confirmed since the central themes of the thermodynamic and the molecular–kinetic approaches, namely surface energy σ and mobility of macromolecular chains s, respectively, appear to be directly interrelated. Indeed, as is shown in Fig. 2.27, there is a linear correlation between the surface energy temperature gradient $d\sigma/dT$ and the equilibrium flexibility s. Another example of the interrelation of the two themes involves the following considerations. By our definition, the proposed general criterion α is of energetic origin as, according to eq. (354) it is given as a ratio between the cohesion energy and the surface energy; however, it may be easily transformed so that the true (van der Waals') volume becomes the major variable of the corresponding formula. This property (a molecular–kinetic parameter) then determines the completeness of the interfacial contact area between an adhesive and a substrate.

The inherent uniformity of the approach taken, as applied to the analysis of polymer adhesion, is also supported by the fact that eq. (475), derived from molecular–kinetic concepts to describe the strength of adhesive joints appears to be, on the one hand, a function of the binding energy between the substrate and the segments of adhesive macromolecules, which can be identified with the Helmholtz' free energy, while, on the other hand, it is a function of the ratio between the numbers of the bonded and the unbonded segments, which is identical to Frenkel's surface entropy. The implications of such interrelationships are quite clear.

In undertaking the analysis we were trying to progress from generalizations relying on the most unambiguous data concerning adhesion in terms of the minimal number of the most general concepts. It is our hope that the results presented were convincing enough to prove the validity (and usefulness) of this approach. If so, it gives us the right to claim that the present state of the art in the science of adhesion was properly summarized and the basis for subsequent generalizations was indeed created. We believe this also to be an important product of this work.

When examining future possibilities there is always the danger that the reader might come across the 'good news' some few years after the publication. Nevertheless, we have ventured to express some thoughts concerning the future of the theory and the practice of polymer adhesion.

First, recalling the chronology of the development of adhesion involving a pre-historical stone age of using any available products as adhesives [1842], the golden age associated with the promotion of a variety of theories [1843], crusades for theoretical uniformity [6, 10, 1844, 1845], the middle age of accumulating empirical data [100–103] and an analytical period associated with the revival of the achievements of the golden age [1], we believe that this latter must be followed by a quiet reflective review of the accumulated data so vast that it resulted in a specialized terminological handbook [1846] being issued. Proper account must be taken of the achievements in modern instrumentation and in the application of fundamental concepts of molecular physics, polymer physics and chemistry, surface science, and physico-chemical biology. One should expect a certain evolution of theoretical concepts resulting in parameters of thermodynamic and molecular–kinetic origin being combined within the framework of uniform complex criteria directly related to the final target. Such comprehension of the nature of adhesion and consequently, of the factors which lead to its control would inevitably elevate it to a higher status. Speaking of particular features, we assume it to be especially important to examine the relationships between the structure and properties of the transition and boundary layers in polymers, on the one hand, and the strength of adhesive joints on the other. It should be acknowledged that these characteristics are affected by the simultaneous effect of two factors, surface energy (thermodynamic parameter) and macromolecular mobility (molecular–kinetic parameter), which influences both the kinetics of adhesive joint formation and its ultimate strength.

From the practical point of view, it should be anticipated that, along with the interpreting function, the prognostic function of the theory would be expanded, primarily, to the area concerning the efficiency of interfacial interaction [1847]. Two trends are possible here. The first involves the planning of physico-chemically sound routes for the development of the new generation of materials for adhesion purposes befitting the technological requirements of the twenty first century [1848] and the second involves the creation of new bonding technologies. We believe that as with development of new adhesives so with the working out of new technological procedures we should rely on the foundations of both the thermodynamic and molecular–kinetic concepts of adhesion. In the final account, applied research in this direction must be confined to the search for control over the above factors governing adhesive ability of polymers.

Finally, the last but not the least important consideration relates to the materials involved in adhesion interaction. The traditional range of materials encompassing adhesives, sealants and coatings has been expanded in recent times to incorporate composites [1797, 1849]. Of these materials, synthetic and biological composites produced by *in situ* precipitation of the filler [1851] seem to be most promising (at least for further development of the theory of adhesion).

One might claim that the perspectives listed are too general. However, even the daily weather broadcasts are often not definitive. We thought it appropriate to name the markers for the shaping of the basic thinking of tomorrow's specialists in the field, rather than to detail the schedule for subsequent studies. We believe that it is along the described routes that the most impressive results might be anticipated, on the one hand in the theory and practice of adhesion, and on the other, in materials science. It is beyond doubt that people working in various fields, from theoretical physics to chemical engineering and molecular biology, have the opportunity to apply their skills and knowledge in the studies of adhesion. At the same time, it is not only about science that we are speaking; we are rather speaking of the art of controlling the adhesion processes [1850], at least until practical experience obtains its scientific understanding. This is why we consider physical chemistry of polymer adhesion to be one of the most prospective and appealing branches of modern science.

References

[1]   Pritykin, L.M., & Vakula, V.L. In *Chemical Encyclopedia*, Soviet Encyclopedia Publ., Moscow, 1988, vol. 1, p. 30.

[2]   McBain, J.W., & Hopkins, G. *J. Phys. Chem.* **29**, 188 (1925).

[3]   Gurney, W. *Trans. Inst. Rubber Ind.* **18**, 207 (1943).

[4]   Bancroft, M.D. *Applied Colloid Chemistry*. New York, 1926.

[5]   Josefowitz, D., & Mark, H. *Ind. Rubber World* **106**, 33 (1942).

[6]   Bikerman, J.J. *Science of Adhesive Joints*. Academic Press, New York, 1968.

[7]   Gul', V.E., & Koudryashova, L.L. In *Adhesion of Polymers*, USSR Acad. Sci. Publ., Moscow, 1963, p. 134.

[8]   Derjaguin, B.V., & Krotova, N.A. *Adhesion*, USSR Acad. Sci. Publ., Moscow, 1949.

[9]   Moskvitin, N.I. *Bonding of Polymers*, Forest Ind. Publ., Moscow, 1968.

[10] Berlin, A.A., & Basin, V.E. *Fundamentals of Polymer Adhesion*. Chimia Publ., Moscow, 1974.

[11] Mittal, K.L. *J. Adhesion Sci. Technol.* **1**, 247 (1987).

[12] Pritykin, L.M., & Emel'yanova, V.P. In *Adhesives and Adhesive Joints*, Mir Publ., Moscow, 1988, p. 217.

[13] Pritykin, L.M. *J. Adhesion*, **9**, 311, 1978.

[14] Figovsky, O.L., Kozlov, V.V., Sholokhova, A.B., Pritykin, L.M., Kreindlin, Yu. G., & Voitovitch, V.A. *Handbook of Adhesives and Mastics in Building*. Building Publ., Moscow, 1984, p. 221.

[15] Pritykin, L.M., Emel'yanov, Yu.V., Emel.yanova, V.P., & Zjuz', V.T. *J. Adhesion Sci. Technol.* in press.

[16] Vakula, V.L. In *Adhesives and Adhesive Joints*, House of Sci. Techn. Propag. Publ., Moscow, 1967, p. 6.

[17] Cagle, C.V. *Adhesive Bonding (Techniques and Applications)*, McGraw-Hill, New York, 1968, ch. 1.

[18] Hinshelwood, C.N. *Structure of Physical Chemistry*, Clarendon Press, Oxford, 1951, Introduction.

[19] Kirakosyan, Kh.A. In *State, Perspectives, and Development Problems in Polymer Adhesives to 2000th Year*, Inst. Pol. Adh. Press, Kirovakan, 1984, p. 229.

[20] Dahlquist, C.A. *Interdisc. Sci. Rev.* 2, 140 (1977).

[21] Rozenberg, B.A., & Enikolopyan, N.S. *Mendeleev Chem. Soc. USSR J.* 25, 524 (1980).

[22] Vakula, V.L. Thesis, Lomonosov Inst. Fine Chem. Technol., Moscow, 1974.

[23] Mittal, K.L. *J. Vacuum Sci.* 13, 19 (1976).

[24] Lipin, Yu.V., Rogachev, A.V., & Kharitonov, V.V. *Vacuum Metallization of Polymer Materials*, Chimia Publ. Leningrad, 1987.

[25] Kragelsky, I.V. *Friction and Wear*, Mach.-Build. Publ., Moscow, 1968.

[26] Bartenev, G.M., & Lavrent'ev, V.V. *Friction and Wear of Polymers*. Chimia Publ., Moscow, 1972.

[27] Ainbinder, S.B., & Tyunina, E.L. *Introduction to Theory of Polymer Friction*, Zinatne Publ., Riga, 1978.

[28] Golden, J.M. *J. Phys. Math. Gen.* 8A, 966 (1975).

[29] Lavrent'ev, V.V., & Vakula, V.L. In *Advances in Polymer Friction and Wear*, Wiley, New York, 1974, p. 765.

[30] Skelcher, W.L. *Microscopic Aspects of Adhesion and Lubrication*, Elsevier, Amsterdam, 1982, p. 719.

[31] Chernyak, Y.B., & Leonov, A.I. *Wear*, **108**, 105 (1986).

[32] *Surface Effects in Adhesion, Friction, Wear, and Lubrication*. Elsevier, Amsterdam, 1981.

[33] Cherry, B.W. *Polymer Surfaces*, Cambridge University Press, 1981.

[34] *On the Nature of Solid Setting*, Science Publ., Moscow, 1968.

[35] Semenov, A.P. *Friction and Adhesion Interaction of Refractory Materials at High Temperatures*, Science Publ., Moscow, 1972.

[36] Patskevitch, I.R., & Deev, G.F. *Surface Phenomena in Welding Processes*, Metallurgy Publ., Moscow, 1974.

[37] Kazakov, N.F. *Diffusion Welding*, Mach.-Build. Publ., Moscow, 1968.

[38] Dolgov, Yu.S., & Sidokhin, Yu.F. *Problems of Formation of Solder Joints*. Mach-Build. Publ., Moscow, 1973.

[39] Golovanenko, S.A., & Meandrov, L.V. *Manufacture of Bimetals*, Metallurgy Publ., Moscow, 1966.

[40] Klein, W. *Schweissen und Schneiden*, **19**, 4 (1967).

[41] Yakobashvili, S.B. *Surface Properties of Welding Fluxing Agents and Cinders*, Technic Publ., Kiev, 1970.

[42] Vasenin, R.M. *USSR Plastics*, No 6, 13 (1986).

[43] Yonezo, K., Fowkes, F.M. & Vanderhoff, J.W. *Ind. Eng. Chem. Prod. Res. Develop.* 21, 441 (1982).

[44] Zimon, A.D. *What is Adhesion?* Science Publ., Moscow, 1983.

[45] Trinkaus, G.P. *Cells into Organs*, New Jersey, Englewood Cliffs, 1969.

[46] Wakefield, H.F. In *Proc. Symp. Experimental Marine Ecology*. University Press, Rhode Island, 1964, p. 51.

[47] Manly, R.S. (ed) *Adhesion in Biological Systems*. Academic Press, New York, 1970.

[48] Curtis, A.S.G. *Progr. Mol. Biol.* **27**, 317 (1973).
[49] Evgen'eva, T.P. *Intercellular Interactions and Their Role in Evolution*, Medicine Publ., Moscow, 1976.
[50] Van Oss, C.J., Good, R.J., & Neumann, A.W. *J. Electroanal. Chem.* **37**, 387 (1972).
[51] Sherbert, G.V. *Biophysical Characterization of the Cell Surface*, Academic Press, New York, 1978.
[52] Dalin, M.V., & Fish, N.G. *Adhesions of Microorganisms*, Inst. for Sci. -Techn. Information Press, Moscow, 1985, p. 91.
[53] *Adhesion of Microorganisms to Surfaces.* Academic Press, New York, 1979.
[54] Pritykin, L.M. *USSR Biophysics,* **21**, 1059 (1976).
[55] Loewenstein, W.R. *J. Colloid Interface Sci.* **25**, 34 (1967).
[56] Newton, C., Pangborn, W., Nir, S., & Papahadjopoulos, D. *Biochim. Biophys. Acta,* **506**, 281 (1978).
[57] Markin, V.S., & Kozlov, M.M. In *Interaction and Merging of Membranes*, Inst. for Sci. -Techn. Information Press, Moscow, 1984, p. 62.
[58] MacIver, D.J.L. *Physiol. Chem. Phys.* **3**, 289 (1979).
[59] Sundler, R., Düzgünes, N., & Papahadjopoulos, D. *Biochem. Biophys. Acta.* **649**, 751 (1981).
[60] Wilschut, J., Holsappel, M., & Jansen, R. *Biochim. Biophys. Acta.* **690**, 297 (1982).
[61] Curtis, A.S.G. *Cell Surface: Its Molecular Role in Morphogenesis.* Academic Press, London, 1967.
[62] Curtis, A.S.G. *Cell Surface.* CRC Press, Boca Raton, **4**, 151 (1981).
[63] Yamada, K.M., Yamada, S.S., & Rastan, I. *Proc. Nat. Acad. Sci. USA.* **72**, 3158 (1975).
[64] Rinaldini, L.M. *Exp. Cell Res.* **16**, 477 (1959).
[65] Nenashev, V.A. In *Interaction and Merging of Membranes.* Inst. for Sci. - Techn. Information Press, Moscow, 1984, p. 91.
[66] Maroudas, N.G. *Nature.* **254**, 695 (1975).
[67] Laggner, P. *Nature.* **294**, 373 (1981).
[68] Freire, E., & Snyder, B. In *Membranes and Transport.* Plenum Press, New York, 1982, vol. 1, p. 37.
[69] Pagano, R.C., & Takeichi, M. *J. Cell Biol.* **74**, 531 (1977).
[70] Neumann, A.W., Gillman, C.F., & Van Oss, C.J. *J. Electroanal. Chem.* **39**, 393 (1974).
[71] Canningham, R.K., Söderström, T.C., Gillman, C.F., & Van Oss, C.J. *Immunol. Comm.* **4**, 429 (1975).
[72] Van Oss, C.J., Gillman, C.F., & Neumann, A.W. *Phagocytic Engulfment and Cell Adhesiveness.* Marcel Dekker, New York, 1975.
[73] Van Oss, C.J., Good, R.J., Neumann, A.W., Wieser, J.D., & Rosenberg, P. *J. Colloid Interface Sci.* **59**, 505 (1977).
[74] Schrader, M.E. *J. Colloid Interface Sci.* **88**, 296 (1982).
[75] Van Oss, C.J., & Neumann, A.W. *Immunol. Comm.* **6**, 341 (1977).
[76] Sharma, C.P. *J. Sci. Ind. Res.* **39**, 453 (1980).
[77] Baier, R.E. In *Proc. 8th Annu. Meet. Adhesion Soc.* Savannah, 1985, p6. 1, p. 19/a.

[78] Rosen, J.J., & Schway, M.B. In *Adhesion and Adsorption of Polymers*. Plenum Press, New York, 1980, p. 667.

[79] Iordansky, A.L., & Zaikov, G.E. *USSR High Polymers*. **25A**, 451 (1983).

[80] Tanzawa, H., Nagaoka, S., Suzuki, J., Kobayashi, S., Masubichi, Y., & Kikuchi, T. *Amer. Chem. Soc. Polymer Prepr.* **20**, 313 (1979).

[81] Neumann, A.W., Hunn, O.S., Francis, D.W., Zingg, W., & Van Oss, C.J. *J. Biomed. Mater. Res.* **14**, 499 (1980).

[82] Plate, N.A., & Valuev, L.I. *Polymers in Contact with Living Organism*. Knowledge Publ., Moscow, 1987, p. 10.

[83] Sanui, K., Yui, N., Takabashi, Y., Ogata, N., Kataoka, K., Okano, T., & Sakurai, Y. *Kobunshi ronbunshu.* **39**, 213 (1982).

[84] Pritykin, L.M., Kardashov, D.A., & Vakula, V.L. *Monomer Adhesives*. Chimia Publ., Moscow, 1988.

[85] Kataoka, K. *Khyomyon.* **21**, 385 (1983).

[86] Mulvihill, J.N., Cazenave, J.-P., Schmitt, A., Maisonneuve, P., & Pusineri, C. *Colloids and Surfaces.* **14**, 317 (1985).

[87] Fuller, R.A., & Rosen, J.J. *Scientific American.* No 12, 55 (1986).

[88] Weiss, L. *Exp. Cell Res.* **74**, 21 (1972).

[89] Weiss, L., & Chang, M.K. *J. Cell Sci.* **12**, 655 (1973).

[90] Weiss, L. *J. Nat. Cancer Inst.* **50**, 3 (1973).

[91] Stackpole, C.W. In *Progress of Surface and Membrane Science*. Wiley, New York, 1978, p. 1.

[92] Vasil'ev, Yu.M. *Biology of Cancer Growth*. Science Publ., Moscow, 1965, p. 200.

[93] Coman, D.R. *Cancer Res.* **4**, 625 (1944).

[94] Matsumoto, T. *J. Adhesion Soc. Japan.* **16**, 325 (1980).

[95] Evdokimov, Yu.M. *USSR Technology and Science.* (6), 22 (1984).

[96] Evdokimov, Yu.M., & Krestov, D.S. *Problems of Industrial Bionization*. Kazan, 1986, p. 86.

[97] Evdokimov, Yu.M. *Bionics*. Leningrad, 1986, p. 130.

[98] Evdokimov, Yu.M., Krestov. D.S. *Adhesion Bionics*. Chuvash Book Press, Cherboxary, 1988, p. 35.

[99] Raevsky, V.G., & Pritykin, L.M. *USSR Plastics* (2), 7 (1970).

[100] Zimon, A.D. *Adhesion of Dust and Powders*. Chimia Publ., Moscow, 1968, p. 9.

[101] Zimon, A.D. *Adhesion of Liquid and Wetting*. Chimia Publ., Moscow, 1974, p. 7.

[102] Zimon, A.D. *Adhesion of Films and Coatings*. Chimia Publ., Moscow, 1977, p. 12.

[103] Belyi, V.A., Egorenkov, N.I., & Pleskatchevsky, Yu.M. *Adhesion of Polymers to Metals*. Science and Technology Publ., Minsk, 1970.

[104] Rehbinder, P.A. In *Physical Encyclopedic Dictionary*. Soviet Encyclopedia Publ., Moscow, 1960, vol. 1, p. 19.

[105] Rehbinder, P.A. In *Short Chemical Encyclopedia*. Soviet Encyclopedia Publ., Moscow, 1961, vol. 1, p. 27.

[106] Bartenev, G.M. In *Structural Materials*. Mach. -Build. Publ., Moscow, 1963, vol. 1, p. 27.

[107] Voyutskii, S.S. In *Encyclopedia of Polymers*. Soviet Encyclopedia Publ., 1972, vol. 1, p. 22.

[108] Basin, V.E. In *Chemical Encyclopedic Dictionary*. Soviet Encyclopedia Publ., Moscow, 1983, p. 10.

[109] Derjaguin, B.V., Krotova, N.A., & Smilga, V.P. *Adhesion of Solids.* Plenum, New York, 1978.

[110] Kardashov, D.A. *Structural Adhesives.* Chimia Publ., Moscow, 1980, p. 7.

[111] Trizno, M.S., & Moscalev, E.V. *Adhesives and Bonding.* Chimia Publ., Leningrad, 1980, p. 6.

[112] Adamson, A.W. *Physical Chemistry of Surfaces.* Wiley-Interscience, New York, 1976, p. 274.

[113] Derjaguin, B.V., & Toporov, Yu.P. *Herald of USSR Acad. Sci.,* (8), 1739 (1982).

[114] Zherebkov, S.K. *Bonding of Rubber to Metal.* Chimia Publ., Moscow, 1966, p. 50.

[115] Voyutskii, S.S. *Autohesion and Adhesion of High Polymers.* Wiley-Interscience, New York, 1963.

[116] Kaelble, D.H. *Physical Chemistry of Adhesion.* Wiley-Interscience, New York, 1971, p. 3.

[117] Smilga, V.P. Cand. Sci. (Chem.) Thesis, Inst. for Phys. Chem. Moscow, 1961.

[118] Krotova, N.A. *On Bonding and Adhesion.* Acad. Sci. Publ., Moscow, 1956, p. 94.

[119] Epstein, G. *Adhesive Bonding of Metals.* Reinhold, New York, 1954.

[120] Enikolopyan, N.S., & Volfson, S.A. *Chemistry and Technology of Polyformaldehyde.* Chimia Publ., Moscow, 1968, p. 213.

[121] ASTM Standard Committee D-907 (26.02. 1982). In *Annual Book of ASTM Standards.* ASTM Publ., Philadelphia, 1982, vol. 15.06.

[122] Pritykin, L.M. *USSR Polymer Mechanics* (2), 360 (1974).

[123] Good, R.J. *J. Adhesion.* 8, 1 (1976).

[124] Dukes, W.A. *J. Adhesion.* 7, 253 (1975).

[125] Mittal, K.L. *J. Adhesion.* 8, 101 (1976).

[126] Mittal, K.L. *J. Adhesion.* 6, 377 (1974).

[127] Runnels, L.K. *Crit. Rev. Solid Sci.* 4, 2 (1974).

[128] Maradudin, A.A., Montroll, E.W., Weiss, G.H., & Ipatova, I.P. *Theory of Lattice Dynamics in the Harmonic Approximation.* Academic Press, New York, 1971, ch. 8.

[129] Gibbs, J.W. *Collected Works.* Green and Co., New York, 1928, vol. 1.

[130] Lovett, R., DeHaven, P.W., Viecell, J., & Buff, E.P. *J. Chem. Phys.* 58, 1880 (1973).

[131] Good, R.J. *Pure Appl. Chem.* 48, 427 (1976).

[132] Derjaguin, B.V., & Kousakov, M.M. *Herald of USSR Acad. Sci. Chem. Ser.* (7), 1119 (1937).

[133] Veitsman, E.V. *USSR J. Phys. Chem.* 56, 1598 (1982).

[134] Bakker, G. In *Handbuch der Experimentalphysik.* Akad. Verlag, Leipzig, 1928, Bd.6, S.70.

[135] Veitsman, E.V. *USSR J. Phys. Chem.* 55, 817 (1981).

[136] Veitsman, E.V. *USSR J. Phys. Chem.* 61, 1713 (1987).

[137] Hey, M.J., & Wood, D.W. *J. Colloid Interface Sci.* 90, 277 (1982).

[138] Somorjai, G.A. In *Treatise on Solid State Chemistry.* Wiley, New York, 1976, vol. 6A, p. 1.

[139] Datta, S.K. *Indian J. Pure Appl. Phys.* 23, 99 (1985).

[140] Pritykin, L.M., & Vakula, V.L. In *Chemical Encyclopedia.* Soviet Encyclopedia Publ., Moscow, 1989, vol. 2, in press.

[141] Filippov, L.P. *Similarity of Properties in Substances.* Univ. Press, Moscow, 1978, p. 53.

[142] Mouratov, G.N. *USSR J. Phys. Chem.* **56**, 1562 (1982).

[143] Mistura, L. *Phys. Rev. Gen. Phys.* **33**, 1275 (1986).

[144] Defay, R., Prigogine, I., Bellemans, A., & Everett, D.H. *Surface Tension and Adsorption.* Longman, London, 1966.

[145] Shuttleworth, R. *Proc. Phys. Soc.* **63A**, 444 (1950).

[146] Shherbakov, L.M., & Volkova, E.M. In *Problems of Physics of Formation and Phase Transitions.* Univ. Press, Kalinin, 1982, p. 83.

[147] Eriksson, J.C. *Surface Sci.* **14**, 221 (1969).

[148] Guggenheim, E.A. *Trans. Faraday Soc.* **36**, 397 (1940).

[149] Defay, R., & Sanfeld, A. *Ann. Chem.* **71**, 856 (1975).

[150] Defay, R., Prigogine, I., & Sanfeld, A. *J. Colloid Interface Sci.* **58**, 498 (1977).

[151] Gerbacia, W., & Rosano, H.K. *J. Colloid Interface Sci.* **44**, 242 (1973).

[152] Budrugeac, P., & Vass, M.I. *Rev. Roum. Phys.* **24**, 77 (1979).

[153] Rusanov, A.I. *USSR Colloid J.* **39**, 711 (1977).

[154] Rusanov, A.I. *J. Colloid Interface Sci.* **63**, 330 (1978).

[155] Herring, C. In *Structure and Properties of Solid Surfaces.* Univ. Press, Chicago, 1953, p. 5.

[156] Petrov, Yu.I. *USSR Surface* No (7), 1 (1982).

[157] Shebzukhov, A.A., & Karachaev, A.M. *USSR Surface.* No 5, 58 (1984).

[158] Kotchurova, N.N. In *Problems in Thermodynamics of Heterogeneous Systems and Theory of Surface Phenomena.* Univ. Press, Leningrad, 1982, p. 191.

[159] Podstrigatch, Ya.S., & Povstenko, Yu.Z. *Introduction to Mechanics of Surface Phenomena in Deformable Solids.* Naukova Dumka Publ., Kiev, 1985, p. 116.

[160] Sacher, E. *J. Colloid Interface Sci.* **88**, 308 (1982).

[161] Temrokov, A.I. & Zadumkin, S.N. In *Wettability and Surface Phenomena.* Naukova Dumka Publ., Kiev, 1972, p. 151.

[162] Temrokov, A.I. *USSR J. Phys. Chem.* **56**, 1207 (1982).

[163] Kornblit, L., & Ignatiev, A. *Physica.* **141A**, 466 (1987).

[164] Soma, H. In *Interface Chemistry.* Iwanami Shoten Publ., Tokyo, 1980, ch. 2.

[165] Rusanov, A.I. *J. Colloid Interface Sci.* **90**, 143 (1982).

[166] Frenkel, J.I. *Collected Papers.* USSR Acad. Sci. Publ., Moscow–Leningrad, 1959, vol. 2.

[167] Flores, F. *Introduction to the Theory of Solid Surfaces.* Cambridge University Press, 1979.

[168] Cohen, S.M.A., Cosgrove, T., & Vincent, B. *Adv. Colloid Interface Sci.* **24**, 143 (1986).

[169] Hill, T.L. *J. Chem. Phys.* **17**, 520 (1949).

[170] Yates, D.J.C. In *Adv. Catalysis and Related Subjects.* Academic Press, New York, 1960, vol. 12, p. 265.

[171] Crawford, V.A., & Tompkins, F.C. *Trans. Faraday Soc.* **46**, 504 (1950).

[172] Balmer, R.T. *Surface Sci.* **52**, 174 (1975).

[173] Flood, E.A. In *Solid–Gas Interface.* Marcel Dekker, New York, 1967, vol. 1, ch. 2.

[174] Rance, D.G. In *Ind. Adhesion Problems.* Oxford University Press, 1985, p. 48.

[175] Rusanov, A.I. In *Surface Phenomena in Polymers.* Naukova Dumka Publ., Kiev, 1971, p. 4.

[176] Nosanow, L.H., Parish, L.J., & Pinski, F.J. *Phys. Rev.* **11B**, 191 (1975).

[177] Bird, R.B. *J. Rheol.* **26**, 277 (1982).

[178] Verkin, B.I., Serbin, I.A., & Sivokon' B.E. *Reports of USSR Acad Sci.* **250**, 645 (1980).

[179] Serbin, I.A., Sivokon' B.E., & Privalko, V.P. *Prepr. No 32, Phys.* Techn. Inst. of Low Temperatures Press, Karkov, 1980.

[180] Doolittle, A.K. *J. Appl. Polymer Sci.* **25**, 2305 (1980).

[181] Davis, H.T. & Scriven, L.E. *J. Phys. Chem.* **80**, 2805 (1976).

[182] Croucher, M.D. *Makromol. Chem. Rapid. Comm.* **2**, 199 (1981).

[183] Meissner, B. *J. Polymer Sci. Polymer Letter.* **19**, 137 (1981).

[184] Dorsey, N.E. *Sci. Papers of Nat. Bur. Stand.* **21**, 563 (1926).

[185] Young, T. *Phil. Trans. Roy. Soc.* **95**, 65 (1805).

[186] Summ, B.D., & Goryunov, Yu.V. *Physico-Chemical Fundamentals of Wetting and Spreading.* Chimia Publ., Moscow, 1976.

[187] Povstenko, Yu.Z. *Reports of Ukrainian Acad. Sci.* No 10A, 45 (1980).

[188] Povstenko, Yu.Z. *USSR Appl. Math. Mech.* **45**, 919 (1981).

[189] Rusanov, A.I. *USSR Colloid J.* **39**, 704 (1977).

[190] Shherbakov, L.M., & Ryazantsev, P.P. In *Res. Surface Forces.* Acad. Sci. Publ., Moscow, 1964, p. 26.

[191] Melrose, J.C. *Adv. Chem.* Ser. 43, 158 (1964).

[192] Andrade, J.D., Smith, L.M., & Gregonis, D.E. In *Surface and Interfacial Aspects of Biomedical Polymers.* Wiley, New York, 1985, vol. 5, p. 249.

[193] Derjaguin, B.V., & Shherbakov, L.M. *USSR Colloid J.* **23**, 40 (1961).

[194] Eick, J.D., Good, R.J., & Neumann, A.W. *J. Colloid Interface Sci.* **53**, 235 (1975).

[195] Neumann, A.W. *Adv. Colloid Interface Sci.* **4**, 105 (1974).

[196] Fortes, M.A. *J. Chem. Soc. Faraday Trans.* **78**, 101 (1982).

[197] Erofeev, V.S. *Dep. Manuscr. No 285.* Niitekhim Press, Tcherkassy, 1986.

[198] Huntsberger, J.R. In *Proc. 2nd Workshop Adv. Restorative Dental Mater.* Dept. Health Press, New York, 1965, p. 7.

[199] Fainerman, A.E., & Min'kov, V.I. In *Modern Methods in Polymer Research.* Naukova Dumka Publ., Kiev, 1975, p. 29.

[200] Shanahan, M.E. *Int. J. Adhesion and Adhesives.* **4**, 179 (1984).

[201] Fowkes, F.M., McCarthy, D.C., & Mostafa, M.A. *J. Colloid Interface Sci.* **78**, 200 (1980).

[202] Yates, D.J.C. *Proc. Roy. Soc.* **A224**, 526 (1954).

[203] Amberg, C.H., & McIntosh, R. *Canad. J. Chem.* **30**, 1012 (1952).

[204] Hardy, W. *Phil. Mag.* **38**, 49 (1919).

[205] Chang, W.V., Chang, Y.M., Wang, L.J., & Wang, Z.G. *Org. Coat. and Appl. Polymer Sci.* **47**, (1982).

[206] De Gennes, P.-G. *Rev. Mod. Phys.* **57**, 827 (1985).

[207] Raevsky, V.G. Cand. Sci. (Techn.). Thesis, Lomonosov Inst. Fine Chem. Technol., Moscow, 1962.

[208] Vasenin, R.M. In *Adhesive Joints in Machine-Building.* Polytechn. Inst. Publ., Riga, 1983, p. 24.

[209] Horsthemke, A., & Schröder, J.J. *Chem. Eng. Process.* **19**, 277 (1985).

[210] Ponter, A.B., & Yekta-Fard, M.J. *Colloid Interface Sci.* **101**, 282 (1984).

[211] Chadov, A.V., Goryunov, Yu.V., Raud, E.A., & Summ, B.D. *Herald of Moscow Univ. Chem. Ser.* **18**, 691 (1977).

[212] Raud, E.A., Goryunov, Yu.V., Summ, B.D., & Chadov, A.V. *USSR Colloid J.* **47**, 1200 (1985).

[213] Goryunov, Yu.V., Raud, E.A., Summ, B.D., & Chadov, A.V. In *Compatibility and Adhesion Interaction of Melts with Metals.* Naukova Dumka Publ., Kiev, 1978, p. 30.

[214] Summ, B.D., Chadov, A.V., Raud, E.A., & Goryunov, Yu.V. *USSR Colloid J.* **42**, 1010 (1980).

[215] Wenzel, R.N. *Ind. Eng. Chem.* **28**, 988 (1936).

[216] Good, R.G. *J. Amer. Chem. Soc.* **74**, 5041 (1952).

[217] Dettre, R.H., & Johnson, R.E. *Adv. Chem.* Ser. 43, 112 (1964).

[218] Finch, J.A. & Smith, G.W. In *Anionic Surfactants* New York, 1981, p. 317.

[219] Künzer, F.-V., & Bonart, R. *Progr. Colloid Interface Sci.* **66**, 311 (1979).

[220] Kawasaki, K. *J. Colloid Sci.* **15**, 402 (1960).

[221] Bikerman, J.J. *J. Colloid Sci.* **5**, 349 (1950).

[222] Frenkel, J.I. *USSR J. Theor. Exper. Phys.* **18**, 659 (1948).

[223] Johnson, R.E. *J. Phys. Chem.* **17**, 288 (1962).

[224] Olsen, D.A., Soynir, P.A., & Olsen, M.D. *J. Phys. Chem.* **17**, 883 (1962).

[225] Saito, M., & Akagawa, N. *J. Soc. Fiber Sci. Techn. Japan.* **36**, 113 (1980).

[226] Andreu, M., Gilbert, Y., & Roques-Carmes, C. *Mater. Techn.* **75**, 147 (1987).

[227] Fortes, M.A. *Phys. Chem. Liquids.* **9**, 285 (1980).

[228] Neumann, A.W., & Good, R.J. *J. Colloid Interface Sci.* **38**, 341 (1972).

[229] Penn, L.S., & Miller, B. *J. Colloid Interface Sci.* **78**, 238 (1980).

[230] Bayramli, E., Van de Ven, T.G.M., & Mason, S.G. *Colloids and Surfaces.* **3**, 279 (1981).

[231] Shanahan, M.E.R., Carre, A.M.S., & Schultz, J. *J. Chim. Phys. et Phys. -Chim. Biol.* **84**, 199 (1987).

[232] Chen. W.Y., & Andrade, J.D. *J. Colloid Interface Sci.* **110**, 468 (1986).

[233] Tolman, R.C. *J. Chem. Phys.* **17**, 333 (1949).

[234] Brodskaya, E.N., & Rusanov, A.I. *USSR Colloid J.* **39**, 646 (1977).

[235] Tabor, D. *J. Colloid Interface Sci.* **58**, 2 (1977).

[236] Israelashvili, J.N. *Faraday Disc. Chem. Soc.* No 65, 20 (1978).

[237] Israelashvili, J.N., Perez, E., & Tandon, R.K. *J. Colloid Interface Sci.* **78**, 260 (1980).

[238] Rusanov, A.I. In *Structure and Properties of Polymer Surface Layers*, Naukova Dumka Publ., Kiev, 1972, p. 3.

[239] Rusanov, A.I. *USSR Colloid J.* **37**, 678 (1975).

[240] Ovrutskaya, L.A., Akopyan, L.A. & Rusanov, A.I. In *Thermodynamic and Structural Properties of Polymers.* Naukova Dumka Publ., Kiev, 1976, p. 15.

[241] Rusanov, A.I. *USSR Colloid J.* **37**, 688 (1975).

[242] Rusanov, A.I. In *Physics of Interface Phenomena.* University Press, Naltchick, 1980, p. 26.

[243] Maijgren, B., & Odberg, L. *J. Colloid Interface Sci.* **88**, 197 (1982).

[244] Berenshtein, G.V., D'yachenko, A.M., & Rusanov, A.I. *USSR Colloid J.* **47**, 236 (1985).

[245] Berenshtein, G.V., D'yachenko, A.M., & Rusanov, A.I. *USSR Colloid J.* **47**, 9 (1985).

[246] Yuk, S.H., & Jhon, M.S. *J. Colloid Interface Sci.* **110**, 252 (1986).

[247] Martynov, G.A., Starov, V.M., & Churaev, N.V. *USSR Colloid J.* **38**, 472 (1977).

[248] Penn, L.S., & Miller, B. *J. Colloid Interface Sci.* **77**, 574 (1980).

[249] Bikerman, J.J. *J. Adhesion.* **6**, 331 (1974).

[250] Bienkowski, R., & Skolnick, M. *J. Colloid Interface Sci.* **48**, 350 (1974).

[251] Birdi, K.S. *J. Colloid Interface Sci.* **88**, 290 (1982).

[252] Akopyan, L.A., Ovrutskaya, L.A., & Rusanov, A.I. In *7th Conf. on Coll. Chem. and Phys. -Chem. Mech.* Science and Technology Publ., Minsk, 1977, pt. A-D, p. 52.

[253] Joanny, J.F., & de Gennes, P.-G. *J. Chem. Phys.* **81**, 552 (1984).

[254] Helmus, M.N., Gibbons, D.F., & Jones, R.D. *J. Colloid Interface Sci.* **89**, 567 (1982).

[255] Bartenev, G.M., & Akopyan, L.A. *USSR High Polymers.* **12**, 395 (1970).

[256] Rusanov, A.I., Ovrutskaya, L.A., & Akopyan, L.A. *USSR Colloid J.* **43**, 685 (1981).

[257] Bartenev, G.M., & Akopyan, L.A. In *Surface Phenomena in Polymers.* Naukova Dumka Publ., Kiev, 1971, p. 89.

[258] Bartenev, G.M., & Akopyan, L.A. In *Macromolecules on Interface.* Naukova Dumka Publ., Kiev, 1971, p. 39.

[259] Larin, V.B., Komarov, V.S., & Danilovitch, S.V. *Herald of Byelorussian Acad. Sci.*: Chem. Ser., No 5, 24 (1984).

[260] Good, R.J., Kvikstad, J.A., & Bailey, W.O. *J. Colloid Interface Sci.* **35**, 314 (1971).

[261] Good, R.J., Kvikstad, J.A., & Bailey, W.O. *Appl. Polymer Symp.* No. 16, 153 (1971).

[262] Chmielewski, H.J., & Bailey, A.I. *Phil. Mag.* **A43**, 739 (1981).

[263] Banerji, B.K. *Colloid Polymer Sci.* **259**, 391 (1981).

[264] Kamusewitz, H., & Possart, W. *Int. J. Adhesion and Adhesives.* **5**, 211 (1985).

[265] Wolfram, E., & Faust, R. *Ann. Univ. Sci. Budapest.* Sec. Chim., **16**, 151 (1980).

[266] Bayramli, E., Van de Ven, T.G.M., & Mason, S.G. *Canad. J. Chem.* **59**, 1954 (1981).

[267] Bayramli, E., & Mason, S.G. *Canad. J. Chem.* **59**, 1962 (1981).

[268] Cox, R.G. *J. Fluid Mech.* **131**, 1 (1983).

[269] Mason, S.G. In *Wetting, Spreading and Adhesion.* Academic Press, New York, 1978, p. 321.

[270] Schwartz, L.W., & Garoff, S. *J. Colloid Interface Sci.* **106**, 422 (1985).

[271] Pomeau, Y., & Vannimenus, J. *J. Colloid Interface Sci.* **104**, 477 (1985).

[272] Geguzin, Ya.E., & Ovcharenko, N.N. *USSR Adv. Phys. Sci.* **76**, 360 (1962).

[273] Herring, C. *Physics of Powder Metallurgy.* Wiley, New York, 1951, p. 143.

[274] Basin, V.E. In *Adhesives and Applications in Technology.* Inst. Pol. Adh. Press, Kirovakan, 1978, p. 3.

[275] Bikerman, J.J. *Phys. Stat. Sol.* **10**, 3 (1965).

[276] Frumkin, A.N. *USSR J. Phys. Chem.* **12**, 337 (1938).

[277] Everett, D.H. *Pure Appl. Chem.* **53**, 2181 (1981).

[278] Safonov, V.P. In *Surface Phenomena in Melts*. Naukova Dumka Publ., Kiev, 1968, p. 351.

[279] Dupre, A. *Ann. Chim. Phys.* **11**, 194 (1867).

[280] Briscoe, B.J. In *Polymer Surfaces*. Wiley, Chichester, 1978, p. 40.

[281] Briscoe, B.J., & McClune, C.R. *J. Colloid Interface Sci.* **61**, 485 (1977).

[282] Gent, A.N., & Schultz, J. *J. Adhesion.* **3**, 281 (1972).

[283] Donnet, J.B. *Bulg. Chem. Ind.* No. 3, 130 (1982).

[284] Platonov, M.P., & Frenkel, S.Ya. *USSR High Polymers.* **20A**, 522 (1978).

[285] Platonov, M.P., Grigorov, A.O., & Kuznetsov, A.V. *USSR High Polymers.* **23B**, 93 (1981).

[286] Platonov, M.P. *USSR High Polymers.* **18B**, 821 (1976).

[287] Tchestyunin, V.P., & Safonov, V.P. In *Problems of Physics of Formation and Phase Transitions*. University Press, Tula, 1973, p. 52.

[288] Good, R.J. In *Treatise on Adhesion and Adhesives*. Marcel Dekker, 1967, vol. 1, p. 56.

[289] Naiditch, Yu.V. In *Physical Chemistry of Surface Phenomena at High Temperatures*. Naukova Dumka Publ., Kiev, 1971, p. 26.

[290] Arslanov, V.V., Tcherezov, A.A., & Ogarev, V.A. In *Adhesives and Their Applications in Technology*. Inst. Pol. Adh. Press, Kirovakan, 1978, p. 18.

[291] Neumann, A.W., & Good, R.J. In *Surface and Colloid Science*. Plenum Press, New York, 1979, vol. 11, p. 31.

[292] Jiang, T.-S., Oh, S.-S., & Slattery, J.C. *J. Colloid Interface Sci.* **69**, 74 (1979).

[293] Hoffman, R.L. *J. Colloid Interface Sci.* **50**, 228 (1975).

[294] Bates, R. In *Adhesion and Adsorption of Polymers*. Plenum Press, New York, 1980, p. 497.

[295] Gershkovitch, V.M., & Borodina, N.N. *USSR Polymer Build. Mater.* N 48, 75 (1978).

[296] Antonoff, G.N. *J. Chim. Phys.* **5**, 372 (1907).

[297] Antonoff, G.N. *Phil. Mag.* **1**, 1258 (1926).

[298] Abramzon, A.A. *USSR Colloid J.* **32**, 475 (1970).

[299] Chizhenko, D.L., Tsvetkova, E.N., & Time, A.V. *USSR Ind. Synth. Rubber.* No 8, 6 (1977).

[300] Mitchell, J.W., & Elton, G.A.N. *J. Chem. Soc.* 839 (1953).

[301] Neumann, F.E. *Vorlesungen über die Theorie der Kapillarität*. Teubner Verlag, Leipzig, 1894, S. 161.

[302] Rowlinson, J.S., & Widom, B. *Molecular Theory of Capillarity*. Clarendon Press, Oxford, 1982, ch. 8.3.

[303] Summ, B.D., & Abramzon, A.A. *USSR J. Appl. Chem.* **53**, 2545 (1980).

[304] Wu, S. *J. Macromol. Chem.* **10C**, 1 (1974).

[305] Rhee, S.K. *Mater. Sci. Eng.* **16**, 45 (1974).

[306] Tovbin, M.V. *Ukrain. Chem. J.* **23**, 13 (1957).

[307] Avraamov, Yu.S., Gvozdev, A.G., Semenov, V.M., & Levin, I.Ya. *Herald of USSR Acad. Sci.* Ser. Ferr. Met., **11**, 99 (1968).

[308] Berthelot, O. *Compt. Rend.* **126**, 1703 (1898).

[309] Fowkes, F.M. *J. Colloid Interface Sci.* **28**, 493 (1968).

[310] Rayleigh, J.W. *Scientific Papers*. Cambridge University Press, 1900, vol. 2, p. 231.

[311] Good, R.J., & Elbing, E. *Ind. Eng. Chem.* **62**, 3, 54 (1970).

[312] Schwartz, A. *J. Polymer Sci.* Phys. Ed. **12**, 1195 (1974).

[313] Good, R.J. In *Surface and Colloid Science.* Plenum Press, New York, 1979, vol. 11, p. 1.

[314] Good, R.J., & Girifalco, C.A. *J. Phys. Chem.* **64**, 561 (1960).

[315] Phillips, M.C., & Riddiford, A.C. *J. Chem. Soc.*, No 8A, 978 (1966).

[316] Driedger, O., Newman, A.W., & Sell, P.-J. *Koll. Z. und Z. Polym.* **201**, 52 (1965).

[317] Lipatov, Yu.S., & Feinerman, A.E. *Adv. Colloid Interface Sci.* **11**, 195 (1979).

[318] Lipatov, Yu.S. *Interfacial Phenomena in Polymers.* Naukova Dumka Publ., Kiev, 1980.

[319] Neumann, A.W., Good, R.J., Hope, C.J., & Sejpal, M. *J. Colloid Interface Sci.* **49**, 291 (1974).

[320] Neumann, A.W., Absolom, D.R., Francis, D.W., & Van Oss, C.J. *Separ. Purif. Meth.* **9**, 69 (1980).

[321] Pittman, A. In *Fluoropolymers.* Wiley, NewsYork, 1972, ch. 3.

[322] Adam, N.K. *Adv. Chem.* Ser. 43, 52 (1964).

[323] Lee, M.C.H. In *Adhesives Chemistry.* Plenum Press, New York, 1984, p. 95.

[324] Bakovets, V.V. *Reports of USSR Acad. Sci.* **228**, 1132 (1976).

[325] Wulff, G. *Z. Kristallogr.* **34**, 449 (1901).

[326] Good, R.J. *Adv. Chem.* Ser. 43, 74 (1964).

[327] Kirkwood, J.G., & Buff, F.P. *J. Chem. Phys.* **17**, 388 (1949).

[328] Davis, B.W. *J. Colloid Interface Sci.* **52**, 150 (1975).

[329] Huntsberger, J.R. *Adh. Age.* **21**, No 12, 23 (1977).

[330] Good, R.J. *J. Colloid Interface Sci.* **59**, 398 (1977).

[331] Girifalco, C.A., & Good, R.J. *J. Phys. Chem.* **61**, 904 (1957).

[332] Adamson, A.W. *Adv. Chem.* Ser. 43, 57 (1964).

[333] Good, R.J. *J. Colloid Interface Sci.* **52**, 308 (1975).

[334] Becher, P. *J. Colloid Interface Sci.* **59**, 429 (1977).

[335] Chashhin, I.P., & Shvab, N.S. In *Methods of Aerodynamics and Heat-Mass Exchange in Technological Processes.* University Press, Tomsk, 1984, p. 133.

[336] Cutowski, W.S. In *8th Annu. Meet. Adhesion Soc.* Savannah, 1985, p. 17/a.

[337] Toyama, M. *Adhes. Seal. Japan.* **30**, 313 (1986).

[338] Cooper, W., & Nuttal, W. *J. Agr. Sci.* **7**, 219 (1915).

[339] Evdokimov, Yu.M., Mamedov, V.Sh., & Krestov, D.S. *Recommendations for Use of Polymer Anti-Transpirants.* Agr. -Ind. Publ., Baku, 1986, p. 20.

[340] Harkins, W.D., & Ewing, W.W. *J. Amer. Chem. Soc.* **42**, 239 (1920).

[341] Huntsberger, J.R. In *Treatise on Adhesion and Adhesives.* Marcel Dekker, New York, 1981, vol. 5, p. 1.

[342] Harkins, W.D. *Physical Chemistry of Surface Films.* Reinhold, New York, 1952, p. 41.

[343] Krotov, V.V., & Rusanov, A.I. *USSR Colloid J.* **34**, 81 (1972).

[344] Pritykin, L.M., & Dranovsky, M.G. In *Diffusion, Phase Transformations, Mechanical Properties of Metals and Alloys.* Mach. Inst. Press, Moscow, 1976, p. 201.

[345] Lasoski, S.W., & Kraus, G. *J. Polymer Sci.* **18**, 359 (1955).

[346] Kaelble, D.H. *J. Adhesion.* **1**, 102 (1969).

[347] Bragole, R.A. *Elastoplastics.* **4**, 226 (1972).

[348] Ausserre, D., Picard, A.M., & Leger, L. *Phys. Rev. Lett.* **57**, 2671 (1986).

[349] Sakai, T. *J. Adhesion Soc. Japan.* **21**, 441 (1985).

[350] Halperin, A., & de Gennes, P.-G. *J. Phys.* **47**, 1243 (1986).

[351] Joanny, J.F., & de Gennes, P.-G. *Compt. rend.* **299B**, 279 (1984).

[352] Hocking, L.M. *Quart. J. Mech. Appl. Math.* **34**, 37 (1981).

[353] Hocking, L.M., & Rivers, A. *J. Fluid Mech.* **121**, 425 (1982).

[354] Dussan, V.E. *Ann. Rev. Fluid Mech.* **11**, 371 (1979).

[355] Marmur, A. *Adv. Colloid Interface Sci.* **66**, 1790 (1981).

[356] Lelah, M., & Marmur, A. *J. Colloid Interface Sci.* **82**, 518 (1981).

[357] Tanner, L. *J. Phys.* **12D**, 1473 (1979).

[358] Popel', S.I. *Adhesion of Melts and Solder. Mater. USSR.* (1), **3** (1976).

[359] Popel', S.I. In *Theory of Metallurgical Processes.* Inst. Sci. -Techn. Information Publ., Moscow, 1978, p. 108.

[360] Cherry, B.W., & Holmes, C.M. *J. Colloid Interface Sci.* **29**, 174 (1969).

[361] Blake, T.D., & Haunes, J.M. *J. Colloid Interface Sci.* **30**, 421 (1969).

[362] Popel', S.I., Pavlov, V.V., & Zakharova, T.V. In *Adhesion of Melts.* Naukova Dumka Publ., Kiev, 1974, p. 53.

[363] Friz, G. *Z. Angew. Phys.* **19**, 374 (1965).

[364] De Gennes, P.-J. *Compt. rend.* **288B**, 219 (1979).

[365] Neogi, P., & Miller, C.A. *J. Colloid Interface Sci.* **86**, 525 (1982).

[366] Kraynik, A.M., & Schowalter, W.R. *J. Rheol.* **25**, 95 (1981).

[367] Burton, R.H., Folkes, M.J., Narh, K.A., & Keller, A. *J. Mater. Sci.* **18**, 315 (1983).

[368] De Gennes, P.-G. *Scaling Concepts in Polymer Physics.* Cornell University Press, Ithaca, 1979.

[369] Khlynov, V.V., Pastukhov, B.A., Furman, E.L., & Boxer, E.L., In *Surface Properties of Melts.* Naukova Dumka Publ., Kiev, 1982, p. 151.

[370] Missol, W. *Surface Energy of Interfaces in Metals.* PWT Publ., Warszawa, 1975.

[371] Ikada, Y., & Matsunaga, T. *J. Adhesion Soc. Japan.* **14**, 427 (1978); **15**, 91 (1979).

[372] Jacobasch, H.-J., & Freitag, K.-H. *Acta Polym.* **30**, 453 (1979).

[373] Huntsberger, J.R. *J. Adhesion.* **12**, 3 (1981).

[374] Pugatchevitch, P.P., Beglyarov, E.M., & Lavygin, I.A. *Surface Phenomena in Polymers.* Chimia Publ., Moscow, 1982.

[375] Jhon, M.S., & Oh, Y. In *Surface and Interfacial Aspects of Biomedical Polymers.* Wiley, New York, 1985, vol. 1, p. 395.

[376] Halsey, G.D. In *Gas–Solid Interface.* Marcel Dekker, New York, 1967, vol. 1, p. 508.

[377] Obreimoff, I.W. *Proc. Roy. Soc.* **A127**, 290 (1930).

[378] Sawai, J., & Nishida, M. *Z. anorg. Chem.* **190**, 375 (1930).

[379] Khokonov, Kh.B., Zadumkin, S.N., & Shebzukhova, I.G. *USSR Pat.* 408198 (1973).

[380] Preston, F. *Trans. Opt. Soc.* **23**, 3 (1921–1922).

[381] Barbour, J., Charbonnier, F.M., Dolan, W.W., Dyke, W.P., Martin, E.E., & Trolan, J.K. *Phys. Rev.* **117**, 1452 (1960).

[382] Allen, B.C. *Trans. AJME.* **245**, 1621 (1969).

[383] Lipsett, S.G., Johnson, F.M.G., & Maass, O. *J. Amer. Chem. Soc.* **49**, 925 (1927).

[384] Benson, G.C., Schreiber, H.P., & VanZeggeren, F. *Canad. J. Chem.* **34**, 1553 (1959).

[385] Bailey, G.L., & Watkins, H.C. *Proc. Phys. Soc.* **B63**, 950 (1950).

[386] Fox, H.W., & Zisman, W.A. *J. Colloid Sci.* **7**, 109 (1952).

[387] Sell, P.-J., & Neumann, A.W. *Z. phys. Chem.* **41**, 191 (1964).

[388] Elton, G.A.N. *J. Chem. Phys.* **10**, 1066 (1951).

[389] Hybart, F.J., & White, T.R. *J. Appl. Polymer Sci.* **3**, 118 (1960).

[390] Fainerman, A.E., Lipatov, Yu.S., & Koulick, V.M. *USSR Colloid J.* **31**, 140 (1969).

[391] Schonhorn, H., Ryan, F.W., & Sharpe, L. *J. Polymer Sci.* **4**, 538 (1964).

[392] Sugden, S. *J. Chem. Soc.* **125**, 32 (1924).

[393] Hildebrand, J.H., & Scott, R.L. *Solubility of Non-electrolytes.* Reinhold, New York, 1950, p. 431.

[394] Davis, B.W. *J. Colloid Interface Sci.* **59**, 420 (1977).

[395] Pritykin, L.M. *USSR High Polymers.* **23A**, 757 (1981).

[396] McLeod, D.B. *Trans. Faraday Soc.* **19**, 38 (1923).

[397] Wu. S. In *Polymer Blends.* Academic Press, New York, 1978, vol. 1, p. 243.

[398] Roe, R.-J. *J. Chem. Phys.* **72**, 2013 (1968).

[399] Wu. S. *J. Phys. Chem.* **74**, 632 (1968).

[400] Wu. S. *J. Polymer Sci.* **34C**, 19 (1971).

[401] Gielen, H.L., Verbeke, O.B., & Thoen, J. *J. Chem. Phys.* **81**, 6154 (1984).

[402] Wu. S. *J. Colloid Interface Sci.* **31**, 153 (1969).

[403] Bender, G.W., & Gaines, G.L. *Macromolecules.* **3**, 128 (1970).

[404] McLaclan, D.J. *Acta Metall.* **5**, 111 (1957).

[405] Zadumkin, S.N., & Karashev, A.A. In *Surface Phenomena in Melts and Emerging Solid Phases.* University Press, Naltchick, 1965, p. 85.

[406] Mezei, L.Z., Marton, D., & Giber, Y. *USSR J. Phys. Chem.* **52**, 850 (1978).

[407] Boyer, R.F. *J. Polymer Sci.* (14C), 3 (1966).

[408] Bunn, C.R. *Trans. Faraday Soc.* **35**, 482 (1939).

[409] Tsvetkov, V.N., Eskin, V.E., & Frenkel', S.Ya. *Structure of Macromolecules in Solutions.* Science Publ., Moscow, 1964, p. 288.

[410] Pritykin, L.M., & Wakula, W.L. *Adhesion.* **27**, (12), 14 (1983).

[411] Fisher, C.H. *Amer. Chem. Soc. Div. Polym. Mater. Sci. Eng.* **53**, 746 (1985).

[412] Fox, T.G., & Flory, P.J. *J. Polymer Sci.* **14**, 315 (1954).

[413] Starkweather, H.W. *Amer. Chem. Soc. Polymer Prepr.* **5**, 360 (1964).

[414] Edwards, H. *J. Appl. Polymer Sci.* **12**, 2213 (1968).

[415] Nose, T. *Polymer J.* **3**, 1 (1972).

[416] Le Grand, D.C., & Gaines, G.L. *J. Colloid Interface Sci.* **50**, 272 (1975).

[417] Le Grand, D.C., & Gaines, G.L. *J. Colloid Interface Sci.* **31**, 162 (1969).

[418] Poser, C.I., & Sanchez, I.C. *J. Colloid Interface Sci.* **69**, 539 (1979).

[419] Gaines, G.L. *Colloids and Surfaces.* **15**, 161 (1985).

[420] Golubkov, Yu.V. *USSR J. Phys. Chem.* **59**, 905 (1985).

[421] Suryanarayana, C.G., & Pugazhendhi, P. *Indian. J. Pure Appl. Phys.* **24**, 406 (1986).

[422] Friend, J.N. *Nature.* **150**, 432 (1942).

[423] Golubkov, Yu.V., & Nisel'son, L.A. *USSR J. Phys. Chem.* **59**, 648 (1985).

[424] Golubkov, Yu.V., & Nisel'son, L.A. *USSR J. Phys. Chem.* **60**, 1742 (1986).

[425] Van der Berwe, J.H., & van der Berg, N.G. *Surface Sci.* **32**, 1 (1972).

[426] Buckley, D. *Surface Phenomena on Adhesion and Friction Contact*. Wiley, New York, 1983, ch. 1.3, 3.2.
[427] Sageman, D.R., & Burnet, G. *J. Inorg. Nucl. Chem.* **36**, 1105 (1974).
[428] Allen, B.C. *Liquid Metals*. Marcel Dekker, New York, 1972.
[429] Siegel, E. In *Physics and Chemistry of Liquids*. Gordon & Breach, New York, 1976, p. 29.
[430] Barton, D., Mezei, L.Z., & Giber, Y. *USSR J. Phys. Chem.* **52**, 847 (1978).
[431] Papazian, H.A. *Int. J. Thermophys.* **6**, 533 (1985).
[432] Fridman, Ya.B. In *Mechanical Properties of Metals, Mechanical Testing, and Strength*. Mach. -Build. Publ., Moscow, 1974.
[433] Troyanovsky, E.A. *Proc. Electrotechn. Inst.* (213), **44** (1975).
[434] Gilman, J.J. *J. Appl. Phys.* **31**, 2208 (1960).
[435] Sorokin, G.M., Albagachiev, A.Yu., & Medelyaev, I.A. *Dep. Manuscr.* No 3995, Inst. Sci. -Techn. Information Publ., Moscow, 1985.
[436] Gel'bers, V.P. In *Welding*. Inst. Sci. -Tech. Information Publ., Moscow, 1970, p. 1.
[437] Kounin, L.L. *Reports USSR Acad. Sci.* **79**, 93 (1951).
[438] Kounin, L.L. In *Theory of Metallurgical Processes*. Metallurgia Publ., Moscow, 1965, p. 67.
[439] Zadumkin, S.N., & Egiev, V.G. *USSR Phys. Metal.* **22**, 121 (1967).
[440] Demchenko, V.V. *USSR Phys. Metal.* **21**, 634 (1966).
[441] Demchenko, V.V., & Khomutov, N.E. *Proc. Mendeleev Chem. -Techn. Inst.*, No 39, 115 (1962).
[442] Zadumkin, S.N. *USSR J. Phys. Chem.* **27**, 502 (1953).
[443] Egiev, V.G. *USSR J. Phys. Chem.* **45**, 2871 (1971).
[444] Zadumkin, S.N., Ibragimov, Kh.I., & Khokonov, Kh.B. *USSR J. Phys. Chem.* **51**, 133 (1977).
[445] Martin, J.R., Johnson, J.F., & Cooper, A.R. *J. Macromol. Sci.* **8C**, 57 (1972).
[446] Fedors, R.F. *Polymer.* **20**, 1055 (1979).
[447] Nazin, G.I., & Arinshtein, E.A. In *Surface Phenomena in Liquids*. University Press, Leningrad, 1975, p. 57.
[448] Turusov, R.A., Korotkov, V.N., & Rozenberg, B.A. In *Surface Phenomena in Polymers*. Naukova Dumka Publ., Kiev, 1982, p. 110.
[449] Privalko, V.P. *Molecular Structure and Properties of Polymers*. Chimia Publ., Leningrad, 1986.
[450] Rostiashvili, V.G., Irzhack, V.I., & Rozenberg, B.A. *Glass-transition of Polymers*. Chimia Publ. Leningrad, 1987.
[451] McLeod, D.B. *Trans. Faraday Soc.* **19**, 41 (1923).
[452] Kleeman, R.D. *Phil. Mag.* **21**, 99 (1911).
[453] Bachinski, A.I. *Short Lectures on Gas Theory*. State Publ., Moscow, 1922.
[454] Wright, F.J. *J. Appl. Chem.* **11**, 193 (1961).
[455] Kasemura, T., & Hata, T. *Japan High Polymers.* **33**, 195 (1976).
[456] Kasemura, T., Yamashita, N., Suzuki, K., Kondo, T., & Hata, T. *Japan High Polymers.* **35**, 263 (1978).
[457] Barton, A.F. *J. Adhesion.* **14**, 33 (1982).
[458] Michaels, A.S. *ASTM Spec. Techn. Publ.* **340**, 3 (1962).
[459] Naghizadeh, J., & Ailawade, N.K. *J. Phys. Math. Gen.* **10**, 59 (1977).

[460] Yagnyatinskaya, S.M., Voyutskii, S.S., & Kaplunova, L.Ya. *USSR Colloid J.* **34**, 132 (1972).

[461] Bonn, R., & van Aartsen, J.J. *Europ. Polymer J.* **8**, 1055 (1972).

[462] Schonhorn, H. *J. Chem. Phys.* **43**, 2041 (1965).

[463] Mason, E.A. *Amer. J. Phys.* **34**, 1193 (1966).

[464] Toyama, M. *Japan Adhes. Seal.* **29**, 120 (1985).

[465] Gardon, J.L. *J. Colloid Interface Sci.* **59**, 308 (1977).

[466] Lee, L.-H. *Adv. Chem.* Ser. 87, 106 (1968).

[467] Beriketov, A.S., Belousov, V.N., & Mikitaev, A.K. *Dep. Manuscr.* No 1054, Niitekhim Press, Tcherkassy, 1982.

[468] Small, P.A. *J. Appl. Chem.* **3**, 71 (1953).

[469] Wu, S. *J. Phys. Chem.* **72**, 3332 (1968).

[470] Godfrey, J.C. *J. Chem. Educ.* **36**, 140 (1959).

[471] Slonimsky, G.L. Askadsky, A.A., & Kitaigorodsky, A.I. *USSR High Polymers.* **12A**, 494 (1970).

[472] Zhdanov, G.S. *Physics of Solids*, University Press, Moscow, 1962, p. 160.

[473] Askadsky, A.A. *Pure Appl. Chem.* **46**, 19 (1976).

[474] Askadsky, A.A., Kitaigorodsky, A.I., & Slonimsky, G.L. *USSR High Polymers.* **17A**, 2293 (1975).

[475] Askadsky, A.A., Matevosyan, M.S., & Slonimsky, G.L. *Dep. Manuscr.* No 6949, Inst. Sci. -Techn. Information Publ., Moscow, 1985.

[476] Askadsky, A.A., Kolmakova, L.K., Tager, A.A., Slonimsky, G.L., & Korshak, V.V. *Reports USSR Acad. Sci.* **226**, 857 (1976).

[477] Askadsky, A.A., Kolmakova, L.K., Tager, A.A., Slonimsky, G.L., & Korshak, V.V. *USSR High Polymers.* **19A**, 1004 (1977).

[478] Askadsky, A.A., & Matveev, Yu.I. *Chemical Structure and Physical Properties of Polymers.* Chimia Publ., Moscow, 1983.

[479] Askadsky, A.A. *Structure and Properties of High Temperature Resistant Polymers.* Chimia Publ., Moscow, 1981.

[480] Askadsky, A.A. *Principle of Additivity in Physico-Chemistry of Polymers.* Knowledge Publ., Moscow, 1987, p. 26.

[481] Askadsky, A.A. In *Synthesis and Development of Properties of Optical Sensitive Materials.* Moscow, 1987, p. 22.

[482] Pritykin, L.M. In *Proc. 30th IUPAC Congr.* Roy. Chem. Soc. Publ., London, 1985, prepr. 2P.A7.

[483] Vasilova, O.I., Zaitseva, V.V., & Koutcher, R.V. *USSR High Polymers.* **29B**, 912 (1987).

[484] Shilov, G.I., Ovchinnikov, Yu.V., & Kronman, A.G. *USSR High Polymers.* **23B**, 199 (1981).

[485] Pritykin, L.M., Zjuz', V.T., & Onushko, A.I. *USSR High Polymers.* **25B**, 271 (1983).

[486] Pritykin, L.M. In *31st IUPAC Macromol. Symp.* IUPAC Publ., Merseburg, 1987, pt.IV–V, p. 57.

[487] Beriketov, A.S., & Kogotyzheva, M.Kh. In *Polycondensation Processes and Polymers.* University Press, Naltchick, 1985, p. 119.

[488] Askadsky, A.A., Matevosyan, M.S., & Slonimsky, G.L. *USSR High Polymers.* **29A**, 753 (1987).

[489] Shits, L.A. In *Encyclopedia of Polymers*. Soviet Encyclopedia Publ., Moscow, 1972, vol. 1, p. 1045.

[490] Van Krevelen, D.W. *Properties of Polymers: Correlations with Chemical Structure*. Elsevier, Amsterdam, 1972.

[491] Rymuza, Z. *Tribol. lubrif.* **16**, 93 (1981).

[492] Wiesner, L. *Chem. prům.* **26**, 532 (1976).

[493] Pritykin, L.M., & Dranovsky, M.G. In *Adhesive Joints in Machine-Building*. Polytechn. Inst. Press, Riga, 1983, p. 48.

[494] Kreibich, U.T., & Batzer, H. *Angew. Makromol. Chem.* **83**, 57 (1979).

[495] Grinenko, E.Yu., Roginsky, S.L., Uloukhanov, A.G., Kanovitch, M.Z., & Nikitina, G.S. *USSR Plastics.* No 10, 27 (1985).

[496] Pritykin, L.M., & Shalakhman, Yu.G. *USSR Plastics.* No 8, 19 (1986).

[497] Oh, Y., & Jhon, M.S. *J. Colloid Interface Sci.* **73**, 467 (1980).

[498] Owens, N.F., Richmond, P., & Mingins, J. In *Wetting, Spreading and Adhesion*. Academic Press, New York, 1978, p. 127.

[499] Beevers, R.B. *J. Polymer Sci.* **12**, 1407 (1974).

[500] Shalaeva, L.F. *USSR Plastics.* No 2, 65 (1968).

[501] Doladugina, V.S., Nizhin, A.M., & Bryakova, L.S. *USSR Opt. -Mech. Ind.* No 6, 36 (1971).

[502] Akhmedov, A.G. *USSR J. Phys. Chem.* **58**, 1807 (1984).

[503] Papazian, H.A. *J. Amer. Chem. Soc.* **93**, 5634 (1971).

[504] Pugachevitch, P.P., & Zhalsabon, B.V. *USSR J. Phys. Chem.* **56**, 764 (1982).

[505] Gambill, W.R. *Chem. Eng.* **64**, 146 (1958).

[506] Pritykin, L.M. *USSR High Polymers.* **20B**, 277 (1978).

[507] Askadsky, A.A., Prozorova, S.N., & Slonimsky, G.L. *USSR High Polymers.* **18A**, 636 (1976).

[508] Blanks, R.F., & Prausnitz, J.M. *Ind. Eng. Chem.* **3**, 1 (1964).

[509] *Handbook on Polymer Chemistry*. Naukova Dumka Publ., Kiev, 1971.

[510] Joffe, B.V. *Refractometric Methods of Chemistry*. Chimia Publ., Leningrad, 1974, p. 339.

[511] Eijkman, J.F., *Rec. Trav. Chim.* **14**, 185 (1895).

[512] Stolyarov, E.A., & Orlova, N.G. *Calculation of Physico Chemical Properties for Liquids*. Chimia Publ., Leningrad, 1976, p. 18.

[513] Gold, P.J., & Ogle, G. *Chem. Eng.* **76**, 170 (1969).

[514] Pritykin, L.M. *J. Colloid Interface Sci.* **112**, 539 (1986).

[515] Ionesawa, M., & Iuki, J. *Japan J. Soc. Synth. Org. Chem.* **30**, 823 (1972).

[516] Shafrin, E.G. In *Handbook of Adhesives*. Van Nostrand Reinhold, New York, 1977, p. 67.

[517] Kopytin, V.S. Cand. Sci. (Chem.) Thesis, Lomonosov Inst. Fine Chem. Techn., Moscow, 1987.

[518] Prokoptchuck, N.R. *Herald Byelorussian Acad. Sci.* Ser. Phys. -Techn. Sci., No 4, 37 (1984).

[519] Geczy, I. *Kem. Tud. Oszt. Közl.* **23**, 285 (1965).

[520] Geczy, I. *Textilpari Kut. Int.* No 11, 3 (1967).

[521] Geczy, I., Pritykin, L.M., Jemel'janov, Ju.V., & Sztesenko, N.I. *Kolor. Ertes.* **25**, 66 (1983).

[522] Weissler, A. *J. Amer. Chem. Soc.* **70**, 1634 (1948).

[523] Menin, J.P., & Roux, R. *J. Polymer Sci.* **10A**, 855 (1972).

[524] Lagemann, R.T., Woolf, W.E., Evans, J.S., & Underwood, N. *J. Amer. Chem. Soc.* **70**, 2994 (1948).

[525] Völker, T., Neumann, A., & Baumann, U. *Makromol. Chem.* **63**, 186 (1963).

[526] Trekoval, J. *Coll. Czech. Chem. Comm.* **38**, 3769 (1973).

[527] Geczy, I., & Gyorgy, K. *Acta Chim. Acad. Sci. Hung.* **112**, 227 (1983).

[528] *Encyclopedia of Polymer Science and Technology.* Wiley-Interscience, New York, 1967, vol. 6, p. 173.

[529] Pliev, T.N., Glavati, O.L., & Popovitch, T.D. *USSR High Polymers.* **12A**, 31 (1970).

[530] Pritykin, L.M., Emel'yanov, Yu.V., Steshenko, N.I., & Geczy, I. *USSR High Polymers.* **26B**, 438 (1984).

[531] Vilesov, F.I., Zagroubsky, A.A., & Sukhov, D.A. *USSR Solid Phys.* **11**, 3049 (1969).

[532] Dimun, M., & Zeman, S. *Petrochemia.* **26**, 137 (1986).

[533] Rogachev, A.G., & Krasovsky, A.M. *Herald Byelorussian Acad. Sci.* Ser. Phys. -Techn. Sci., No 1, 49 (1979).

[534] Loushheikin, G.A. *Polymer Electrets.* Chimia Publ., Moscow, 1984, p. 11.

[535] Rogachev, A.V. *USSR High Polymers.* **24A**, 1108 (1982).

[536] Rogachev, A.V., & Krasovsky, A.M. In *Coating of Metals by Plastics and Metallization of Plastics in Machine-Building.* Sci. -Techn. Soc. Publ., Moscow, 1980, pt. 1, p. 29.

[537] Sears, G.B. *J. Appl. Phys.* **21**, 721 (1950).

[538] Imoto, M. *Japan Techn. Adhes. Seal.* **32**, 1 (1988).

[539] Kagan, D.F., Prokopenko, V.V., Malinsky, Yu.M., & Bakeev, N.F. *USSR High Polymers.* **22A**, 110 (1980).

[540] Fainerman, A.E., Min'kov, V.I., & Tremuth, V.M. In *Physical Chemistry of Polymer Compositions.* Naukova Dumka Publ., Kiev, 1974, p. 46.

[541] Fox, H.W., & Zisman, W.A. *J. Colloid Sci.* **5**, 514 (1950).

[542] Shafrin, E.G., & Zisman, W.A. *J. Phys. Chem.* **71**, 1309 (1967).

[543] Rhee, S.K. *Mater. Sci. Eng.* **11**, 311 (1973).

[544] Elban, W.L. *J. Mater. Sci.* **14**, 1008 (1979).

[545] Zisman, W.A. *Ind. Eng. Chem.* **55**, No 10, 19 (1964).

[546] Bernett, M.K., & Zisman, W.A. *J. Colloid Interface Sci.* **28**, 243 (1968).

[547] Bernett, M.K., & Zisman, W.A. *J. Colloid Interface Sci.* **29**, 413 (1969).

[548] Wu, S. *J. Colloid Interface Sci.* **71**, 605 (1979).

[549] Wu, S. In *Adhesion and Adsorption of Polymers.* Plenum Press, New York, 1980, p. 53.

[550] Kammer, H.-W., & Gräfe, F. *Acta Polym.* **36**, 378 (1985).

[551] Bartenev, G.M., Akopyan, L.A., & Rusanov, A.I. *USSR High Polymers.* **28A**, 207 (1986).

[552] Johnson, R.E., & Dettre, R.H. *J. Colloid Interface Sci.* **21**, 610 (1966).

[553] Rozenboim, N.A., Ovchinnikov, V.N., & Pestov, S.S. *USSR Rubber.* (8), **27** (1987).

[554] Pritykin, L.M. *USSR Rubber.* (3), **33** (1988).

[555] Jensen, W.B. *Rubber Chem. Techn.* **55**, 881 (1982).

[556] Fowkes, F.M. In *Ind. Appl. Surface Anal.* Amer. Chem. Soc. Publ., New York, 1982, p. 69.

[557] Fowkes, F.M. *J. Adhesion Sci. Techn.* **1**, 7 (1987).

[558] Fowkes, F.M. *J. Adhesion.* **4**, 155 (1972).

[559] Drago, R.S., & Nozari, M.S. *J. Amer. Chem. Soc.* **92**, 7086 (1970).

[560] Drago, R.S., Vogel, G.C., & Needham, T.E. *J. Amer. Chem. Soc.* **93**, 6014 (1971).

[561] Wu. S. *J. Adhesion.* **5**, 35 (1973).

[562] Janczuk, B., & Bialopiotrowicz, T. *Przem. chem.* **61**, 407 (1982).

[563] Spelt, J.K., Absolom, D.R., & Neumann, A.W. *Langmuir.* **2**, 620 (1986).

[564] Spelt, J.K., & Neumann, A.W. *Langmuir.* **3**, 588 (1987).

[565] Moy, E., & Neumann, A.W. *J. Colloid Interface Sci.* **119**, 296 (1987).

[566] De Bruyne, N.A. *Aircraft Eng.* (12), **53** (1939).

[567] Erb, R.A. *J. Phys. Chem.* **69**, 1306 (1965).

[568] Brewis, D. *Polymer Eng. Sci.* **7**, 17 (1967).

[569] Michel, M. *Adhesion.* **27**, (11), **11** (1983).

[570] Korolev, A.Ya. In *Adhesives and Joints on Its Base.* House Sci. Techn. Propag. Publ., Moscow, 1970, pt. 1, p. 15.

[571] Dirska, B. *Sesz. Nauk. Akad. Techn. Bydgoszczy.* Chem. Techn., No 62, 113 (1979).

[572] Birdi, K.A., & Jeppesen, J. *Coll. Polymer Sci.* **256**, 261 (1978).

[573] Busscher, H.J., & Arends, J. *J. Colloid Interface Sci.* **81**, 75 (1981).

[574] Fowkes, F.M. *J. Phys. Chem.* **66**, 382 (1962).

[575] Kaelble, D.H. In *Proc. 23rd IUPAC Int. Congr.* Elsevier, Amsterdam, 1971, vol. 8, p. 265.

[576] Dynes, P.J., & Kaelble, D.H. *J. Adhesion.* **6**, 195 (1974).

[577] Neumann, A.W., & Tanner, W. In *Proc. 5th Int. Congr. Surface Activity.* Butterworth, London, 1968, vol. 2, p. 727.

[578] Sherriff, M. *J. Adhesion.* **7**, 257 (1976).

[579] Sacher, E. *J. Colloid Interface Sci.* **83**, 649 (1981).

[580] Busscher, H.J., Van Pelt, A.W.J., De Jong, H.P., & Arends, J. *J. Colloid Interface Sci.* **95**, 23 (1983).

[581] El-Shimi, A., & Goddard, E.D. *J. Colloid Interface Sci.* **48**, 242 (1974).

[582] Escoubes, M., Berticat, P., Charbert, B., Chauchard, J., & Sage, D. *Makromol. Chem.* **183**, 3041 (1982).

[583] Janczuk, B., & Chibowski, E. *J. Colloid Interface Sci.* **95**, 268 (1983).

[584] Zorll, U. *Gummi–Asbest–Kunstst.* **30**, 634 (1977).

[585] Dann, J.R. *J. Colloid Interface Sci.* **32**, 302 (1970).

[586] Kloubek, J. *J. Adhesion.* **6**, 293 (1974).

[587] Bantysh, A.N., Klepikov, E.S., Kyzin, I.S., & Savel'ev, V.B. In *Surface Phenomena in Polymers.* Naukova Dumka Publ., Kiev, 1982, p. 11.

[588] Bantysh, A.N., Klepikov, E.S., Kyzin, I.S., & Savel'ev, V.B. *USSR Compos. Polym. Mater.* (21), **43** (1984).

[589] Dahlquist, C.A. In *Aspects of Adhesion.* University Press, London, 1969, vol. 5, p. 183.

[590] Hamilton, W.C. *J. Colloid Interface Sci.* **47**, 672 (1974).

[591] Andrade, J.D., Ma, S.M., King, R.N., & Gregonis, D.E. *J. Colloid Interface Sci.* **72**, 488 (1979).

[592] Fowkes, F.M. In *Chemistry and Physics of Interfaces*. Marcel Dekker, Washington, 1965, ch. 1.

[593] Tamai, Y., Makuuchi, K., & Suzuki, M. *J. Phys. Chem.* **71**, 4176 (1967).

[594] Shanahan, M.E.R., Cazeneuve, C., Carre, A., & Schultz, J. *J. Chim. Phys. et Phys. Chim. Biol.* **79**, 241 (1982).

[595] Matsunaga, T., & Ikada, Y. *J. Colloid Interface Sci.* **84**, 8 (1981).

[596] Shiomi, T., Nishioka, S., Tezuka, Y., & Imai, K. *Polymer.* **26**, 429 (1985).

[597] Janczuk, B., & Chibowski, E. *Pol. J. Chem.* **59**, 1251 (1985).

[598] Tamai, Y., Matsunaga, T., & Horiuchi, K. *J. Colloid Interface Sci.* **60**, 112 (1977).

[599] Dann, J.R. *J. Colloid Interface Sci.* **32**, 321 (1970).

[600] Matsunaga, T. *J. Appl. Polymer Sci.* **21**, 2847 (1977).

[601] Aveyard, R. *J. Colloid Interface Sci.* **52**, 621 (1975).

[602] Matsunaga, T., Tamai, Y., & Suzuki, K. In *7th Int. Congr. Surfactants*. Foreign Commerce Publ., Moscow, 1976, pt.Bm p. 89.

[603] Cazeneuve, C., Donnet, J.B., Schultz, J., & Shanahan, M.E.R. In *Int. Congr. Adhesion and Adhesives*. University Press, London, 1980, p. 19.1.

[604] Schultz, J., Cazeneuve, C., Shanahan, M.E.R., & Donnet, J.B. *J. Adhesion.* **12**, 221 (1981).

[605] Feynman, R.P. *Statistical Mechanicss* Academic Press, New York, 1975.

[606] Navascues, G. *J. Colloid Interface Sci.* **72**, 150 (1979).

[607] Pletnev, M.Yu., & Tereshhenko, N.B. *Ukrainian Chem. J.* **52**, 427 (1986).

[608] Nag, Z., & Mingins, J. *Polym. Comm.* **25**, 269 (1984).

[609] Kasemura, T., Inagaki, M., & Hata, T. *Kobunshi ronb.* **44**, 131 (1987).

[610] Yasuda, H., Sharma, A.K., & Yasuda, T. *J. Polymer Sci.* **19**, 1285 (1981).

[611] Suzuki, K., Christie, A.B., & Howson, R.P. *Vacuum.* **34**, 181 (1984).

[612] Thomas, G.E., Van der Ligt, G.O., Lippits, G.J., & Van der Hei, G.M.M. *Appl. Surface Sci.* **6**, 204 (1980).

[613] Cognard, J. *J. Chim. Phys. et Phys. Chim. Biol.* **84**, 357 (1987).

[614] Andrade, J.D., & Chen, W.-Y. *Surf. Interf. Anal.* **8**, 253 (1986).

[615] Carre, A., & Schultz, J. *J. Adhesion.* **18**, 171 (1986).

[616] Metsick, M.S. *USSR J. Techn. Phys.* **28**, 109 (1958).

[617] Orman, S., Kerr, C. In *Aspects of Adhesion*. University Press, London, 1970, vol. 6, p. 64.

[618] Shhukin, E.D., Bryukhanova, L.S., Andreeva, I.A., Petrova, I.V., & Rehbinder, P.A. *USSR Colloid J.* **35**, 828 (1973).

[619] Efimov, A.G., Muhammed, Y., Shitov, N.A., Volynsky, A.L., Kozlov, P.V., & Bakeev, N.F. *USSR High Polymers.* **24B**, 433 (1982).

[620] Kalnin', M.M., Karlivan, V.P., & Brakere, R.R. *USSR High Polymers.* **10A**, 2513 (1970).

[621] Malers, L.Ya., & Kalnin', M.M. In *Modification of Polymer Materials*. Polytechn. Inst. Press, Riga, 1978, p. 12.

[622] Malers, L.Ya., & Kalnin', M.M. In *Modification of Polymer Materials*. Polytechn. Inst. Press, Riga, 1972, p. 39.

[623] Vinogradov, G.V., Kurbanaliev, M.K., Dreval, V.E., & Malkin, A.Ya. *Reports USSR Acad. Sci.* **257**, 386 (1981).

[624] Vinogradov, G.V., Kurbanaliev, M.K., Dreval, V.E., & Malkin, A.Ya. *Polymer.* **23**, 100 (1982).

[625] Lobanov, Yu.E. Cand. Sci. (Techn.) Thesis, Lomonosov Inst. Fine Chem. Techn., Moscow, 1969.

[626] Garnish, E.W. In *Adhesion*. Appl. Science, London, 1978, vol. 2, p. 35.

[627] Nechiporenko, N.A., Yakhnin, E.D., & Dembovskaya, Yu.V. *USSR High Polymers*. **20B**, 917 (1978).

[628] Hammermesh, C.L., & Crane, L.W. *J. Appl. Polymer Sci.* **22**, 2395 (1978).

[629] Iwao, M., & Fujio, T. *J. Adhesion Soc. Japan.* **4**, 196 (1968).

[630] Egorenkov, N.I., & Tishkov, N.I. *Reports Byelorussian Acad. Sci.* **18**, 813 (1974).

[631] Egorenkov, N.I., Mlynsky, V.L., & Belyi, V.A. *Reports Byelorussian Acad. Sci.* **17**, 329 (1973).

[632] Brown, H.R., & Kramer, E.J. *Polymer*. **22**, 687 (1981).

[633] Cherezov, A.A. Cand. Sci(Chem.) Thesis, Inst. Phys. Chem., Moscow, 1979.

[634] Dynes, P.J., & Kaelble, D.H. In *Proc. 5th Conf. Compos. Mater.* Philadelphia, 1979, p. 566.

[635] Owens, D.K. *J. Appl. Polymer Sci.* **19**, 265 (1975).

[636] Shmourack, I.L., & Myasnikova, N.M. *USSR Colloid J.* **36**, 412 (1974).

[637] Nenakhov, S.A., Chalykh, A.E., Kreitus, A.E., Metra, A.Ya., & Kalnin', M.M. In *Modification of Polymer Materials*. Polytechn. Inst. Press, Riga, 1975, p. 179.

[638] Kinloch, A.J., Dukes, W.A., & Gledhill, R.A. In *Adhesion Science and Technology*. Plenum Press, New York, 1975, p. 597.

[639] Mikhailova, S.S., Freidin, A.S., & Sokolnikova, I.N. *USSR Compos. Polym. Mater.* No 6, 42 (1980).

[640] Arslanov, V.V., Gevorkyan, O.M., & Ogarev, V.A. *USSR Colloid J.* **43**, 952 (1981).

[641] Schultz, J., & Carre, A. *J. Appl. Polymer Sci.* Appl. Polymer Symp., No 39, 103 (1984).

[642] Krivoshei, V.N., & Iovanovitch, K.S. In *Diffusion Phenomena in Polymers*. Inst. Chem. Phys. Press, Chernogolovka, 1985, p. 91.

[643] Hayes, L.J., & Dixon, D.D. *J. Appl. Polymer Sci.* **23**, 1907 (1979).

[644] Avotin'sh, Ya.Ya., Yurtaeva, A.G., Kalnin', M.M. In *Modification of Polymer Materials*. Polytechn. Inst. Press, Riga, 1985, p. 75.

[645] Avotin'sh, Ya.Ya., Vyatere, E.F., & Kalnin', M.M. *Diffusion Phenomena in Polymers*. Inst. Chem. Phys. Press, Chernogolovka, 1985, p. 89.

[646] Pinchouck, L.S. *Reports Byelorussian Acad. Sci.* **22**, 150 (1978).

[647] Tikhonova, N.I., Kazantseva, V.V., Teyes-Akun'ya, G., Rudakova, T.E., Askadsky, A.A., Moiseev, Yu.V., & Zaikov, G.E. *USSR High Polymers*. **20A**, 1543 (1978).

[648] Porchkhidze, A.D., Rudakova, T.E., Moiseev, Yu.V., Kazantseva, V.V., & Askadsky, A.A. *USSR High Polymers*. **22B**, 783 (1980).

[649] Volynsky, A.L., & Bakeev, N.F. *USSR High Polymers*. **17A**, 1610 (1975).

[650] Volynsky, A.L., Loginov, V.S., & Bakeev, N.F. *USSR High Polymers*. **22B**, 484 (1980).

[651] Volynsky, A.L., Loginov, V.S., & Bakeev, N.F. *USSR High Polymers*. **23B**, 371 (1981).

[652] Volynsky, A.L., Aleskerov, A.G., Zavarova, T.B., Skorobogatova, A.E., Arzhakov, S.A., & Bakeev, N.F. *USSR High Polymers*. **19A**, 845 (1977).

[653] Sinevitch, E.A., & Bakeev, N.F. *USSR High Polymers*. **22B**, 485 (1980).

[654] Gul', V.E., & Zadoya, M.A. *USSR Plastics.* No 3, 25 (1980).

[655] Comyn, J. In *Developments in Adhesion.* Englewood, London, 1981, vol. 2, p. 279.

[656] Kalnin', M.M., Reikhmanis, P.K., & Dzenis, M.A. *Herald Latvian Acad. Sci.* Ser. Sci. Rev., No 2, 116 (1980).

[657] Baun, W.L. *J. Adhesion.* **12**, 81 (1981).

[658] Nakamura, K., Maruno, T., & Sasaki, S. *J. Metal. Finish. Soc. Japan.* **36**, 160 (1985).

[659] Brewis, D.M., Comyn, J., Cope, B.C., & Moloney, A.C. *Polymer Eng. Sci.* **24**, 797 (1981).

[660] Nenakhov, S.A., Chalykh, A.E., & Kreitus, A.E. In *Modification of Polymer Materials.* Polytechn. Inst. Press, Riga, 1975, p. 179.

[661] Comyn, J., Brewis, D.M., & Tegg, J.L. In *Proc. 26th Int. Symp. Macromol.* IUPAC Publ., Mainz, 1979, vol. 2, p. 1136.

[662] Sargent, J.P., & Ashbee, K.H.G. *Polym. Compos.* **1**, 93 (1980).

[663] Brewis, D.M., Comyn, J., Moloney, A.C., & Tegg, J.L. *Europ. Polymer J.* **17**, 127 (1981).

[664] Schulte, R.L., & De Iasi, R.J. *IEEE Trans. Nuclear Sci.* **28**, 1841 (1981).

[665] Brewis, D.M., Comyn, J., Shalash, J.A., & Tegg, J.L. *Polymer.* **21**, 357 (1980).

[666] Comyn, J. *Plast. Rubb. Process. Appl.* **3**, 201 (1983).

[667] Pesetsky, S.S., Egorenkov, N.I., & Shherbakov, S.V. *USSR Colloid J.* **43**, 992 (1981).

[668] Haraga, K. *J. Adhesion Soc. Japan.* **15**, 568 (1979).

[669] Brockmann, W. In *Proc. Int. Conf. Adhesion and Adhesives.* University Press, London, 1980, p. 9.1.

[670] Inagaki, N., & Yasuda, H. *J. Appl. Polymer Sci.* **26**, 3333 (1981).

[671] Andrews, E.H., Pingsheng, H., & Vlachos, C. *Proc. Roy. Soc. A.* **381**, 745 (1982).

[672] Skvortsov, A.M., Gorbunov, A.A., Zhulina, E.B., & Birshtein, T.M. *USSR High Polymers.* **20A**, 278 (1978).

[673] Schonhorn, H. *J. Appl. Polymer Sci.* **23**, 687 (1979).

[674] Veselovsky, R.A. *Regulation of Adhesion Strength of Polymers.* Naukova Dumka, Publ., Kiev, 1988, p. 153.

[675] Ruhsland, K. *ZIS-Mitt.* **22**, 1172 (1980).

[676] Allen, K.W., & Alsalim, H.S. *J. Adhesion.* **8**, 183 (1977).

[677] Mittal, K.L. In *Adhesion Science and Technology.* Plenum Press, New York, 1975, p. 129.

[678] Huntsberger, J.R. In *Proc. Int. Conf. Adhesion and Adhesives.* University Press, London, 1980, p. 18.1.

[679] Miedema, A.R., & Nieuwenhuys, B.E. *Surface Sci.* **104**, 491 (1979).

[680] Pritykin, L.M., & Dranovsky, M.G. In *Diffusion, Phase Transformations, Mechanical Properties of Metals and Alloys.* Mach. Inst. Press, Moscow, 1976, p. 206.

[681] Schonhorn, H. *Macromolecules.* **1**, 145 (1968).

[682] Raisin, I.B. Cand. Sci(Techn.) Thesis Karpov Phys. -Chem. Inst., Moscow, 1975..

[683] Veselovskij, R.A. In *VIII Konf. lepeni kovov.* CVTS Press, Bratislava, 1981, s. 28.

[684] Pritykin, L.M. In *Polymer Adhesives in Modern Engineering.* Novosibirsk, 1978, p. 44.

[685] Toyama, M., Ito, T., Moriguchi, H. *J. Appl. Polymer Sci.* **14**, 2039 (1970).

[686] Ito, T., & Kitazaki, Y. *J. Adhesion Soc. Japan.* **12**, 1006 (1976).

[687] Fukuzawa, K. *J. Adhesion Soc. Japan.* **16**, 230 (1980).

[688] Ostrovskaya, N.B. In *Progressive Structural Films Materials and Assembly of Goods.* House Sci. -Techn. Propaga. Publ., Moscow, 1988, p. 37.

[689] Shalygin, G.F., Gribkova, N.Ya., & Kozlov, P.V. *USSR High Polymers.* **23B**, 426 (1981).

[690] Korshak, V.V., Polyakova, A.M., Mager, K.A., & Semyantsev, V.N., *USSR Plastics.* (3), **54** (1968).

[691] Korshak, V.V., Polyakova, A.M., Mager, K.A., Semyantsev, V.N., & Askadsky, A.A. *USSR Plastics.* (1), **44** (1970).

[692] Pritykin, L.M., Vakula, V.L., Kardashov, D.A., Polyakova, A.M., & Korshak, V.V. *Reports USSR Acad. Sci.* **304**, 390 (1989).

[693] Cheung, K.-H., Guthrie, J., Otterburn, M.S., & Rooney, J.M. *Amer. Chem. Soc. Polymer Prepr.* **26**, 224 (1985).

[694] Pritykin, L.M., Vakula, V.L., Polyakova, A.M., & Korshak, V.V. In *Perspectives of Formulation and Application of New High Effective Adhesive Materials in National Economy.* Inst. Polymer Adhes. Press, Kirovakan, 1988, p. 60.

[695] Raevsky, V.G., & Pritykin, L.M. In *Int. Rubber Symp.* Gottwaldov, 1971, prepr. V. 4.

[696] Mitrokhina, L.L., Chernikov, O.I., Kachan, A.A., & Nosatch, L.V., *USSR Colloid J.* **45**, 1009 (1983).

[697] Kobets, L.P., & Prigorodov, V.N. *USSR Plastics.* No 2, 12 (1981).

[698] Leger, L. *Ann. Chim.* **12**, 175 (1987).

[699] Scola, D.A. In *Interfaces in Polymer Matrix Composition.* Academic Press, New York, 1974, ch. 7.

[700] Piiroya, E.K. *USSR Factory Labor.* **49**, 678 (1983).

[701] Voyutskii, S.S., & Vakula, V.L. *USSR High Polymers.* **2**, 51 (1960).

[702] Yagnyatinskaya, S.M., Kaplunova, L.Ya., Garetovskaya, N.L., & Voyutskii, S.S. *USSR Rubber.* No 11, 25 (1968).

[703] Voyutskii, S.S., Raevsky, V.G., & Yagnyatinskaya, S.M. *Reports USSR Acad. Sci.* **150**, 1296 (1963).

[704] Kouptsov, Yu.D. Cand. Sci(Chem.) Thesis, Lomonosov Inst. Fine Chem. Technol., Moscow, 1971.

[705] Toyama, M., Ito, T., Nakatsuka, H., & Ikeda, M. *J. Appl. Polymer Sci.* **17**, 3495 (1973).

[706] Ieynger, Y., & Erickson, D.E. *J. Appl. Polymer Sci.* **11**, 2311 (1967).

[707] Derjaguin, B.V., Zherebkov, S.K., & Medvedeva, A.M. *USSR Colloid J.* **18**, 404 (1956).

[708] Pritykin, L.M., & Demidenko, L.G. *USSR High Polymers.* **24B**, 89 (1982).

[709] Matsumoto, T., Okubo, M., & Shimao, M. *J. Adhesion Soc. Japan.* **10**, 153 (1974).

[710] Kim, Y.-H., Walker, G.F., Kim, J., & Park, J. *J. Adhesion Sci. Techn.* **1**, 331 (1987).

[711] Leidheister, H., Music, S., & Simmens, G.W. *Nature.* **297**, 667 (1982).

[712] Bauer, A., Bischof, C., Kapelle, R., Possart, W., & Sander, B., *Potsdam. Forsch.* (25), **71** (1981).

[713] Ukrainets, A.M., Koval'chuk, E.P., Aximent'eva, E.I. In *Surface Phenomena in Polymers*. Naukova Dumka Publ., Kiev, 1982, p. 112.

[714] Chang, R.-J., & Gent, A.N. *J. Polymer Sci.* **19**, 1619 (1981).

[715] Gent, A.N. In *Proc. 28th Macromol. Symp.* IUPAC Publ., Amherst, 1982, pt. 1, p. 544.

[716] Braun, E.V., Kruglikov, S.S., & Simonov-Emel'yanov, I.D. *USSR Plastics.* No 1, 29 (1983).

[717] Borodulin, V.N. In *Proc. Lenin Electrotechn. Inst.* (340), 100 (1977).

[718] Adamson, A.W., *Amer. Chem. Soc. Polymer Prepr.* **20**, 673 (1979).

[719] Ikada, Y. *Japan Eng. Mater.* **33**, 3 (1985).

[720] Canova, L., Garbassi, F., & Occhiello, E. *J. Adhesion Sci. Techn.* **1**, 319 (1987).

[721] Bolger, J.C. In *Adhesion Aspects of Polymeric Coatings*. Plenum Press, New York, 1983, p. 3.

[722] Bichwalter, L.P. *J. Adhesion Sci. Techn.* **1**, 341 (1987).

[723] Cross, J.A. In *Proc. Conf. Elec. Math. Mach. Form. Coat.* London, 1975, p. 46.

[724] Briscoe, B.J. In *Polymer Surfaces*. Wiley, Chichester, 1978, p. 31.

[725] Gorbatkina, Yu.A., & Shaidurova, N.K. In *Surface Phenomena in Polymers*. Naukova Dumka Publ., Kiev, 1982, p. 30.

[726] Derjaguin, B.V., Toporov, Yu.P., Mouller, V.M., & Aleinikova, I.N. In *Proc. Polytechn. Inst.* Frunze, (97), **15** (1976).

[727] Jacobs, P., & Tompkins, F. In *Chemistry of Solid State*. Butterworths, London, 1955, ch. 4.4.

[728] Yoshito, M., Kazuo, F., & Yoshinosuke, N. *J. Text. Mach. Sci. Japan.* **13**, 113 (1967).

[729] Ebnett, H., & Klimashewski, A. *Galvano.* **37**, 167 (1968).

[730] Barquins, M., & Cognard, J. *Gold Bull.* **19**, 82 (1986).

[731] Snowden, W.E., & Aksay, I.A. In *Surfaces and Interfaces in Ceramics and Ceramic–Metal Systems*. xxxxxxxxxxxxxxx New York, 1981, p. 651.

[732] Hammermesh, C.L., & Dynes, P.J. *J. Polymer Sci. Lett.* **13**, 663 (1975).

[733] Saito, M., & Yabe, A. *J. Pharm. Soc. Japan.* **28**, 902 (1979).

[734] Matsunaga, T., & Tamai, Y. *J. Appl. Polymer Sci.* **22**, 3525 (1978).

[735] Toshio, H. *Japan Chem.* **6**, 281 (1968).

[736] Anand, J.N., & Balwinski, R.Z. *J. Adhesion.* **1**, 24 (1969).

[737] Litt, M.H., & Matsuda, T. *Amer. Chem. Soc. Polymer Prepr.* **16**, 58 (1975).

[738] Yamaishi, K., Kumazawa, H., & Sanuki, H., *J. Japan Soc. Fiber Sci. Techn.* **32**, 62 (1976).

[739] Ko. Y.C., Ratner, B.D., & Hoffman, A.S. *J. Colloid Interface Sci.* **82**, 25 (1981).

[740] Barbarisi, M.J. *Nature.* **215**, 383 (1967).

[741] Boucher, E.A., *Nature.* **215**, 1071 (1967).

[742] Aleshina, E.A. In *Physics and Chemistry of Cellulose*. Tashkent, 1982, vol. 1, p. 53.

[743] Skalicky, C., & Milichovsky, M. *Sb. ved. praci.* Vys. sk. chem. techn., Pardubice, No 37, 41 (1977).

[744] Arslanov, V.V., & Ogarev, V.A. *USSR Colloid J.* **40**, 841 (1978).

[745] Arslanov, V.V., & Ogarev, V.A. In *Problems of Polymer Composite Materials*. Naukova Dumka Publ., Kiev, 1979, p. 76.

[746] Dzhaparidze, P.N. *Physico-chemical and Energetical Characteristic of Formation Process of Condensed Surface Substances.* Metsniereba Publ., Tbilisi, 1976, p. 33.

[747] Akhmatov, A.S. *Molecular Physics of Boundary Friction.* Phys. -Math. Publ., Moscow, 1963.

[748] Tarazona, P., & Vicente, L. *Mol. Phys.* **56**, 557 (1985).

[749] Ono, S., & Kondo, S. In *Handbuch der Physik.* **10**, T.1, S.13.

[750] Adam, N.K. *Physics and Chemistry of Surfaces.* Oxford Press, London, 1941.

[751] Livshits, I.M., & Geguzin, Ya.E. *USSR Phys. Solid.* **7**, 3 (1965).

[752] Solunsky, V.N. *Reports USSR Acad. Sci.* **257**, 160 (1981).

[753] Cahn, I.W., & Hillard, I.E. *J. Chem. Phys.* **28**, 258 (1958).

[754] Lebovka, N.I., Ovcharenko, F.D., & Mank, V.V. *Reports USSR Acad. Sci.* **285**, 392 (1985).

[755] Borisov, V.I., & Soushkov, V.I. *USSR High Polymers.* **24A**, 437 (1982).

[756] Soushkov, V.I., Gousev, S.S., Perepelkin, A.N., Minyailo, S.A., & Bablyuck, E.B. *USSR High Polymers.* **26A**, 2291 (1984).

[757] White, J.M., & Heidrich, P.F. *Appl. Opt.* **15**, 151 (1976).

[758] Starovoitov, L.E., Gousev, S.S., & Frigin, V.F. *USSR High Polymers.* **29A**, 2042 (1987).

[759] Altoiz, B.A., & Popovsky, Yu.M. In *Problems of Physics of Formation and Phase Transitions.* University Press, Kalinin, 1982, p. 3.

[760] Matveenko, V.N., & Kirsanov, E.A. *USSR Adv. Chem.* **55**, 1319 (1986).

[761] Krasikov, N.N., & Lavrov, I.S. *Herald USSR High School Chem. Techn.* **25**, 428 (1982).

[762] Derjaguin, B.V., & Churaev, N.V. *Wetting Films.* Nauka Publ., Moscow, 1984, p. 20–28.

[763] Paschel, G., & Aldfinger, K.H. *Coll. Polym. Sci.* **254**, 1011 (1976).

[764] Kondrashov, O.F., & Markhasin, I.L. *USSR J. Phys. Chem.* **52**, 1052 (1978).

[765] Popovsky, Yu.M. In *Research on Surface Forces.* Nauka Publ., Moscow, 1967, p. 148.

[766] Derjaguin, B.V., Churaev, N.V., & Mouller, V.M. *Surface Forces.* Nauka Publ., Moscow, 1985, p. 129.

[767] Derjaguin, B.G., & Zakhavaeva, N.N. In *Research on High Polymers.* USSR Acad. Sci. Publ., Moscow, 1949, p. 223.

[768] Derjaguin, B.V., & Samygin, M.M. In *Conf. Viscosity of Liquids and Colloid Solutions.* USSR Acad. Sci. Publ., Moscow, 1941, vol. 1, p. 159.

[769] Derjaguin, B.V., & Karasev, V.V. *USSR Adv. Chem.* **57**, 1110 (1988).

[770] Derjaguin, B.V., Karasev, V.V., & Lavygin, I.A. *Reports USSR Acad. Sci.* **187**, 846 (1969).

[771] Derjaguin, B.V., Karasev, V.V., Lavygin, I.A., Skorokhodov, I.I., & Khromoma, E.N. *Polymer Compos. Mater.* (10), 51 (1981).

[772] Cohen, Y., & Reich, S. *J. Polymer Sci.* **19**, 599 (1981).

[773] Machac, I., Doleis, V., Cakl, J., & Jesensky, M. *Chem. prüm.* **31**, 254 (1981).

[774] Chalykh, A.E. In *Surface Phenomena in Polymers.* Naukova Dumka Publ., Kiev, 1982, p. 123.

[775] Pireaux, J.J., Thiry, P.A., Caudano, R., & Pfluger, P. *J. Chem. Phys.* **84**, 6452 (1986).

[776] Pozdn'yakov, O.F., & Redkov, B.P. *USSR High Polymers.* **28B**, 852 (1986).

[777] Regel', V.R. *USSR Mech. Compos. Mater.* No. 5, 999 (1979).

[778] Tsarev, P.K., & Lipatov, Yu.S. *USSR High Polymers.* **17A**, 717 (1975).

[779] Tsarev, P.K., & Lipatov, Yu.S. *USSR High Polymers.* **12A**, 282 (1970).

[780] Curro, J.G., Lagasse, R.R., & Simha, R. *Macromolecules.* **15**, 1621 (1982).

[781] Egami, T., Maeda, K., & Vitek, V. *Phys. Rev. Phil. Mag.* **41A**, 883 (1980).

[782] Srolovitz, D., Egami, T., & Vitek, V. *Phys. Rev.* **24B**, 6936 (1981).

[783] Ratner, B.D. In *Surface and Interfacial Aspects of Biomedical Polymers.* Wiley, New York, 1985, vol. 1, p. 373.

[784] Cherkasov, A.N., Vitovskaya, M.G., & Boushin, S.V. *USSR High Polymers.* **18A**, 1628 (1976).

[785] Grishhenko, A.E., Rouchin, A.E., Koroleva, S.G., Skazka, V.S., Bogdanova, L.M., Irzhack, V.I., Rozenberg, B.A., & Enikolopyan, N.S. *Reports USSR Acad. Sci.* **269**, 1384 (1983).

[786] Sarkisyan, V.A., Airapetyan, G.A., & Dadivanyan, A.K. *USSR High Polymers.* **26B**, 860 (1984).

[787] Sarkisyan, V.A., & Dadivanyan, A.K. *Herald Armenian Acad. Sci. Phys. Ser.* **22**, 45 (1987).

[788] Bogdanova, L.M., Grizhhenko, A.E., Irzhack, V.I., Nikolaev, V.Ya., Rozenberg, B.A., & Tourov, S.V. *USSR High Polymers.* **29A**, 1588 (1987).

[789] Yuferov, E.A., & Yuferov, A.M. In *Synthesis, Structure and Properties of Network Polymers.* Zvenigorod, 1988, p. 201.

[790] Malinsky, Yu.M. Thesis. Karpov Phys. -Chem. Inst., Moscow, 1970.

[791] Balandin, A.A. *Modern State of Multiplet Theory of Heterogeneous Catalysis.* Nauka Publ., Moscow, 1968, p. 202.

[792] Lipatov, Yu.S. *Physical Chemistry of Filled Polymers.* Chimia Publ., Moscow, 1977.

[793] Lipatov, Yu.S. *Colloid Chemistry of Polymers.* Naukova Dumka Publ., Kiev, 1984.

[794] Fabulyack, F.G. *Molecular Mobility of Polymers in Surface Layers.* Naukova Dumka Publ., Kiev, 1983.

[795] Privalko, V.P. *Handbook of Physical Chemistry of Polymers.* Naukova Dumka Publ., Kiev, 1984, vol. 2, p. 115, 215.

[796] Lebedev, E.G. In *Physico-Chemistry of Multicomponent Polymer Systems.* Naukova Dumka Publ., Kiev, 1986, vol. 2, p. 74.

[797] Lipatov, Yu.S., & Babitch, V.F. *USSR Mech. Compos. Mater.* (1), 17 (1987).

[798] Theorcaris, P.S. In *Mech. Charact. Load Bear.* Fibre Compos. Laminates, London, 1985, p. 55.

[799] Gorbatkina, Yu.A. *Adhesion Strength in Polymer-Fibre Systems.* Chimia Publ., Moscow, 1987.

[800] Lipatov, Yu.S., & Sergeeva, L.M. *Adsorption of Polymers.* Naukova Dumka Publ., Kiev, 1972.

[801] Silberberg, A. *Amer. Chem. Soc. Polymer Prepr.* **27**, 28 (1987).

[802] De Gennes, P.-G. *Ann. Chim.* **77**, 389 (1987).

[803] Klein, J., & Pincus, P. *Macromolecules.* **15**, 1129 (1982).

[804] Kobershtein, J.T., Morra, B., & Stein, R.S. *J. Appl. Crystallogr.* **13**, 34 (1980).

[805] Wendorf, J.H. *J. Polymer Sci.* **18**, 439 (1980).

[806] Cosgrove, T. *Chem. and Ind.* (2), 45 (1988).

[807] Shilov, V.V., Lipatov, Yu.S., & Tsoukrouck, V.V. In *Physico-Chemistry of Multi-component Polymer Systems*. Naukova Dumka Publ., Kiev, 1986, vol. 2, p. 101.

[808] Bogdanova, L.M., Irzhack, V.I., Rozenberg, B.A., & Enikolopyan, N.S. *Reports USSR Acad. Sci.* **268**, 1139 (1983).

[809] Nouzhdina, Yu.A., Ul'berg, Z.R., & Nizhnick, Yu.V. In *Surface Phenomena in Polymers*. Naukova Dumka Pugl., Kiev, 1982, p. 80.

[810] Lipatov, Yu.S., Philippovitch, A.Yu., & Veselovsky, R.A. *Reports USSR Acad. Sci.* **275**, 118 (1984).

[811] Lipatov, Yu.S., Philippovitch, A.Yu., & Veselovsky, R.A. *USSR High Polymers*. **28A**, 2259 (1986).

[812] Lipatov, Yu.S. *USSR Adv. Chem.* **50**, 355 (1981).

[813] Clark, A., Lal, M., & Turpin, M. *Faraday Disc. Chem. Soc.* No 59, 189 (1975).

[814] Gaylord, R.J., Paisner, M.J., & Lohse, D.J. *J. Macromol. Sci.* **17B**, 473 (1980).

[815] Burstrand, J.M. *J. Appl. Phys.* **52**, 4795 (1981).

[816] Lipatov, Yu.S. In *Adhesion and Adsorption of Polymers*. Plenum Press, New York, 1980, p. 601.

[817] Gourinovitch, L.N., Kovriga, V.V., & Lourye, E.G. *USSR Mech. Compos. Mater.* **(6)**, 974 (1980).

[818] Lipatov, Yu.S., Moisya, E.G., & Semenovitch, G.M. *USSR High Polymers*. **19A**, 125 (1977).

[819] Lipatov, Yu.S., Souslo, S.A., & Fabulyack, F.G. *Reports Ukrainian Acad. Sci.* **(5B)**, 834 (1979).

[820] Prokopenko, V.V., Titova, O.K., & Malinsky, Yu.M. In *Problems of Polymer Composite Materials*. Naukova Dumka Publ., Kiev, 1979, p. 40.

[821] Goudova, E.G., Lipatov, Yu.S., & Todosiichouck, T.T. *USSR High Polymers*. **29B**, 539 (1987),

[822] Lipatov, Yu.S. *USSR Mech. Compos. Mater.* (5), 808 (1980).

[823] Kirakosyan, Kh.A., Kiselev, M.R., & Zoubov, P.I. *Reports USSR Acad. Sci.* **251**, 1160 (1980).

[824] Malinsky, Yu.M., & Titova, O.K. *USSR High Polymers*. **18B**, 259 (1976).

[825] Maurer, F.H.J., Kosfeld, R., & Uhlenbroich, T. *Coll. Polym. Sci.* **263**, 624 (1985).

[826] Kaspersky, A.B. In *Physics of Condensed State*. Naukova Dumka Publ., Kiev, 1980, p. 81.

[827] Bronnikov, S.V., Vettegren', V.I., & Korzhavin, L.N. *USSR High Polymers*. **23B**, 97 (1981).

[828] Flourney, P.A. *Spectrochim. Acta.* **22**, 15 (1966).

[829] Ionina, N.V., & Nel'son, K.V. *USSR J. Appl. Spectrosc.* **35**, 329 (1981).

[830] Bronnikov, S.V., & Vettegren', V.I. *USSR High Polymers*. **22B**, 430 (1980).

[831] Kuraeva, L.N., Mikhailova, N.V., & Zolotarev, V.M. *USSR High Polymers*. **19B**, 918 (1977).

[832] Nachinkin, O.I., Lexovskaya, N.P., Karchmartchick, O.S., & Direnko, L.Yu. In *3rd Int. Symp. Chem. Fibres*. Kalinin, 1981, vol. 1, p. 231.

[833] Lipatov, Yu.S., Privalko, V.P., Demtchenko, S.S., & Titov, G.V. *Reports USSR Acad. Sci.* **284**, 651 (1985).

[834] Saidov, G.V. *USSR High Polymers*. **29B**, 453 (1987).

[835] Elliott, A. *Infra-red Spectra and Structure of Organic Long-chain Polymers*. Edward Arnold, London, 1969.

[836] Barbanel', L.Yu., Zhuravleva, I.P., & Saidov, G.V. *USSR High Polymers*. **26A**, 837 (1984).

[837] Babayants, V.D., & Sannikov, S.G. *USSR Colloid J.* **50**, 558 (1988).

[838] Stas'kov, N.I. *Herald Byelorussian Acad. Sci.* (3), **105**, (1986).

[839] Semenovitch, G.M., Dubrovina, L.V., & Lipatov, Yu.S. *USSR High Polymers*. **29A**, 452 (1987).

[840] Rogachev, A.V., Bui, M.V. In *Processes and Equipment for Manufacture of Polymer Materials, Methods and Equipment for Its Processing*. Moscow, 1986, vol. 2, p. 68.

[841] Vixne, A.V., Kalnin', M.M., Krauya, U.E., Upitis, Z.T., Ozolin'sh, Yu.L., & Toutans, M.Ẏa. *USSR Mech. Compos. Mater*. No 6, 1099 (1983).

[842] Bogdanova, L.M., Grishhenko, A.E., Irzhack, V.I., & Rosenberg, B.A. *USSR High Polymers*. **26A**, 1400 (1984).

[843] Bogdanova, L.M., Ponomareva, T.I., Irzhack, V.I., & Rozenberg, B.A. *USSR Compos. Polymer Mater*. (24), **14** (1985).

[844] Alexandrov, V.N., Bogdanova, L.M., Naidovskaya, V.I., Tarasov, A.I., Irzhack, V.I., & Rozenberg, B.A. *USSR High Polymers*. **27B**, 914 (1985).

[845] Pleskachevsky, Yu.M., Smirnov, V.V., & Dubova, E.B. *Herald Byelorussian Acad. Sci*. Phys. -Techn. Ser. (4), **40** (1979).

[846] Smourougov, V.A., & Delikatnaya, I.O. *Herald Byelorussian Acad. Sci*. Phys. -Techn. Ser. (4), **37** (1987).

[847] Smourougov, V.A., & Delikatnaya, I.O. *USSR Polymer Mater. Res*. (18), 194 (1988).

[848] Douglass, D.C., McBrierty, N.J., & Weber, T.A. *Macromolecules*. **10**, 178 (1977).

[849] Zhbankov, R.G., & Tretinnikov, O.N. *USSR High Polymers*. **26B**, 146 (1984).

[850] Zhbankov, R.G., & Tretinnikov, O.N. *USSR J. Appl. Spectrosc*. **40**, 99 (1984).

[851] Tretinnikov, O.N., & Zhbankov, R.G. *USSR High Polymers*. **30B**, 259 (1988).

[852] Kouznetsov, G.K., & Irgen, L.A. In *Thermodynamic and Structural Properties of Polymer Boundary Layers*. Naukova Dumka Publ., Kiev, 1976, p. 94.

[853] Shilov, V.V., Tsoukrouck, V.V., & Lipatov, Yu.S. *USSR High Polymers*. **26A**, 1347 (1984).

[854] Moisya, E.G., Menzheres, G.Ya., & Lipatov, Yu.S. *USSR High Polymers*. **21A**, 333 (1979).

[855] Pesetskii, S.S., Starzynskii, V.E., & Shcherbakov, S.V. In *Adhesive Joints: Formation, Characterization, and Testing*. ASTM Publ., New York, 1982, p. 195.

[856] Pennings, J.F.M. In *Physicochemical Aspects of Polymer Surfaces*. Plenum Press, New York, 1983, vol. 2, p. 1199.

[857] Langevini, D. *J. Colloid Interface Sci*. **80**, 412 (1981).

[858] Kawaguchi, M., Sano, M., Chen, Y.-L., Zografi, G., & Yu, H. *Macromolecules*. **19**, 2606 (1986).

[859] Maleev, I.I., & Opainitch, I.E. *Herald Lvov Univ. Chem. Ser*. (28), 33 (1987).

[860] Allain, C., Ausserre, D., & Rondelez, F. *Phys. Rev. Lett*. **49**, 1694 (1982).

[861] Leary, H.J., & Campbell, D.S. *Amer. Chem. Soc. Polymer Prepr*. **21**, 147 (1980).

[862] Hammond, J.S. *Amer. Chem. Soc. Polymer Prepr*. **21**, 149 (1980).

[863] Thomas, H.R., & O'Malley, J.J. *Amer. Chem. Soc. Polymer Prepr*. **21**, 144 (1980).

[864] Thomas, H.R., & O'Malley, J.J. *Macromolecules*. **14**, 1316 (1981).

[865] O'Malley, J.J., & Thomas, H.R. In *Contemporary Topics of Polymer Science*. Wiley, New York, 1979, vol. 3, p. 215.

[866] Croucher, M. In *26th Int. Macromol. Symp.* IUPAC Publ., Mainz, 1979, vol. 3, p. 1205.

[867] Schreiber, H.P., & Croucher, M.D. *J. Appl. Polymer Sci.* **25**, 1961 (1980).

[868] Voronin, I.A., & Lavrent'ev, V.V. *USSR High Polymers.* **21A**, 1742 (1979).

[869] Krzeminski, J.L., & Wiechowicz-Kowalska, E. *Polymer Eng. Sci.* **21**, 594 (1981).

[870] Hahn, O., & Kötting, G. *Kunstst.* **74**, 238 (1984).

[871] Kalnin', M.M., & Ozolin'sh, Yu.L. *USSR Mech. Compos. Mater.* (2), 201 (1984).

[872] Lipatov, Yu.S., Shifrin, V.V., Besklubenko, Yu.D., Demchenko, S.S., & Privalko, V.P. *Reports Ukrainian Acad. Sci.* No 8B, 48 (1984).

[873] Papanicolaou, G.C., & Theocaris, P.S. *Coll. Polymer Sci.* **257**, 239 (1979).

[874] Sarkisyan, V.A., Asratyan, M.G., Mkhitaryan, A.A., Katrijyan, K.Kh., & Dadivanyan, A.K. *USSR High Polymers.* **27A**, 1331 (1985).

[875] Lal, M., Turpin, M., Ricardson, K., & Spencer, D. *Amer. Chem. Soc. Symp.* Ser. 8, 16 (1975).

[876] Gähde, J. In *Polymer Composites*. IUPAC Publ., Berlin, 1986, p. 431.

[877] Besklubenko, Yu.D., Lipatov, Yu.S., & Privalko, V.P. *USSR Compos. Polym. Mater.* (16), 72 (1983).

[878] Rusanov, A.I. In *Modern Theory of Capillarity*. Chimia Publ., Leningrad, 1980, p. 13.

[879] Elmgren, H. *J. Polymer Sci.* **18**, 339 (1980).

[880] Babitch, V.F. In *New Methods for Polymer Research*. Naukova Dumka Publ., Kiev, 1975, p. 118.

[881] Theocaris, P.S., Spatnis, G., & Kefalas, B. *Coll. Polymer Sci.* **260**, 837 (1982).

[882] Arslanov, V.V. *USSR High Polymers.* **29A**, 130 (1987).

[883] Aniskina, T.A., Efremova, A.I., Zolotoukhin, S.P., Ivanova, L.L., Ponomareva, T.I., Shteinberg, V.G., Dudina, L.A., Irzhack, V.I., & Rozenberg, B.A. *USSR Mech. Comp. Mater.* No 6, 1115 (1984).

[884] Kouleshov, I.V., & Emel'yanov, Yu.V. *USSR Plastics.* (5), 18 (1987).

[885] Bartenev, G.M., Kouleshov, I.V., Kalnin' M.M., & Kaibin, S.I. *USSR Compos. Polym. Mater.* No 23, 9 (1984).

[886] Kouleshov, I.V., Bartenev, G.M., Kalnin', M.M., & Kaibin, S.I. *Reports USSR Acad. Sci.* **272**, 1418 (1983).

[887] Bartenev, G.M., Kulesov, I.W., Kalnin, M.M., & Kajbin, S.I. *Plaste und Kautsch.* **31**, 133 (1984).

[888] Kouleshov, I.V., Kalnin', M.M., Avotin'sh, Ya.Ya., & Kaibin, S.I. *Herald Latvian Acad. Sci.* (2), 186 (1982).

[889] Arslanov, V.V., Cherezov, A.A., Chalykh, A.E., & Ogarev, V.A. *USSR High Polymers.* **25B**, 437 (1983).

[890] Arslanov, V.V., Vavkushevski, A.A., Ogarev, V.A., Schulzs, R.-D, Göschel, U., & Bischof, C. *Acta Polym.* **36**, 637 (1985).

[891] Theocaris, P.S. *Coll. Polym. Sci.* **262**, 929 (1984).

[892] Lavrent'ev, V.V., Gorshkov, M.M., & Vakula, V.L. *Reports USSR Acad. Sci.* **214**, 352 (1974).

[893] Lipatov, Yu.S., Shilov, V.V., & Gomza, Yu.P. *Reports USSR Acad Sci.* **252**, 1393 (1980).

[894] Voyutskii, S.S., Raevsky, V.G., & Yagnyatinskay, S.M. In *Advances in Colloid Chemistry*. Nauka Publ., Moscow, 1973, p. 339.

[895] Raevsky, V.G. *Adv. Colloid Interface Sci.* **8**, 1 (1977).

[896] Dobkowski, Z. In *Polymer Blends: Processing, Morphology, and Properties*. New York, 1984, vol. 2, p. 85.

[897] Lipatov, Yu.S., Babitch, V.F., & Svyatnenko, G.P. *USSR Compos. Polym. Mater.* (19), 65 (1983).

[898] Lipatov, Yu.S. *Mendeleev Chem. Soc. USSR J.* **31**, 35 (1986).

[899] Lukas, J., & Jesek, B. *Coll. Czech. Chem. Comm.* **48**, 2909 (1983).

[900] Briggs, D. In *Surface Analysis and Pretreatment of Plastics and Metals*. Appl. Sci. Publ., London, 1982, ch. 4.

[901] Briggs, D. In *Practical Surface Analysis by Auger and X-ray Photoelectron Spectroscopy*. Wiley, New York, 1983, ch. 9.

[902] Grigorov, L.N. *USSR High Polymers*. **27A**, 1098 (1985).

[903] Pritykin, L.M., & Vakula, V.L. *USSR High Polymers*. **25A**, 1887 (1983).

[904] *Handbook of Rubber Man*. Chimia Publ., Moscow, 1971, p. 64.

[905] Dvoretskaya, N.M., Erokhina, R.A., Solov'ev, A.M., & Ogarkov, V.A. *Herald USSR High School*. Chem. Techn. **25**, 996 (1982).

[906] Pritykin, L.M. In *Int. Adhesion Conf.* Chameleon Press, London, 1984, p. 12.1.

[907] Tverdokhlebova, I.I. *Conformation of Macromolecules*. Chimia Publ., Moscow, 1981, p. 118.

[908] Poddoubnyi, I.Ya., Erenburg, E.G., & Eremina, M.A. *USSR High Polymers*. **10A**, 1381 (1968).

[909] Pritykin, L.M., Askadsky, A.A., Gal'pern, E.G., & Korshak, V.V. *Reports USSR Acad. Sci.* **273**, 1424 (1983).

[910] Pritykin, L.M., Askadsky, A.A., Gal'pern, E.G., & Korshak, V.V. *USSR High Polymers*. **27A**, 24 (1985).

[911] Pritykin, L.M., Askadsky, A.A., & Korshak, V.V. *USSR High Polymers*. **27A**, 1663 (1985).

[912] Pritykin, L.M. *Probl. Chem. and Chem. Technol.* (85), 78 (1987).

[913] Poddoubnyi, I.Ya., & Podalinsky, A.V. *USSR High Polymers*. **14A**, 780 (1972).

[914] Gorshkov, M.M., Lavrent'ev, V.V., Naumenko, V.Yu., & Sadov, B.D. In *Surface Phenomena in Polymers*. Naukova Dumka Publ., Kiev, 1976, p. 36.

[915] Lavrent'ev, V.V. In *Methods for Increasing Polymer Adhesion*. USSR Chem. Soc. Publ., Moscow, 1977, p. 3.

[916] Toyama, M., Ito, T., Nukatsuka, H., & Ikeda, M. *J. Appl. Polymer Sci.* **17**, 3495 (1973).

[917] Buckley, D.H. *J. Colloid Interface Sci.* **58**, 36 (1977).

[918] Bikerman, J.J. *USSR Adv. Chem.* **41**, 1431 (1972).

[919] Bikerman, J.J. In *Adhesion Measurement of Thin Films, Thick Films, and Bulk Coatings*. ASTM Publ., Philadelphia, 1978, p. 30.

[920] Lipatov, Yu.S. *USSR High Polymers*. **17A**, 2358 (1975).

[921] Lipatov, Yu.S. *Pure Appl. Chem.* **43**, 273 (1975).

[922] Lipatov, Yu.S., Privalko, V.P., Sharov, A.N. *Reports of USSR Acad. Sci.* **263**, 1403 (1982).

[923] Lipatov, Yu.S., Semenovitch, G.M., Sergeeva, L.M., Karabanova, L.V., & Skiba, S.I. *USSR High Polymers*. **29B**, 530 (1987).

[924] Deng, Z., & Cao, J. *Chem. J. Chin. Univ.* **7**, 177 (1986).

[925] Gaines, G.L. *Macromolecules.* **14**, 208 (1981).

[926] Schmitt, R.L., Gardelia, J.A., & Salvati, L. *Macromolecules.* **19**, 648 (1986).

[927] Lipatov, Yu.S., Fainerman, A.E., & Shifrin, V.V. *Reports USSR Acad. Sci.* **231**, 381 (1976).

[928] Lipatov, Yu.S., Fainerman, A.E., & Anokhin, O.V. *Reports USSR Acad. Sci.* **234**, 596 (1977).

[929] Lipatov, Yu.S., Sergeeva, L.M., Nesterov, A.E., & Todosiichouck, T.T. *Reports USSR Acad. Sci.* **259**, 1132 (1981).

[930] Maloshhouck, Yu.S., Titarenko, A.T., Raevsky, V.G., & Voyutskii, S.S. *USSR Mech. Polym.* (6), 857 (1966).

[931] Raevsky, V.G., Voyutskii, S.S., Moneva, I., & Kamensky, A.N. *Herald USSR High School.* Chem. Techn. **3**, 305 (1965).

[932] Hinrichsen, G. *Polymer.* **10**, 718 (1969).

[933] Bikerman, J.J., & Marshall, S.W. *J. Appl. Polymer Sci.* **7**, 1031 (1963).

[934] Gul', V.E., Dvoretskaya, N.M., & Erokhina, R.A. *Reports USSR Acad. Sci.* **248**, 409 (1979).

[935] Gul', V.E., & Dvoretskaya, N.M. *USSR Mech. Compos. Mater.* (5), 858, (1980).

[936] Gul', V.E., Dvoretskaya, N.M., Popova, G.G., & Raevsky, V.G. *Reports USSR Acad. Sci.* **172**, 637 (1967).

[937] Dvoretskaya, N.M., Zaitseva, I.A., Novikov, V.U., Tougov, I.I., & Gul', V.E. *USSR Plastics.* (6), 33 (1981).

[938] Kanthos, M. *J. Appl. Polymer Sci.* **16**, 381 (1972).

[939] Gracheva, N.I., Kornev, A.E., Potapov, E.E., & Glagolev, V.A. *USSR Rubber.* (5), 28 (1981).

[940] Korsoukov, V.E., Vettegren', V.I., Novak, I.I., & Chmel', A.E. *USSR Mech. Compos. Mater.* (5), 621 (1972).

[941] Vettegren', V.E., & Chmel', A.E. *USSR High Polymers.* **18B**, 521 (1976).

[942] Bershtein, V.A., Nikitin, V.V., & Razgoulyaeva, L.G. In *Thermodynamic Properties and Structure of Polymer Layers.* Naukova Dumka Publ., Kiev, 1978, p. 66.

[943] Vettegren', V.I., Novak, I.I., & Jalilov, F. *USSR Compos. Polym. Mater.* (21), 16 (1984).

[944] Vettegren', V.I., & Chmel', A.E. *USSR Mech. Compos. Mater.* (3), 512 (1976).

[945] Zhurkov, S.N., Kouxenko, V.S., & Frolov, D.I. *USSR Phys. Solid.* **16**, 2201 (1974).

[946] Vettegren', V.I., & Chmel', A.E. *Europ. Polymer J.* **12**, 856 (1976).

[947] Vettegren', V.I., Kouxenko, V.S., Frolov, D.I., & Chmel', A.E. *USSR Mech. Compos. Mater.* (5), 771 (1979).

[948] Lexovskaya, N.P., Nachinkin, O.I., Doudrova, A.G., Shoubina, T.G., & Rouban, I.G. In *Surface Phenomena in Polymers.* Naukova Dumka Publ., Kiev, 1982, p. 66.

[949] Vettegren', V.I. Dr. Sci(Chem.) Thesis, Joffe Phys. -Techn. Inst. Leningrad, 1987.

[950] Vettegren', V.I., & Fridlyand, K.Yu. *USSR Opt. Spectrosc.* (5), 521 (1975).

[951] Novak, I.I., Korsoukov, V.E., & Banzhhikov, A.G. *Reports USSR Acad. Sci.* **224**, 1297 (1975).

[952] Jalilov, F., & Vettegren', V.I. *USSR High Polymers.* **22B**, 886 (1983).

[953] Orlov, L.G., Lexovsky, A.M., & Regel', V.R. *USSR Phys. Chem. Processing Mater.* (3), 140 (1980).

[954] Gabaraeva, A.D., Lexovsky, A.M., & Orlov, L.G. *USSR Mech. Compos. Mater.* (1), 16 (1980).

[955] Popov, A.A., Krisyuck, B.E., & Zaikov, G.E. *USSR High Polymers.* **22A**, 1366 (1980).

[956] Yamamoto, S., Hayashi, M., & Inoue, T. *J. Appl. Polymer Sci.* **19**, 2107 (1975).

[957] Tourousov, R.A., & Vuba, K.T. *USSR Phys. Chem. Proc. Mater.* (5), 87 (1979).

[958] Regel', V.R., Pozdnyakov, O.F., & Redkov, B.P. In *Active Surface of Solids.* Nauka Publ., Moscow, 1976, p. 230.

[959] Pozdnyakov, O.F., Amelin, V.V., Mal'chevsky, V.A., Podolsky, A.F., Regel', V.R., Redkov, B.P., & Shalimov, V.V. In *Proc. Sem. Physics of Composite Materials Strength.* Phys. -Techn. Inst. Press, Leningrad, 1978, p. 134.

[960] Pozdnyakov, O.F., Kouzenko, E.N., Poulatov, A.A., & Shalimov, V.V. *USSR Compos. Polym. Mater.* (25), 44 (1985).

[961] Pozdnyakov, O.F., Regel', V.R., Amelin, A.V., & Redkov, B.P. *Reports 7th Symp. Mechanoemission and Mechanochemistry of Solids.* Tashkent, 1982, vol. 2, p. 39.

[962] Pozdnyakov, O.F., & Redkov, B.P. In *Mechanisms of Damage and Strength of Heterogeneous Materials.* Leningrad, 1985, p. 40.

[963] Possart, W., Yudin, Y.S., Redkov, B.P., Ziegler, H.-J., Pozdnyakov, O.F., & Bischof, C. *Acta polym.* **36**, 631 (1985).

[964] Pozdnyakov, O.F., & Regel', V.R. In *Modern Physical Methods for Polymer Research.* Chimia Publ., Moscow, 1982, p. 169.

[965] Pozdnyakov, O.F. In *Kinetics of Deformation and Fracture of Composite Materials.* Leningrad, 1983, p. 55.

[966] Pozdnyakov, O.F., Regel', V.R., Kortov, V.S., Redkov, B.P., & Shalimov, V.V. In *2nd Symp. Active Surface of Solids.* University Press, Tartu, 1977, p. 7.

[967] Tabarov, S.Kh., Amelin, A.V., Pozdnyakov, O.F., & Regel', V.R. In *Proc. Sem. Physics. of Composite Materials Strength*, Phys. -Techn. Inst. Press, Leningrad, 1978, p. 124.

[968] Pozdnyakov, O.F., Redkov, B.P., & Tabarow, S.C. *Plaste und Kautsch.* **29**, 148 (1982).

[969] Pozdnyakov, O.F., Kouzenko, E.N., Poulatov, A.A., & Shalimov, V.V. In *Surface Phenomena in Polymers.* Naukova Dumka Publ., Kiev, 1982, p. 87.

[970] Ma, L., Middlemiss, K.M., Torrie, G.M., & Whittington, S.G. *J. Chem. Soc. Faraday Soc.* **74**, 721 (1978).

[971] Birshtein, T.M. *Macromolecules.* **12**, 715 (1979).

[972] Grosberg, A.Yu. *USSR High Polymers.* **24A**, 1194 (1982).

[973] Skvortsov, A.M., Gorbounov, A.A., Zhoulina, E.B., & Birshtein, T.M. *USSR High Polymers.* **20A**, 816 (1978).

[974] Lemaire, B., & Bothorel, P. *Macromolecules.* **13**, 311 (1980).

[975] Komarov, V.M., Vainshtein, E.F., Entelis, S.G., & Orlov, V.A. In *Adhesion of Polymers and Adhesive Joints in Machine Building.* Sci. -Techn. Soc. Press, Moscow, 1976, vol. 2, p. 142.

[976] Lal, M., & Stepto, R.F.T., *J. Polymer Sci.* **61C**, 401 (1977).

[977] Komarov, V.M. Cand. Sci(Chem.) Thesis, Inst. Chem. Phys., Moscow, 1979.

[978] Griffith, A.A. *Phil. Trans. Roy. Soc. A.* **221**, 180 (1920).

[979] Smith, J. *J. Adhesion.* **11**, 243 (1980).

[980] Regel', V.R., Sloutsker, A.I., & Tomashevsky, E.E. *Kinetic Nature of Solids Strength.* Nauka Publ., Moscow, 1974.

[981] Sanzharovsky, A.T. *Physicochemical Properties of Polymer and Paint Coatings.* Chimia Publ., Moscow, 1978, p. 33.

[982] Ol'khovick, O.E., & Goldman, A.Ya. *USSR Mech. Polym.* (2), 163 (1973).

[983] Tsygankov, S.A., & Goldman, A.Ya. *USSR High Polymers.* **21A**, 294 (1979).

[984] Sytov, V.A. *USSR Factory Labor.* **46**, 1126 (1980).

[985] Pritykin, L.M., & Dranovsky, M.G. *Diffusion, Phase Transformations, Mechanical Properties of Metals and Alloys.* **12**, pt. 2, 140 (1975).

[986] Pritykin, L.M., Dranovsky, M.G., & Parcksheyan, C.R. *Adhesives and Applications in Electrotechnic.* Energoatom Publ., Moscow, 1983.

[987] Gorbatkina, Yu.A., & Ivanova-Mumghieva, V.G. *Rheol. Acta.* **13**, 789 (1974).

[988] Regel', V.R., Lexovsky, A.M., & Kireenko, O.F. *USSR Mech. Polym.* (4), 544 (1977).

[989] Pegoraro, M., Pagani, G., Clerici, P., & Penati, A. *Fibre Sci. Techn.* **10**, 263 (1977).

[990] Ratner, S.B., Yartsev, V.P., & Andreeva, E.K. *USSR High Polymers.* **24B**, 563 (1982).

[991] Bershtein, V.A., Nikitin, V.V., Stepanov, A.V., & Shamrai, L.M. *USSR Phys. Solid.* **15**, 3260 (1973).

[992] Regel', V.R., Savitsky, A.V., & Sanfirova, T.P. *USSR Mech. Polym.* (5), 1002 (1976).

[993] Shtarkman, B.P., Voyutskii, S.S., & Kargin, V.A. *USSR High Polymers.* **7**, 135 (1965).

[994] Voyutskii, S.S., & Zamazii, V.M. *Reports SSR Acad. Sci.* **81**, 85 (1951).

[995] Voyutskii, S.S., & Zamazii, V.M. *USSR Colloid J.* **15**, 407 (1953).

[996] Shtarkman, B.P., Voyutskii, S.S., & Kargin, V.A. *USSR High Polymers.* **7**, 141 (1965).

[997] Voyutskii, S.S., & Shtarkh, B.V. *Reports USSR Acad. Sci.* **90**, 573 (1953).

[998] Voyutskii, S.S., & Shtarkh, B.V. *USSR Colloid J.* **16**, 3 (1954).

[999] Gul', V.E., & Fomina, L.L. *USSR High Polymers.* **7**, 45 (1965).

[1000] Voyutskii, S.S., Markin, Yu.I., Gorchakova, V.M., & Gul', V.E. *USSR J. Phys. Chem.* **37**, 2027 (1963).

[1001] Kanamaru, K. *Koll. -Z. und Z. Polym.* **192**, 51 (1963).

[1002] Hofrichter, C.H., & McLaren, A.D. *Ind. Eng. Chem.* **40**, 329 (1948).

[1003] Lavrent'ev, V.V., & Konstantinova, N.A. *USSR Mech. Polym.* (6), 329 (1971).

[1004] Kammer, H.-W., & Piglowski, J. In *Polymer Blends: Processing, Morphology and Properties.* New York, 1984, vol. 2, p. 19.

[1005] Regel', V.R. *USSR High Polymers.* **19A**, 1915 (1977).

[1006] Regel', V.R., Lexovsky, A.M., Orlov, L.G., Lexovskaya, N.A., Mazo, A.I., & Perepelkin, K.E. *USSR Mech. Polym.* (3), 815 (1976).

[1007] Bershtein, V.A., & Glikman, L;A. *USSR Phys. Solid.* **5**, 2270 (1963).

[1008] Regel', V.R., Lexovsky, A.M., Abdoumanonov, A., & Orlov, L.G. In *Proc. Sem. Physics of Composite Materials Strength.* Phys. -Techn. Inst. Press, Leningrad, 1978, p. 107.

[1009] Briscoe, B.J., & Kremnitzer, S.L. *J. Phys.* **12D**, 505 (1979).

[1010] Dudina, L.A., Aliev, A.D., Djavadyan, E.A., Zspinock, G.S., Ivanova, L.L., Il'in, M.I., Chalykh, A.E., Rozenberg, B.A., & Enikolopyan, N.S. *Reports USSR Acad. Sci.* **257**, 670 (1981).

[1011] Kamensky, A.N., Hexel, L., Markov, D.A., Kiselev, V.Ya., & Toutorsky, I.A. In *Surface Phenomena in Polymers*. Naukova Dumka Publ., Kiev, 1982, p. 50.

[1012] Kamensky, A.N., Hexel, L., Kiselev, V.Ya., & Toutorsky, I.A. *Herald USSR High School*. Chem. Techn., **28**, 77 (1985).

[1013] Lederer, D.A., Kear, K.E., & Kuhls, G.H. *Rubb. Chem. Techn.* **55**, 1482 (1982).

[1014] Kamensky, A.N. Cand. Sci(Chem.) Thesis, Lomonosov Inst. Fine Chem. Technol. Moscow, 1969.

[1015] Aivazov, A.B. Cand. Sci.(Chem.) Thesis, Karpov Phys. -Chem. Inst., Moscow, 1970.

[1016] Takahashi, A., Kawaguchi, M. In *Behaviour Macromol.* Berlin-e.a., 1982, p. 1.

[1017] Saito, T. *J. Adhesion Soc. Japan.* **22**, 9 (1986).

[1018] Lipatov, Yu.S., & Fainerman, A.E. *Reports USSR Acad. Sci.* **271**, 896 (1983).

[1019] Voyutskii, S.S., Derjaguin, B.V., & Raevsky, V.G. *Reports USSR Acad. Sci.* **161**, 377 (1965).

[1020] Voyutskii, S.S., & Vakula, V.L. *Herald USSR High School*: Chem. Techn., **3**, 186 (1960).

[1021] Voyutskii, S.S., & Vakula, V.L. *USSR High Polymers*. **2**, 51 (1960).

[1022] Voyutskii, S.S., & Vakula, V.L. *J. Appl. Polymer Sci.* **7**, 475 (1963).

[1023] Voyutskii, S.S., & Vakula, V.L. *Rubb. Chem. Technol.* **37**, 1153 (1964).

[1024] Antchack, V.K., Chourakova, I.K., & Berestneva, Z.Ya. *Mendeleev Chem. Soc. USSR J.* **15**, 581 (1970).

[1025] Antchack, V.K., Chourakova, I.K., & Shhekatourova, G.Yu. *Mendeleev Chem. Soc. USSR J.* **21**, 234 (1976).

[1026] Jud, K., & Kausch, H.H. *Polymer Bull.* **1**, 697 (1979).

[1027] Kausch, H.H., & Jud, K. In *26th Int. Macromol. Symp.* IUPAC Press, Mainze, 1979, vol. 3, p. 1426.

[1028] Miller, G.W. In *Treatise on Adhesion and Adhesives*. Marcel Dekker, New York, 1973, vol. 3, p. 123.

[1029] Flory, P.J. *J. Amer. Chem. Soc.* **62** 1057 (1940).

[1030] Eyring, H., & Kauzmann, W. *J. Amer. Chem. Soc.* **62**, 3113 (1940).

[1031] Voyutskii, S.S., & Vakula, V.L. *USSR Adv. Chem.* **28**, 205 (1964).

[1032] Calvert, P.D., & Billingham, N.C. *J. Appl. Polymer Sci.* **24**, 357 (1979).

[1033] Gent, A.N., & Tobias, R.H. *J. Polymer Sci.* **22**, 1483 (1984).

[1034] Ueno, H., Hasegawa, H., & Kishimoto, A. *J. Adhesion Soc. Japan.* **14**, 5 (1978).

[1035] Ueno, H., Otsuka, S., Taira, K., & Kishimoto, A. *Japan High Polymers*. **35**, 759 (1978).

[1036] Ellul, M.D., & Gent, A.N. *J. Polymer Sci.* **22**, 1953 (1984).

[1037] Voyutskii, S.S., Gul', V.E., Jahn-In-si, & Vakula, V.L. *USSR High Polymers*, **4**, 285 (1962).

[1038] Kouleznev, V.N. Dr.Sci(Chem.) Thesis, Lomonosov Inst. Fine Chem. Technol., Moscow, 1973.

[1039] Chalykh, A.E. *Diffusion in Polymer Systems*. Chimia Publ., Moscow, 1987.

[1040] Maklakov, A.I., Skirda, V.D., & Fatkoullin, N.F. *Autodiffusion in Solutions and Melts of Polymers*. University Press, Kazan, 1987.

[1041] Di Benedetto, A.T. *J. Polymer Sci.* **1A**, 3459 (1963).

[1042] Pace, R.J., & Datyner, A. *J. Polymer Sci.* **17**, 437 (1979).

[1043] Koulezney, V.N. In *Composite Polymer Materials*. Naukova Dumka Publ., Kiev, 1975, p. 93.

[1044] Helfand, E. *Acc. Chem. Res.* **8**, 295 (1975).

[1045] Tagami, Y. *J. Chem. Phys.* **73**, 5354 (1980).

[1046] Weber, T., & Helfand, E. *Macromolecules.* **9**, 311 (1976).

[1047] Anastasiadis, S.H., Chen, J.K., Kobershtein, J.T., Sohn, J.E., & Emerson, J.A. *Polymer Eng. Sci.* **26**, 1410 (1986).

[1048] Helfand, E., & Tagami, Y. *J. Chem. Phys.* **56**, 3592 (1972).

[1049] Helfand, E., & Sapse, A.M. *J. Chem. Phys.* **65**, 1327 (1975).

[1050] Nakao, K. *Japan High Polymers.* **25**, 104 (1975).

[1051] Kamensky, A.N., Fodiman, N.M., & Voyutskii, S.S. *USSR High Polymers.* **7**, 696 (1965).

[1052] Kamensky, A.N., Fodiman, N.M., & Voyutskii, S.S. *USSR High Polymers.* **11**, 394 (1969).

[1053] Kamensky, A.N., Fodiman, N.M., & Voyutskii, S.S. In *Surface Phenomena in Polymers*. Naukova Dumka Publ., Kiev, 1969, p. 80.

[1054] Kamensky, A.N., Maltsev, V.F., Fodiman, N.M., & Voyutskii, S.S. *USSR High Polymers.* **12**, 574 (1970).

[1055] Sapozhnikova, I.N., & Chalykh, A.E. In *Surface Phenomena in Polymers*. Naukova Dumka Publ., Kiev, 1982, p. 93.

[1056] Chalykh, A.E., Aliev, A.D., & Roubtsov, A.V. *USSR High Polymers.* **25A**, 2217 (1983).

[1057] Rafailovich, M.H., Sokolov, J., Jones, R.A.L., Krausch, G., Klein, J., & Mills, R. *Europhys. Lett.* **5**, 657 (1988).

[1058] Bueche, F., Cashin, W., & Debye, P. *J. Chem. Phys.* **20**, 1956 (1952).

[1059] Bresler, S.E., Zakharov, G.M., & Kirillov, S.V. *USSR High Polymers.* **3**, 1072 (1961).

[1060] Morozova, L.P., & Krotova, N.A. *USSR Colloid J.* **20**, 59 (1958).

[1061] De Gennes, P.-J. *Hebd. Seances Acad. Sci.* **291FB**, 218 (1980).

[1062] Kim, Y.H., & Wool, R.P. *Macromolecules.* **16**, 1115 (1983).

[1063] Roland, C.M., & Böhm, G.G.A. *Macromolecules.* **18**, 1310 (1985).

[1064] Kamensky, A.N., Markov, D.A., Perneker, T., Kiselev, V.Ya., & Toutorsky, I.A. In *Adhesive Joints in Machine Building*. Polytechn. Inst. Press, Riga, 1983, p. 68.

[1065] Bothe, L., & Rehage, G. *Angew. makromol. Chem.* **100**, 39 (1981).

[1066] Olabisi, O., & Robeson, L.M. *Polymer–Polymer Miscibility*. Academic Press, New York, 1979.

[1067] Koulezney, V.N. *Blends of Polymers*. Chimia Publ., 1980.

[1068] Koulezney, V.N., Krokhina, L.S., & Dogadkin, B.A. *USSR Colloid J.* **29**, 170 (1967).

[1069] Roe, R.-J. *J. Colloid Interface Sci.* **31**, 228 (1969).

[1070] Champpetier, D.C. *Amer. Chem. Soc. Polymer Prepr.* **5**, 363 (1964).

[1071] Sholokhovitch, T.A. Cand.Sci(Chem.) Thesis, Univ., Sverdlovsk, 1975.

[1072] Krause, S. *Macromolecules.* **11**, 1288 (1978).

[1073] Koulezney, V.N. *USSR Colloid J.* **39**, 407 (1977).

[1074] Fowler, M.E., Barlow, J.W., & Paul, D.R. *Polymer.* **28**, 2145 (1987).

[1075] Krause, S. *Pure Appl. Chem.* **58**, 1553 (1986).

[1076] Roland, C.M. *J. Polymer Sci.* **26B**, 839 (1988).
[1077] Babitch, V.F., Nesterov, A.E., Nizelsky, Yu.N., Rosovitsky, V.F., Semenovitch, G.M., Todosiichouck, T.T., Fainerman, A.E., Shifrin, V.V., & Shoumsky, V.F. In *Physicochemistry and Modification of Polymers.* Naukova Dumka Publ., Kiev, 1987, p. 116.
[1078] Roe, R.-J. *J. Chem. Phys.* **62**, 490 (1975).
[1079] Scheutjens, J.M.H.M., & Fleer, G.J. *Macromolecules.* **18**, 1882 (1985).
[1080] Kovarsky, A.L., Bourkova, S.G., Vasserman, A.M., & Morozov, Yu.L. *Reports USSR Acad. Sci.* **196**, 383 (1971).
[1081] Kaelble, D.H., & Cirlin, E.H. *J. Polymer Sci. Symp.* (43), 131 (1973).
[1082] Silberberg, A. *J. Macromol. Sci.* **18B**, 677 (1980).
[1083] Northolt, M.G. *J. Mater. Sci.* **16**, 2025 (1981).
[1084] Barlow, J.W., & Paul, D.R. *Polym. Eng. Sci.* **27**, 1482 (1987).
[1085] Polyakov, M.L. *USSR High Polymers.* **29A**, 2212 (1987).
[1086] Lipatov, Yu.S., Bezrouck, L.I., & Lebedev, A.E. *USSR Colloid J.* **37**, 481 (1975).
[1087] Kouleznev, V.N., Krokhina, L.S., Oganesov, Yu.G., & Zlatsen, L.M., *USSR Colloid J.* **33**, 98 (1971).
[1088] Shoutilin, Yu.F., & Dankovtsev, V.A. In *Blends of Polymers.* Ivanovo, 1986, p. 52.
[1089] Voyutskii, S.S., & Vakula, V.L. *USSR Mech. Polym.* (3), 455 (1969).
[1090] Chalykh, A.E., Mikhailov, Yu.M., Avdeev, N.N., & Lotmentsev, Yu.M. *Reports USSR Acad. Sci.* **247**, 890 (1979).
[1091] Voyutskii, S.S. *USSR High Polymers.* **1**, 230 (1959).
[1092] Kouleznev, V.N., Oganesov, Yu.G., Evreinov, Yu.V., Voyutskii, S.S., Filippovitch, L.D., & Gil'man, I.M. *USSR Colloid J.* **34**, 863 (1972).
[1093] Adam, G., & Gibbs, J.H. *J. Chem. Phys.* **43**, 139 (1965).
[1094] Cohen, M.H., & Turnball, D. *J. Chem. Phys.* **31**, 1164 (1959).
[1095] Kästner, S. *Faserforsch. Textiltechn.* **28**, 399 (1977).
[1096] Kästner, S. *Wiss. Z. Techn. Hochsch. Leuna-Merseburg.* **20**, 187 (1978).
[1097] Hurai, N., & Eyring, H. *J. Polymer Sci.* **37**, 51 (1959).
[1098] Boyer, R.F. *Rubb. Chem. Techn.* **36**, 1303 (1963).
[1099] Simha, R., & Boyer, R.F. *J. Chem. Phys.* **37**, 1003 (1962).
[1100] Lipatov, Yu.S. *Reports Ukrainian Acad. Sci.* (3B), 363 (1972).
[1101] Parizenberg, M.D., & Boukhina, M.F. *USSR Rubber.* (10), 43 (1979).
[1102] Koloupaev, B.S. *USSR High Polymers.* **24B**, 105 (1982).
[1103] Ponomareva, T.I., Irzhack, V.I., & Rozenberg, B.A. *USSR High Polymers.* **20A**, 597 (1978).
[1104] Petryaev, S.V., Blyakhman, E.M., Pilipenock, D.A., Gvirts, E.M., & Gofman, P.E. *USSR High Polymers.* **14A**, 1624 (1972).
[1105] Ponomareva, T.N., Efremova, A.I., Smirnov, Yu.N., Irzhack, V.I., Oleinick, E.F., & Rozenberg, B.A. *USSR High Polymers.* **22A**, 1958 (1980).
[1106] Privalko, V.P., & Lipatov, Yu.S. *USSR High Polymers.* **13A**, 2733 (1971).
[1107] Campion, R.P. *J. Adhesion.* **7**, 1 (1974).
[1108] Curro, J.G., Lagasse, R.R., & Simha, R. *Macromolecules.* **15**, 1621 (1982).
[1109] Oleynik, E.F., Bulatov, V.V., & Gusev, A.A. *Macromol. Chem. Suppl.* **6**, 305 (1984).

[1110] Regel', V.R., & Slutsker, A.I. In *Physics Today and Tomorrow*. Nauka Publ., Leningrad, 1973, p. 90.

[1111] Piglowski, A., Kammer, H.-W., & Defer, G. *Acta Polym.* **32**, 87 (1981).

[1112] Toyama, M. *Japan Adhes. Seal.* **30**, 387 (1986).

[1113] Wu. S. *Polymer Eng. Sci.* **27**, 335 (1987).

[1114] Gutowski, W. *Int. J. Adh. Adhesives.* **7**, 189 (1987).

[1115] Ruckenstein, E., & Gourisankar, S.V. *J. Colloid Interface Sci.* **101**, 436 (1984).

[1116] Sharma, C.P. *J. Colloid Interface Sci.* **110**, 292 (1986).

[1117] Ruckenstein, E., & Gourisankar, S.V. *J. Colloid Interface Sci.* **110**, 293 (1986).

[1118] Andrade, J.D., Gregonis, D.E., & Smith, L.M. In *Surface and Interfacial Aspects of Biomedical Polymers*. Plenum Press, New York–London, 1985, vol. 1, p. 15.

[1119] Kimura, H. *Japan High Polymers.* **43**, 177 (1986).

[1120] Zhbankov, R.G., Tretinnikov, O.N., & Tretinnikova, G.K. *USSR High Polymers.* **26B**, 104 (1984).

[1121] Bogue, R., Gamet, D., & Schreiber, H.P. In *8th Annu. Meet. Adhesion Soc.* Savannah, 1985, p. 16a.

[1122] Takahara, A., & Kajiyama, T. *J. Chem. Soc. Japan.* Chem. Ind. Chem. (6), 1293 (1985).

[1123] Izmailova, V.N., Yampol'skaya, G.P., & Summ, B.D. *Surface Phenomena in Protein Systems*. Chimia Publ., Moscow, 1988, pp. 52, 86, 221.

[1124] Reardon, J.P., & Zisman, W.A. *Macromolecules.* **7**, 920 (1974).

[1125] Gul', V.E., Vakula, V.L., Khe-Yuan-tsui, & Voyutskii, S.S. In *Res. on Surface Forces*. USSR Acad. Sci. Publ., Moscow, 1961, p. 55.

[1126] Vasenin, R.M. In *Adhesion of Polymers*. USSR Acad. Sci. Publ., Moscow, 1963, p. 17.

[1127] Wool, R.P. *Rubb. Chem. Technol.* **54**, 307 (1984).

[1128] Wool, R.P. *J. Elastom. Plast.* **17**, 106 (1985).

[1129] Vasenin, R.M., Gromov, V.K., Vakula, V.L., & Voyutskii, S.S. In *Adhesion of Polymers*. USSR Acad. Sci. Publ., Moscow, 1963, p. 52.

[1130] Markin, Yu.I., & Voyutskii, S.S. In *Adhesion of Polymers*. USSR Acad. Sci. Publ., Moscow, 1963, p. 23.

[1131] Voyutskii, S.S., & Markin, Yu.I. *USSR High Polymers.* **4**, 926 (1962).

[1132] Vasenin, R.M. *USSR High Polymers.* **3**, 679 (1961).

[1133] Vasenin, R.M. In *Adhesion: Fundamentals and Practice*. McLaren & Son, London, 1969, p. 29.

[1134] Pritykin, L.M., Emel'yanov, Yu.V., & Steshenko, N.I. *USSR Plastics.* (3), 13 (1984).

[1135] Prokopchouck, N.R. *Herald Byelorussion Acad. Sci.* Ser. Phys. -Techn. Sci. (4), 37 (1984).

[1136] Anastasiadis, S.H., & Kobershtein, J.T. *Amer. Chem. Soc. Polymer Prepr.* **28**, 24 (1987).

[1137] Kasemura, T., Yamaguchi, S., & Hata, T., *Kobunshi Romb.* **44**, 657 (1987).

[1138] Kasemura, T., Yamaguchi, S., Hattori, K., & Hata, T. *Kobunshi Romb.* **45**, 63 (1988).

[1139] Pritykin, L.M., Emel'yanov, Yu.V., & Vakula, V.L. In *New Adhesives, Technology of Bonding, and Applications*. House Sci. -Techn. Propag. Publ., Moscow, 1989, p. 86.

[1140] Bothe, L., & Rehage, G. *Rubb. Chem. Technol.* **55**, 1308 (1982).

[1141] Kozlov, G.V., Shetov, R.A., & Mikitaev, A.K. *USSR High Polymers.* **28B**, 643 (1986).

[1142] Bershtein, V.A., Egorov, V.M., & Emel'yanov, Yu.A. *USSR High Polymers.* **27A**, 2451 (1985).

[1143] Skolnick, J., & Helfand, E. *J. Chem. Phys.* **72**, 5489 (1980).

[1144] Bershtein, V.A., & Egorov, V.M. *USSR High Polymers.* **27A**, 2440 (1985).

[1145] Bershtein, V.A., Egorov, V.M., Emel'yanov, Yu.A., & Stepenov, V.A. *Polymer Bull.* **9**, 98 (1983).

[1146] Bershtein, V.A., Emel'yanov, Yu.A., & Stepanov, V.A. *USSR High Polymers.* **26A**, 2272 (1984).

[1147] Bershtein, V.A., Egorov, V.M., & Stepanov, V.A. *Reports USSR Acad. Sci.* **269**, 627 (1983).

[1148] Shalygin, G.F., Grobkova, N.Ya., & Kozlov, P.V. *USSR Mech. Compos. Mater.* (5), 771 (1981).

[1149] Tager, A.A., & Blinov, V.S. *USSR Adv. Chem.* **56**, 1004 (1987).

[1150] Derjaguin, B.V., Zherebkov, S.K., & Medvedeva, A.M. *USSR Colloid J.* **18**, 404 (1956).

[1151] Pritykin, L.M., & Vakula, V.L. In *Adhesive Joints in Machine Building*. Polytechn. Inst. Press, Riga, 1983, p. 3.

[1152] Akopyan, L.A., Ovrutskaya, L.A., & Nikiforov, V.P. *USSR Pat. No.* 657314 (1979).

[1153] Pritykin, L.M. In *Int. Adhesion Conf.* Chameleon Press, London, 1984, p. 11.1.

[1154] Pritykin, L.M. *USSR Leath. -Shoe Ind.* (7), 40 (1981).

[1155] Shimbo, M. *J. Adhesion Soc. Japan.* **17**, 331 (1981).

[1156] Berg, J.C. In *Composite Systems from Natural and Synthetic Polymers*. Elsevier, Amsterdam, 1986, p. 23.

[1157] Pritykin, L.M. *USSR Plastics.* (6), 63 (1988).

[1158] Usachev, S.V., Zakharov, N.D., & Vetoshkin, A.B. *USSR Rubber.* (7), 11 (1981).

[1159] Usachev, S.V., Zakharov, N.D., & Vetoshkin, A.B. *USSR Rubber.* (8), 10 (1981).

[1160] Netzel, D.A., Hoch, G., & Marx, T.I. *J. Colloid Sci.* **19**, 774 (1964).

[1161] Arslanov, V.V., Ivanova, T.N., & Ogarev, V.A. *Reports USSR Acad. Sci.* **198**, 1113 (1971).

[1162] Ogarev, V.A. Dr.Sci(Chem.) Thesis, Inst. Phys. Chem., Moscow, 1975.

[1163] Vyrodov, I.P. In *Physical Chemistry of Surface Phenomena at High Temperatures*. Naukova Dumka Publ., Kiev, 1971, p. 154.

[1164] Joanny, J.-F. *J. Chim. Phys. et Phys. -Chim. Biol.* **84**, 197 (1987).

[1165] Schonhorn, H., Frisch, H.L., & Kwei, T.K. *J. Appl. Phys.* **37**, 4967 (1966).

[1166] Newman, S. *J. Colloid Interface Sci.* **26**, 209 (1968).

[1167] Smith, D.E. In *29th Pittsburgh Conf. on State of the Art Analysis in Chemistry and Applied Spectroscopy*. Monroeville, 1978, p. 117.

[1168] Rymuza, Z. *J. Colloid Interface Sci.* **112**, 221 (1986).

[1169] Van Oene, H.V., Chang, Y.F., & Newman, S. *J. Adhesion.* **1**, 54 (1969).

[1170] Dettre, R.H., & Johnson, R.E. *J. Adhesion.* **2**, 62 (1970).

[1171] Kaelble, D.H. *J. Macromol. Sci.* **6C**, 85 (1971).

[1172] Yin, T.P. *J. Phys. Chem.* **73**, 2413 (1969).

[1173] Ruckenshtein, E., & Dann, M. *J. Colloid Interface Sci.* **59**, 137 (1976).

[1174] Noolandi, J. *Amer. Chem. Soc. Polymer Prepr.* **28**, 8 (1987).

[1175] Ahmad, J., & Hansen, R.S. *J. Colloid Interface Sci.* **38**, 601 (1972).

[1176] Sucio, D.G., Smigelski, O., & Ruckenshtein, E. *J. Colloid Interface Sci.* **33**, 520 (1970).

[1177] Summ, B.D., Raud, E.A., & Shhukin, E.D. *Reports USSR Acad. Sci.* **209**, 164 (1973).

[1178] Summ, B.D., Pinter, Y., Wolfram, E., & Stergiopulos, H. *USSR Colloid J.* **43**, 601 (1981).

[1179] Gouloyan, Yu.A. *USSR J. Appl. Chem.* **50**, 1692 (1977).

[1180] Kanamaru, H. *Japan High Polymers.* **12**, 294 (1963).

[1181] Anand, J.N., & Karam, H.J. *J. Adhesion.* **1**, 16 (1969).

[1182] Gent, A.N., & Petrich, R.P. *Proc. Roy. Soc. A.* **310**, 433 (1969).

[1183] Connelly, R.W., Parsons, W.F., & Pearson, G.N. *J. Rheol.* **25**, 315 (1981).

[1184] Kleinert, H. *USSR Plastics.* (9), 63 (1987).

[1185] Eckert, R. Thesis, Techn. Univ., Dresden, 1981.

[1186] Eckert, R., Kleinert, H., & Blume, F. *Wiss. Z. Techn. Univ. Dresden.* **32**, 193 (1983).

[1187] Kleinert, H., & Hammer, G. *Schweisstechnik.* **34**, 260 (1984).

[1188] Albring, W. *Angewandte Strömungslehre.* Steinkopf Verl., Dresden, 1966, p. 282.

[1189] Kleinert, H., & Richter, J. *Plaste und Kautsch.* **30**, 38 (1983).

[1190] Vavkushevsky, A.A., Arslanov, V.V., & Ogarev, V.A. *USSR Colloid J.* **46**, 1076 (1984).

[1191] Lelah, M.D., & Marmur, A. *J. Colloid Interface Sci.* **82**, 518 (1981).

[1192] Ogarev, V.A., Timonina, T.N., Arslanov, V.V., & Trapeznikov, A.A. *J. Adhesion.* **6**, 337 (1974).

[1193] Arslanov, V.V., & Ogarev, V.A. *Herald USSR Acad. Sci. Ser. Chem.* (8), 1795 (1974).

[1194] Torbin, I.D. Thesis, Opt. Inst. Leningrad, 1972.

[1195] Torbin, I.D. *USSR High Polymers.* **13B**, 605 (1971).

[1196] Kleinert, H. In *VIII Konf. lepeni kovov.* CSVTS Press, Bratislava, 1981, p. 11.

[1197] Wake, W.C. In *Adhesion.* University Press, London, 1961, p. 191.

[1198] Vavkushevsky, A.A., & Vavkushevskaya, I.N. In *Adhesive Joints in Machine Building.* Polytechn. Inst. Press, Riga, 1983, p. 131.

[1199] Imoto, T. *Bessatsu Kagaku kogyo.* **10**, 207 (1966).

[1200] Klingenfuss, H. *Adhesion.* **25**, 179 (1981).

[1201] Loutfy, R.O., & Law, K.Y. *J. Phys. Chem.* **84**, 2803 (1980).

[1202] Loutfy, R.O. *J. Polymer Sci.* **20**, 825 (1982).

[1203] Loutfy, R.O., & Teegarden, D.M. *Macromolecules.* **16**, 452 (1983).

[1204] Sheppard, N.F., Galverick, S.L., Day, D.R., & Sentury, S.D. In *26th SAMPE Symp.* 1981, p. 65.

[1205] Levy, R.L., & Ames, D.P. In *Adhesive Chemistry: Developments and Trends.* Plenum Press, New York, 1984, p. 245.

[1206] Koike, T., Fukuzawa, T., & Saga, M. *J. Adhesion Soc. Japan.* **22**, 292 (1986).

[1207] Akiyama, S., Ushiki, H., Kano, Y., & Kitazaki, Y. *Europ. Polymer J.* **23**, 327 (1987).

[1208] Tirrell, M., Aubert, J.H., Adolf, D., & Davis, H.T. In *Proc. 28th Macromol. Symp.* IUPAC Publ., Amherst, 1982, p. 716.

[1209] Bogdanov, A.A., Listopad, M.I., & Uklisty, A.E. In *Rheology in Processes and Equipment of Chemical Technology*. Polytechn. Inst. Press, Volgograd, 1975, p. 171.

[1210] Sarkisyan, Z.G., Ovanesov, G.T., & Kabalyan, Yu.K. *USSR Rubber.* (1), 19 (1980).

[1211] Vostroknoutov, E.G., Kamensky, B.Z., & Malkina, Kh.E. *Retreaded Repair of Tyres*. Chimia Publ., Moscow, 1974.

[1212] White, R.P. *Polymer Eng. Sci.* **14**, 50 (1974).

[1213] Khvostick, L.I., Vostroknoutov, E.G., Prozorovskaya, N.V., & Smirnova, N.I. *USSR Rubber.* (6), 21 (1979).

[1214] Porter, D. *Polymer.* **28**, 1051 (1987).

[1215] Porter, D. *Polymer.* **28**, 1056 (1987).

[1216] Porter, D. *Polymer.* **28**, 1652 (1987).

[1217] Hayashi, M., & Sakai, T. *J. Adhesion Soc. Japan.* **17**, 407 (1981).

[1218] Kalnin', M.M., & Kapishnikov, Yu.V. *USSR Mech. Compos. Mater.* (6), 1106 (1980).

[1219] Porter, R.S., & Johnson, J.F. *J. Polymer Sci.* **15C**, 373 (1966).

[1220] *Thermophysical and Rheological Characteristics of Polymers*. Naukova Dumka Publ., Kiev, 1977, p. 75.

[1221] Privalko, V.P., & Lipatov, Yu.S. *J. Polymer Sci.* Phys. Ed. **14**, 1725 (1976).

[1222] Eremenko, V.N., Lesnick, N.D., Pestoun, T.S., & Ryabov, V.R. In *Wettability and Surface Properties of Melts and Solids*. Naukova Dumka Publ., Kiev, 1972, p. 38.

[1223] Landau, L.D., & Livshits, E.M. *Mechanics of Continuous Medium*. State Techn. Publ., Moscow, 1953.

[1224] Khlynov, V.V., Pastoukhov, B.A., Boxer, E.L., & Serkov, I.P. *Adhesion of Melts and Solder. Mater. USSR* (4), 38 (1979).

[1225] Gul', V.E., Genel', S.V., & Fomina, L.L. *USSR Mech. Polym.* (2), 203 (1970).

[1226] Minaev, Yu.V., Ixanov, B.A., & Zhukhovitsky, A.A. In *Physical Chemistry of Condensed Phases, Superhard Materials, and Their Interfaces*. Naukova Dumka Publ., Kiev, 1975, p. 51.

[1227] Lincoln, B. *Nature.* **172**, 169 (1953).

[1228] Bowden, F.P., & Tabor, D. *Friction and Lubrication of Solids*. Oxford University Press, 1950.

[1229] Archard, I.E. *Proc. Roy. Soc.* **243A**, 190 (1957).

[1230] Area of Actual Contact of Coupled Surfaces. USSR Acad. Sci. Publ., Moscow, 1963.

[1231] Lopovock, T.S. *Wavy Surfaces and Measurement*. Standard Publ., Moscow, 1973, p. 27.

[1232] Barwell, F.T., Jones, M.H., & Probert, S.D. In *Mechanical Contact of Deformed Bodies*. Delft, 1975, p. 304.

[1233] Thomas, T.R. *Wear.* **33**, 205 (1975).

[1234] Khusu, A.P., Vitenberg, Yu.R., & Pal'mov, V.A. *Roughness of Surfaces (Theoretico–Statistical Approach).* Nauka Publ., Moscow, 1975.

[1235] Demkin, N.B. *Contacting of Roughed Surfaces.* Nauka Publ., Moscow, 1970.

[1236] Rudzit, Ya.A. *Microgeometry and Contact Interaction of Surfaces.* Zinatne Publ., Riga, 1975.

[1237] Ainbinder, S.B., & Tyunina, E.L. *USSR Mech. Polym.* (2), 241 (1977).

[1238] Kragelsky, I.V., Bely, V.A., & Sviridyonok, A.I. In *Adv. Polymer Friction and Wear.* Plenum Press, New York, 1974, vol. B, p. 729.

[1239] Ainbinder, S.B., & Tyunina, E.L. *USSR Mech. Polym.* (5), 651 (1972).

[1240] Kombalov, V.S. In *Contact Interaction of Solids and Calculation of Forces of Friction and Wear.* Nauka Publ., Moscow, 1971, p. 89.

[1241] Rabinowicz, E. *Prod. Eng.* **36**, 95 (1965).

[1242] Gul', V.E., Vakhroushina, L.A., & Dvoretskaya, N.M. *USSR High Polymers.* **18A**, 122 (1976).

[1243] Konstantinova, N.A. Cand. Sci(Chem.) Thesis, Pedagogic Inst., Moscow, 1967.

[1244] Gribanov, V.N., Lavrent'ev, V.V., & Voyutskii, S.S. *Proc. Inst. Fine Chem. Techn.* **5**, 133 (1975).

[1245] Harrick, N.J. *Ann. New York Acad. Sci.* **101**, 928 (1963).

[1246] Bartenev, G.M. *Reports USSR Acad. Sci.* **96**, 1161 (1954).

[1247] Sviridyonok, A.I., Kalmykova, T.F., Petrokovets, M.I., & Belyi, V.A. *USSR Mech. Polym.* (6), 710 (1974).

[1248] Stouchebryukov, S.D., Vavkushevsky, A.A., & Rudoy, V.M. *Reports USSR Acad. Sci.* **267**, 82 (1982).

[1249] Plisko, L.F. Cand. Sci(Chem.) Thesis, Lomonosov Inst. Fine Chem. Techn., Moscow, 1972.

[1250] Demkin, N.B. *Actual Contact Area of Solid Surfaces.* USSR Acad. Sci. Publ., Moscow, 1962.

[1251] Gribanov, V.N., Lavrent'ev, V.V., & Voyutskii, S.S. In *Adhesion of Polymers and Adhesive Joints in Machine Building.* Sci. -Techn. Soc. USSR Publ., Moscow, 1976, vol. 2, p. 137.

[1252] Bartenev, G.M. *Reports USSR Acad. Sci.* **103**, 1017 (1955).

[1253] Bartenev, G.M., & Lavrent'ev, V.V. *Herald USSR Acad. Sci.* Div. Techn. Sci. **9**, 126 (1958).

[1254] Voevodsky, V.S. Cand. Sci(Techn.) Thesis, Pedagogic Inst., Moscow, 1971.

[1255] Voyutskii, S.S., Shapovalova, A.I., & Pisarenko, A.P. *USSR Colloid J.* **18**, 485 (1956).

[1256] Vakula, V.L., & Voyutskii, S.S. *USSR Colloid J.* **23**, 672 (1961).

[1257] Miyashita, I., Hara, K., & Imoto, T. *Koll. Z. und Z. Polym.* **221**, 108 (1967).

[1258] Sviridyonok, A.I., Petrokovets, M.I., Kalmykova, T.F., & Ken'ko, V.M. *USSR Mech. Polym.* (5), 866 (1977).

[1259] Lavrent'ev, V.V., Ostreiko, K.K., & Gorshkov, M.M. In *Adhesives and Joints.* House Sci. -Techn. Propag., Moscow, 1970, vol. 1, p. 49.

[1260] Kaelble, D.H. In *Treatise on Adhesion and Adhesives.* Marcel Dekker, New York, 1966, vol. 1, p. 169.

[1261] Bartenev, G.M., & Lavrent'ev, V.V. *Reports USSR Acad. Sci.* **141**, 334 (1961).

[1262] Bartenev, G.M., Lavrent'ev, V.V., & Konstantinova, N.A. *USSR Mechn. Polym.* (5), 726 (1967).

[1263] El'kin, A.I. Thesis, Pedagogic Inst., Moscow, 1970.

[1264] Mikhailov, V.K., & Nikolaev, V.N. In *Some Problems in Physical Kinetics of Solids*. University Press, Saransk, 1975, p. 131.

[1265] Levit, M.Z., Zakharov, N.D., Chekanova, A.A., & Aref'ev, N.V. *USSR Rubber* (10), 57 (1980).

[1266] Voyutskii, S.S., Markin, Yu.I., Gorchakova, V.M., & Gul', V.E. *USSR J. Phys. Chem.* **37**, 2027 (1963).

[1267] Huntsberger, J.R. *J. Polymer Sci.* **1A**, 2241 (1963).

[1268] Schonhorn, H., & Ryan, F.W. *Adv. Chem.* Ser. 87, 121 (1968).

[1269] McLaren, A.D., & Seiler, C.J. *J. Polymer Sci.* **4**, 63 (1949).

[1270] Meissner, H.P., & Merrill, E.W. *ASTM Bull.* (151), 80 (1948).

[1271] Rayatskas, V.L. *Mechanical Strength of Adhesive Joints of Leather–Shoe Materials*. Light Ind. Publ., Moscow, 1976.

[1272] Pekarskas, V.-P.V., & Rayatskas, V.L. *USSR Mech. Polym.* (6), 937 (1974).

[1273] Narkevichyus, L.M. Thesis, Polytechn. Inst., Vilnius, 1974.

[1274] Pekarskas, V.-P.V., & Rayatskas, V.L. *USSR Mech. Polym.* (1), 168 (1974).

[1275] Egorenkov, A.I., & Egorenkov, N.I. *USSR Paint Mater. Appl.* (4), 41 (1978).

[1276] Bright, W.M. In *Fundamentals and Practice on Adhesion and Adhesives*. Wiley, New York, 1954, p. 130.

[1277] Egorenkov, N.I. *USSR High Polymers.* **22A**, 83 (1980).

[1278] Schonhorn, H., & Sharpe, L. *Polymer Lett.* **2**, 719 (1964).

[1279] Lavrent'ev, V.V. *Reports USSR Acad. Sci.* **175**, 125 (1967).

[1280] Ivanova-Chumakova, L.V., & Rehbinder, P.A. *Reports USSR Acad. Sci.* **81**, 219 (1951).

[1281] Gul', V.E., & Genel', S.V. In *Adhesion and Strength of Adhesive Joints*. House Sci. -Techn. Propag. Publ., Moscow, 1968, vol. 1, p. 30.

[1282] Greenwood, J.A., & Williamson, J.B.P. *Proc. Roy. Soc. A.* **292**, 300 (1966).

[1283] Bhushan, B. *Trans. ASME: J. Tribol.* **160**, 26 (1984).

[1284] Rabotnov, Yu.N. *Elements of Hereditary Mechanics of Solids*. Nauka Publ., Moscow, 1977.

[1285] Rabotnov, Yu.N. *Mechanics of Deformed Solid*. Nauka Publ., Moscow, 1988.

[1286] Hatfield, M.R., & Rathmann, G.B. *J. Phys. Chem.* **60**, 357 (1956).

[1287] Brunt, N.A. *J. Appl. Polymer. Sci.* **6**, 548 (1962).

[1288] Bartenev, G.M. *Reports USSR Acad. Sci.* **90**, 819 (1953).

[1289] Lavrent'ev, V.V. Dr. Sci(Chem.) Thesis, Karpov Phys. -Chem. Inst., Moscow, 1969.

[1290] Duszyk, M. *Rheol. Acta.* **25**, 199 (1986).

[1291] Vakula, V.L., Mishoustin, V.I., & Voyutskii, S.S. *USSR Plastics.* No 3, 32 (1969).

[1292] Plisko, L.F., Vakula, V.L., Lavrent'ev, V.V., & Voyuskii, S.S. *Proc. Inst. Fine Chem. Techn.* **2**, 87 (1972).

[1293] Plisko, L.F., Lavrent'ev, V.V., Vakula, V.L., & Voyutskii, S.S. *USSR High Polymers.* **14A**, 2131 (1972).

[1294] Ostreiko, K.K. Thesis, Karpov Phys. -Chem. Inst., Moscow, 1970.

[1295] Konstantinova, N.A., Lavrent'ev, V.V., & Vakula, V.L. *USSR High Polymers.* **14B**, 672 (1972).

[1296] Lavrent'ev, V.V., Vakula, V.L., Plisko, L.F., & Voyutskii, S.S. *USSR Colloid J.* **40**, 145 (1978).

[1297] Bartenev, G.M., & Styran, Z.E. *USSR High Polymers.* **1**, 7 (1959).

[1298] Williams, M.L., Landel, R.F., & Ferry, J.D. *J. Amer. Chem. Soc.* **77**, 3701 (1955).

[1299] Plazek, D.Y., Vanken, M.N., & Berge, Y.W. *Trans. Soc. Rheol.* **2**, 39 (1958).

[1300] Brown, G.M., & Tobolsky, A.V. *J. Polymer Sci.* **6**, 165 (1951).

[1301] Dahlquist, C.A., & Hatfield, M.R. *J. Colloid Sci.* **7**, 253 (1952).

[1302] Bartenev, G.M. *USSR High Polymers.* **6**, 335 (1964).

[1303] Lavrent'ev, V.V., Plisko, L.F., Vakula, V.L., & Sadov, B.D. *Reports USSR Acad. Sci.* **205**, 632 (1972).

[1304] Plisko, L.F., Ostreiko, K.K., Lavrent'ev, V.V., Vakula, V.L., & Voyutskii, S.S. *USSR High Polymers.* **15A**, 2579 (1973).

[1305] Alexandrov, A.P., & Lazourkin, Yu.S. *USSR. J. Techn. Phys.* **9**, 1250 (1939).

[1306] Bartenev, G.M., & Ermilova, N.V. In *Phys. Chem. Mechanics of Dispersed Structures.* Nauka Publ., Moscow, 1964, p. 371.

[1307] Bartenev, G.M., & Styran, Z.E. *Reports USSR Acad. Sci.* **121**, 87 (1958).

[1308] Lavrent'ev, V.V., & Sadov, B.D. In *5th Conf. Phys. Chem. Mechanics.* USSR Acad. Sci. Publ., Ufa, 1971.

[1309] Sadov, B.D., Lavrent'ev, V.V., & Konstantinova, N.A. In *2nd Conf. Relaxation Phenomena in Polymers.* USSR Acad. Sci. Publ., Baku, 1974, vol. 2, p. 28.

[1310] Evstigneeva, E.V., Maumenko, V.Yu., Lavrent'ev, V.V., & Toutorsky, I.A. *USSR High Polymers.* **23B**, 650 (1981).

[1311] Evstigneeva, E.G., Maloushhouck, Yu.S., & Markina, G.V. *USSR Rubber* No 4, 10 (1978).

[1312] Fridman, B.S., Dobren'kov, G.A., & Zilberman, A.B. *Dep. Manuscr.* (776), NIITEKHIM Press, Tcherkassy, 1983.

[1313] Zilberman, A.B. Thesis, Chem. -Techn. Inst., Kazan, 1987.

[1314] Sokolova, L.V., Nikolaeva, A.O., & Shershnev, V.A. *USSR High Polymers.* **26A**, 1544 (1984).

[1315] Fridman, B.S., Khramov, Yu.V., Zilberman, A.B., Gouseva, M.S., & Efimov, M.A. In *Galvanotechnics.* Chem. Soc. USSR Press, Kazan, 1987, p. 263.

[1316] Zilberman, A.B., Dobren'kov, G.A., & Fridman, B.S. *J. Adhesion Sci. Techn.* in press.

[1317] Pritykin, L.M., Zilberman, A.B., & Vakula, V.L. In *New Adhesives, Technology of Bonding and Applications.* House Sci. -Techn. Propag. Publ., Moscow, 1989, p. 7.

[1318] Pritykin, L.M., Vakula, V.L., Silberman, A.B., & Silberman, I.I. *Plaste und Kautsch.* **36**, 33 (1989).

[1319] Pritykin, L.M., Silberman, A.B., Silberman, I.I., & Vakula, V.L. *Plaste und Kautsch.* In press.

[1320] Tamm, I.E. *Phys. Z. Sow.* **1**, 733 (1932).

[1321] Lee, Y.S., & Kertesz, M. *Int. J. Quant. Chem.* Quant. Chem. Symp. (21), 163 (1987).

[1322] Ladik, J., Suhai, S., & Seel, M. *Amer. Chem. Soc. Polymer Prepr.* **21**, 118 (1980).

[1323] Welsh, W.J. In *Polymer Materials*. Science and Engineering, Washington, 1986, p. 134.

[1324] Perelygina, T.K., & Tekoucheva, I.A. *USSR High Polymers*. **20A**, 1471 (1978).

[1325] Duke, C.B. *Amer. Chem. Soc. Polymer Prepr*. **21**, 117 (1980).

[1326] Kiess, H., & Rehwald, W. *Coll. Polym. Sci*. **258**, 241 (1980).

[1327] Enikolopyan, N.S., Berlin, Yu.A., Beshenko, S.I., & Zhorin, V.A. *Reports USSR Acad. Sci*. **258**, 1400 (1981).

[1328] Kasowski, R.V., Caruthers, E., & Hsu, W.Y. *Phys. Rev. Lett*. **44**, 676 (1980).

[1329] Karl, N. *Adv. Sol. Stat. Phys*. **14**, 261 (1974).

[1330] Bauser, H. *Elektrostatische Aufladung*. Chemie Verl., Weinheim, 1974, p. 11.

[1331] Lewis, T.J. In *Polymer Surfaces*. Wiley, Chichester, 1978, p. 65.

[1332] Fabish, T.J., Saltsburn, H.M., & Hair, M.L. *J. Apply. Phys*. **47**, 940 (1976).

[1333] Garton, C.G. *J. Phys*. **7D**, 1814 (1974).

[1334] Belyi, V.A., Goldade, V.A., Neverov, A.S., & Pinchuk, L.S. *Reports USSR Acad. Sci*. **245**, 132 (1979).

[1335] Belyi, V.A., Goldade, V.A., Neverov, A.S., & Pinchuk, L.S. *J. Polymer Sci*. **17**, 3193 (1979).

[1336] Mizutani, T., Takai, Y., Osawa, T., & Ieda, M. *J. Phys*. **9D**, 2253 (1976).

[1337] Vol'kenshtein, F.F. *Electronic Theory of Catalysis on Semiconductors*. Phys.-Math. Publ., Moscow, 1963.

[1338] Davison, S.G., & Levine, J.D. *Surface States*. Academic Press, New York, 1970.

[1339] Schokley, W. *Phys. Rev*. **56**, 317 (1939).

[1340] Morrison, S.R. *Chemical Physics of Surfaces*. Plenum Press, New York, 1977.

[1341] Tomasek, M., & Koutecky, J. In *Electronic Phenomena in Adsorption and Catalysis for Semiconductors*. Mir Publ., Moscow, 1969, p. 71.

[1342] Levine, J. *Surface Sci*. **34**, 90 (1973).

[1343] Pritykin, L.M. In *Proc. 7th Vacuum Congr., 3rd Int. Conf. Solid Surfaces*. Techn. Univ. Press, Vienna, 1977, vol. 1, p. 675.

[1344] Pritykin, L.M., & Dranovsky, M.G. In *Methods for Increase of Polymer Adhesion*. USSR Chem. Soc. Publ., Moscow, 1977, p. 21.

[1345] Prasad, C.V., & Sundaram, K. *Int. J. Quant. Chem*. **20**, 613 (1981).

[1346] Pritykin, L.M. In *Adhesives and Adhesive Joints*. Mir Publ., Moscow, 1988, p. 203.

[1347] Mark, H.F. In *Adhesion and Cohesion*. Elsevier, Amsterdam, 1961, p. 271.

[1348] Mark, H.F. *Adh. Age*. **22**, (7), 35 (1979).

[1349] Mark, H.F. *Adh. Age*. **22** (9), 45 (1979).

[1350] Morgounov, N.N., Kokhanova, E.A., & Kozhevnikov, A.N. *USSR Plastics*. (3), 56 (1984).

[1351] Lipatov, Yu.S. In *Adhesion-7*. Sci. Press, London, 1983, p. 149.

[1352] Lipatov, Yu.S., Filippovitch, A.Yu., Veselovsky, R.A., Khranovsky, V.A., & Fedorchenko, E.I. *USSR High Polymers*. **28B**, 668 (1986).

[1353] Ramharack, R., & Nguen, T.H. *J. Polymer Sci*. Lett. Ed. **25C**, 93 (1987).

[1354] Iwatsuki, S., Kondo, A., & Sakai, Y. *J. Chem. Soc. Japan. Chem. Ind*. (10), 1884 (1985).

[1355] Pritykin, L.M., Dranovsky, M.G., Bougaeva, S.V., & Eremov, Ya.E. *USSR Plastics*. (6), 49 (1975).

[1356] Pritykin, L.M., Zaichenko, G.A., & Fokin, I.Ya., *Probl. Chem. Techn. USSR* (60), 30 (1980).

[1357] Pritykin, L.M. In *Modern Problems in Synthesis of Rubber*. Chem. Soc. USSR Publ., Dnepropetrovsk, 1980, p. 164.

[1358] Kotlyar, I.V., & Zilberman, E.N. In *Advances in Chemistry and Physics of Polymers*. Chimia Publ., Moscow, 1973, pp. 260, 271.

[1359] Misono, A., & Uchida, Y. *J. Polymer Sci.* **5B**, 401 (1967).

[1360] Pritykin, L.M., & Raevsky, V.G. USSR Pat. No 358935 (1975).

[1361] Yokota, K., & Hirabayashi, T. *Polymer J.* **13**, 813 (1981).

[1362] Pritykin, L.M., Raevsky, V.G., Borzenko, E.M., & Orlov, V.A., USSR Pat. No 357203 (1972).

[1363] Pritykin, L.M., & Orlov, V.A. In *Modern Problems in Synthesis of Rubber*. Chem. Soc. USSR Publ., Dnepropetrovsk, 1980, p. 163.

[1364] Trostyanskaya, E.B., & Babevsky, P.G. *USSR High Polymers*. **10A**, 288 (1968).

[1365] Pritykin, L.M., & Raevsky, V.G. USSR Pat. No. 383377 (1975).

[1366] Raevsky, V.G., & Pritykin, L.M. USSR Pat. No. 303323 (1975).

[1367] Pritykin, L.M., Borzenko, E.M., & Raevsky, V.G. USSR Pat. No 452210 (1975).

[1368] Pritykin, L.M., Borzenko, E.M., & Raevsky, V.G. USSR Pat. No 437391 (1975).

[1369] Pritykin, L.M., Borzenko, E.M., & Raevsky, V.G. USSR Pat. No 444411 (1975).

[1370] Pritykin, L.M., Borzenko, E.M., & Raevsky, V.G. USSR Pat. No 444412 (1975).

[1371] Raevsky, V.G., & Pritykin, L.M. USSR Pat. No 309605 (1975).

[1372] Pritykin, L.M., Borzenko, E.M., & Raevsky, V.G. USSR Pat. No 447047 (1975).

[1373] Bell, V. *J. Polymer Sci.* **2A**, 5291 (1964).

[1374] Marvel, C.S., Keiner, O.E., & Wessel, E.D. *J. Amer. Chem. Soc.* **81**, 4694 (1959).

[1375] Marvel, C.S., & Keiner, O.E. *J. Polymer Sci.* **61**, 311 (1962).

[1376] Pritykin, L.M. In *31st Macromol. Symp.* IUPAC Publ., Merseburg, 1987, vol. 1, p. 62.

[1377] Ingold, C.K. *Structure and Mechanism in Organic Chemistry*. Cornell University Press, Ithaca, 1969, ch. III.3b.

[1378] Zaitsev, B.A. In *Carbon-chain Polymers*. Nauka Publ., Moscow, 1977, p. 139.

[1379] Pritykin, L.M. In *Adhesion '87*. PRI, London, 1987, p.M/1.

[1380] Pritykin, L.M., Borzenko, E.M., Orlov, V.A., & Ravesky, V.G. USSR Pat. No 426490 (1975).

[1381] Pritykin, L.M., & Dranovsky, M.G. USSR Pat. No 611920 (1978).

[1382] Ohmori, S., Ito, S., Onogi, Y., & Nishijima, Y. *Polymer J.* **19**, 1269 (1987).

[1383] Pritykin, L.M., Borzenko, E.M., Orlov, V.A., & Raevsky, V.G. USSR Pat. No 421259 (1975).

[1384] Pritykin, L.M., Steshenko, N.I., & Androushkiv, I.I. *USSR Light Ind.* (1), 51 (1977).

[1385] Raevsky, V.G., & Pritykin, L.M. USSR Pat. No 376394 (1973).

[1386] Raevsky, V.G., & Pritykin, L.M. USSR Pat. No 401158 (1975).

[1387] Pritykin, L.M., & Dranovsky, M.G. In *New Adhesives and Technology of Bonding*. House Sci. -Techn. Propag. Publ., Moscow, 1976, p. 19.

[1388] Pritykin, L.M. In *Adhesion '87*, PRI Publ., London, 1987, p. 9/1.

[1389] Nazarov, I.N., & Kouznetsova, A.I. *Herald USSR Acad. Sci.* Div. Chem. Sci. (4), 431 (1941).

[1390] Bragina, M.N., Iltina, V.M., Utyansky, Z.S., & Iotkovskaya, L.A. In *Proc. Plastics Inst.* Chimia Publ., Moscow, 1971, p. 58.

[1391] Raevsky, V.G., Kazaryan, S.A., Obidin, E.A., & Pritykin, L.M. USSR Pat. No 401163 (1975).

[1392] Pritykin, L.M., & Kazaryan, S.A. In *Proc. Sci. Papers.* Polytechn. Inst. Press, Kirovakan, 1971, p. 3.

[1393] Pritykin, L.M., Raevsky, V.G., & Obidin, E.A. *USSR Plastics.* No 7, 14 (1974).

[1394] Pritykin, L.M., Raevsky, V.G., Steshenko, N.I., & Obidin, E.A. USSR Pat. No 523924 (1976).

[1395] Pritykin, L.M., Raevsky, V.G., Steshenko, N.I., & Obidin, E.A. USSR Pat. No 603653 (1978).

[1396] Pritykin, L.M., Obidin, E.A., & Dranovsky, M.G. *USSR Plastics.* No 10, 29 (1977).

[1397] Pritykin, L.M., & Dranovsky, M.G. *USSR Machine-building.* (4), 32 (1976).

[1398] Bruck, H. Dissertation, Stuttgart, 1985.

[1399] Lipatov, Yu.S., Yakovenko, A.G., & Gorichko, E.Ya. *USSR Compos. Polym. Mater.* (5), 3 (1980).

[1400] Villoutreix, J., Nogues, P., & Berlot, R. *Europ. Polymer J.* **22**, 147 (1986).

[1401] Lipatov, Yu.S., Sergeeva, L.M., Gorichko, E.Ya., & Brovko, A.A. *USSR High Polymers.* **26B**, 643 (1984).

[1402] Kaneko, N., Takahashi, A., & Ohtsuka, Y. *J. Appl. Polymer Sci.* **27**, 4365 (1982).

[1403] Polyakov, A.M., & Krotova, N.A. In *Research in Surface Forces.* Nauka Publ., Moscow, 1967, p. 461.

[1404] Andreev, V.K., Lipson, A.G., Shabanova, S.A., Klyuev, V.A., & Toporov, Yu.P. *USSR High Polymers.* **27B**, 440 (1985).

[1405] Novak, I., & Romanov, A. *Text. chem.* **11**, 11 (1981).

[1406] Aubrey, D.W., & Ginosatis, S. *J. Adhesion.* **12**, 189 (1981).

[1407] Yang, Z., Huang, S., Zhang, X., & Chen, D. *J. Cent. China Norm. Univ. Natur. Sci.* **22**, 51 (1988).

[1408] Uspenskaya, Z.R., Tyazhlo, N.I., Trofimova, N.V., Knyazheva, T.V., Lavrova, N.V., & Arkhipova, I.N. In *Modification of Structure and Properties of Polymeric Plastics.* Chimia Publ., Leningrad, 1981, p. 60.

[1409] Roumyantsev, V.D., Mouzykantov, A.I., Borisov, V.A., Bykova, I.P., & Bolikhova, V.D. In *Synthesis, Properties and Processing of Polyolefins.* Chimia Publ., Leningrad, 1984, p. 95.

[1410] Efremov, G.A., & Kostikov, Yu.P. *USSR High Polymers.* **24B**, 813 (1982).

[1411] Efremov, G.A., Maroushhack, N.V., Ol'khova, M.B., & Strykanov, V.C. *USSR High Polymers.* **26B**, 750 (1984).

[1412] Efremov, G.A., & Strykanov, V.S. In *3rd Conf. Oligomer Chemistry.* Nauka Publ., Chernogolovka, 1986, p. 57.

[1413] Mao, T.J., & Reegen, S.L. In *Adhesion and Cohesion.* Elsevier, Amsterdam, 1962, p. 209.

[1414] Ryskina, Yu.P., Mamedova, E.S., & Mekhtiev, S.I. *USSR Plastics.* No 5, 33 (1985).

[1415] Rantell, A. *Trans. Inst. Metal Finish.* **47**, 197 (1969).

[1416] Kim, M.W., Peiffer, D.G., & Pincus, P. *J. Phys. Lett.* (France) **45**, L-953 (1984).

[1417] Rudoy, V.M., & Ogarev, V.A. In *Modern Physical Methods for Research on Polymers.* Chimia Publ., Moscow, 1982, p. 197.

[1418]   Yamamoto, H. *J. Chem. Soc. Japan Chem. Ind.* (1), 90 (1986).

[1419]   Roland, C.M., & Böhm, G.G.A. *J. Appl. Polymer Sci.* **29**, 3803 (1984).

[1420]   Hamed, G.R., & Shieh, C.-H. In *28th Macromol. Symp.* IUPAC Publ., Amherst, 1982, vol. 1, p. 559.

[1421]   Ozolin'sh, Yu.L., & Sirmatch, A.I. In *Modification of Polymer Materials.* Polytechn. Inst. Press, Riga, 1985, p. 94.

[1422]   Kouznetsova, V.M., Yakovleva, R.A., Lebedev, V.S., Doubinskaya, B.G., & Efanova, V.V. *USSR Plastics.* (8), 23 (1986).

[1423]   Kakhramanov, N.T., Abbasov, A.M., Bouniyat-zade, A.A., & Abbasov, K.A. *USSR High Polymers.* **27B**, 227 (1985).

[1424]   Baraban, O.P. *USSR Rubber.* (9), 8 (1971).

[1425]   Chang, R.-J., & Gent, A.N. *J. Polymer Sci.* **19**, 1635 (1981).

[1426]   Shagov, V.S. *USSR Rubber.* (12), 9 (1981).

[1427]   Dogadkin, B.A. *Chemistry of Elastomers.* Chimia Publ., Moscow, 1972, p. 133.

[1428]   Baratsevitch, E.N., Beresneva, N.K., Kalaous, A.E., & Sabourova, T.S. *USSR Rubber.* (9), 7 (1976).

[1429]   Petrov, G.N., & Lykin, A.S. *USSR High Polymers.* **20A**, 1203 (1978).

[1430]   Brossas, J., & Clonet, G. *Makromol. Chem.* **175**, 3067 (1974).

[1431]   Shitov, V.S. *USSR Paint Mater. Appl.* (4), 49 (1977).

[1432]   Reznichenko, S.V., Kazakova, L.N., & Kanaouzova, A.A. *USSR Ind. Tyres, Rubber- and Asbestos-Technic Goods.* (12), 2 (1980).

[1433]   Zimnitskaya, E.A., Trostyanskaya, I.I., & Tourgenevskaya, Sh.B. *Proc. Technol. Inst. Leningrad* (5), pt. 1, 62 (1975).

[1434]   Shmourack, I.I. *USSR Rubber* (3), 21 (1975).

[1435]   Karapetyan, N.G., Boshnyakov, I.S., Nokogosyan, M.G., Mkryan, G.M., Papazyan, N.A., Yutoudjyan, K.K., Sarkisyan, R.R., & Belyakova, V.I. USSR Pat. No 451331 (1974).

[1436]   Souproun, A.P., Zhoulin, V.M., Klimentova, N.V., Soboleva, T.A. & Akopyan, A.N. USSR Pat. No 418050 (1974).

[1437]   Ginzburg, L.V., & Pol'sman, G.S. *USSR High Polymers.* **14A**, 1667 (1972).

[1438]   Shmourack, I.L. Dr. Sci(Techn.) Thesis, Lomonosov Inst. Fine Chem. Technol. Moscow, 1978.

[1439]   Glagolev, V.A., Il'in, N.S., Orlov, V.A., & Starkova, N.A. In *Adhesives and Joints.* House Sci. -Techn. Propag. Publ., Moscow, 1970, vol. 2, p. 18.

[1440]   Glagolev, V.A., Lyusova, P.R., Krokhina, L.S., & Kouleznev, V.N. *USSR Rubber* (4), 33 (1979).

[1441]   Yamakawa, S. In *Adhesion and Adsorption of Polymers.* Plenum Press, New York, 1980, pp. 473, 580.

[1442]   Lin, D.G., Vinogradova, T.B., & Poukshansky, M.D. *USSR Plastics* (1), 29 (1979).

[1443]   Sirota, A.G., Sergeeva, N.N., Gol'denberg, A.L., Kouznetsova, E.P., & Karkozova, G.F. *USSR J. Appl. Chem.* **54**, 1558 (1981).

[1444]   Vader, F.v.V., & Dekker, H. *J. Adhesion.* **7**, 73 (1974).

[1445]   Cagnon, D.R., & McCarthy, T.J. *J. Appl. Polymer Sci.* **29**, 4335 (1984).

[1446]   Gaines, J.L., & Bender, G.W. *Macromolecules.* **5**, 82 (1972).

[1447]   Allan, A.J.G. *J. Colloid Sci.* **14**, 206 (1959).

[1448]   Owens, D.K. *J. Appl. Polymer Sci.* **14**, 185 (1970).

[1449] Zhigletsova, S.K., Rudoy, V.M., & Ogarev, V.A. *USSR High Polymers.* **24B**, 446 (1982).

[1450] Kouz'mina, R.P., Zaev, N.E. *Proc. USSR Inst. Electromech.* No 40, 188 (1974).

[1451] Dontsov, A.A., Tarasova, G.I., & Lapshova, A.A. *USSR High Polymers.* **24A**, 1895 (1982).

[1452] Pritykin, L.M., Shoumetov, V.G., Babayan, M.S., Navasardyan, T.N., & Karadjyan, K.N. *Armenian Chem. J.* **24**, 878 (1971).

[1453] Pritykin, L.M., Raevsky, V.G., Borzenko, E.M., & Sklyarsky, L.S. USSR Pat. No 445346 (1975).

[1454] Pritykin, L.M., & Dranovsky, M.G. *Diffusion, Phase Transformations, Mechanical Properties of Metals and Alloys.* **12**, pt. 2, 146 (1975).

[1455] Pritykin, L.M., *USSR Plastics* (2), 69 (1974).

[1456] Pritykin, L.M., Raevsky, V.G., Steshenko, N.I., & Borzenko, E.M. USSR Pat. No 538562 (1976).

[1457] Pritykin, L.M., Raevsky, V.G., Stremock, T.M., Cherkasova, L.A., Apoukhtina, N.P., Vinogradova, L.A., & Morozova, L.P. USSR Pat. No 418083 (1975).

[1458] Pritykin, L.M., Raevsky, V.G., Stremock, T.M., & Morozova, L.P. *USSR Plastics* (3), 72 (1972).

[1459] Pritykin, L.M., Stremock, T.M., Raevsky, V.G., Sklyarsky, L.S., Morozova, L.P., & Karmazina, E.N. *USSR Leather-shoe Ind.* (6), 31 (1972).

[1460] Pritykin, L.M., Raevsky, V.G., & Semenyuck, S.P. *USSR Light Ind.* (6), 19 (1973).

[1461] Tourousov, R.A., Vuba, K.T., Freidin, A.S., Soulyaeva, Z.P., Babitch, V.F., & Andreevskaya, G.D. In *Thermodynamic and Structural Properties of Polymer Boundary Layers.* Naukova Dumka Publ., Kiev, 1976, p. 88.

[1462] Soukhareva, L.A., Zemtsov, A.I., Kiselev, M.R., & Zoubov, P.I. *USSR Colloid J.* **36**, 991 (1974).

[1463] Pritykin, L.M., & Dranovsky, M.G. In *New Polymer Composite Materials in Machine Building.* USSR Sci. -Techn. Soc. Publ., Moscow, 1978, p. 228.

[1464] Epshtein, L.G., Nagoumanova, E.A., Voskresensky, V.A., & Kiseleva, A.S. *USSR Compos. Polym. Mater.* (23), 51 (1984).

[1465] Markov, V.V., Finogenova, E.N., Zakharov, V.P., & Zachesova, G.N. *USSR Rubber* (12), 18 (1982).

[1466] Denisov, V.A., Gerenroth, V.G., Solntseva, A.V., & Pritykin, L.M. USSR Pat. No 658158 (1979).

[1467] Pritykin, L.M., Gerenroth, V.Gl, & Oleinikova, V.V. *USSR Ind. Domest. Chem. Goods* (5), 21 (1975).

[1468] Veselovsky, R.A. In *Polymers-80.* Naukova Dumka Publ., Kiev, 1980, p. 121.

[1469] Foltyniwicz, Z., Yamaguchi, K., Czajka, B., & Regen, S.L. *Macromolecules.* **18**, 1394 (1985).

[1470] Veselovsky, R.A. In *Intermetalbond '81.* CSVTS Press, Bratislava, 1981, p. 28.

[1471] Veselovsky, R.A., Vysotskaya, G.V., Ishhenko, S.S., Fedorchenko, E.I., & Sheinina, L.S. In *Physicochemistry and Modification of Polymers.* Naukova Dumka Publ., Kiev, 1987, p. 146.

[1472] Nesterov, A.E., & Ignatova, T.D. *USSR High Polymers.* **28B**, 804 (1986).

[1473] Baradene, V.A., & Rayatskas, V.L. In *Problems of Adhesive Joint Strength.* Polytechn. Inst. Publ., Kaunas, 1971, p. 66.

[1474] Pritykin, L.M., Vakula, V.L., & Dranovsky, M.G. *USSR Rubber* (3), 18 (1985).

[1475] Brewis, D.M., & Briggs, D. *Polymer.* **22**, 7 (1981).

[1476] Packham, D.E. In *Adhesion Aspects of Polymeric Coatings.* Plenum Press, New York, 1983, p. 19.

[1477] Erykalova, T.A., & Vladychina, S.V. In *New Elaborations in Manufacture of Artificial Leathers and Film Materials.* Light Ind. Publ., Moscow, 1983, p. 58.

[1478] Yasuda, H.K., Cho, D.L., & Yeh, Y.-S. In *Polymer Surfaces and Interfaces.* Wiley, Chichester, 1987, p. 149.

[1479] Povstougar, V.I., Kodolov, V.I., & Mikhailova, S.S. *Structure and Properties of Surfaces of Polymer Materials.* Chimia Publ., Moscow, 1988, p. 91.

[1480] Vol'kenshtein, F.F. *Physicochemistry of Semiconductor Surfaces.* Nauka Publ., Moscow, 1977.

[1481] Orlov, V.A., & Pritykin, L.M. In *New Adhesives and Technology of Bonding,* House Sci. -Techn. Propag. Publ., Moscow, 1973, p. 18.

[1482] Pritykin, L.M. In *Proc. Int. Congr. Surface Technology.* VDE Verl., Berlin, 1985, S. 59.

[1483] Soshko, A.I. *USSR Phys. -Chem. Mech. Mater.* **6**, 74 (1970).

[1484] Dwight, D.W., & Riggs, W.M. *J. Colloid Interface Sci.* **47**, 650 (1974).

[1485] Khlebnikov, B.M., Kouz'menko, N.N., & Ponomareva, T.M. USSR Pat. No 615123 (1978).

[1486] Dake, S.B., Bhoraskar, S.V., Patil, P.A., & Narasimhan, N.S. *Polymer.* **27**, 910 (1986).

[1987] Von Brecht, H., Mayer, F., & Binder, H. *Makromol. Chem.* **33**, 89 (1973).

[1488] Baumhardt-Neto, R., Galembeck, S.E., Jockes, I., & Galembeck, F., *J. Polymer Sci.* **19**, 819 (1981).

[1489] Dahm, R.H., Brewis, D.M., Gribbin, J.D., Barker, D.J., & Hoy, L.R.J. In *Int. Conf. Adhesion and Adhesives.* Science, Technology, and Applications, London, 1980, p. 3.1.

[1490] Minford, J.D. *Adh. Age.* **17**, (7), 24 (1974).

[1491] Kimura, K. *J. Adhesion Soc. Japan.* **23**, 443 (1987).

[1492] Popovska, N., Ljubtscheva, M., Ljubschev, L., Mladenov, T., & Mladenov, I. *Acta Polym.* **37**, 563 (1986).

[1493] Fonseca, C., Pereha, J.M., Fatou, J.G., & Bello, A. In *Int. Macromol. Symp.* IUPAC Publ., Bucharest, 1983, Sect. 6–7, p. 239.

[1494] Suganuma, A., & Kani, M. Japan Pat. No 54-144473 (1979).

[1495] Seregin, V.I., & Artem'ev, V.A., USSR Pat. No 757558 (1980).

[1496] Rodchenko, D.A., Parshikova, Z.V., Barkan, A.I., & Sorokoletova, V.A. USSR Pat. No 653274 (1979).

[1497] Popov, V.I., Egorov, Yu.V., Pouzako, V.D., & Betenekov, N.D. USSR Pat. No 588230 (1977).

[1498] Sugita, K. *Japan Met.* **24**, 371 (1986).

[1499] Piiroja, E.K., Lippmaa, H.V., Metlitskaya, O.F., & Dankovics, A. *Europ. Polymer J.* **16**, 641 (1980).

[1500] Kapachyaouskene, Ya.P., & Shalkauskas, M.I. *Proc. Lithuanian Acad. Sci.* (100B), 43 (1977).

[1501] Briggs, D. *Appl. Surface Sci.* **6**, 188 (1980).

[1502] Briggs, D., Zichy, V.J.I., & Brewis, D.M. *Surface Interface Anal.* **2**, 107 (1980).

[1503] Piiroja, E.K. *USSR Plastics* (1), 61 (1988).
[1504] Balis, P., Carlsson, D.J., Csullog, G.W., & Wiles, D.M. *J. Colloid Interfaces Sci.* **47**, 636 (1974).
[1505] Piiroja, E.K., & Payula, S.F. *Proc. Tallinn Polytechn. Inst.* (471), p. 39 (1979).
[1506] Piiroja, E.K., & Dankovics, A. *Acta Polym.* **33**, 200 (1982).
[1507] Sokolov, V.D., Smirnov, E.P., Kol'tsov, S.I., & Aleskovsky, V.B. *Reports USSR Acad. Sci.* **256**, 1443 (1981).
[1508] Kopylov, V.B., Tsetkova, M.N., Pack, V.N., Malygin, A.A., & Kol'tsov, S.I. *USSR J. Appl. Chem.* **54**, 293 (1981).
[1509] Golander, C.-G., & Sultan, B.-A. *J. Adhesion Sci. Techn.* **2**, 125 (1988).
[1510] Piiroja, E.J. *Dep. Manuscr.* No 4E, Inst. Inf. Publ., Tallinn, 1983.
[1511] Kaufman, M.H. USA Pat. No 4243771 (1981).
[1512] Delamar, M., Zeggane, S., & Dubois, J.E. In *23rd Annu. Conf. Adhesion and Adhesives.* London, 1986, p. 57.
[1513] Kosicki, J. *Zesz. nauk. Pozn. Masz. robocze pojazdy.* No 24, 133 (1984).
[1514] Courduvellis, C.I., & Del Gobbo, A.R. USA Pat. No 4629636 (1986).
[1515] Redick, H.E., & Barnes, R.D. USA Pat. No 4608402 (1986).
[1516] Drzal, L.T., Rich, M.J., & Lloyd, P.E. *Amer. Chem. Soc. Polymer Prepr.* **22**, 199 (1981).
[1517] Shoul', G.S., Shkirkova, L.M., Shhoukina, L.A., & Gorbatkina, Yu.A. *USSR Mech. Compos. Mater.* (4), 600 (1987).
[1518] Kobets, L.P. In *Composite Materials.* Moscow, 1981, p. 201.
[1519] Kobets, L.P., Prigorodov, V.N., & Gounyaev, G.M. In *Fibrous and Dispersed-strengthened Composite Materials.* Nauka Publ., Moscow, 1976, p. 159.
[1520] Kobets, L.P., & Gounyaev, G.M. *USSR Mech. Polym.* (3), 445 (1977).
[1521] Kobets, L.P., Gounyaev, G.M., & Kouznetsova, M.A. *USSR Aircraft Mater.* (2), 74 (1977).
[1522] Dorn, L., & Bischof, R. *Maschinenmarkt.* **93**, 64 (1987).
[1523] Emanuel', N.M. *USSR High Polymers.* **21A**, 2624 (1979).
[1524] Zaikov, G.E., & Razoumovsky, S.D. *USSR High Polymers.* **23A**, 513 (1981).
[1525] Piiroja, E.K., & Dankovics, A. *USSR Plastics* (7), 35 (1982).
[1526] Vakabayashi, K. *Japan Technol. Adhes.* **32**, 52 (1988).
[1527] Piiroja, E.K., & Lippmaa, H.V. *Acta Polym.* **35**, 669 (1984).
[1528] Piiroja, E.K. *Dep. Manuscr.* No 5E, Inst. Inf. Publ., Tallinn, 1983.
[1529] Esumi, K., Maguro, K., Schwartz, A.M., & Zettlemoyer, A.C. *Bull. Chem. Soc. Japan.* **55**, 1649 (1982).
[1530] Kato, K. Japan Pat. No 59-129234 (1983).
[1531] Goldblatt, R.D., Park, J.M., White, R.C., Matienzo, L.J., Johnson, J.F., & Huang, S.J. *Amer. Chem. Soc. Polymer Prepr.* **28**, 60 (1987).
[1532] Peeling, J., Courval, G., & Jazzar, S.M. *J. Polymer Sci.* **22**, 419 (1984).
[1533] Kiyushkin, S.G., Dalinkevitch, A.A., Shemarov, F.V., & Shlyapnikov, Yu.A. *USSR Plastics.* No 8, 14 (1984).
[1534] Gatechair, L.R., & Wostratzky, D. In *Adhesive Chemistry; Developments and Trends.* Plenum Press, New York, 1984, p. 409.
[1535] Tanaka, H., & Otani, S. Japan Pat. No 56-18607 (1981).
[1536] Dalinkevitch, A.A., Kiryushkin, S.G., Shemarov, F.V., & Shlyapnikov, Yu.A. *USSR Plastics.* No 8, 21 (1988).

[1537] Bragole, R.A. *Rubber Age.* **106**, 53 (1974).
[1538] Carter, A.R. In *Int. Conf. Adhesion and Adhesives.* Science, Technology, and Applications, London, 1980, p. 26.1.
[1539] Wake, W.C. *Plast. Rubb. Age.* **5**, 157 (1980).
[1540] Bragole, R.A., & Weidman, R.A. USA Pat. No 4321307 (1982).
[1541] Angert, L.G., Kouznetsova, M.N., & Shourygina, T.B. In *Surface Phenomena in Polymers.* Naukova Dumka Publ., Kiev, 1982, p. 4.
[1542] Gaske, J.E. USA Pat. No 3925349 (1975).
[1543] Bartolomew, R.F., & Davidson, R.S. *J. Chem. Soc.* 2342 (1971).
[1544] Slobodetskaya, E.M. *USSR Adv. Chem.* **49**, 1594 (1980).
[1545] Malers, Yu.Ya., & Kalnin', M.M. *Herald Latvian Acad. Sci. Ser. Chem.* (6), 654 (1979).
[1546] Malers, L.Ya., & Kalnin', M.M. *USSR Modif. Polym. Mater.* (6), 19 (1976).
[1547] Tsvetkova, M.N., Skorick, Yu.I., Kouchaeva, S.K., Tallier, Yu.A., & Malygin, A.A. *USSR J. Appl. Chem.* **58**, 1420 (1985).
[1548] Egorenkov, N.I,. In *Metal-polymer Materials and Goods.* Chimia Publ., Moscow, 1979, p. 19.
[1549] Egorenkov, N.I., & Kouzavkov, A.I. *USSR High Polymers.* **26A**, 518 (1984).
[1550] Egorenkov, N.I. In *Composite Polymer Materials.* Naukova Dumka Publ., 1975, p. 124.
[1551] Malers, L.Ya., Zeltserman, G.A., Vixne, A.V., Kalnin', M.M., & Karlivan, V.P. *USSR High Polymers.* **18A**, 551 (1976).
[1552] Kalnin', M.M., & Malers, Yu.Ya. *USSR High Polymers.* **27A**, 793 (1985).
[1553] Kalnin', M.M., & Malers, Yu.Ya. *Herald Latvian Acad. Sci. Ser. Chem.* No 2, 166 (1985).
[1554] Pesetsky, S.S., Alexandrova, O.N., & Shherbakov, S.V. *Herald Byelorussian Acad. Sci. Ser. Chem. Sci.* No 1, 45 (1987).
[1555] Rekhmanis, P.K., & Kalnin', M.M. *USSR Modif. Polym. Mater.* No 3, 46 (1972).
[1556] Kalnin', M.M. In *Synthesis and Physicochemistry of Polymers.* Naukova Dumka Publ. (23), 100 (1978).
[1557] Bel'tenas, R.A., Bal'tenene, Ya.Yu., & Kevyalaite, Z.K. *USSR High Polymers.* **23A**, 1466 (1981).
[1558] Kalnin', M.M., & Ozolin'sh, Yu.L. *USSR Mech. Compos. Mater.* (2), 201 (1984).
[1559] Kapishnikov, Yu.V. *USSR Modif. Polym. Mater.* (14), 27 (1986).
[1560] Kalnin', M.M., & Malers, Yu.Ya. *USSR Modif. Polym. Mater.* (10), 24 (1981).
[1561] Malers, L.Ya., & Kalnin', Yu.Ya. *USSR Modif. Polym. Mater.* (6), 32 (1976).
[1562] Egorenkov, N.I., Lin, D.G., & Rouzavkov, A.I. *Reports Byelorussian Acad. Sci.* **20**, 417 (1976).
[1563] Malers, L.Ya., Yansons, A.B., & Matisane, L.G. *USSR Modif. Polym. Mater.* (14), 63 (1986).
[1564] Kalnin', M.M., Vixne, A.V., Toupoureina, V.V., & Malers, Yu.Ya. In *Adhesion of Polymers and Adhesive Joints in Machine Building.* USSR Sci. -Techn. Soc. Publ., Moscow, 1976, vol. 1, p. 50.
[1565] Egorenkov, N.I., Lin, D.G., & Belyi, V.A. *J. Adhesion.* **7**, 269 (1976).
[1566] Chang, R.J., Gent, A.N., Hsu, C.C., & Sengai, K.C. *J. Appl. Polymer Sci.* **25**, 163 (1980).

[1567] Egorenkov, N.I., Kouzavkov, A.I., & Doktorova, V.A. *USSR High Polymers.* **28A**, 1525 (1986).

[1568] Rapoport, N.Ya., & Shlyapnikov, Yu.A. *USSR High Polymers.* **17A**, 738 (1975).

[1569] Sokolovsky, A.A., Ukhova, E.M., Bandourina, V.A., & Kouz'minsky, A.S., *USSR High Polymers.* **20B**, 142 (1978).

[1570] Roudakova, T.E., & Zaikov, G.E. *USSR High Polymers.* **29A**, 3 (1987).

[1571] Rapoport, N.Ya., & Zaikov, G.E. *USSR Adv. Chem.* **52**, 1568 (1983).

[1572] Melkoumov, A.N., Proutkin, V.P., Bronovitsky, V.E., & Tevshouzhsky, L.I. *USSR Mech. Polym.* (2), 353 (1975).

[1573] Rapoport, N.Ya., Livanova, N.M., & Miller, V.B. *USSR High Polymers.* **18A**, 2045 (1976).

[1574] Bol'shakova, S.I., Sokolovsky, A.A., & Kouz'minsky, A.S. *USSR High Polymers.* **23B**, 391 (1981).

[1575] Yasuda, E., Kimura, S., & Shibusa, Y. *Trans. Japan Soc. Compos. Mater.* **6**, 14 (1980).

[1576] Krisyuck, B.E., Popov, A.A., & Denisov, E.T. In *Proc. Sem. Chem. Ozone*, Tbilisi, 1981, p. 18.

[1577] Brewis, D.M., Konieczko, M.B., & Briggs, D. In *Adhesion*. University Press, London, 1978, vol. 2, p. 77.

[1578] Carley, J.F. In *37th Ann. Techn. Conf. Plastic Eng.* Greenwich, 1979, p. 728.

[1579] Eisby, V. *Japan Pap. Plast.* **8**, (10), 15 (1980).

[1580] Dalinkevitch, A.A., Shemarov, F.V., Kiryuskkin, S.G., & Shlyapnikov, Yu.A. *USSR Plastics* (12), 13 (1986).

[1581] Piiroja, E.K., Rayalo, G.Yu., Ki'yanen, I.R., Oidram, R.A., & Ebber, A.V. *Dep. Manuscr.* (3), Tallinn Inst. Inf. Publ., 1983.

[1582] Kodokian, G.K.A., & Kinloch, A.J. *J. Mater. Sci. Lett.* **7**, 625 (1988).

[1583] Stradal, M., & Goring, D.A.I. *Canad. J. Chem. Eng.* **53**, 427 (1975).

[1584] Briggs, D., & Kendall, C.R. *Polymer.* **20**, 1053 (1979).

[1585] Kagan, D.F., Prokopenko, V.V., & Stepanova, O.N. *USSR Plastics* (1), 21 (1983).

[1586] Takahashi, N., Rault, J., Goldman, A., & Goldman, M. In *2nd Int. Conf. Conduct. Breakdown Solid Diel.* New York, 1986, p. 179.

[1587] Catoire, B. *Polym. Mater. Sci. Eng.* **53**, 875 (1985).

[1588] Takahashi, N., Aouinti, A., Rault, J., Goldman, A., & Goldman, M. *C. r. Acad. Sci.* **305**, Ser. 2, 81 (1987).

[1589] Carley, J.F., & Kitze, P.T. *Polymer Eng. Sci.* **18**, 326 (1978).

[1590] Briggs, D., & Kendall, C.R. *Int. J. Adhesion Adhesives.* **2**, 13 (1982).

[1591] Kruger, R., & Potente, H. *J. Adhesion.* **11**, 113 (1980).

[1592] Artemenko, S.E., Lougovets, N.V., Gorbatkina, Yu.A., & Kopytin, V.S. *USSR Plastics* (6), 26 (1981).

[1593] Fitzer, E., Geigl, K.-H., Hüttner, W., & Weiss, R. *Carbon.* **18**, 389 (1980).

[1594] Leclercq, B., Sotton, M., Baszkin, A., & Ter-Minassian-Saraga, L. *Polymer.* **18**, 675 (1977).

[1595] Briggs, D., Rance, D.G., Kendall, C.R., & Blythe, A.R. *Polymer.* **21**, 895 (1980).

[1596] Owens, D.K. *J. Appl. Polymer Sci.* **19**, 3315 (1975).

[1597] Demuth, O., Rouzbehi, F., & Goldman, M. In *1st Int. Conf. Conduct. Breakdown Solid Diel.* New York, 1983, p. 93.

[1598] Luc, J. USA Pat. No 4024038 (1977).

[1599] Tourkina, E.S., Yakhnin, E.D., & Balog, I.I. In *Proc. USSR Inst. Polygr.* **30**, (3), 9 (1981).

[1600] Tourkina, E.S., Balog, I.I., & Yakhnin, E.D. *USSR High Polymers.* **22B**, 794 (1980).

[1601] Stradal, M., & Goring, D.A.I. *J. Adhesion.* **8**, 57 (1976).

[1602] Tsudzi, O., & Ishii, T. *J. Adhesion Soc. Japan.* **15**, 46 (1979).

[1603] Inagaki, N. *Japan Dyeing Ind.* **35**, 298 (1987).

[1604] Brovikova, I.N. *Dep. Manuscr.* No 559, Niitekhim Publ., Tcherkassy, 1988.

[1605] Yasuda, H., March, H.C., Brandt, S., & Reilley, C.N. *J. Polymer Sci.* **15**, 991 (1977).

[1606] Gerenser, L.J. *J. Adhesion Sci. Techn.* **1**, 303 (1987).

[1607] Hudis, M., & Prescott, L.E. *J. Appl. Polymer Sci.* **19**, 451 (1975).

[1608] Hamdam, S., & Evans, J.R.G. *J. Adhesion Sci. Techn.* **1**, 281 (1987).

[1609] Evans, J.R.G., Bulpett, R., & Ghezel, M. *J. Adhesion Sci. Techn.* **1**, 291 (1987).

[1610] Rose, P.W., & Liston, E.M. *Plast. Age.* **43**, No 10, 41 (1985).

[1611] Allred, R.E., Merrill, E.W., & Roylance, D.K. In *Molecular Characterisation of Composite Interfaces.* New York, 1985, p. 333.

[1612] Hiroshima, M., & Igarashi, T. Japan Pat. No 62-91535 (1987).

[1613] Mme T.n.T.T.N.M., Dissertation, P.et M.Curie Univ., Paris, 1980.

[1614] Mikame, H., Akai, I., & Hirose, M. Japan Pat. No 56-147832 (1981).

[1615] Gheorghiu, M., Turcu, D., Heinisch, P., Holban, V., & Mihaescu, A. In *Int. Macromol. Symp.* IUPAC Publ., Bucharest, 1983, Sect. 6–7, p. 92.

[1616] Fukuoka, S., Yasuda, H., & Sugiyama, H. Japan Pat. No 61-241330 (1986).

[1617] Fudzii, Y., Isioka, K., & Umedzawa, M., Japan Pat. No 61-157536 (1986).

[1618] Bourgeanu, G., Grunichievici, E., & Borgeanu, M. *Roum. Rev. chim.* **30**, 859 (1985).

[1619] Wakida, T., Kawamura, H., Song, J.C., Goto, T., & Takagishi, T. *Japan J. Soc. Fiber Sci. Techn.* **43**, 384 (1987).

[1620] Bartos, K. *Pr. Inst. wlok.* **35**, 5 (1987).

[1621] Kuriyama, D. *Japan Pap. Plast.* **9**, (4), 12 (1981).

[1622] Nuzzo, R.G., & Smolinsky, G. *Macromolecules.* **17**, 1013 (1984).

[1623] Zharov, V.A., & Solov'eva, O.N. *USSR Electron. Process. Mater.* (5), 49 (1986).

[1624] Isaka, T., & Nagano, T. Japan Pat. No 59-136333 (1984).

[1625] Isaka, T., & Nagano, T. Japan Pat. No 59-191740 (1984).

[1626] Yamakawa, S., & Yamamoto, F. *J. Appl. Polymer Sci.* **25**, 41 (1980).

[1627] Takass, M., Fritsch, R., Wiswanathan, N.S., & Patsis, A.V. *Amer. Chem. Soc. Polymer Prepr.* **21**, 141 (1980).

[1628] Anand, M., Cohen, R.E., & Baddour, R.F. *Polymer.* **22**, 361 (1981).

[1629] Klausner, M., Loh, I.H., Baddour, R.F., & Cohen, R.E. *Polymer Mater. Sci.* **56**, 227 (1987).

[1630] Momose, Y., Nishiyama, H., Noguchi, M., & Okazaki, S. *Japan J. Chem. Soc. Chem. Ind.* (10), 1876 (1985).

[1631] Strobel, M., Thomas, P.A., & Lyons, C.S. *J. Polymer Sci.* **25A**, 3343 (1987).

[1632] Kokoma, M., Moriwaki, T., Takahashi, K., & Okadzaki, Y. Japan Pat. No 61-133239 (1986).

[1633] Wakida, T., Kawamura, H., Song, J.C., Goto, T., & Takagishi, T. *Chem. Express.* **1**, 507 (1986).

[1634] Dorn, L., Gärtner, J., & Rasche, M. *Kunstst.* **76**, 249 (1986).

[1635] Haque, Y., & Ratner, B.D. *J. Appl. Polymer Sci.* **32**, 4369 (1986).

[1636] Hirotsu, T., & Ohnisi, S. *J. Adhesion.* **11**, 57 (1980).

[1637] Pederson, L.A., *J. Electrochem. Soc.* **129**, 205 (1982).

[1638] Vasilets, V.N., Tikhomirov, L.A., & Ponomarev, A.N. *USSR Chem. High Energ.* **12**, 442 (1978).

[1639] Tsudzi, O., Dodo, H., Nisino, H., & Kamio, H. Japan Pat. No 53-794 (1978).

[1640] Kaplan, S.I., & Rose, P.W. *Plast. Eng.* **44**, (5), 77 (1988).

[1641] Rogachev, A.V. Thesis, Inst. Mech. Metal-polymer Syst., Gomel, 1979.

[1642] Kolotyrkin, V.M., Kozlov, V.T., Khan, A.A., Gil'man, A.B., & Tounitsky, N.N. *USSR High Polymers.* **14B**, 742 (1972).

[1643] Kozlov, V.T., Khan, A.A., Gil'man, A.B., & Orlov, V.A. *USSR Rubber* No 1, 44 (1973).

[1644] Kolotyrkin, V.M., Kozlov, V.T., Khan, A.A., Gil'man, A.B., Orlov, V.A., & Tounitsky, N.N. *USSR High Polymers.* **17A**, 1319 (1975).

[1645] Ueda, Y., Fukutomi, Y., & Ashida, M. *Japan High Polymers.* **38**, 717 (1981).

[1646] Ishikawa, Y., Honda, K., Sasakawa, S., Hatada, K., Kobayashi, H., Soeda, F., Yoshimura, K., & Igaki, H. *Japan High Polymers.* **38**, 709 (1981).

[1647] Vonsyatsky, V.A., Karmazina, L.V., Gorbatkina, Yu.A., Ivanov-Moumzhieva, V.G., Pochinock, V.Ya., Dorokhovitch, V.P., & Kopytin, V.S. *USSR Mech. Compos. Mater.* No 1, 9 (1982).

[1648] Westerlind, B., Larsson, A., & Rigdahl, M. *Int. J. Adhesion Adhesives.* **7**, 141 (1987).

[1649] Amoroux, J., Goldman, M., & Revoil, M.F. *J. Polymer Sci.* **20**, 1373 (1982).

[1650] Poll, H.-U. DDR Pat. No 143082 (1980).

[1651] Yasuda, H. *J. Macromol. Sci.* **10A**, 383 (1976).

[1652] Gil'man, A.B., Khan, A.A., Shifrina, R.R., Kolotyrkin, V.M., Kozlov, V.P., & Orlov, V.A. *USSR High Polymers.* **21B**, 220 (1979).

[1653] Sharma, A.K., & Yasuda, H. *J. Vacuum Sci. Techn.* **21**, 994 (1982).

[1654] Washo, B.D. *Amer. Chem. Soc. Polymer Prepr.* **16**, 98 (1975).

[1655] Gil'man, A.B., Goldshtein, D.V., Potapov, V.K., & Shifrina, R.R. *USSR Chem. High Energ.* **22**, 465 (1988).

[1656] Westerdahl, C.A.L., Hall, J.R., & Levi, D.W. *Amer. Chem. Soc. Polymer Prepr.* **19**, 538 (1978).

[1657] Legeay, G., Epaillard, F., & Brosse, H.C. In *2nd Annu. Int. Conf. Plasma Chem. Techn.* Lancaster, 1986, p. 29.

[1658] Eakes, D.W., Newton, J.M., Watts, J.F., & Edgell, M.J. *Surface Interface Anal.* **10**, 416 (1987).

[1659] Friedrich, J., Wittrich, H., Gähde, J., & Richter, K. *Acta Polym.* **32**, 337 (1981).

[1660] Friedrich, J., Wittrich, H., & Gähde, J. *Acta Polym.* **31**, 59 (1980).

[1661] Freidrich, J., Gähde, J., Frommelt, H., & Wittrich, H. *Faserforsch. Textiltechn.* **27**, 604 (1976).

[1662] Garton, A., Sturgeon, P.Z., Carlsson, D.J., & Wiles, D.M. *J. Mater. Sci.* **13**, 2205 (1978).

[1663] Simionescu, C.I., Neamtu, I., Ioanid, G., Ioanid, A., Chiriac, A., & Rusan, V. *Roum. Mater. plast.* **24**, 212 (1987).

[1664] Chow, B.L., & Whittle, D.J. *J. Adhesion Sci. Techn.* **2**, 363 (1988).

[1665] Lipatov, Yu.S., Nesterov, A.E., Artemenko, S.E., & Andreeva, V.V. *Reports Ukrainian Acad. Sci.* No 9B, 51 (1981).

[1666] Mascia, L., Carr, G.E., & Kember, P. *Plast. Rubb. Process. Appl.* **9**, 133 (1988).

[1667] Boukhgalter, V.I., Belova, R.I., Evdokimova, N.V., & Gol'denberg, A.L. *USSR Plastics.* (2), 59 (1981).

[1668] Skorobogatov, A.A., Alexeev, O.V., & Ben'kova, L.F. *USSR Plastics* (10), 39 (1983).

[1669] Ivanov, S.I., & Pechenyakova, V.P. *Boulgar. Phys. J.* **11**, 340 (1984).

[1670] Ide, F. *J. Text. Mach. Soc. Japan.* **38**, 1 (1985).

[1671] Rasche, M. *Adhäsion.* **30** (3), 25 (1986).

[1672] Vonsyatsky, V.A., Roter, E.A., Tetersky, V.A., & Tynny, A.N. *USSR Phys. -Chem. Mech. Mater.* **12**, 100 (1976).

[1673] Inagaki, N., & Yamamoto, H. *J. Chem. Soc. Japan. Chem. Ind.* (11), 2031 (1987).

[1674] Petrov. A.K., Vlasova, L.F., & Bagryansky, V.A. *Herald Siberian Div. USSR. Acad. Sci. Ser. Chem. Sci.* (12/4), 118 (1988).

[1675] Wakida, T., Takeda, K., Kawamura, H., Tanaka, I., & Takagischi, T. *Chem. Express.* **2**, 711 (1987).

[1676] Pelzbauer, Z., & Lednicky, F. *J. Polymer Sci.* **12**, 2173 (1974).

[1677] Anishhenko, L.M., Kouznetsov, S.E., & Lavrenyuck, S.Yu. *USSR Phys. Chem. Process Mater.* No 6, 124 (1985).

[1678] Saka, N., Yee, G.Y., & Suh, N.P. In *35th Annu. Techn. Conf. Chem. Plast. Eng.* Montreal, 1977, p. 337.

[1679] Plate, N.A., Alieva, E.D., & Kalachev, A.A. *USSR High Polymers.* **23A**, 640 (1981).

[1680] Matsuo, T., & Igarashi, T., Japan Pat. No 62-91534 (1987).

[1681] Ikada, Y., Uyama, R., & Gen, S. Japan Pat. No 62-104843 (1987).

[1682] Hirotsu, T., & Ohnishi, T. Japan Pat. No 55-131026 (1980).

[1683] Rudoy, V.M., Sidorova, L.P., & Kabanov, V.Ya. *USSR High Polymers.* **30A**, 398 (1988).

[1684] Schultz, J., Carre, A., & Mazeau, C. *Int. J. Adhesion Adhesives.* **4**, 163 (1984).

[1685] Sugama, T., Kukacka, L.E., Clayton, C.R., & Hua, H.C. *J. Adhesion Sci. Techn.* **1**, 265 (1987).

[1686] Friedrich, J., Loeschke, I., & Gähde, J. *Acta Polym.* **37**, 687 (1986).

[1687] Fudjita, S., & Sudo, Y. Japan Pat. No 54-28872 (1979).

[1688] Hoffman, A.S., Keller, T.S., Miyake, A., Ratner, B.D., & McElroy, B.J. In *3rd Pacific Chem. Eng. Congr.* Seoul, 1983, vol. 2, p. 54.

[1689] Dias, A.J., & McCarthy, T.J. *Macromolecules.* **20**, 2068 (1987).

[1690] Yamashita, Y. *Japan Eng. Mater.* **33** (12), 46 (1985).

[1691] Sekamoto, I., Akaki, T., & Yamauti, S. Japan Pat. No 61-89236 (1986).

[1692] Nakamoto, H., Seidzai, F., & Fukushima, H. Japan Pat. No 61-233024 (1986).

[1693] Valiev, K.A., Makhviladze, T.M., & Sarychev, M.E. *USSR Surface* (3), 70 (1988).

[1694] Yamakawa, S. *J. Appl. Polymer Sci.* **20**, 3057 (1976).

[1695] Sage, D., Barticat, P., & Valiet, G. *Angew. makromol. Chem.* **54**, 151 (1976).

[1696] Orlov, V.A., Gerasimova, V.P., Kotova, I.P., Shashkov, A.S. *USSR Rubber* (8), 23 (1978).

[1697] Toutorsky, I.A., Dontsov, A.A., Filippovitch, B.V., & Orlov, V.A. *USSR Rubber* (5), 26 (1979).

[1698] Yanagihara, E. Japan Pat. No 59-142224 (1984).

[1699] Carter, A.R. In *2nd SATRA Int. Conf.* Kettering, 1972, p. 20/2.

[1700] Choulkov-Kovalev, S.A., Chesounova, A.G., Kourdoubov, Yu.F., & Leonov, V.V. In *Improving of Methods of Design of Leather Goods.* Light Ind. Publ., Moscow, 1983, p. 60.

[1701] Petravichyus, A.V. In *Materials Science and Technology of Leather Goods,* Polytechn. Inst. Press, Kaunas, 1979, p. 13.

[1702] Kapko, J., & Jakubowski, K. Poland Pat. No 111386 (1981).

[1703] Kinstle, J.F., & Watson, S.L. *Amer. Chem. Soc. Polymer Prepr.* **16**, 137 (1975).

[1704] Kitchens, J.D., & Novak, L.R. USA Pat. No 4567241 (1986).

[1705] Sakurai, Y., & Osiro, A. Japan Pat. No 61-85447 (1986).

[1706] Sinohara, D. Japan Pat. No 61-76535 (1986).

[1707] Hood, J.L.L. In *6th Int. Conf. Gas Discharg. Appl.* London, 1981, pt. 1, p. 86.

[1708] Yamamoto, H., & Komazu, M. Japan Pat. No 61-204238 (1986).

[1709] Kasai, D. Japan Pat. No 62-121736 (1987).

[1710] Chew, A., Brewis, D.M., Briggs, D., & Dahm, R.R. *21st Annu. Conf. Adhesion and Adhesives.* London, 1984, p. 97.

[1711] Glebko, Yu.I., Rastorguev, I.N., Zagorouyko, E.D., Knaus, O.G., Pavlova, E.A., & Rylova, T.G. USSR Pat. No 907049 (1982).

[1712] Voskanyan, E.S., & Melkonyan, N.K. *USSR Rubber* (8), 4 (1982).

[1713] Novikova, E.A., Zhmaeva, I.V., Tokarev, A.G., Zhigoun, I.G., Yunikova, T.G., & Mishenznikov, G.E. In *3rd Conf. Young Special on Mech. Compos. Mater.* Inst. Mech. Polym. Press, Riga, 1981, p. 29.

[1714] Kondo, S. Japan Pat. No 61-136525 (1986).

[1715] Kondo, S. Japan Pat. No 61-136543 (1986).

[1716] Ogino, K., Ohtsuka, T., & Ishikawa, N. *Kobunshi Romb.* **43**, 377 (1986).

[1717] Orlov, V.A., Rostovtseva, E.E., Gerasimova, V.P., & Malyshev, A.I. In *Adhesives and Joints.* House Sci. -Techn. Propag. Publ., Moscow, 1970, vol. 2, p. 3.

[1718] Filippovitch, B.V., Shmyrev, I.K., Orlov, V.A., & Toutorsky, I.A., *USSR Ind. Tyres, Rubber- and Asbestos- Technic Goods* (5), 8 (1979).

[1719] Doubinovsky, M.Z., Pritykin, L.M., Takoev, N.N., Stremock, T.M., Steshenko, N.I., & Dranovsky, M.G. *USSR Leather-shoe Ind.* No 9, 44 (1976).

[1720] Schonhorn, H. *J. Polymer Sci.* **1A**, 2343, 3523 (1969).

[1721] Vinogradova, L.M., & Korolev, A.Ya. *USSR High Polymers.* **11B**, 126 (1969).

[1722] Skeist, I. *Rev. Polymer Techn.* **1**, 19 (1972).

[1723] Kitamura, Y., Hata, H., & Ohira, T. Japan Pat. No 60-55073 (1985).

[1724] Yano, E., Komai, T., Kawasaki, T., Kaifu, K., Atsuka, T., Kubo, Y., & Fujiwara, Y. *J. Biomed. Mater. Res.* **19**, 863 (1985).

[1725] Tesoro, G.C., Rajendran, G.P., Park, C., & Uhlmann, D.R. *J. Adhesion Sci. Techn.* **1**, 39 (1987).

[1726] Pritykin, L.M., Genel', L.S., Shapiro, A.B., Gerenroth, V.G., & Vakula, V.L. *USSR Plastics* (9), 36 (1981).

[1727] Pritykin, L.M., Gerenroth, V.G., & Shapiro, A.B. USSR Pat. No 825573 (1981).

[1728] Pritykin, L.M., Gerenroth, V.G., Shapiro, A.B., & Pavlikov, V.V. USSR Pat. No 883132 (1981).

[1729] Donskikh, V.I., & Khar'kov, S.N. USSR Pat. No 1199783 (1985).

[1730] Nakamura, Y., Mori, K., & Tamura, K. *J. Adhesion Soc. Japan.* **17**, 308 (1981).

[1731] Ahagon, A., & Gent, A.N. *J. Polymer Sci. Phys. Ed.* **13**, 1285 (1975).

[1732] Vixne, A.V., Avotin'sh, Ya.Ya., & Ketsel'man, Ya.V. *USSR Modif. Polym. Mater.* (10), 59 (1981).

[1733] Dreyfuss, R., Eckstein, J., Lien, Q.S., & Dollwet, H.H. *J. Polymer Sci.* **19**, 427 (1981).

[1734] Runge, M.L., & Dreyfuss, R. *J. Polymer Sci. Chem. Ed.* **17**, 1067 (1979).

[1735] Lee, M.C.H. *J. Appl. Polymer Sci.* **33**, 2479 (1987).

[1736] Azrak, R.G. *Chem. Techn.* **4**, 683 (1974).

[1737] Tsukada, T. *Sekkei Seidzu.* **18**, 251 (1983).

[1738] Tsukada, T. *J. Adhesion Soc. Japan.* **24**, 266 (1988).

[1739] Malpass, B.W., Packham, D.E., & Bright, K. *J. Appl. Polymer Sci.* **18**, 3249 (1974).

[1740] Henneman, O.-D., & Brockmann, W. *J. Adhesion.* **12**, 297 (1981).

[1741] Takashi, M., Ogawa, K., & Kunio, T. In *17th Japan Congr. Mater. Res.* Kyoto, 1974, p. 192.

[1742] Bischof, C., & Bauer, A. *Plaste und Kautsch.* **34**, 111 (1987).

[1743] Breznick, M., Banbaji, J., Guttmann, H., & Marom, G. *Polymer Comm.* **28** (2), 55 (1987).

[1744] Mervinsky, R.I., & Avramenko, V.L. *USSR Phys. -Chem. Mech. Mater.* **15**, 70 (1979).

[1745] Roudakova, T.E., Moiseev, Yu.V., Astrina, V.I., Razoumova, L.L., Vlasov, S.V., & Zaikov, G.E. *USSR High Polymers.* **17A**, 1791 (1975).

[1746] Durning, C.J. Weigmann, H.D., Rebenfeld, I., & Russel, W.B. In *Polymer Fibers and Elastomers.* Washington, 1984, p. 309.

[1747] Astrina, V.I., Razoumova, L.L., Shatalova, O.V., Vlasov, S.V., Sagalaev, G.V., & Zaikov, G.E. *USSR High Polymers.* **20A**, 342 (1978).

[1748] Astrina, V.I., Vlasov, S.V., Razoumova, L.L., Shatalova, O.V., Goumen, R.G., & Zaikov, G.E. *USSR High Polymers.* **21B**, 505 (1979).

[1749] Chotorlishvili, L.S. *USSR Plastics* (9), 53 (1988).

[1750] Srinavasan, R., & Lazare, S. *Polymer.* **26**, 1297 (1985).

[1751] Lazare, S., & Srinavasan, R. *J. Phys. Chem.* **90**, 2124 (1986).

[1752] Hennemann, O.-D. In *3rd Int. Congr. Surface Technology.* VDE Verl. Berlin, 1985, S.41.

[1753] Vixne, A.V., Rentse, L.K., Koryukin, V.M., & Bereza, M.P. *USSR Plastics* (9), 31 (1988).

[1754] Zorll, U. *Metalloberfläsche.* **32**, 257 (1978).

[1755] Wartusch. J., & Saure, M. BRD Pat. No 2950589 (1981).

[1756] Loupinovitch, L.N. In *Intensification of Bonding Process*. House Sci. -Techn. Propag. Publ., Leningrad, 1987, p. 50.

[1757] Gilbert, Y. *Composites*. **25** (2), 40 (1985).

[1758] Goloubkov, G.E., & Savel'eva, L.N. USSR Pat. No 638611 (1979).

[1759] Pol'sman, G.S., & Ginzburg, L.B. *USSR High Polymers*. **18B**, 319 (1976).

[1760] Zaporozhskaya, E.A., Ginzburg, L.V., & Dontsov, A.A. *USSR High Polymers*. **25A**, 371 (1983).

[1761] Tikhonova, N.P., Ginzburg, L.V., & Dontsov, A.A. *USSR Rubber* (1), 14 (1987).

[1762] Zaporozhskaya, E.A. Cand.Sci(Chem.) Thesis, Lomonosov, Inst. Fine Chem. Techn. Moscow, 1983.

[1763] Bechhold, H., & Neumann, S. *Angew. Chem.* **37**, 225 (1924).

[1764] Zakrevsky, V.A., & Korsoukov, V.E. *USSR High Polymers*. **14A**, 955 (1972).

[1765] Polyakov, A.M., Kourdoubov, Yu.F., Baramboim, N.K., & Krotova, N.A. *Reports USSR Acad. Sci.* **175**, 72 (1967).

[1766] Poll. H.U., & Kleeman, R. *Wiss. Z. Techn. Hochsch. Karl-Marx-Stadt.* **20**, 819 (1978).

[1767] Boutyagin, Yu.P. *USSR Adv. Chem.* **40**, 1935 (1970).

[1768] Smourougov, V.A. *France Mec. mater. elec.* (374), 83 (1981).

[1769] Gerashhenko, E.I., & Prokazova, E.V. *USSR Plastics* (12), 17 (1987).

[1770] Kobets, L.P., Polyakova, N.V., & Kouznetsova, M.A. *USSR Mech. Polym.* (4), 579 (1978).

[1771] Juhacz, Z., & Opoczky, L. In *Int. Conf. Colloid Interface Sci.* Akad. Kiado Publ., Budapest, 1975, vol. 1, p. 57.

[1772] Evdokimov, Yu.M. *Proc. Moscow Forest-Techn. Inst.* (81), 5 (1975).

[1773] Lerchenthal, C.H., Brenman, M., & Yits'haq, N. *J. Polymer Sci. Chem. Ed.* **13**, 737 (1975).

[1774] Lerchenthal, C.H., Brenman, M., & Yits'haq, N. *Polymer Eng. Sci.* **16**, 760 (1976).

[1775] Lerchenthal, C.H., Brenman, M., & Yits'haq, N. *J. Appl. Polymer Sci. Symp.* No 35, 537 (1979).

[1776] Genel', L.S., Vakula, V.L., & Fokine, M.N. USSR Pat. No 622831 (1978).

[1777] Genel', L.S., Vakula, V.L., Akoutin, M.S., & Lokshin, R.F. USSR Pat. No 763432 (1980).

[1778] Ruhsland, K. In *Adhesive Joints: Formation, Characterization, and Testing*. New York, 1982, p. 257.

[1779] Genel', L.S., Gasyuck, O.V., Mouromtsev, V.I., Vakula, V.L., Krasnosel'skaya, N.S., & Galkina, V.V., USSR Pat. No 1002338 (1983).

[1780] Genel', L.S., Wakula, W.L., & Kestelman, W.N. *Plaste und Kautsch.* **29**, 604 (1982).

[1781] Genel', L.S., Wakula, W.L., & Kestelman, W.N. *Oberfläche Surf.* **23**, 433 (1982).

[1782] Genel', L.S., & Vakula, V.L. *USSR Herald Mach. Build.* (5), 71 (1978).

[1783] Genel', L.S., Vakula, V.L., & Akoutin, M.S. In *New Polymer Composite Materials in Machine Building*. USSR Sci. -Tech. Soc. Publ., Moscow, 1978, p. 210.

[1784] Genel', L.S., Vakula, V.L., & Akoutin, M.S. *USSR Plastics* (10), 27 (1980).

[1785] Genel', L.S., & Vakula, V.L. In *Coating of Metals by Plastics and Metallization of Plastics in Machine-Building.* USSR Sci. -Techn. Soc. Publ., Moscow, 1980, pt. 2, p. 163.

[1786] Genel', L.S., & Vakula, V.L. In *Coating of Metals by Plastics and Metallization of Plastics in Machine-Building.* USSR Sci. -Techn. Soc. Publ., Moscow, 1980, pt. 2, p. 168.

[1787] Genel', L.S., Gasyuck, O.V., Mouromtsev, V.N., Vakula, V.L., Kaizer, M.F., Korenev, S.E., & Galkina, V.V. USSR Pat. No 876695 (1981).

[1788] Genel', L.S. In *Epoxide Oligomers and Paint Materials.* USSR Sci. -Techn. Soc. Publ., Tcherkassy, 1980, p. 22.

[1789] Genel', L.S., & Vakula, V.L. In *Adhesives and Applications in Technology.* USSR Polym. Adh. Ints. Press, Kirovokan, 1978, p. 36.

[1790] Pritykin, L.M., & Orlov, V.A. USSR Pat. No 600162 (1978).

[1791] Pritykin, L.M., & Dranovsky, M.G. In *Adhesion of Polymers and Adhesive Joints in Machine-Building.* USSR Sci. -Techn. Soc. Publ., Moscow, 1976, vol. 1, p. 55.

[1792] Gouseva, M.S., Liakoumovitch, A.G., & Fridman, B.S. *USSR Plastics* (1), 31 (1988).

[1793] Danyushina, G.A., Kagan, E.Sh., Smirnov, V.A., & Pritykin, L.M. *USSR Plastics* (11), 36 (1982).

[1794] Danyushina, G.A., Pritykin, L.M., Smirnov, V.A., Kagan, E.Sh., Mikhailov, V.M., & Sysoeva, V.P. USSR Pat. No 732405 (1980).

[1795] Pritykin, L.M., Genel', L.S., Shapiro, A.B., Rozantsev, E.G., Vakula, V.L., Akoutin, M.S., & Dranovsky, M.G. USSR Pat. No 950743 (1982).

[1796] Voyutskii, S.S., Raevsky, V.G., & Yagnyatinskaya, S.M. *USSR Rubber* (7), 16 (1964).

[1797] Voyutskii, S.S., Raevsky, V.G., & Yagnyatinskaya, S.M. In *Advances of Colloid Chemistry.* Nauka Publ., Moscow, 1973, p. 339.

[1798] Filyanov, E.M., Telegina, E.B., Tarakanov, O.G., & Demina, A.I. *USSR Compos. Polym. Mater.* (4), 16 (1979).

[1799] Papirer, E. In *1st Int. Conf. Compos. Interfaces.* New York, 1986, p. 203.

[1800] Hoffmann, F. *Polym. Paint Colour J.* **171**, 722 (1981).

[1801] Kryszewski, M., & Jeszka, J.K. In *1st Int. Conf. Compos. Interfaces.* New York, 1986, p. 81.

[1802] Williams, J.W., & Shang, S.W. In *194th Amer. Chem. Soc. Meet.* ACS Press, Washington, 1987, p. 1042.

[1803] Schröder, J., & Honigmann, B. *Farbe und Lack.* **87**, 176 (1981).

[1804] Mamin, V.N., Gromov, A.N., & Lyong-Dyk-tuan. In *Fillers of Polymer Materials.* House Sci. -Techn. Propag. Publ., Moscow, 1977, p. 170.

[1805] Loupinovitch, L.N., & Garshin, A.P. *USSR Plastics* (5), 43 (1986).

[1806] Lyapina, L.A., & Gorshkova, L.M. In *Formulation and Use of Modified Types of Carbon Blacks.* Moscow, 1981, p. 61.

[1807] Novikova, O.A., Sergeev, V.P., & Litvinov, V.F. *USSR Plastics* (3), 19 (1985).

[1808] Messick, D.L., Progar, D.J., & Wightman, J.P. In *15th Nat. SAMPE Techn. Conf.* Azusa, 1983, p. 170.

[1809] Rassokha, V.N., & Avramenko, V.L. *USSR Plastics* (5), 63 (1987).

[1810] Shemyakin, V.A., Loukasick, V.A., Ogrel', A.M., Shvetsov, V.A., Antsoupov,

Yu.A., Boukalov, V.P., Sizov, S.Yu., & Moiseev, B.S. USSR Pat. No 410032 (1974).

[1811] Tanaka, Y., Murakami, N., & Bo, D. Japan Pat. No 55-69658 (1980).

[1812] Kolesnikov, V.I., Volkov, A.V., Chebotarev, S.I. In *Properties and Applications of Glass Fiber and Glass Reinforced Plastics*. Chimia Publ., Moscow, 1988, p. 81.

[1813] Zakharov, N.D., Izyumova, V.I., Mel'nikov, M.Ya., Shakh-Paron'yants, A.M., & Kostrykina, G.I. *USSR High Polymers*. **26A**, 2082 (1984).

[1814] Deroun, S.I., Pritykin, L.M., Chouiko, A.A., Pavlov, V.V., Khaber, N.V., Dranovsky, M.G., & Kostylev, Yu.S. USSR Pat. No 806713 (1981).

[1815] Deroun, S.I., Pritykin, L.M., Chouiko, A.A., Pavlov, V.V., Khaber, N.V., Dranovsky, M.G., Vorob'eva, T.N., & Agamalyan, S.G., USSR Pat. No 806712 (1981).

[1816] Deroun, S.I., Pritykin, L.M., Chouiko, A.A., Pavlov, V.V., Khaber, N.V., Dranovksy, M.G., & Kostylev, Yu.S. USSR Pat. No 789548 (1980).

[1817] Pritykin, L.M., Obidin, E.A., & Deroun, S.I. In *New Materials Based on Epoxide Resins, Properties and Applications*. House Sci. -Techn. Propag. Publ., Leningrad, 1974, vol. 2, p. 53.

[1818] Raevsky, V.G., Pritykin, L.M., & Sarkisyan, M.S. USSR Pat. No 376414 (1973).

[1819] Ravesky, V.G., Pritykin, L.M., & Borzenko, E.M. USSR Pat. No 380674 (1973).

[1820] Raevsky, V.G., Pritykin, L.M., & Sarkisyan, M.S. USSR Pat. No 357208 (1972).

[1821] Pritykin, L.M., Obidin, E.A., & Deroun, S.I. In *2nd USSR Conf. Epoxide Monomers and Resins*. University Press, Dnepropetrovsk, 1974, p. 278.

[1822] Pritykin, L.M., Deroun, S.I., & Dranovsky, M.G. In *Advances in Formulation and Applications of Adhesives in Industry*. House Sci. -Techn. Propag. Publ., Moscow, 1983, p. 29.

[1823] Pritykin, L.M., & Dranovsky, M.G. USSR Pat. No 607836 (1978).

[1824] Pritykin, L.M., & Dranovsky, M.G. In *New Polymer Composite Materials in Machine Building*. USSR Sci. -Techn. Soc. Publ., Moscow, 1978, p. 224.

[1825] Pritykin, L.M., Dranovsky, M.G., & Krasheninnikov, A.I. USSR Pat. No 981331 (1982).

[1826] Pritykin, L.M., & Dranovksy, M.G. In *Modern Problems in Synthesis of Rubbers*. USSR Chem. Soc. Publ., Dnepropetrovsk, 1980, p. 162.

[1827] Pritykin, L.M., Zjuz', V.T., Onouskko, A.I., Dranovsky, M.G., & Krasheninnikov, A.I. *USSR Rubber* (9), 18 (1983).

[1828] Modyanova, E.A. *USSR Cytology*. **12**, 35 (1970).

[1829] Modyanova, E.A. *USSR Cytology*. **15**, 183 (1973).

[1830] Bocharova, O.A., & Modyanova, E.A. *USSR Probl. Oncology*. **28**, 58 (1982).

[1831] Modyanova, E.A., Bocharova, O.A., & Ushakov, V.F. *USSR Bull. Exp. Biol.* **4**, 459 (1980).

[1832] Malenkov, A.G., & Modyanova, E.A. *USSR Biophysics*. **32**, 1033 (1987).

[1833] Modyanova, E.A., & Malenkov, A.G. *USSR Cytology*. **17**, 1155 (1975).

[1834] Modyanova, E.A., Kasatkina, N.K., & Malenkov, A.G. In *1st USSR Biophys. Congr.* Moscow, 1982, vol. 2, p. 119.

[1835] Modyanova, E.A. *USSR Ontogenesis*. **5**, 198 (1974).

[1836] Malenkov, A.G., Modyanova, E.A., & Yamskova, V.P. *USSR Biophysics*. **22**, 156 (1977).

[1837] Modyanova, E.A., Bocharova, O.A., & Malenkov, A.G. *USSR Exp. Oncology.* **5**, 39 (1983).

[1838] Bocharova, O.A., Modyanova, E.A., & Malenkov, A.G. In *1st USSR Biophys. Congr.* Moscow, 1982, vol. 2, p. 119.

[1839] Duchesne, J. *C.r. Acad. Sci.* **293**, XXXIX (1975).

[1840] Emanuel', N.M., Konovalova, N.P., D'yachkovskaya, R.F., & Vasil'eva, L.S. In *Free-Radical States and Their Role in Radiation Sickness and Malignant Growth.* Moscow, 1971, p. 102.

[1841] Konovalova, N.P. Dr.Sci(Biol.) Thesis, Inst. Chem. Phys., Moscow, 1975.

[1842] Schultz, J., & Carre, A. *Inf. Chim.* (232), 101 (1982).

[1843] Allen, K.W. *J. Adhesion.* **21**, 261 (1987).

[1844] Basin, V.E., & Berlin, A.A. *USSR Mech. Polym.* (2), 295 (1972).

[1845] Raevsky, V.G. *J. Adhesion.* **5**, 203 (1973).

[1846] *Dictionary-Handbook on Welding and Bonding of Plastics.* Naukova Dumka Publ., Kiev, 1988.

[1847] Mittal, K.L. *Pure Appl. Chem.* **52**, 1295 (1980).

[1848] Nakajima, T. *J. Adhesion Soc. Japan.* **20**, 395 (1984).

[1849] Kardos, J.L. *Chemtech.* **14**, 431 (1984).

[1850] Lee, L.-H. In *Adhesion Science and Technology.* Plenum Press, New York, 1975, vol. B, p. 711.

[1851] Calvert, P., & Mann, S. *J. Mater. Sci.* **23**, 3801 (1988).

Index